Alumina Ceramics

Related titles

Bioceramics and Their Clinical Applications
(ISBN: 9781845692049)

Cambier and De Barra, Advances in Ceramic Biomaterials
(ISBN: 9780081008812)

Biomimetic Biomaterials
(ISBN: 9780857094162)

Woodhead Publishing Series in Biomaterials

Alumina Ceramics
Biomedical and Clinical Applications

Andrew Ruys

WOODHEAD
PUBLISHING
An imprint of Elsevier

Woodhead Publishing is an imprint of Elsevier
The Officers' Mess Business Centre, Royston Road, Duxford, CB22 4QH, United Kingdom
50 Hampshire Street, 5th Floor, Cambridge, MA 02139, United States
The Boulevard, Langford Lane, Kidlington, OX5 1GB, United Kingdom

Copyright © 2019 Elsevier Ltd. All rights reserved.

No part of this publication may be reproduced or transmitted in any form or by any means, electronic or mechanical, including photocopying, recording, or any information storage and retrieval system, without permission in writing from the publisher. Details on how to seek permission, further information about the Publisher's permissions policies and our arrangements with organizations such as the Copyright Clearance Center and the Copyright Licensing Agency, can be found at our website: www.elsevier.com/permissions.

This book and the individual contributions contained in it are protected under copyright by the Publisher (other than as may be noted herein).

Notices

Knowledge and best practice in this field are constantly changing. As new research and experience broaden our understanding, changes in research methods, professional practices, or medical treatment may become necessary.

Practitioners and researchers must always rely on their own experience and knowledge in evaluating and using any information, methods, compounds, or experiments described herein. In using such information or methods they should be mindful of their own safety and the safety of others, including parties for whom they have a professional responsibility.

To the fullest extent of the law, neither the Publisher nor the authors, contributors, or editors, assume any liability for any injury and/or damage to persons or property as a matter of products liability, negligence or otherwise, or from any use or operation of any methods, products, instructions, or ideas contained in the material herein.

Library of Congress Cataloging-in-Publication Data
A catalog record for this book is available from the Library of Congress

British Library Cataloguing-in-Publication Data
A catalogue record for this book is available from the British Library

ISBN: 978-0-08-102442-3 (print)
ISBN: 978-0-08-102443-0 (online)

For information on all Woodhead publications visit our website at https://www.elsevier.com/books-and-journals

Publisher: Matthew Deans
Acquisition Editor: Laura Overend
Editorial Project Manager: Joshua Bayliss
Production Project Manager: Joy Christel Neumarin Honest Thangiah
Cover Designer: Greg Harris

Typeset by SPi Global, India

Transferred to Digital Printing in 2018

Contents

About the author		ix
Author biography		xiii
Preface		xv
The beginning		xxi

1 Introduction to alumina ceramics — 1
- 1.1 Introduction to alumina ceramics — 1
- 1.2 Natural alumina — 13
- 1.3 A brief history of alumina — 17
- 1.4 The alumina ceramics manufacturing industry — 31
- 1.5 Companies for which alumina is a high-volume platform technology — 37

2 Bauxite: The principal aluminum ore — 39
- 2.1 Bauxite, alumina refining, and aluminum smelting: A mega-industry — 42

3 Refining of alumina: The Bayer process — 49
- 3.1 Overview of the Bayer process for refining bauxite into alumina — 49
- 3.2 A century of Bayer refining: Review of problems and lessons learned — 52
- 3.3 Autoclave digestion of bauxite — 53
- 3.4 Silica problems: Dissolution, removal, and scale — 55
- 3.5 Precipitation of the gibbsite — 61
- 3.6 The sodium oxalate problem (bauxite organic content) — 62
- 3.7 Red mud — 63
- 3.8 Alternatives to Bayer refining — 68

4 Processing, structure, and properties of alumina ceramics — 71
- 4.1 A brief history of alumina ceramics — 72
- 4.2 Alumina powder preparation — 77
- 4.3 Forming of alumina ceramics — 81
- 4.4 Sintering of alumina — 93
- 4.5 Structure and crystallography of alumina — 103
- 4.6 Overview of the properties of alumina ceramics — 104

5 Dental, tissue scaffold, and other specialized biomedical applications of alumina — 123

5.1 The early history of alumina in biomedical engineering — 124
5.2 Alumina in dental technology — 126
5.3 Alumina bone tissue scaffolds — 134
5.4 Other biomedical uses for alumina — 137

6 Alumina bearings in orthopedics: Origin and evolution — 139

6.1 Introduction to alumina in hip replacements — 139
6.2 Introduction to the anatomy of the hip and hip joint pathologies — 143
6.3 A brief history of the hip replacement — 146
6.4 Alumina bearings in hip replacements: The alumina hip — 161
6.5 Other ceramic orthopedic bearings — 168

7 Alumina bearings in orthopedics: Present and future — 179

7.1 Zirconia-toughened alumina — 179
7.2 Wear of orthopedic bearings — 188
7.3 Other alumina-relevant issues arising from hip replacement — 200
7.4 Commercialization of the alumina hip: The transition from research to commercialization — 210

8 Alumina in bionic feedthroughs: The pacemaker — 225

8.1 Alumina: A key enabling technology for implantable bionic medical devices — 225
8.2 A brief history of early pacemaker technology — 228
8.3 The alumina/titanium hermetic feedthrough: Technology overview — 232
8.4 The concept of the alumina feedthrough — 240
8.5 The specifics of the alumina/titanium feedthrough innovation — 240
8.6 Feedthrough evolution post-1970 — 247
Appendix 8.1: Feedthrough Terminology — 252

9 Alumina in bionic feedthroughs: The bionic ear — 255

9.1 Bionic ear: A global race in the 1960s, 1970s, and 1980s — 255
9.2 A brief history of the bionic ear — 257
9.3 Development of the alumina feedthrough technology for the bionic ear — 261

10 Alumina in bionic feedthroughs: The bionic eye and the future — 283

10.1 Alumina in bionics today — 284
10.2 The bionic eye — 296
10.3 Quantum leap 3 in alumina feedthrough technology: Alumina retinal implant for the bionic eye — 304
10.4 The future for alumina in bionics — 315

11 Alumina in lightweight body armor — 321

- 11.1 Ballistic armor—The sociocultural context — 322
- 11.2 A brief history of body armor — 323
- 11.3 The origin of alumina ceramic composite body armor — 333
- 11.4 Ceramic armor design principles — 345
- 11.5 The state of the art in alumina and other ceramic body armor systems — 351
- 11.6 A brief summary of vehicle armor — 362
- 11.7 Notable armor ceramic companies — 366

12 Alumina as a wear-resistant industrial ceramic — 369

- 12.1 Tribology — 371
- 12.2 Wear-resistant alumina in industry — 389
- 12.3 Concluding remarks — 411

13 Alumina as an electrical insulator — 413

- 13.1 The electrical properties of alumina — 415
- 13.2 Electrical insulator applications of alumina: Alumina macro-insulators — 426
- 13.3 Electrical insulator applications of alumina: Alumina substrates in microelectronic technology — 431
- 13.4 Alumina substrate technology — 436
- 13.5 Microwave-industry electrical applications of alumina — 441
- 13.6 Alumina as an electrical insulating ceramic. Concluding remarks — 444

14 Alumina-metal bonding for electrical feedthroughs — 447

- 14.1 Evolution of the hermetic feedthrough concept — 449
- 14.2 Evolution of the alumina feedthrough — 454
- 14.3 Alumina-metal bonded feedthroughs: General discussion — 462
- 14.4 Multilayer micro-feedthroughs: Cofired alumina-substrate systems — 468
- 14.5 Concluding remarks — 469

15 Refractory and other specialist industrial applications of alumina — 473

- 15.1 The refractories industry: A brief overview — 474
- 15.2 The contemporary refractories industry — 478
- 15.3 Contemporary refractory uses of alumina ceramics — 481
- 15.4 High-purity alumina ceramics for heat containment: Labware and industrial uses — 484
- 15.5 Alumina machining tools — 486
- 15.6 Architectural applications for alumina — 490
- 15.7 Aerospace applications for alumina — 493
- 15.8 Oxygen sensors — 495
- 15.9 Specialist industrial applications of alumina not discussed elsewhere — 498

16 Alumina: The future — 501

16.1 Economic impact of alumina—2020 — 503
16.2 Social impact of alumina—2020 — 504
16.3 Alumina in the future — 505
16.4 Conclusions: Alumina in the year 2050 — 507
16.5 Closing comments — 508

References — 509
Index — 541

About the author

Andrew J. Ruys
Founding Director of Biomedical Engineering, University of Sydney, Sydney, NSW, Australia

Andrew J Ruys is a professor of Biomedical Engineering and has served as director of Biomedical Engineering at the University of Sydney since 2003. He holds a bachelor's degree (Honours 1—1987) and doctorate (1992) in Ceramic Engineering.

Professor Ruys has over 200 academic publications, and has been awarded over 30 competitive research grants totalling over $5 million. He has worked in bioceramics and advanced ceramics research for over 30 years, in the complementary disciplines of Biomedical Engineering and Ceramic Engineering, and has been an active participant as researcher and educator, for this entire time. He also has extensive experience as an industrial consultant in the world-changing applications of advanced ceramics in wear-resistant linings in mineral processing, anticorrosion-resistant vessels, high-temperature gas seals, bioceramics, ballistic ceramics, and numerous other world-changing applications. The most important of the advanced ceramics, and the bioceramics, is alumina, the topic of this book. Professor Ruys has been involved with alumina for the last three decades.

Professor Ruys serves on three editorial boards, and is a reviewer for 24 scientific journals. He has been teaching bioceramics, industrial ceramics, biomaterials, medical device technology, dental materials, chemistry, physics, and general engineering, a total of 27 different university units of study in the last 30 years.

His experience is unique, spanning as it does three decades in ceramics, bioceramics, and industrial advanced ceramics, with an ongoing association with alumina for that entire time.

He leads a large multidisciplinary research team in bioceramics, and has supervised 25 PhD students to completion as well as about 200 Honours students, during his academic career. His academic research has covered a broad range of areas, intersecting with almost every area covered in this book.

Section A: Alumina as a ceramic

Professor Ruys has been involved in alumina ceramics research since 1987 (Chapter 4):

- Published the seminal paper on alumina micro-slip casting, and subsequent papers on this theme [636–640].

- Published the seminal paper on thixotropic casting, and subsequent papers on this theme [117–124].
- Gelcasting [116].
- Tapecasting [375,412–414,459].
- 3D printing [641].
- Hydrostatic shock forming [133].
- Zirconia-toughened alumina (ZTA) [116,278–281].
- Metal-ceramic functionally graded materials, including alumina [629–632].
- Metal fiber-reinforced alumina and other ceramics [642–650].

Section B: Biomedical

Professor Ruys has been involved in alumina biomaterials since the 1990s, and biomaterials in general since the 1980s.

- Published the seminal paper on porous high-strength alumina tissue scaffolds doped with bioactive ions (Chapter 5 [174–177]).
- ZTA for orthopedics (Chapters 6 and 7 [278–281]).
- Bionics: FES—functional electrical stimulation (Chapters 8–10 [651–655]).
- Bionic ear: numerous collaborative projects with Cochlear the bionic ear company in the last two decades (Chapter 9 [656–670]).
- In the last decade, Professor Ruys has been involved in the bionic eye alumina microfeedthrough development program of the Suaning group (Chapter 10). The Suaning group set the world record of 1145 platinum channels in a retinal implant in 2012 [375,412]. Professor Ruys was a coauthor on the paper documenting the 1145 channel alumina feedthrough [375]. The next highest globally was 360 channels.
- In the 1980s, the author published the seminal paper on silicon-doped hydroxyapatite for optimal bioactivity in 1988 [671–673]. Though not an alumina technology, this concept is one for which the author is renowned in the biomaterials realm. It has become a global paradigm in bone tissue scaffolds and was later commercialized by Cambridge University through start-up company Apatech, which was sold to Baxter Healthcare for $300 Million in 2010.
- Transparent alumina orthodontic brackets (Chapter 5 [674]).
- The author spent much of the 1990s developing a metal microfiber-reinforced biomaterial for biocompatible orthopedic load bearing (Chapters 6 and 7 [639–647]). Currently, the focus of Professor Ruys in the biomedical alumina realm is in the development of high-toughness metal-reinforced alumina for the ceramic knee and ceramic hip resurfacing applications that require higher toughness than ZTA is capable of (Chapters 6 and 7).

Professor Ruys has specialized in the ceramic armor realm as a technical consultant since 1988 (Chapter 11):

- 1988: represented ESK (now 3M-ESK: Europe's first company to commercialize ceramic armor).
- 1992: became a consultant to Signal 1 International, Australia's first company to introduce alumina ceramic body armor for the civilian sector.
- 1996: became a director of MC^2 (Modern Ceramics Company). Alumina ceramics were a part of the MC^2 technology portfolio. In their day, MC^2 were one of the world's two leading manufacturers of reaction sintered SiC (RSSC) body armor, along with M-Cubed of the United States [505]. MC^2 supplied ceramic armor breastplates to numerous countries globally.

Section C: Industrial

Professor Ruys had a long involvement with alumina ceramics in the industrial realm in his role as an industrial consultant, dating back to the 1980s, and increasing in scope in the 1990s through his (then) role as director of MC^2—Modern Ceramics Company. This particularly involved wear-resistant ceramics (Chapter 12) and aerospace ceramics (Chapter 15), and in this capacity Professor Ruys has also had an involvement with the United States in the wear, armor, and aerospace realm, specifically, AFRL, NASA, and various ceramic companies in California, Arizona, and Texas. The electrical alumina activities of Professor Ruys (Chapters 13 and 14) have focussed on bionic feedthroughs, as discussed above.

In addition to his research and consultancy activities, Professor Ruys has taught 27 different undergraduate and master's courses since 1987. Of particular relevance to this book are the following:

- Physical ceramics
- Chemical ceramics
- Ceramic process principles
- Ceramic engineering 2
- Refractories
- Materials science for dentistry
- Design with brittle materials
- Materials 1
- Materials 4
- Phase equilibrium
- Biomedical design and technology
- Biomechanics and biomaterials

Author biography

Andrew J. Ruys is a professor of Biomedical Engineering and has served as director of Biomedical Engineering at the University of Sydney since 2003. He holds a bachelor's degree (Honours 1—1987) and doctorate (1992) in Ceramic Engineering.

He has over 200 academic publications, and has been awarded over 30 competitive research grants totalling over $5 million. He has worked in bioceramics and advanced ceramics research for 30 years, and has been an active participant as researcher and educator, for this entire time. He also has extensive experience as an industrial consultant in the world-changing applications of advanced ceramics in wear-resistant linings in mineral processing, anticorrosion-resistant vessels, high-temperature gas seals, bioceramics, ballistic ceramics, and numerous other world-changing applications. The most important of the advanced ceramics, and the bioceramics, is alumina, the topic of this book. He has been involved with alumina for the entire 30 years while pursuing a career in the interconnected disciplines of Biomedical Engineering and Ceramic Engineering.

He serves on three editorial boards, and is a reviewer for 24 scientific Journals. He has been teaching bioceramics, industrial ceramics, biomaterials, medical device technology, dental materials, chemistry, physics, and general engineering, a total of 27 different university units of study in the last 30 years.

His experience is unique, spanning as it does three decades in ceramics, bioceramics, and industrial advanced ceramics, with an ongoing association with alumina for that entire time.

Preface

I am pleased to publish the first book on alumina since 1984. I can still recall using the 1984 alumina book (Dorre and Hubner) as a reference source during the 1980s, as I was pursuing research in the emerging new fields of biomaterials, biomedical engineering, and advanced ceramics. I did not imagine at that time that three decades later I would be writing its sequel. I have been involved in alumina in its many important applications, ever since 1984, and therefore I am pleased to be the one to receive the baton from my distinguished predecessors of so long ago: Erhard Dorre and Heinz Hubner. Erhard Dorre was a pioneer of the early CeramTec (then called Feldmuhle) alumina hip bearing.

The fledgling alumina industry began in the 1950s. In 1984, 30 years on, it was slowly moving along the pathway from boutique technology toward industrial significance. Fast forward another 34 years to 2018, the world of alumina has changed almost beyond recognition. Refractories and emerging developments in electronics and wear resistance were the focus in 1984. A new niche application in orthopedics got a few pages in the 1984 book, as did another niche application in body armor. The recently invented alumina bionic feedthrough did not feature at all.

Alumina ceramics have today become a multibillion dollar industry that has changed the world over the last few decades. Alumina uniquely combines low cost with extreme hardness, extreme electrical resistivity, extreme corrosion resistance, high refractoriness, and it is the most biocompatible material in current clinical use. This book is about an extraordinary material, and its extraordinary applications, combining science, engineering, technology history, and a look to the exciting future.

Bionics: Millions of lives have been saved by a bionic implant known as the cardiac pacemaker. What is not so well known is that alumina is the key enabling technology for the bionic implant. Early pacemakers were clad in epoxy resin and had a very short lifespan in the body. The invention of the alumina bionic feedthrough was the key event that made the pacemaker a mass-market technology. Spin-offs have included the defibrillator, bionic ear, bionic eye, deep brain stimulator, spinal cord stimulator, and many more. Indeed there are now 14 spin-off bionic implant types that make up the global $25 billion pacemaker industry today. This book details the entire bionic feedthrough evolution, from the invention of the alumina bionic feedthrough in the late 20th century through to the future: the brain-computer interface (BCI). The BCI is already prototyped using thought control to move a robotic arm. The future for the alumina bionic implant is an exciting one indeed. I was part of the team that set the world record for the largest channel number in an alumina feedthrough in 2012: 1145 channels for a bionic eye retinal implant. Consequently, this book can offer unique perspectives on this breakthrough. Chapters 8–10 outline the fascinating development of the alumina bionic feedthrough.

Hip replacements: At the turn of the century, almost all hip replacements used CobaltChrome-on-Polyethylene bearings, which deposit 2 trillion polyethylene wear particles in the hip joint per year, with all the problems that flow from that. One year's wear with alumina bearings is equivalent to one day's wear with a CobaltChrome-on-Polyethylene bearing. Consequently, alumina hip bearings have shown a meteoric rise since the turn of the century from obscurity to now over 50% of the 1.3 million hip replacements implanted annually. The author has been involved in this field since the 1990s. This incredible alumina hip replacement technology is discussed in Chapters 6 and 7.

Lightweight body armor: Millions of lives have been protected and many saved since the late 1990s thanks to lightweight ceramic body armor. Ever since the gun made metal armor obsolete in the 15th century, soldiers, police, and security guards were routinely sent into combat situations unprotected against hails of bullets. Our society came to accept this grotesque state of affairs as normal. Alumina lightweight body armor changed all this. Developed by the US military in secret between 1945 and 1963, and rarely seen by anyone but elite US special forces until the late 20th century, in 1996, the US military deployed it to their regular troops, which opened the floodgates globally, and in the early 21st century virtually every country in the world has been feverishly uparmoring their military, police, and security forces with the miracle of lightweight ceramic body armor.

I have been a consultant in this field since the early days (1988) and I was a director of one of the two leading RSSC armor ceramic companies in the world from the 1990s. This book is also the first published document I am aware of that traces the origin of ceramic body armor to its original inventor, whose name appears in a classified 1945 US Military report (now declassified) and was lost to history until now. His discovery has protected millions of lives worldwide since 1996. Life-saving alumina body armor is the focus of Chapter 11.

Wear-resistant alumina linings: The global mining industry is a $500 billion industry. Excavation and mineral processing cause huge amounts of erosion of mineral processing equipment. Alumina has revolutionized this industry since the late 20th century, because of its unique combination of low cost and outstanding wear resistance. Chapter 12 not only explores the science behind the extraordinary wear resistance of alumina, but also has a detailed industry case study profiling the comprehensive alumina technology of a company (Taylor Ceramic Engineering) that has been a pioneer in industrial alumina innovation for nearly half a century.

Alumina in electronics: In the 1980s, the semiconductor industry was just $12 billion. Today it is a $500 billion industry. Also in that time, microchip nanocomponent density has increased 1 millionfold. Alumina uniquely combines extreme electrical resistivity, with high-thermal conductivity and low cost. This makes it the market workhorse material for microchip heatsink substrates in the $50 billion silicon on insulator (SOI) microchip market sector that is particularly important to the MEMS market (e.g., accelerometers and gyroscopes for smart phones, tablets, and cameras, plasma TV drivers). Alumina also has a uniquely low dielectric loss tangent, which makes it extremely important to the microwave industry. The microwave industry is huge globally: from the billion ovens globally to the millions of microwave generators in telecommunications, cellphone towers, radar, and other industrial technology. Alumina has many other vital roles in electronics, for example, it dominates the $3 billion spark plug industry. Chapters 13 and 14 comprehensively cover alumina in electronics.

Refractories: The global heat containment industry (refractories) is a $50 billion industry in which alumina has a key role, as outlined in Chapter 15.

The role of alumina in other important fields such as dentistry (transparent orthodontic brackets, implants, etc.), aerospace, architectural ceramics, machine tools, oxygen sensors, and other industrial uses is also outlined in this book.

This book also has four chapters on alumina fundamentals. Its history, an industry overview, a detailed review on Bauxite and Bayer refining, and a detailed overview of alumina manufacturing techniques, its structure, and properties. Chapter 1 overviews the alumina industry. Chapter 2 is a short review of the bauxite industry. Chapter 3 is a detailed review of the Bayer process for refining alumina. The global alumina industry produces a staggering 100 million tonnes of alumina a year, most of which is used for aluminum smelting, but this huge industrial scale means that high-quality alumina powder is cheap and plentiful. This is one of the reasons for the success of alumina. Outstanding properties combined with a capacity for pressureless sintering in air to full density, and plentiful and cheap raw material. It is a triumvirate of industrial advantage. In Chapter 4, I draw on my extensive experience of three decades in alumina fabrication technologies.

I have been involved in alumina ceramics as a researcher and as an industrial consultant, for three decades. My career has uniquely straddled both bioceramics for medical devices, for which alumina is the premier bioinert and wear-resistant candidate (a significant focus of my academic research career), and technical ceramics for industrial and general engineering applications for which alumina is the premier cost-effective wear-resistant material of choice (a major focus of my long career as an industrial consultant). There is almost not an area covered in this book that I have not at some stage been involved in. The last three decades have been an interesting journey in the world of advanced ceramics and biomaterials.

Moreover, I believe the time is now right for a book like this one to bring us up to date on the extraordinary progress in alumina ceramic technology of the last few decades. This book is not just a scholarly update on alumina research, it has a dual focus on the research and the commercial applications. In that regard it contains an important industrially relevant case study from a company with a stellar track record of over 40 years of innovation in industrial alumina ceramics: Taylor Ceramic Engineering—Alyssa and Julie Taylor. The book also benefits from the input of key pioneers from the biomedical/industrial alumina realm: Bill Walter (orthopedics), David Cowdery (pacemaker), Peter Crowhurst and Dawn Kanost of Ceramic Oxide Fabricators, Paul Carter (bionic ear), and Gregg Suaning (bionic eye).

The intention is that this will be a useful reference work for engineers, medical personnel, students and scholars, as well as those in the diversity of industries that depend on alumina ceramics: orthopedics, bionics, dentistry, armor, mineral processing, paper making, textile production, chemical processing, the electrical and microelectronics industries, and refractories.

This is not just a book about the science. It is also about the engineering, and the commercial side—the $Million and $Billion markets for alumina.

The book documents all the applications of alumina, both those that are currently significant and those that are currently insignificant, always with a fourfold focus:

- History and evolution of the technology. Without understanding the past, how can we understand the present or envisage the future?
- The underlying science and engineering principles.
- The applications from both a technological and commercial perspective.
- The future: where to from here?

The chapter structure is designed to focus on the most commercially significant applications, but secondary applications were also addressed. Minor applications of today may be significant tomorrow. For example, bionics is arguably the number one commercial application for alumina ceramics in the world today, the focus of three chapters of this book, but it did not even appear in the last alumina book in 1984, at which time it was a seemingly insignificant niche application. The book is divided into three sections:

Section A: Alumina as a ceramic
    (1) Introduction to alumina
    (2) Bauxite
    (3) Bayer refining
    (4) Processing structure and properties
Section B: Alumina in biomedical engineering
    (5) Dental and biomedical niche applications
    (6) Alumina in orthopedics: Origin and evolution
    (7) Alumina in orthopedics: Present and future
    (8) Alumina in bionic implants: Pacemaker
    (9) Alumina in bionic implants: Bionic ear
    (10) Alumina in bionic implants: Bionic eye and the future
    (11) Alumina in lightweight body armor
Section C: Alumina as an industrial ceramic
    (12) Alumina as a wear-resistant ceramic
    (13) Alumina as an electrical insulator
    (14) Alumina-metal bonding for electrical feedthroughs
    (15) Refractory and industrial niche applications

I do not wish to prolong this preface with a detailed exposition on each chapter. The title of each is self-explanatory. However, it is important to note that the proportion of the book allocated to each alumina ceramic application is approximately equal to the level of its commercial significance in the world today. However, this is not to be taken as a statement regarding either its technical significance, or its significance in the world of the future. No application, whether significant or niche, is overlooked in this treatise.

While this book has a technical and industrial focus, I am mindful of the importance of history, and where appropriate, incorporate history into my review sections. We must never forget, as we move onward, and ever upward, that we stand on the shoulders of giants to do so. There is much to learn from the footsteps of our predecessors. I would therefore like to finish this preface with a quote from the 1831 book of Thomas Thomson, who chronicled the slow process of the discovery of alumina by a series of visionary pioneers during the 1700s.

Andrew J. Ruys
Biomedical Engineering, University of Sydney, Sydney, NSW, Australia

The beginning

Alum is a salt which was known many centuries ago, and employed in dyeing, though its component parts were unknown. The alchymists discovered that it is composed of sulfuric acid and an earth; but the nature of this earth was long unknown. Stahl and Neuman supposed it to be lime; but in 1729 Geoffroy, junior, proved this to be a mistake, and demonstrated that the earth of alum constitutes a part of a clay. In 1754, Margraff showed that the basis of alum is an earth of a peculiar nature, different from any other; an earth which is an essential ingredient in clays, and gives them their peculiar properties. Hence this earth was called argil; but Morveau afterwards gave it the name of alumina, because it is obtained in the state of greatest purity from alum. The properties of alumina were still farther examined by Macquer in 1758 and 1762, by Bergman in 1767 and 1771, and by Scheele in 1776: not to mention several other chemists who have contributed to the complete investigation of this substance. A very ingenious treatise on it was published by Saussure, junior, in 1801.

T. Thomson, A System of Chemistry of Inorganic Bodies, Baldwin & Cradock/William Blackwood, London/Edinburgh, 1831 (Quoting from Volume 1, page 451).

Introduction to alumina ceramics

1.1 Introduction to alumina ceramics

The word "alumina" conveys images of a pearly white ceramic. However, alumina ceramics are much more than that. In the biomedical sector, alumina ceramics have delivered important life-saving and life-enhancing technologies to millions of people in the world in the last few decades. Moreover, in the industrial sector, alumina underpins some of the world's largest industries: mining and mineral processing ($500 billion), semiconductors ($500 billion), and refractories ($50 billion).

1.1.1 Live-saving alumina technologies

Implantable bionic pacemaker (Chapter 8): All cardiac pacemakers use an alumina feedthrough, a key enabling platform technology that made the bionic implant possible. Millions of implantable bionic pacemakers have kept arrhythmia sufferers alive in recent decades (the great majority since the turn of the century), and more still since the advent of the implantable bionic defibrillator and pacemaker/defibrillator.

Alumina ceramic lightweight body armor (Chapter 11): Millions of body armor insert plates have protected lives in recent decades, the great majority since the turn of the century when ceramic lightweight body armor went mainstream.

Many lives have been saved as a result, and I consider myself fortunate to have been able to serve the community by being active in both these technology sectors in my long career.

1.1.2 Life-enhancing alumina technologies

Orthopedics (Chapters 6 and 7): Alumina bearings bring greatly enhanced implant longevity to about 700,000 or so hip replacement recipients a year, i.e., 55% of the 1.3 million hip replacement recipients, the half who are lucky enough to receive the "deluxe" alumina version of the hip replacement, rather than the regular CobaltChrome hip replacement, and its problems with 2 trillion polyethylene wear particles released per year, and cobalt and chromium release into the joint cavity through galvanic corrosion and fretting wear.

Other bionics (Chapters 9 and 10): While it is the pacemaker that saves lives, all bionic implants use an alumina feedthrough. Today there is a wide range of bionic implants on the market that treat drug-resistant Parkinson disease, dystonia, obsessive-compulsive disorder (OCD), depression, chronic pain, sensorineural deafness, blindness, and incontinence, to name but a few of the 25 or so significant disease states now treated by bionic implants.

1.1.3 Community enhancing

Mining industry (Chapter 12): As a low-cost highly wear-resistant ceramic, alumina has significantly enhanced the economic viability of the $500 billion mining and mineral processing industry, one of the world's largest industries.

Textile industry (Chapter 12): This is a significant employer in developing countries, where the affordability of alumina and the longevity it brings to textile processing equipment is of significant economic and social importance.

In addition, alumina is important in many other large global industries such as electronics, and refractories.

One of the most unique aspects of the global alumina industry is how little interaction is there between those involved in each market sector. There are a number of large global markets for alumina, but those involved in each market have virtually no interaction with those from the other markets. This is much more pronounced for those involved in the biomedical applications (orthopedic bearings, feedthroughs for implantable bionics, lightweight body armor, dental implants, and tissue scaffolds) than for those involved in the industrial applications (wear-resistant linings, refractories, microchips, insulators, and electrical feedthroughs). Those involved in the industrial applications are commonly aware of the other industry sectors, and even to some extent, the biomedical sectors. However, in my long career in bioceramics and industrial ceramics since the 1980s, in which I have had an involvement in all of the above sectors, I have never come across such a book or review that has overviewed all biomedical and industrial applications in one document.

The most recent monograph written on alumina, the 1984 book of Dorre and Hubner [1], came the closest to achieving this. However, the rudimentary world of alumina in 1984 is so far removed from the world of alumina today, that this excellent book written 35 years ago has mainly historical value today. Bionics did not feature in the 1984 book at all, and lightweight body armor received only a very brief mention. Orthopedics covered several pages by virtue of the fact that the principal author Erhard Dorre was a pioneer in orthopedic alumina. Dorre was from Feldmuhle, the German company that pioneered the commercialization of the alumina hip bearing in the late 20th century, and became CeramTec some years later. The name CeramTec is now synonymous with alumina hips, and has a global monopoly on the alumina orthopedic bearings industry today.

It is interesting to reflect that now, 35 years after the publication of the book of Dorre and Hubner [1], implantable bionics is the largest alumina-dependent biomedical industry at $25 billion and has three chapters in this book. Hip replacement is the second largest alumina-dependent biomedical industry at $7 billion, and has two chapters in this book. Lightweight body armor is the third largest alumina-dependent biomedical industry, in the multiple hundred million dollar range, and has one dedicated chapter in this book.

In the industrial realm, wear resistance, electrical applications, and refractories are all in the multiple hundred million dollar (wear) to billion dollar range (electrical and refractories), with electrical applications the largest sector of the three, receiving two dedicated chapters in this book. The global semiconductor industry is now $500 billion a year, with the alumina SOI (silicon on insulator) sector $50 billion. Wear resistance and refractories received a chapter each. Even in the industrial sector, the world of 2018 is very different to 1984. The now $50 billion refractories industry has changed almost beyond recognition since 1984. The fledgling semiconductor industry of 1984 is now a $500 billion industry, and the fledgling wear-resistant industry of 1984 is now a vital component of the global $500 billion mining industry.

Thus the time is ripe for a book that overviews all of the contemporary commercial uses of alumina, combining a historical context with a focus on current research and future directions.

1.1.4 Scope of this book

This book is focussed on *monolithic alumina ceramics*, where the assumption is that the ceramic is alumina in the 85%–100% purity range.

1.1.4.1 Alumina topics WITHIN the scope of this book

Monolithic alumina ceramic, in the 85%–100% purity range: While most commercial applications for alumina involve purities above 94%, there are significant commercial uses for alumina down to 85% pure.

Thin and thick film alumina wafers: Alumina wafers in the electronics industry are included in this book, since they are monolithic alumina ceramics, however thin they may be.

Zirconia toughened alumina (ZTA): Strictly speaking this is an alumina matrix composite, as it involves up to 20% zirconia microparticles in an alumina matrix. The ZTA is within the scope of this book. This is because ZTA is proving to be a very commercially important modern manifestation of alumina in orthopedic, and dental, applications, and a discussion of the biomedical applications of alumina would be incomplete without comprehensive coverage of ZTA. Moreover, ZTA is not an advanced alumina composite (unlike FGMs, SiC whisker-reinforced alumina, DIMOX alumina, etc.) but is essentially processed and used just like a conventional aluminaceramic that has up to 20% zirconia added.

1.1.4.2 Alumina topics BEYOND the scope of this book

Powdered alumina commercial technologies: Powdered alumina is used on a very large scale. The abrasives industry is currently $40 billion. There are also many other commercially important uses for powdered alumina in polishing powders, catalysts, chromatography, dehydrating agents, alumina cements, manufacture of zeolites, fire retardant/smoke suppressant, as an additive in glass and other ceramics, and numerous other minor uses. The scope of the powdered alumina industry is very large and deserves to be the topic of a dedicated book. Moreover, these commercial applications for powdered alumina are not part of the alumina ceramics industry, which concerns advanced ceramics made of pure-alumina or high-alumina formulations.

Alumina coatings: This book will not discuss alumina coatings, neither thin film nor thick film coatings. Coating is an industry in its own right with a very broad scope of its own.

Alumina-based advanced composite materials: This book is not about advanced alumina-based composite ceramics such as fiber-reinforced alumina, alumina-reinforced glass-matrix composites, calcium-alumina composites, DIMOX (directed-metal-oxidation) alumina composites, or alumina-based functionally graded materials (FGMs). With all their specialized manufacturing technologies and extraordinary properties and applications, advanced composites represent a scope so broad and so different in application to pure alumina that this deserves to be the topic of a dedicated book.

1.1.4.3 Focus of this book

This book is not so much about the science of alumina ceramics as it is about the engineering of alumina ceramics. This is an important distinction. Engineering is about applications. It is about taking scientific discoveries and using them to develop and commercialize applications, so as to build industry, and bring technological benefits to the society. A great deal of quality alumina research was done in the mid-to-late 20th century. That science has come to fruition in the 21st century. Nonetheless, given that it is a third of a century since the last alumina monograph [1], this book will also review the science of alumina, but always, the focus will be on the engineering applications: where the science has taken us, in the world-changing engineering applications of alumina that we benefit from in the world of 2018 (Fig. 1.1).

Alumina was virtually unknown until the post-World War II era. The alumina ceramics industry began around 1950, but this promising material remained the focus of scientific study and engineering obscurity for many years, while the scientific knowledge slowly gathered momentum up until the mid-to-late 20th century, when

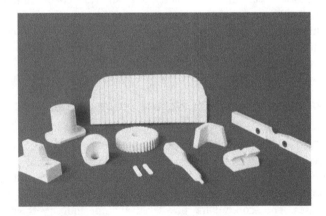

Fig. 1.1 Specialty alumina ceramics.
Manufactured by Taylor Ceramic Engineering (Image courtesy of Taylor Ceramic Engineering).

the last monograph updating progress in alumina was published in 1984 [1]. The year 1984 was a time point midway between now and the beginning of the alumina ceramics industry in around 1950. In the 35 years since 1984, alumina science has given way to alumina engineering. In 1984 it could be said that alumina ceramics were beginning to be noticed. Today, it can be said that alumina ceramics have changed the world.

The primary commercial applications for alumina as a ceramic arise from its combination of low cost for an advanced ceramic, with a set of outstanding mechanical/chemical/thermal properties commensurate with much more expensive materials. The key properties are as follows:

- low production cost—both in raw materials and in manufacture
- excellent size and shape capability
- capable of being pressureless sintered in air to high density
- with sintering aids the grainsize and sintering temperature can be greatly reduced
- hard and wear resistant (second only to diamond in the Mohs scale)
- high strength and stiffness
- outstanding electrical insulator, combined with high thermal conductivity making it an outstanding heatsinking insulator
- extremely low dielectric loss tangent up to GHz frequencies
- high chemical inertness and corrosion resistance
- outstanding biocompatibility
- refractoriness
- high-temperature oxidation resistance

In short, alumina ceramics have a unique combination of industrially valuable properties, combined with an unusually low production cost for an advanced ceramic. This means that alumina can deliver outstanding properties for a much lower price than its competitor ceramics.

This is why alumina has now become the premier material of choice as the bearing material in hip replacements, as the feedthrough material for bionic implants, as an outstanding heatsinking insulator for microelectronics and microchips, as the hardlining material for mining and mineral processing equipment that supports the global mining industry, as the lightweight hard ceramic component in body armor, and in many specialist applications that require extreme hardness, extreme bioinertness and chemical durability, and affordability, or combinations thereof. Moreover, outstanding biocompatibility and electrical resistivity of alumina have enabled its widespread use in electronics and bionic implants. The most important commercial applications of alumina are as follows.

Bionics: Alumina ceramics in bionic implants was a fledgling technology in 1984. Now bionic implants, which exclusively use alumina feedthroughs, are a $25 billion industry (Fig. 1.2).

Orthopedics: Well over 10 million alumina-bearing components have been implanted in hip prostheses in the last decade or so, in the $7 billion PA global hip replacement industry. The total number of alumina bearing components implanted annually is currently approaching a million. Until the turn of the century, alumina bearings in orthopedic hip replacements were still seen as a research activity (Fig. 1.3).

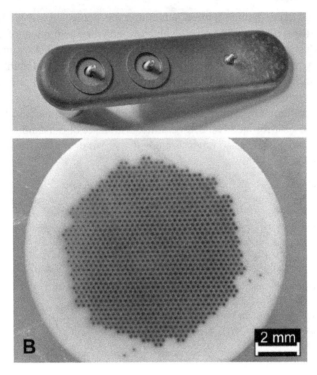

Fig. 1.2 Alumina bionic feedthroughs. Above: 1980s era two-channel alumina/titanium-titanium bionic feedthrough—9-mm alumina cylinder with 1.5-mm titanium conductor (courtesy of David Cowdery). Below: Current world-record breaking 1145 channel feedthrough alumina-platinum co-fired feedthrough 7-mm diameter region containing the 1145 channels (courtesy Elsevier Journal "Biomaterials"). Shown ready for brazing into a titanium casing for a bionic eye retinal implant. Invented by Gregg Suaning, with principal researcher Thomas Guenther, and other team members including Rylie Green, Charlie Kong, Hong Lu, Martin Svehla, Nigel Lovell, Christoph Jeschke, Amandine Jaillon, Jin Yu, Wolfram Dueck, William Lim, William C. Henderson, Anne Vanhoestenberghe, and Andrew Ruys (for more details about the author's role in this breakthrough see "About the Author") [2]. The progress in alumina bionic feedthroughs since the 1980s is remarkable.
Image courtesy of David Cowdery.

Lightweight body armor: Lightweight alumina ceramic armor was veiled in military secrecy for many years after the first use of alumina armor in 1965. It remained an almost unknown concept in the public domain until the 1990s when alumina armor (and ceramic armor in general) made the move from a low-tech concept used only in a few niche applications to a high-tech concept that today sees widespread use in infantry and law enforcement situations. Today alumina ceramic body armor and vehicle armor are mass-market products representing an industry in the multiple hundred million dollar range (Fig. 1.4).

Wear-resistant applications: While alumina ceramics have been used in wear-resistant applications since the 1950s, progress was very gradual for a long time. In the recent decades, the use of alumina has become very widespread in the $500 billion global mining

Introduction to alumina ceramics

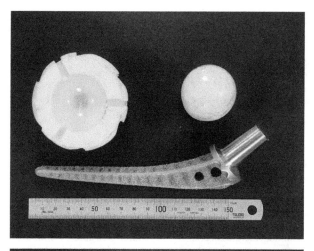

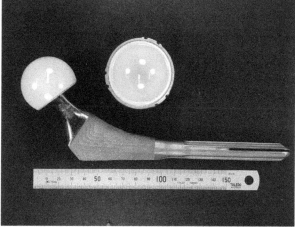

Fig. 1.3 Above: 1980s era experimental alumina ball-socket hip replacement. Failure by fracture rates in service in this era were sometimes over 1%. Below: Contemporary alumina (zirconia toughened) micropolished ball-socket hip replacement. Failure by fracture rates in service are <0.001%, a 1000-fold improvement in mechanical reliability of alumina orthopedic ceramics since then.
Exhibit provided courtesy of Professor Bill Walter. Specialist Orthopedic Group. Wollstonecraft NSW.

and mineral processing industry. It is also widely used in many wear-resistant applications in other industries such as the paper industry and the textile industry, representing a global industry in the multiple hundred million dollar range (Fig. 1.5).

Electrical applications: Electrical applications for alumina were the first of the large markets for alumina ceramics to arise, beginning with early alumina sparkplug development trials in the 1930s (now a $3 billion industry). In the postwar era, three $multibillion markets have arisen: alumina-kovar electrical feedthroughs beginning in the 1960s, alumina substrates for

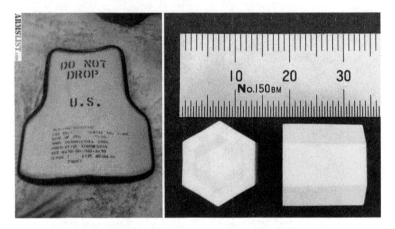

Fig. 1.4 Left: Alumina-fiberglass ceramic body armor breastplate technology available in the 1980s [3]. Right: Contemporary precision-engineered (hexagonal, perimeter lip) miniature alumina armor mosaic tile for lightweight multihit armor panel.

Fig. 1.5 Contemporary alumina wear-resistant pump components. From rudimentary beginnings in the 1960–80s, a huge diversity of precision wear-resistant alumina ceramics are now on the market: large, small, complex shape, complex radii, and tight tolerances. Manufactured by Taylor Ceramic Engineering (Image courtesy of Taylor Ceramic Engineering).

thick-film multilayer microelectronics beginning in the 1970s, and thin-film SOI alumina substrates for microchips beginning in the 1980s. The microchip industry has seen meteoric growth since 1980 when the fledgling microchip industry was around $12 billion. Today it has grown to $500 billion, with a million-fold increase in nanocomponent density since the 1980s (Fig. 1.6).

Refractory applications: This industry was already well established in the 1980s, but has been dramatically revolutionized since then, with the switch to monolithics. The refractories

Introduction to alumina ceramics 9

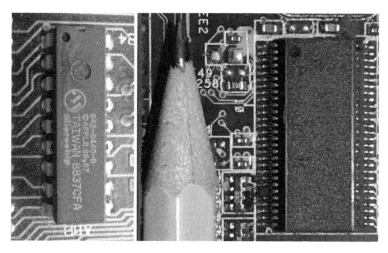

Fig. 1.6 Alumina thin-film substrates are important components in the microchip industry. Left 1988 computer microprocessor from one of the early personal computers. Right: Contemporary microprocessor, with a pencil for scale (note this microprocessor is substantially more miniaturized). The microchip industry has grown from $12 Billion to $500 billion since 1980, with an associated million-fold increase in nanocomponent density.
Image Author Binarysequence [4]. Reproduced from Wikipedia (Public Domain Image Library).

industry has certainly grown since the 1980s: valued at $50 billion today. The primary role for high-purity alumina ceramics is in the glass-contact refractories subsector, a multiple hundred million dollar industry, while alumina as an ingredient is also an important component in a large proportion of the entire refractories industry.

Each of these developments (orthopedics, bionics, armor, wear resistance, electrical, refractories) involves a subsector of the alumina ceramics market in hundreds of millions to billion dollar order of magnitude in its own right.

The following list compiles all the significant uses for alumina in the world today, each of which will be discussed in this book. The focus will be on the more important applications in terms of dollar value and humanitarian benefit, but all these applications compiled below will be discussed in this book.

Traditional industrial applications and biomaterials applications for alumina are easily categorized as per the list below. Lightweight body armor uses of alumina, and its spin-offs in lightweight vehicle armor, are neither a traditional industrial application, nor a biomaterials application. They do involve protecting the human body, combining materials science and the science of traumatology, and have delivered an enormous humanitarian benefit to society with millions of lives protected, and many saved. Therefore, body armor and its spin-offs have been categorized in this book as belonging to the biomedical engineering category, though not biomaterials.

In summary, the following applications for alumina ceramic will be discussed in this book:

1.1.4.4 Section B: Biomedical engineering applications for alumina (Chapters 5–11)

$Billion biomaterials markets for alumina

- Orthopedic hip bearings (Chapter 6 and 7). $7billion Hip replacement industry
- Bionic feedthroughs (Chapters 8–10) $25 billion bionic implants industry

Niche biomaterials markets for alumina (Chapter 5)

- white or translucent orthodontic brackets
- dental crowns
- dental bridges
- dental abutments
- dental implants
- tissue scaffolds
- prosthetic eyeball
- osteotomy spacers

$Multimillion biomedical engineering markets for alumina—Armor realm (Chapter 11)

- lightweight body armor
- lightweight vehicle/aircraft armor
- synthetic sapphire bulletproof windows

1.1.4.5 Section C: Traditional industrial applications for alumina (Chapters 12–15)

Multiple hundred $Million wear- and corrosion-resistance market (Chapter 12)

- hydrocyclone components and linings (mineral processing)
- spigots (wear and corrosion)
- straight and curved piping (wear and corrosion)
- reducers (wear and corrosion)
- orifice and baffle plates (wear, corrosion, papermaking)
- dart valve plugs and Seats (wear and corrosion)
- tiles for ore chutes (standard, engineered, and weldable)
- knife-edge blades (conveyors, wine industry)
- mill linings and milling media (wear and corrosion)
- thread guides (textiles)
- wire drawing step cones and dies (wire)
- brick and heavy clayware (brick cores, sleeves, die boxes, and shaper caps)
- augers and trough liners (numerous industries)
- nozzles (wear and corrosion)
- water-faucet valves
- rotary seals
- bearings and sleeves
- cylinders
- rods and disks

- spacers, tubes, and washers
- tiles (standard, engineered, and weldable)

$Billion electrical applications market (Chapters 13 and 14)

- electrical insulators: high voltage,
- electrical insulators: high temperature
- electrical insulators: high frequency
- sparkplug insulators
- thin-film SOI microelectronic substrates
- thick-film microelectronic substrates
- microelectronic insulating heatsinks
- hermetic electrical feedthroughs
- fuse bodies
- heating-coil formers
- oxygen sensors

$Billion heat containment applications market (Chapter 15)

- general refractories ($50 billion industry of which alumina is a key component)
- $1.3 billion fuse-cast glass-contact refractories sector, of which alumina services a significant proportion
- crucibles
- furnace tubes
- labware

Other niche industrial applications (Chapter 15)
- machining/cutting tools
- architectural
- aerospace applications
- gas sensors
- gas laser tube
- laboratory instrument tube
- Kiln furniture
- thermocouple sheath
- microwave window
- agricultural-plow tine tips
- alumina gauge
- jewel bearings in watches
- mortar and pestle

This book will tie together the world-changing research and industrial progress on alumina ceramics since alumina ceramics became a fledgling commercial product in around 1950 in the postwar era, and particularly since the 1980s, and bring us up to date from both a technical and commercial perspective on the developments with this extraordinary ceramic material alumina, and the million dollar and billion dollar markets it has generated in the recent decades.

This book, in addition to all the necessary background material on science, structure, and properties of alumina, will particularly examine the developments that have taken place in recent decades with regard to the fabrication and optimization of alumina ceramics, and with regard to the numerous applications for alumina globally. Not just the large dollar-volume markets in orthopedics, bionics, body armor, mineral processing, electronics, and refractories, but also the other boutique applications.

Finally, this book will not overlook the historical context of the early developmental period, where the knowledge and groundwork were laid back in the 18th, 19th, and 20th centuries. This monograph is intended as a full update, spanning the full 70-year history of alumina ceramics, and its prehistory.

The concluding chapter, Chapter 16, will look at what the future holds for this premier high-performance ceramic. From electronics to orthopedics to implantable bionics for the brain-computer interface, alumina looks set to remain on the center stage in the realm of advanced ceramic technology.

1.1.5 Outline and structure of this book

The book is structured in three sections, each comprising several chapters, as follows:

- Alumina as a ceramic
- Biomedical applications of alumina ceramics
- Industrial applications of alumina ceramics

The chapter breakdown is designed to present alumina in a logical sequence, from the basic ceramic technology, moving to the biomedical applications, and finally exploring the industrial applications. This chapter structure does not mesh with the chronology of the development of alumina, but rather it meshes with the commercial relevance of alumina in the world today.

This is a technical monograph with an application and commercial focus, and a seamless transition between the scientific overview and the commercial overview aspects of the work. This book is overviewing a single material, alumina ceramic, but this material has evolved independently, with little cross-pollination, in several distinct and disconnected industries, and therefore, the science and commercial application research focus differs for each application, in terms of the timing of the developments, and the current commercial status. The actual chronology of alumina technology is as follows:

Electrical and refractory applications (Chapters 13 and 14): The original applications for alumina, which arose in the early to mid-20th century, were refractory and electrical applications. These evolved quite independent of one another in approximately the same time scale. The public domain scientific research underpinning these developments was clustered in the mid-20th century, beginning with a small trickle from 1910 to the 1940s, and with early commercialization beginning in the 1950s, and evolving throughout the late 20th century. Since the 1990s, the focus has been on commercial activity, with most of the research in the form of confidential corporate in-house applications and process-focussed technology development, while the focus of pure published science has moved on to new fields of ceramic research.

Wear and corrosion resistance applications (Chapter 12): The next application to arise was wear-resistant and corrosion-resistant applications. The scientific research underpinning these developments was clustered in the mid-to-late 20th century, from the 1950s to the

1990s, with commercialization beginning in the 1950s in the textiles industry, then moving on to many other fields. Since the 2000s, the research focus has shifted from public-domain scientific research to proprietary fine-tuning, with an application focus, in the industrial domain, while the focus of pure published science has moved on to new fields of ceramic research.

Ballistic armor (Chapter 11): The next major application for alumina to arise was ballistic armor (Chapter 16). This had its origins in a secret US military program from 1945 to 1963, and very slowly began to trickle into the public domain from the 1980s. It didn't become a mainstream technology until the turn of the century. Much intensive research has taken place in this field as military research since the 1960s, and public domain research much later, reaching its zenith in the first two decades of the 21st century. Scientific papers exploring new ground in this area remain commonplace today, but this is just the tip of a huge, and largely secret, industrial/military iceberg in terms of the current ongoing developments in the field.

Bionic implants (Chapters 8–10): Rudimentary bionic implants arose around the same time as ballistic armor, with their origins in the late 1950s. Implantable bionics depend entirely on alumina feedthrough technology. This field did not begin until the 1970s with the invention of the hermetic alumina/titanium feedthrough for the world's first hermetic implantable bionic cardiac pacemaker in 1970 (Chapter 8). An explosion of private scientific and industrial activity followed, which continues intensively today, with now 14 different bionic implant technologies on the market treating over 25 disease states (Chapters 9 and 10). The zenith of the global research effort in bionic alumina technology is yet to be reached as we move into new and exciting futuristic applications in smart bionic implants, as well as the recently invented bionic eye and the brain-computer interface bionic implant which is currently undergoing development. The scientific literature is highly active in this field today. The future holds great promise.

Orthopedics (Chapter 6 and 7): The orthopedic realm for alumina was the last to arise. With its scientific origins in the 1970s, and a commercial launch in the mid-1990s, but real momentum only developed in the 21st century. We stand at the early years of an exciting era with orthopedic applications of alumina, and the scientific literature is most active of all alumina technologies in this field today.

Alumina—One material, multiple independent industries: Thus alumina is one ceramic, but it has been commercialized essentially in a set of completely independent industries, which have shown very limited cross-pollination. Each is following its own trajectory in terms of the scientific development, and the subsequent commercial development, and mass-marketing activity. If this book achieves one thing, my hope is that it serves as an impetus for cross-pollination and all industries would benefit from this, and ultimately society would benefit.

1.2 Natural alumina

Aluminum is the world's third most abundant element, comprising 8% of the Earth's crust, behind oxygen (nonmetal) and silicon (metalloid), making aluminum the most abundant metal in the Earth's crust. Paradoxically, in spite of its abundance, aluminum was not discovered until surprisingly recently, just two centuries ago in 1808 by British chemist Humphry Davy.

Aluminum is found in a wide range of minerals, and is present in trace quantities in nearly all minerals. Aluminum-bearing minerals have been used by humanity since before the dawn of civilization, particularly clay. Also aluminum has seen long use in a purer chemical form, for example, aluminum hydroxide has been used since the time of ancient Egypt in textiles as a mordant (die fixing agent), and interestingly, it was in servicing this application, in 1887, that Dr. Karl Josef Bayer invented the now famous Bayer process, which launched both the global aluminum and alumina industries, and for well over a century has been the dominant global process for refining alumina. The importance of the Bayer process is so pivotal to both the global aluminum and alumina industries that it forms the basis of Chapter 3.

Aluminum is almost always found in an oxidized state in nature. Due to its strong susceptibility to oxidation, natural geological occurrences of aluminum are extremely rare, but have been found where extreme reducing conditions prevail. Only the noble metals such as gold are commonly found in their metallic state in nature.

Alumina is the ceramic name for aluminum oxide, and is aluminum in its highest oxidation state. In its pure form, alumina is aluminum oxide with the stoichiometry Al_2O_3, known geologically by its mineral name of corundum. Alumina is the only oxide of aluminum. In the aluminum-oxygen phase diagram, Al_2O_3 is essentially a point compound with no solid solubility region. It is important to make a clear distinction at this point between:

(1) Aluminum the metal (Al), also sometimes known as aluminum in the United States.
(2) Alumina the ceramic (Al_2O_3), also known as aluminum oxide, and aluminum oxide. Best known to the public by its gemmological names sapphire or ruby, it is also known by its mineral name corundum. The term corundum is usually used only in the context of powdered abrasives, and sapphire in the context of both gemmology and transparent synthetic alumina. Ruby is used exclusively in the gemmological context.

The terminology conventions of alumina need to be clarified at this point:

(1) Alumina is the name for advanced ceramic shapes and components that are pure (or near-pure) Al_2O_3.
(2) Chemical alumina is powdered Al_2O_3 that is used to make high-purity alumina ceramics, and also used in a wide range of chemical applications, such as catalysts and desiccants.
(3) Metallurgical alumina is powdered Al_2O_3 designated for aluminum smelting.
(4) Aluminum oxide is the name used in the abrasives industry for synthetic abrasive grit made of Al_2O_3.
(5) Corundum is a rarely used term, but when used it primarily refers to abrasive grit made of Al_2O_3 sourced naturally as the mineral corundum.
(6) Sapphire is used to refer both (1) synthetic transparent (usually single crystal) Al_2O_3; (2) the natural gemmological form of Al_2O_3 when colored (any color but clear or red, red is called ruby, clear is called corundum).
(7) Alumina cement refers to calcium aluminate hydraulic cement, widely used in bonding refractories.

This book concerns item 1, *alumina the ceramic*.

From time to time aluminum the metal, and aluminum oxide or corundum the abrasive grits are mentioned. Probably the greatest relevance of aluminum the metal to this book is the well-known fact that aluminum is much less susceptible to corrosion than steel. It is not because aluminum is more oxidation resistant than iron, indeed both oxidize rapidly when bare metal is exposed to the air. It is because the thin (several nanometres) oxide scale formed on aluminum is aluminum oxide—alumina in other words. It is a durable oxide, and well bonded to the parent metal aluminum. Thus the durability of aluminum is because of the thin durable well-bonded surface oxide coating of alumina the ceramic. Aluminum oxide the abrasive grit is relevant because chemically it is alumina, but not a ceramic, it is a powder.

The commercial importance of alumina derives from three main factors:

(1) Alumina is the raw material used for the smelting of aluminum. Over 60 million tonnes of aluminum are produced every year globally from alumina.
(2) Alumina powder is an important industrial chemical in a wide range of applications from hydrocarbon desiccant to catalyst to abrasive.
(3) Alumina ceramics have a unique combination of extreme hardness, outstanding biocompatibility, refractoriness, electrical resistivity, thermal conductivity, and low production cost. In the last few decades they have changed the world, as this book outlines.

While elemental aluminum is abundant in nature, alumina itself as the mineral corundum is relatively uncommon. Corundum, being pure crystalline alumina, is very hard with a hardness of Mohs 9, which makes it an excellent abrasive. Corundum (discovered in 1799) is mined from a number of locations worldwide for use as an abrasive, including the United States, India, Sri Lanka, Russia, and Zimbabwe [5]. There is also a lower-grade alumina mineral called emery, a relatively uncommon mineral mined in Greece and Turkey [6]. Emery is an impure alumina mineral incorporating significant quantities of noncorundum softer minerals, with the result that it is a softer abrasive of around Mohs 8. Emery is also used as an industrial abrasive, but is less commonly used than corundum. The global abrasives industry is very large, and only a very small portion of it is serviced by corundum and emery, with a wide variety of other synthetic industrial abrasives in current use.

Corundum is a transparent visually attractive mineral, commonly found in large crystals in nature. These can be very large, crystals up to half a meter in size have been recorded [7]. At the time of its discovery in 1799, corundum was seen to be a new mineral, hard, crystalline, and not attractively colored. However, corundum has been known since antiquity to the general public by its gemmological names sapphire (traditionally the blue manifestation, but in practice any color but red or clear) and ruby (red manifestation).

Corundum can be found in many colors, depending on impurities present, and when red and of high quality it is called ruby, and when blue and of high quality it is called sapphire. Ruby and sapphire have high intrinsic value as *gem*stones. Strictly speaking, sapphire is merely good quality blue corundum, and ruby is good quality red corundum. Gemmological value is mainly about appearance and color. When sapphire is red it is called ruby and when pink, it may be considered a pink sapphire or a pink ruby depending on who is consulted for an opinion. Clarity, transparency, perfection, and size of the *gem* are also very important in determining the value (Fig. 1.7).

Fig. 1.7 Alumina in the form of naturally occurring single crystals. Left: Sapphire in classic "sapphire blue" color. Middle: Ruby in classic "blood red" color. Right: Corundum crystals, rough, uncut, and of various colors.
Left: Image Author LesFacettes [8]. Middle: Image Author Humanfeather [9]. Right: Image Author Ra'ike [10]. Reproduced from Wikipedia (Public Domain Image Library).

While corundum in nature is found in many colors, generally only pure clear "sapphire blue" and pure clear "blood red" fetch high prices as gemstones, as sapphire and ruby, respectively. There is a large industry dedicated to heat treating (under a controlled atmosphere) low-grade sapphires, such as a yellow, orange, and green sapphire to transform them to high-grade blue sapphire. A large proportion of gem grade sapphires are heat treated to improve them before sale. While there is a moral obligation, and in some jurisdictions a legal obligation, to disclose heat treatment, it is very difficult to determine if a sapphire has been heat treated.

In current practice, heat treatment of sapphires remains something of a modern form of alchemy. Some sapphire alchemists, through their secret knowledge, and a combination of luck, skill, and long experience, have made their fortune in this way. It requires a deep knowledge of the raw sapphire and the heat-treatment process. The same process will not give the same result for sapphires from a different location.[1] The heat-treatment process is done to change the color by changing the oxidation state of key trace impurities, such as iron, titanium chromium, and vanadium, through high-temperature oxidation/reduction and solid-state diffusion, and also to enhance clarity.

To a certain extent color is predetermined. The amounts and identities of these trace impurities are easily determined by elemental analysis of actual gemstones. Blue "sapphire" contains about 1% titania and 2% iron oxide. Red "ruby" contains about 1%–7% chromic oxide. Green "sapphire" contains 8% or more chromic oxide. The skill lies in successfully heat treating a sapphire so as to engineer this outcome. Control of temperature, furnace atmosphere, and other proprietary techniques are involved, and the process is stone-specific: sapphires from one location can behave completely differently when heat treated to sapphires from another location. Heat treatment can also remove small regions of opacity (imperfection) known as inclusions.

---

[1] The author was involved in two sapphire heat-treatment projects in the early 1990s, one sourcing sapphires from Inverell in Australia, and from China.

Low-grade sapphires can be bought in kilogram, while high-quality gem-grade sapphires can be sold for thousands of dollars per gram. Gemstones are sold in carats (1 carat = 0.2 g) and a fine quality sapphire will typically fetch from $1000 to $10,000 per carat, depending on the size, appearance, and quality. The world record was set in 2007: $135,000 per carat ($675,000 per gram) for a 22.66-carat sapphire [11]. Rubies are similar in value. In the gemstone hierarchy, sapphire and ruby are considered to be just below diamond in value, and in fact in some special cases, a high-quality sapphire or ruby can surpass a diamond in cost per carat.

Corundum is a relatively rare mineral. However, aluminum is abundant in nature in many less pure mineral forms. The most common of these is clay, a nanoparticulate aluminosilicate mineral. Aluminum is a major elemental component of clay. Clay is one of the most common minerals in nature. There are a large number of other natural minerals that contain significant amounts of aluminum, and this is how aluminum is predominantly found in nature, as a diffuse component in many minerals. The best of these, as an aluminum source for commerce, is bauxite. Virtually all alumina and aluminum produced in the world today comes from bauxite. Thus, due to the importance of bauxite, it forms the basis of Chapter 2 (Fig. 1.8).

1.3 A brief history of alumina

The history of the discovery of alumina, and methods for its chemical synthesis, was a slow evolution of understanding by a number of chemists working in the 18th century, prefaced by the earlier work of unknown alchemists. The discovery of alumina, and its name, both derive from the mineral alum ($KAl(SO_4)_2 \cdot 12H_2O$). Alum was well known and widely used throughout antiquity, the ancient Romans, Greeks, Egyptians, and Babylonians used it for a range of medicinal and industrial purposes, some of which

Fig. 1.8 A piece of Bauxite with US penny for scale.
Author unknown [12]. Reproduced from Wikipedia (Public Domain Image Library).

continued until modern times. Alum was also an important mineral to the alchemists, both Eastern and Western. The early history and discovery of alumina is linked to the study of alum, and is very well described by Thomas Thomson, in his 1831 book "A System of Chemistry of Inorganic Bodies," quoting from Volume 1 page 451 [13]:

> "Alum is a salt which was known many centuries ago, and employed in dyeing, though its component parts were unknown. The alchymists discovered that it is composed of sulphuric acid and an earth; but the nature of this earth was long unknown. Stahl and Neuman supposed it to be lime; but in 1729 Geoffroy, junior, proved this to be a mistake, and demonstrated that the earth of alum constitutes a part of a clay. In 1754, Margraff showed that the basis of alum is an earth of a peculiar nature, different from any other; an earth which is an essential ingredient in clays, and gives them their peculiar properties. Hence this earth was called argil; but Morveau afterwards gave it the name of alumina, because it is obtained in the state of greatest purity from alum. The properties of alumina were still farther examined by Macquer in 1758 and 1762, by Bergman in 1767 and 1771, and by Scheele in 1776: not to mention several other chemists who have contributed to the complete investigation of this substance. A very ingenious treatise on it was published by Saussure, junior, in 1801."

Thomson then goes on to describe the earliest documented process for refining alumina, the prequel to the 1855 Le Chatelier process from which the famous Bayer process was derived in 1887–1892:

> "Alumina may be obtained by the following process: Dissolve alum in water, and add to the solution an excess of carbonate of soda, and digest the mixture for some time to deprive the precipitated alumina of all the sulphuric acid with which it was united. Wash the precipitate with a sufficient quantity of water, dissolve it in muriatic acid, and precipitate the alumina from this solution by carbonate of ammonia. Digest the precipitate for some time in carbonate of ammonia. Then wash it and dry it. Pure alumina is easily obtained from ammoniacal alum, simply by exposing it to a strong red heat."

The German chemist Andreas Sigismund Marggraf, who is credited with discovering alumina in 1754, is actually well known as the father of the modern sugar industry for his 1747 discovery of sugar in beets, and the isolation thereof. Baron Louis-Bernard Guyton de Morveau, credited with naming alumina in 1761, was a distinguished French Chemist who developed the first system of chemical nomenclature.

1.3.1 The discovery of alumina refining

Neither the alumina ceramics industry nor the aluminum smelting industry could come into being until a commercially viable industrial process for synthesizing high-purity alumina powder was developed. The first step toward this goal was the "Le Chatelier" process for refining alumina from bauxite, invented by French Mining Engineer Louis Le Chatelier in 1855—not to be confused with his famous son Henri Louis Le Chatelier, the originator of "Le Chatelier's principle of chemical

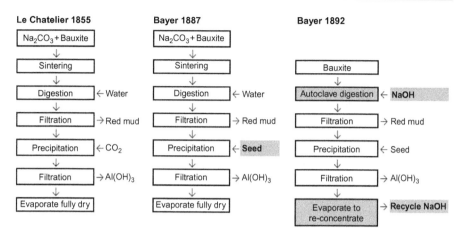

Fig. 1.9 The evolution in alumina refining from Le Chatelier (1855) through Bayer's 1887 innovation (seeding with Al(OH)$_3$) and Bayer's 1892 innovations (NaOH digestion and recycling the Bayer liquor). Very little has changed in the Bayer process since 1892. Bayer's three innovations are highlighted in gray on the schematic.

equilibrium." The Le Chatelier process, shown in Fig. 1.9, was not a very commercially viable process. It involved bauxite dissolution by sintering at very high temperatures with sodium carbonate, followed by dissolution in water, then aluminum hydroxide precipitation by the rather inefficient method of carbon dioxide infusion through the solution.

The big breakthrough in alumina refining came from Dr. Karl Josef Bayer in 1887, in the Tentelev Chemical Plant in St Petersburg, Russia. This plant was making aluminum hydroxide for the textile industry. Aluminum hydroxide was used as a fixing agent (mordant) for binding the die to the fibers of the textile in the textile dyeing process [14]. Bayer, an Austrian Chemist working at the Tentelev plant, made the discovery that high-purity aluminum hydroxide (Al(OH)$_3$ or gibbsite) could be precipitated much more effectively from Le Chatelier sodium aluminate solutions by seeding the solution with aluminum hydroxide particles, rather than the carbon dioxide process of Le Chatelier. Bayer filed British, United States, and German patents on this innovation in 1887 which were granted in 1888 [15].

The next big breakthrough came 4 years later when Bayer discovered that bauxite dissolution could be achieved by digesting bauxite in sodium hydroxide in an autoclave. Moreover Bayer discovered that the sodium hydroxide liquor could be reused after the aluminum hydroxide precipitation step. He filed a patent for this innovation in 1892, which was granted in 1894 [16]. Refining bauxite by NaOH digestion at the Bayer temperatures of 150–250°C consumes approximately a third of the energy of a sintering process like that of Le Chatelier. Refining bauxite by sintering involves temperatures of 1200°C or so.

In short, Bayer's first innovation (seeding) greatly improved the production yield, and his second innovation (NaOH digestion) greatly improved the energy efficiency and dramatically lowered the production cost. These innovations were a "game changer" in terms of bauxite refining.

The Bayer process, as we know it today, is a result of these two innovations of Bayer. This is shown schematically in Fig. 1.9 in the context of the original Le Chatelier process. The Bayer process enables high-purity aluminum hydroxide to be separated from all the silica, iron, organic, and other impurities found in bauxite in a large-scale commercially viable mass-production processes with the main process outputs being the pure aluminum hydroxide precipitate and the waste sediment called "red mud." It is a simple matter then to calcine aluminum hydroxide and convert it into high-purity alumina powder, which can then be used as feedstock for aluminum smelting, and as a source of alumina powder for making alumina ceramics.

Bayer's innovation was perfect timing in that the Hall-Héroult process, used almost universally ever since for aluminum smelting, had just been invented in 1886, the year before Bayer's invention, and therefore the fledgling aluminum smelting industry was just beginning at that time. While the Le Chatelier process may have been adequate for supplying the modest needs of the textile dyeing industry, the global aluminum industry as it is today producing over 60 million tonnes of aluminum PA, depends almost entirely on the much more energy efficient Bayer process for its alumina feedstock.

Therefore, while these two innovations of Bayer may not sound like a dramatic improvement on the Le Chatelier process, and at the time the Bayer process was slow to catch on globally, history has proven the importance of Bayer's innovation, particularly in terms of its energy efficiency. Bayer processing is typically three times more energy efficient than the sinter process of Le Chatelier. The global aluminum industry today, which processes over 100 million tonnes of Bauxite a year, uses almost exclusively the Bayer process. The energy savings when processing 100 million tonnes a year at one- third the energy cost are huge. Likewise the global alumina industry, though still huge, is an order of magnitude smaller in material volume than aluminum smelting.

Bayer eventually returned to Austria his homeland, developed the first bauxite mine in Austria, and set up an alumina refining plant. However, he was unable to raise sufficient investment capital and the venture ended in failure. There is a salutary lesson in the Bayer saga about IP. Bayer was wise to patent his inventions, but it didn't help him greatly. The Bayer process has generated revenues in the last century of multiple $trillions ($120 billion revenue in 2017 alone), but when he died in 1904, Bayer was not a wealthy man, and left a widow and six children in financial difficulties. After his death, all but two aluminum companies stopped paying royalties to his widow, who was in no position to sue them. Pro-bono lawyers were not an option in that era. Other aluminum companies had simply continued using the Le Chatelier process until 1911 when Bayer's patent expired, and then they switched to the Bayer process.

Bayer has received many posthumous accolades, particularly in 1987, the centenary of his invention. There is no doubt that the scale and economic viability of the aluminum smelting industry today is due, in no small part, to the inventive genius of Dr. Karl Josef Bayer. One could speculate that someone else might have made this discovery had he not, but that is purely hypothetical. Given its central importance to the alumina ceramic industry (not to mention the aluminum smelting industry), the Bayer process for refining bauxite into high-purity alumina will form the basis of Chapter 3. Without the source of plentiful low-cost high-purity Bayer alumina, the alumina industry would be much less commercially viable, if viable at all in some sectors.

Smelting aluminum from alumina is not a simple process either. It involves dissolving the alumina in molten cryolite (Na_3AlF_6, sodium hexafluoroaluminate) in a carbon-lined "pot" (cathode) and then using electrolysis via an immersion anode to produce aluminum by carbothermic reduction. This is known as the Hall-Héroult process, named after its inventors Hall and Héroult who invented the process in 1886, around the same time as Bayer's invention, in the decade that laid the foundations for the aluminum industry. It is interesting to note that in addition to inventing the Bayer process, Bayer went on to not only develop the first bauxite deposit in Austria, but he also developed a method for the manufacture of synthetic cryolite [14].

1.3.2 The first alumina ceramics

After the world-changing inventions of Bayer in the 1880s, the focus for some time was on aluminum refining. This is unsurprising given the industrial importance of aluminum. Two decades were to elapse before the scientific literature records the first rudimentary attempts at manufacturing a pure alumina ceramic.

Alumina is a nonoxide nonplastic ceramic. Thus powdered alumina, unlike clay-based ceramic bodies, has no inherent cohesiveness. In essence alumina has the cohesiveness of sand. The range of methods by which alumina can be formed is extensive. A look at the literature shows that it began with slip casting (crucibles) by Graf (Count) Botho Schwerin in 1910 [17], the first documented pure alumina ceramic in the literature. In the 1930s, the second method to be documented was isostatic pressing by Homer Daubenmeyer of the Champion Spark Plug Company [18]. Powder injection molding (spark plugs) followed in the 1940s with the innovation of Karl Schwartzwalder [19]. In addition to these methods, die pressing as a forming method, as always, is ubiquitous. Tapecasting has long been used in the electronics industry. The great majority of alumina ceramics made in the world today are still produced by these five methods, to which can be added green-machining and 3D printing. Alumina forming methods are discussed in Chapter 4.

1.3.3 Alumina ceramics come of age

From 1910 to the 1940s, the literature contains a mere trickle of publications on alumina. Essentially one could say that alumina as a ceramic was virtually unknown, other than by a handful of pioneering scientists, prior to the 1950's. In the 1950s, the literature began to contain a slowly growing number of papers on alumina innovation which gradually built momentum over the next 70 years. Most synthetic commercial forms of alumina ceramics are of the pearly white polycrystalline ceramic manifestation. While single crystal alumina (corundum, sapphire, ruby) is found in nature, pearly white polycrystalline alumina is a purely synthetic concept. It is not found in nature unless it is in disposed landfill, and is indistinguishable by casual inspection from conventional pearly white ceramics such as porcelain. Table 1.1

Table 1.1 Timeline of the history of alumina ceramics

Alum (an aluminum salt) widely used through antiquity and alchemy eras	−3000
Margroff is the first to *chemically purify alumina* from the salt alum and calls it "*argil*"	1754
Morveau *renames argil "alumina"*—alumina is named before its parent metal	1761
Davy. *Aluminum* the metal is discovered	1808
Le Chatelier. First alumina industrial *refining* process	1855
Bayer. Modern alumina *refining*	1887
Schwerin. *First alumina ceramic* (slipcasting patent)	1910
Rock. First mention in the literature that alumina could have potential as a biomaterial	1933
Daubenmeyer. First *isostatic pressing* of Alumina (spark plug insulator). *Birth of electrical alumina ceramics*	1934
Webster. Discovery of ceramic composite armor concept	1945
Schwartzwalder First *injection molded* Alumina	1949
First commercial *wear-resistant* Alumina. *Birth of wear resistant alumina ceramics*	1950
Cook. First Alumina ceramic *body armor* patent. *Birth of alumina lightweight body armor*	1963
First deployment of alumina ceramic body armor (US Helicopter Crews-Vietnam)	1965
Sandhaus: First documented biomedical device made of alumina (*Dental implant*)	1965
Cowdery: First Alumina bionic feedthrough (pacemaker). *Birth of implantable bionics*	1970
Boutin: First Alumina *hip implant* trial. *Birth of alumina orthopedic bearings*	1970
Kuzma/Clark/Patrick: First *bionic ear* capable of speech recognition (first co-fired Alumina-platinum bionic feedthrough)	1985
First non-military Alumina ceramic *body armor*	1985
First commercial spinal cord stimulator bionics for *back pain* (Alumina bionic feedthrough)	1989
EU (CE-Mark) Approved *Alumina hip*. Orthopedic alumina becomes a commercial technology in Europe	1994
First commercial vagus nerve stimulator for *epilepsy* (Alumina bionic feedthrough)	1994
First commercial sacral nerve stimulator for *incontinence* (Alumina bionic feedthrough)	1994
ISAPO. First ceramic *body armor* for US troops. Ceramic body armor becomes a global phenomenon	1996
First commercial deep brain stimulation: *Parkinsons disease* (Alumina bionic feedthrough)	1997
FDA Approves *Alumina hip*. Orthopedic alumina becomes a global technology	2003
Zirconia toughened alumina (ZTA) Market Launch as orthopedic bearing (CeramTec)	2003
Suaning: World record—1145 channel (7 mm diameter) Alumina bionic micro-feedthrough (*bionic eye*)	2012
First rudimentary commercial *bionic eye* (Argus 60 channel)	2012
Alumina hip replacements reach 50% Global Market Share	2015

illustrates a timeline of the key dates for alumina ceramics which demonstrates the slow rise of alumina ceramics from the 1950s, and the huge boom in the alumina ceramics industry since the late 20th century.

Polycrystalline alumina products are almost exclusively made by powder forming followed by sintering in a furnace in a simple ambient air atmosphere, in much the same way as conventional oxide ceramics such as terracotta, stoneware, and porcelain. The main differences being that high purity is usually necessary with alumina and the sintering temperatures are much higher than that for conventional oxide ceramics.

We have come a very long way since the first alumina ceramic, just over a century ago: the rudimentary 1910 patent of Schwerin [17], who innovated fine grinding, acid deflocculation, and slip casting as a means of ceramic forming alumina, with the proposed application being crucibles. Alumina is often still made by the same conventional powder forming methods used for conventional ceramics, such as pressing, slip casting, and plastic forming. However, given its advanced applications, there is a range of advanced powder forming techniques also used for alumina. With the use of advanced chemical additives, and building on decades of innovation, there is virtually no traditional (clay-based) ceramic forming method that has not also been used for alumina ceramics. But more importantly, since the three earliest innovations (slip casting in 1910, isostatic pressing in 1933, and injection molding in 1949), a number of advanced innovative alumina ceramic forming and densification technologies have been developed for alumina ceramics that had never been used for traditional ceramics. Since the 1950s, and particularly in the last few decades, these have been optimized, and in recent years, commercialized, on a very large scale in some cases. These are discussed in Chapter 4.

However, in contrast to traditional clay-based ceramics, and more importantly to the family of advanced ceramics, the outstanding combination of hardness, temperature resistance, wear resistance, corrosion resistance, biocompatibility, electrical resistivity, and modest production cost of alumina set it apart from them as the premier oxide ceramic. This is why alumina is one of the leading advanced ceramics in the world today. Moreover, the relatively low production cost of alumina compared to other advanced ceramics, such as silicon carbide, boron carbide, boron nitride, and silicon nitride, set it apart from them as by far the highest performer per unit cost, which is why it sees widespread use in the mining and body armor industries.

The biocompatibility of alumina is the highest of all commercial biomaterials, and this combined with its outstanding hardness and wear resistance have made it the pre-eminent material for bearings in orthopedic joints, particularly hip replacements, where alumina hips now have over 50% market share in the global hip replacement industry. Its biocompatibility, electrical resistivity, and impervious microstructure have made alumina the pre-eminent material in bionic feedthroughs, where it is an essential platform technology in the global implantable bionics industry today. Moreover, its capability of being co-fired with platinum has made alumina the pre-eminent feedthrough for futuristic miniaturized bionic implants, like the retinal implant with its 1000+ platinum channels in a 7-mm diameter alumina feedthrough ceramic, for the bionic eye.

Synthetic sapphire, that is, synthetic transparent alumina, also exists, and is used primarily for high-temperature windows in furnace or aerospace applications, in bulletproof windows, and in microwave and laser technology. It competes with spinel in the application of bulletproof windows. This is a very niche application. In the other applications it is significant.

Some synthetic alumina products are not formed by conventional powder forming. For example, some synthetic alumina products in the electronics industry are prepared by vapor deposition methods. Alumina glass-contact refractories are made by fuse-casting, in the $1.3 billion global market in glass-contact refractories (Chapter 15). There are also some other very high-tech specialty alumina densification techniques used in research and commercially. This will be discussed further in Chapter 4.

While naturally pearly white, often alumina is deliberately colored commercially. For example Sintox alumina armor ceramics were pink, CeramTec alumina biomedical bearings (Biolox forte) are bone colored, and CeramTec ZTA biomedical bearings (Biolox delta) are bright pink. Alumina dental implants are pearly white, or translucent (Fig. 1.10).

1.3.4 Alumina ceramics today

Decades of innovation and development of alumina since the 1950s have ultimately led to the six independent, huge-market alumina ceramic applications as we know them today: bionics, orthopedics, lightweight body armor, electronics, wear, and refractories. In each case, we note the size of the market in 1950 (the beginning of

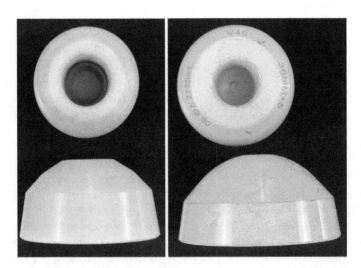

Fig. 1.10 Colored alumina. Left: CeramTec pure alumina biomedical bearings (BIOLOX*forte*) are bone-colored. Right: CeramTec zirconia-toughened alumina biomedical bearings (BIOLOX*delta*) are pink.
Exhibit provided courtesy of Professor Bill Walter. Specialist Orthopedic Group. Wollstonecraft NSW.

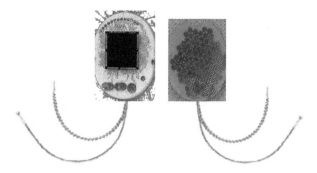

Fig. 1.11 686 Channel co-fired alumina-platinum feedthrough containing 686 platinum micro-conductors laser machined into a 10-mm wide alumina wafer, as a retinal implant for the bionic eye. Left: the "chip side" of the 686 channel feedthrough with some of the 686 platinum tracks clearly visible in the 10-mm wide 686 channel feedthrough and microprocessor chip attached. Right: the "electrode side" with the 686 electrodes in 98 hexagonal clusters of 7 electrodes (six corners and one center), for direct retinal tissue stimulation.
Image courtesy of Hayat Chamtie.

the alumina ceramics industry), in 1984 (the last major update [1]), and in 2020 to give an idea of the trends since 1950. In fact 1984 is midway between 1950, the beginning of alumina ceramic commercialization, and now. In dollar terms, almost all of the alumina industry boom has happened since 1984. Bear in mind the trends are for *alumina ceramics* in these applications, and the significance level is in market dollar terms not in tonnage terms.

(a) *Alumina feedthroughs in bionic implants*
   $25 billion market.
   Exclusive to alumina? Yes.
   (1950 nothing; 1984 fledgling industry; 2020 global dominance $25 billion industry)

The global market in implantable bionics today involves a $25 billion market comprising 14 different devices, treating over 25 disease states, including pacemaker, bionic ear, deep brain stimulator, and bionic eye. All of these devices involve an alumina hermetic feedthrough, a key enabling platform technology, with tens, hundreds, and up to 1000+ platinum wires (channels) passing through the alumina and hermetically bonded to it. It is an extraordinary technology that has evolved substantially since it was invented in 1970. Representing a significant part of the embodied value in a bionic implant, this is a market for alumina in the billion dollar range, and arguably the most important application for alumina ceramics in the world today, in dollar terms and in humanitarian terms. This is the focus of Chapters 8 (pacemaker), 9 (bionic ear), and 10 (bionics: the future). A state-of-the-art retinal implant for the bionic eye, with alumina-feedthrough, is shown in Fig. 1.11.

(b) *Alumina bearings in hip replacements.*
   $7 billion market.
   Exclusive to alumina? No.
   (1950 nothing; 1984 early research; 2020 >55% market share $7 billion industry)

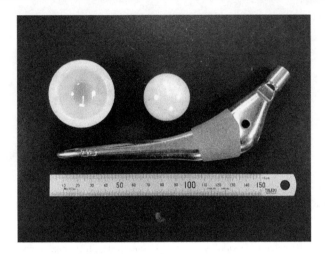

Fig. 1.12 Hip replacement with alumina ball and cup.
Exhibit provided courtesy of Professor Bill Walter. Specialist Orthopedic Group.
Wollstonecraft NSW.

The $7 billion hip replacement industry is greatly enhanced today by a highly value-added niche product: alumina bearings for hip replacements in two components: taper-fit femoral head (ball); taper-fit articulating acetabular cup (socket). First investigated in the 1970s, it wasn't until the mid-1990s that this technology was commercialized, and it went global in 2003. Within a decade or so alumina ceramic hip replacements grew rapidly and have now reached almost a million alumina-bearing components sold per year, the proportion depending on market jurisdiction. This is a very important application for alumina ceramics in the world today, in dollar terms and also in humanitarian terms. This will be discussed in more detail in Chapters 6 and 7 (Fig. 1.12).

(c) *Lightweight alumina body armor*:
   ~*$Hundred millions order of magnitude.*
   Exclusive to alumina? No, but alumina is part of a duopoly with SiC.
   (1950 nothing; 1984 very small; 2020 ~$hundred millions order of magnitude)

Ceramic-composite armor is a life-saving technology that has protected millions of military, law enforcement, security, and VIP personnel in recent decades. This technology slowly evolved over the period 1945–63 into its first working prototype (alumina/fiberglass) in 1963 and was first deployed in 1965, but saw very limited usage until the late 1990s. Alumina was used almost exclusively as the armor ceramic for most of the 20th century, and it was not for 40 years that silicon carbide began to compete significantly on cost with alumina, as reaction sintered silicon carbide (RSSC). Boron carbide has never competed effectively. Ceramic armor is now widely deployed as lightweight body armor (police, security, and military personnel), in military aircraft (particularly helicopters), in private vehicles from VIP cars to cash vans, and in military vehicles from light armored vehicles to tanks. It is a huge industry.

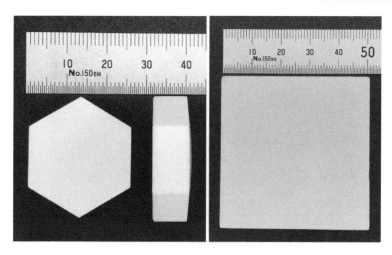

Fig. 1.13 Alumina tiles for ballistic applications.

Vehicle sector is largely serviced by alumina. Body armor is a duopoly of alumina (civilian sector) and SiC (elite military sector), with B_4C also significant in the military sector. This will be discussed in detail in Chapter 11. This is one of the most important applications for alumina ceramics in the world today in humanitarian terms, and also significant in dollar terms (Fig. 1.13).

(d) *Electrical applications of alumina ceramics*:
  $Multiple billion order of magnitude.
  Exclusive to alumina? No, but alumina is significant.
  (1950 just beginning; 1984 large and growing; 2020 $Multibillion order of magnitude)

The use of alumina in electrical applications is one of the very earliest uses of alumina ceramics, beginning with spark plug insulators in 1927. Today, 94% pure alumina insulators are used in virtually all spark plugs, in the global $3 billion spark plug industry. Alumina-kovar electrical feedthroughs were developed in the 1960s to 1980s and underpin a huge and diverse industry that utilizes advanced electrical feedthroughs. High-purity alumina ceramics as are used as SOI insulating substrates in 10% of the $500 billion global semiconductor industry. Alumina-based thick-film microelectronic devices are a huge market globally. Alumina feedthroughs in electronics underpin a large and diverse industry. This will be discussed in more detail in Chapters 13 and 14. Some examples of alumina macro-insulators are shown in Fig. 1.14.

(e) *Alumina wear-resistant ceramics*:
  ~$Hundred millions order of magnitude.
  Exclusive to alumina? No, but alumina is dominant.
  (1950 just beginning; 1984 small but significant; 2020 ~$hundred millions order of magnitude)

Wear-resistant applications for alumina date back to the 1950s. Initially alumina found specialized niche uses in machinery, such as wire drawing, textiles, and paper-making machinery. In more recent years, in mass-market volume, the largest

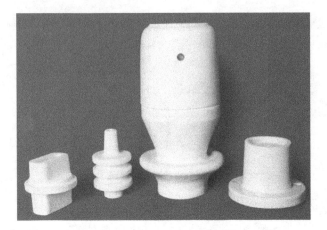

Fig. 1.14 Alumina electrical insulators.
Manufactured by Taylor Ceramic Engineering (Image courtesy of Taylor Ceramic Engineering).

use of alumina has been in wear-resistant ceramics for mining and mineral processing equipment: tiles, pipes, spigots and other hydrocyclone parts, augers, dart valve plugs and seats, bearings, sleeves, knife-edge blades, mill linings, nozzles, orifice and baffle plates, rotary seals, pump components, pipe reducers, spacers, tubes, and washers. Essentially in general plumbing and machine components for high wear environments. Also used in the food and chemical processing industries, and metal smelting equipment (alumina being inert and heat resistant can work well in molten metal contact). The traditional areas of paper making, textiles, and wire drawing remain important. This will be discussed in detail in Chapter 12 (Fig. 1.15).

Fig. 1.15 Alumina wear-resistant ceramic components (spigots) produced by alumina.
Manufactured by Taylor Ceramic Engineering (Image courtesy of Taylor Ceramic Engineering).

Fig. 1.16 Round alumina crucibles of various sizes. A square flat crucible and a long-rectangular furnace boat.
Manufactured by Taylor Ceramic Engineering (Image courtesy of Taylor Ceramic Engineering).

(**f**) *Alumina in refractories*:
  Exclusive to alumina? No, but alumina is significant.
  (1950 established; 1984 and 2020 very large, little change in the alumina technology)

The refractories industry is certainly the oldest of these six main market applications for alumina ceramics, and one of the largest. Alumina refractories form a subset of the larger $50 billion refractories industry. A great many refractories contain alumina. A much smaller, but still significant percentage, is near-pure alumina. The main example of this is high-purity fused-cast alumina ceramics as refractories in the industry in refractories for glass production, which represents a $1.3 billion industry of which a significant proportion is serviced by alumina. Refractories will be discussed in more detail in Chapter 15. Examples of high-purity refractory alumina heat containment ceramics (crucibles) are shown in Fig. 1.16.

In addition to these six large-dollar-value markets for alumina, there are also significant markets in alumina ceramics in the various other applications itemized in Section 1.1.1. Some of the more significant applications are compiled below. Again, note that the trends are for *alumina ceramics* in these applications, and the significance level is in dollar terms not in tonnage terms:

(g) Laboratory refractories: crucibles and furnace tubes
  (1950 established; 1984 and 2020 significant, little change)
(h) Furnace furniture
  (1950 conceptualized; 1984 and 2020 significant, little change)
(i) Labware
  (1950 conceptualized; 1984 and 2020 significant, little change)
(j) Lasers
  (1950 nothing; 1984 small but significant; 2020 substantial)

(k) Dental technology
   (1950 nothing; 1984 experimental; 2020 experimental)
(l) Rotary seals
   (1950 nothing; 1984 significant; 2020 substantial)
(m) Components for specialty slurry pumps
   (1950 nothing; 1984 developing; 2020 substantial)
(n) Synthetic sapphire windows (ballistic and high-temperature applications)
   (1950 nothing; 1984 developing; 2020 substantial)
(o) *Oxygen sensors*
   (1950 nothing; 1984 significant; 2020 substantial)

Technologies g) through to n) will be discussed in more detail in various chapters, of this book where relevant, and those that are not covered there will be covered in the broad-application chapters, Chapters 5 (specialty biomedical) and 15 (specialty industrial).

(p) *Powdered alumina industry*:

Powdered alumina also has many large markets:

Powdered alumina as a functioning powder
- powdered abrasives
- polishing powders
- catalysts, both as alumina itself, and as a support for other catalysts such as platinum
- chromatography
- dehydrating agents, for example, water removal from hydrocarbons in industrial hydrocarbon processing
- fire retardant/smoke suppressant

Powdered alumina as an additive in a ceramic
- component in refractories
- bonded abrasives
- additive in glass and other ceramics
- component in zeolite manufacture
- alumina cements (calcium aluminate)—commonly used in refractory applications, but calcium aluminate cements have many other uses

This book is about *alumina ceramics*, and will not be discussing the uses for powdered alumina, none of which involve alumina ceramic synthesis as such. It should be noted that the above-mentioned use of alumina as a component in the manufacture of zeolites, glass, and alumina cements, ceramics for which alumina is a significant component, do constitute ceramics. However, this book is focussed on monolithic alumina ceramics, where the assumption is that the ceramic is either pure alumina, or predominantly alumina, and the very commercially important material ZTA.

With regard to powdered alumina consumption globally, it should be noted that the use of powdered alumina globally is very large, especially by the abrasives industry. In 1951, at a time when alumina ceramics in dollar terms were an almost nonexistent industry, artificial abrasives constituted 85% of world abrasives production [20]. Today artificial abrasives are closer to 100%. Moreover, in 1951, 97% of all alumina produced in the world that was not used for aluminum smelting was for the abrasives industry. The remaining 3% was used in refractories, cement, crucibles, and electrical resistors [20]. Today, the global abrasives industry still dominates alumina consumption globally. It was $40 billion in 2015, of which a significant proportion was alumina. However, abrasives are not alumina ceramics, they are graded alumina powders, bonded to paper, or wheels, or loose for sandblasting, lapping, and polishing.

We have come a long way since 1951. While the alumina ceramic tonnages are still very much smaller than abrasive alumina tonnages globally, the alumina advanced ceramics industry in dollar terms (bionics, orthopedics, electronics, wear, armor, and refractories) is now on par with the alumina abrasives industry in global dollar terms.

1.4 The alumina ceramics manufacturing industry

There are many small companies around the world that manufacture alumina as sole product or as part of a product portfolio, especially in China. However, the number of significant alumina manufacturers globally is quite small. With many closedowns, mergers, and acquisitions, it is sometimes difficult to keep track of the fast-changing industry. Nonetheless, the following is a useful guide to significant companies in the alumina ceramics realm.

1.4.1 Large broad-portfolio companies with a significant alumina activity

- Saint Gobain (France)
- Kyocera (Japan)
- AGC Group (Japan)

1.4.1.1 Saint Gobain

- France-based multinational (67 countries, 170,000 employees)
- Founded: 1665
- Turnover: US$50 billion

St Gobain is a huge and widely diversified company with a long and turbulent history, dating back to before the French Revolution. It was nationalized from 1971 to 1986, but it has been a private company for the last three decades. In this era it has risen to prominence in the ceramics field. The "Innovative Materials" division comprises approximately 25% of St Gobain today, half of which is glass products, and half of which is high-performance materials [21]. Within this high-performance materials sector is "Ceramic Materials" [22], which has three divisions: powders, abrasives, and ceramic components and systems. The latter concerns a number of ceramics, including alumina. Thus, although Saint Gobain is a huge company, its alumina ceramics activity is a very small proportion of the entire company activity.

In the alumina area, Saint Gobain is strong in refractories and kiln furniture. They also have renown in the ceramic armor area. Saint Gobain provided the very first ceramic body armor protection product in the world (alumina for Vietnam war helicopter crews—see Chapter 11), although today their armor focus is on silicon carbide armor ceramics (Silit, Forceram, Hexoloy, Norbide), with some boron carbide also. Their primary alumina armor ceramic is the SAFirE (synthetic sapphire), a transparent ceramic armor product, used for vehicles and aircraft. Their wear ceramics are primarily silicon carbide (Hexoloy).

1.4.1.2   Kyocera Corporation (Kyoto Ceramic Company, Limited)

- Japan Ceramic and Electronics company
- Founded: 1934
- Employees: 70,000
- Turnover: US$12 billion

Kyocera began life as a ceramics company making ceramic electronic components, and over the years through many acquisitions, it has diversified more and more from its original core business of ceramics and acquired a broad range of electronic and communications product portfolios, launching its first laptop in 1982, cameras and audio equipment in the 1980s, solar cells in the 1990s, and mobile phones in the 2000s, to name but a few diversifications. In ceramics, Kyocera has achieved renown for novelty ceramic products: zirconia knives and other ceramic kitchenware (Kyocera Advanced Ceramics brand), and synthetic gemstones, acquired in the 1980s from Swiss synthetic gemstone inventor Pierre Gilson [23].

Kyocera has a significant footprint in ceramic manufacturing, including alumina. In its Kyocera Fine Ceramics division it manufactures a range of ceramics including zirconia, silicon carbide, aluminum nitride, silicon nitride, and various traditional oxide ceramics. One of these ceramics is alumina, manufactured in grades ranging from 76% to 99.5% pure, and there is also a transparent sapphire alumina product—SA100 99.99% pure [24]. The Kyocera focus with alumina is on electrical and specialty products [25]. A wide range of ceramic manufacturing methods are used and numerous product areas serviced including electrical, wear, biomedical, food processing, and a range of specialty components. Alumina does not service all these needs but is a significant ceramic among the Kyocera portfolio of fine ceramic products. In 2011 Kyocera Medical and Dental products division entered the orthopedic bearings realm, previously the monopoly domain of CeramTec, although CeramTec have retained global market dominance in orthopedic alumina in the intervening decade.

1.4.1.3   AGC group

- Japan
- Employees: 51,000
- Turnover: US$11 billion
- Glass, electronic components, and refractories

Only 2% of the business of AGC is ceramics "other," which includes a small activity in alumina, 93% grade abrasion-resistant ceramics product (pipes and augers), and

Introduction to alumina ceramics 33

they also have a small presence in silicon carbide, zirconium diboride, aluminum titanate, silicon nitride, cordierite, and a novel cermet molybdenum boride/nickel.

1.4.2 Ceramic companies with a substantial alumina activity

It should be noted that the alumina activity of the three companies (Saint Gobain, Kyocera, and AGC) discussed in Section 1.4.1, as a proportion of group general turnover, is much smaller than the following five companies. However, Saint Gobain, Kyocera, and AGC all have turnovers in the $10+ billion range, and so their global impact in alumina is substantial.

In contrast, the alumina activity of the following five companies, as a proportion of group general turnover, is substantial. In the case of TCE, alumina is their sole focus and they are the number 1 alumina manufacturer in the southern hemisphere, while CeramTec and CoorsTek are globally renowned in the alumina area. However, these five companies are much smaller companies ($millions to $\sim$1$billion turnover range) than the three companies discussed in Section 1.4.1. Therefore, overall the alumina impact of all eight companies is substantial.

- Morgan Advanced Materials (UK)
- CoorsTek (USA)
- CeramTec (UK/Germany)
- CUMI Murugappa (India)
- Taylor Ceramic Engineering (Australia)

1.4.2.1 Morgan advanced materials

- UK-based advanced materials company
- Founded: 1856
- Employees: 9000
- Turnover: $\sim$$1 billion GBP (US$1.4 billion)

Morgan began as a graphite crucible manufacturer in 1856, has since diversified broadly, and has a long history of acquisitions in its core business of advanced materials. Among its many activities, Morgan is considered one of the world's leading advanced ceramics manufacturers in the world today. Alumina is only one of the many ceramics in the Morgan advanced materials portfolio. Its best-known alumina product is called Sintox FF [26]. Other alumina products of Morgans include: AL300, Deranox 970, Deranox 995, A950, synthetic rubies/sapphire, AL998, and ZTA [27].

In the areas of greatest relevance to this book: biomedical (Chapters 5–11), electrical (Chapters 13 and 14), wear (Chapter 12), and other industrial applications of alumina, Morgan is a significant global player. Morgan manufactures many alumina-metal brazed products for electrical and biomedical applications (Chapters 8–10, and 14), for example, brazed feedthroughs for active medical implants, and specialty electrical components for medical imaging equipment) [27].

Morgan has a long history in alumina armor ceramics dating back to the late 20th century. The UK was a world leader in ceramic vehicle armor (Chobham Armor) discussed in Chapter 13. Morgan has a strong presence in electrical ceramics and is particularly well known for its presence in the semiconductor industry. CeramTec acquired the Morgan UK electro-ceramics business ("UK Electro-Ceramics"), comprising two manufacturing sites at Ruabon and Southampton, from Morgan Advanced Materials for £47m, in April 2017. Morgan has been strong in the refractories area (Chapter 15) for a very long time [27].

1.4.2.2 CoorsTek

- USA Advanced Ceramics Company
- Founded: 1910
- Employees: 6000
- Turnover: US$1.25 billion

Along with CeramTec, CoorsTek is one of the world's pre-eminent alumina manufacturers, though alumina is just one of its many products. CoorsTek began with glass and porcelain in the early 20th century, and is one of the original alumina ceramic manufacturers in the world, becoming an alumina manufacturer in the 1950s when the alumina era was just beginning. CoorsTek is the originator of the AD85 alumina (85% pure) which is a lightweight alumina armor for nonarmor-piercing applications (see Chapter 11), suitable for civilian uses, as well as higher purity aluminas AD90, AD94, AD96, AD995, AD998, and plasmapure 99.9%+. CoorsTek also makes ZTA which is the pre-eminent wear-resistant biocompatible ceramic (see Chapter 7) and also makes all the other standard advanced ceramics such as zirconia, silicon carbide, boron carbide, tungsten carbide, aluminum nitride, silicon nitride, and various traditional oxide ceramics. CoorsTek uses a wide range of forming methods and finishing/machining techniques for their alumina and other ceramics [28]:

CoorsTek sells armor ceramics (Chapter 11) under the tradename CeraShield, alumina, SiC and B4C, both breastplates and tiles for vehicle armor mosaics. They also sell a wide range of alumina wear products under the tradename CeraSurf including wear-resistant tiles and components of various shapes and sizes. CoorsTek are also renowned in the electrical substrates arena, both alumina and aluminum nitride (Chapter 13).

1.4.2.3 CeramTec

- German Advanced Ceramics Company
- Founded: 1903
- Employees: 3600
- Turnover: EU500 million (US$600 million).

CeramTec has a long and complex corporate history of mergers and acquisitions. Beginning in 1903 as the Thomas Factory in Marktredwitz, which became Hoechst Ceramtec in 1985, and merged with Cerasiv GMBH in 1996 to become CeramTec, which was acquired by Rockwood Specialties, USA, and in 2013 Ceramtec was bought from Rokwood by British private equity firm Cinven. In 2017 CeramTec bought the Electro-Ceramics division from Morgan Advanced Materials. CeramTec operations have remained centered in Plochingen Germany.

CeramTec has revolutionized the world of orthopedics (Chapters 6 and 7) having developed the ceramic hip replacement bearings that are now used in over 50% of hip replacements globally, obtained CE-Mark approval in 1994 and FDA approval in 2003 for Biolox (biomedical-grade alumina brand) and launched the first commercial biomedical-grade ZTA in 2003. CeramTec now supplies all the world's leading orthopedic companies: Stryker, Zimmer-Biomet, Smith & Nephew, and De Puy/J&J, and has retained market dominance through an unrivaled quality standard. This is essential for orthopedic implants where failure is not an option. CeramTec forms the basis of a detailed case study on orthopedic ceramics in Chapter 7. In addition to this, CeramTec is a broad-based ceramics manufacturer, manufacturing not just alumina and ZTA, but also all the usual advanced ceramics including zirconia, silicon carbide, aluminum nitride, silicon nitride, and various traditional oxide ceramics. All the usual advanced ceramic applications are represented in the CeramTec portfolio, from armor to wear to electronics, with a global renown in biomedical applications [29].

1.4.2.4 CUMI Murugappa

- Indian Advanced Ceramics Company
- Founded:1954
- Turnover: US$350 million

Asia's leading manufacturer of high-quality low-cost alumina, the CUMI Murugappa alumina plant in Bangalore, India was set up, and operated by, German specifications, and manufactures a wide range of alumina ceramics for both industrial applications and armor. It is one of the largest dedicated alumina manufacturing plants in the world. CUMI Murugappa makes a very wide range of alumina ceramic products for the electrical, wear, armor, and specialty areas. CUMI Murugappa is one of the premier alumina armor ceramic suppliers in the world [30].

1.4.2.5 Taylor Ceramic Engineering

- Australian Company specializing in high-purity alumina ceramics
- Founded: 1967

Australia is one of the top two largest mining nations in the world, and therefore has a very high consumption of wear-resistant materials. It is inevitable then that Australia has some leading wear-resistant ceramic-manufacturing companies. The most prominent of these is Taylor Ceramic Engineering (TCE) [31]. TCE was founded by ceramic innovators and pioneers David Taylor and Julie Taylor half a century ago in 1967, and is now under the leadership of General Manager Alyssa Taylor. The TCE has been the leading manufacturer of advanced alumina ceramics in the southern hemisphere for four decades, servicing the huge Australian mining industry, and also servicing many other industries (see Chapter 12).

TCE has been a pioneer in many industrial alumina ceramic applications, from mining to steel smelting, to bionics, to armor and was the first company to supply armor ceramics to the Australian military when it made the move to ceramic armor in the late 1990s (see Chapter 11). TCE also supplied high-alumina tiles for the

Sydney Opera House refurbishment program, one of the few examples in the world of alumina as a boutique architectural ceramic (Chapter 15). They are also active in alumina electrical ceramics (Chapters 13 and 14), and heat-containment ceramics (Chapter 15). TCE manufacture their own alumina powders, so as to maintain high and reproducible quality, and purity. They also specialize in very large monolithic alumina ceramics. The product range in advanced alumina ceramics of TCE is so extensive and comprehensive, and the range of industries serviced by TCE industrial alumina so broad, that it forms the basis of a case study on industrial wear/corrosion-resistant alumina ceramics in Chapter 12.

1.4.3 SME (small- to medium-sized companies) with an alumina activity

There are a large number of small- to medium-sized companies with an alumina activity scattered throughout the world, and space does not permit discussing them here. At the low-cost end of the spectrum, there are a number of small Chinese companies that manufacture technical ceramics, including alumina in some cases, but none that are known outside of China. There are numerous other such companies elsewhere globally.

1.4.4 Large ceramic companies ($10 billion+ turnover) with no significant alumina activity

Corning Inc.
  USA
  35,000 people
  $10 billion sales
  Mainly Glass

Vesuvius PLC
  UK
  12,000 people
  $2.4 billion sales
  Refractories

Murata Manufacturing Co., Ltd.
  Japan
  52,000 people
  $8.5 billion sales
  Electronic components

The NSG Group
  Japan
  27,000 employees

$5 billion in sales
Glass

Schott AG
  Germany
  15,000 employees
  $2 billion in sales
  Glass

RHI AG
  Germany
  8000 people
  $2 billion sales
  Refractories

PPG Industries
  USA
  46,000 employees
  $15 billion in sales
  Paints, coatings, and glass

1.5 Companies for which alumina is a high-volume platform technology

In the electronics industry, which is dominated by Asia, microelectronics manufacturing companies use a large quantity of alumina substrates, underpinning the $50 billion SOI industry. Much of this substrate volume is supplied by many SME companies located in Asia, particularly China. Similarly, the numerous refractory manufacturing companies around the world, many of which are SMEs, also produce a significant quantity of alumina-containing refractories, but only those servicing the glass industry work with high-purity alumina ceramic. China has the largest mining industry in the world, and all its wear-resistant alumina is locally sourced. Alibaba Online, the conduit for a vast number of Chinese SMEs, lists a huge number of Chinese SMEs supplying alumina ceramics for most of the industrial applications discussed in this book, though ISO13485-certified biomedical grade alumina products are not significant in Alibaba. Thus, a significant part of the global industrial alumina ceramic activity is serviced by a large number of SME companies, particularly in China and other Asian countries, not by the global giants.

Bauxite: The principal aluminum ore

2

Virtually all alumina and aluminum produced in the world today comes from the aluminum ore bauxite. By far, the largest consumer of bauxite commercially is the aluminum industry. This chapter will overview bauxite the mineral, and the global bauxite industry, in the context of both the global aluminium smelting industry and the global alumina ceramic industry.

The obvious question here is why not use corundum, the purest source of natural alumina? There are two reasons for this. Corundum is a very impractical mineral for use as a raw material in aluminum production. First, it is far too scarce geologically to be able to feed the huge global demand for aluminum production. Second, corundum, being pure crystalline alumina, is very hard with a hardness of Mohs 9, which makes it an excellent abrasive but very expensive to process, causing expensive wear and tear on processing equipment. Bauxite, on the other hand, is much more abundant than corundum, much softer than corundum, and easier to crush and grind, which makes it a very suitable mineral source of aluminum for the aluminum industry.

Bauxite is essentially hydrated alumina (aluminum hydroxide), usually contaminated with clay, silica, and iron oxide. The hydrated alumina is present in various minerals, primarily gibbsite, diaspore, and boehmite, and contaminated with organic material and various other minerals, particularly iron oxide and silica. Bauxite ore can vary from soft red-yellow earth to small round pebbles (pisolites) to hard rock. It derives its name from Las Baux in France, where it was first discovered in 1821.

Diaspore was the first of the hydrated alumina polymorphs to be discovered, in 1801, 2 years after the discovery of corundum, then followed the discoveries of gibbsite (1820), boehmite (1924), bayerite (1925), and nordstrandite (1956). These hydrated alumina polymorphs comprise the commercially useful part of bauxite. The global bauxite industry has now grown to around 260 million tonnes a year (Fig. 2.1).

Global production of alumina is currently around 100 million tonnes a year, almost all of it by the Bayer process, and the majority used in aluminum smelting. Aluminum metal production was about 60 million tonnes a year globally in 2017.

Australia is the largest producer of alumina in the world with 20% of global production [32,33]. In 2015, Australia produced 19.9 million tonnes of alumina, with nearly 90% of that (17.4 million tonnes) being exported [34]. Of this, the majority was metallurgical grade alumina. Chemical grade alumina production in Australia (usable for alumina ceramics) was around half a million tonnes.

Of the 100 million tonnes of alumina manufactured every year globally, approximately 85 million are used for aluminum smelting, approximately 10 million are used for abrasives and other non-alumina-ceramic applications for alumina. The alumina ceramics industry uses approximately 4 million tonnes per year of alumina [35]. While

Fig. 2.1 A piece of Bauxite with US penny for scale.
Author unknown [12]. Reproduced from Wikipedia (Public Domain Image Library).

4 million tonnes may not sound like a large amount compared to the 85 million smelted into aluminum each year, looking at dollar values rather than tonnages presents a different picture. The value of aluminum metal averages $2/kg, making the 60 million tonnes of aluminum per year a $120 billion dollar industry. In contrast, the value of alumina ceramics can vary from $15/kg (industrial) up to $10,000/kg (biomedical), as discussed below. In practice, the bulk of the tonnage is at the low price end of the scale, but this still puts the size of the alumina ceramics industry in the $10 billion order of magnitude.

Because alumina ceramics are used in such a diversity of applications, in many cases as a platform technology such as in microelectronics and refractories, it is difficult to get a true estimate of the dollar value of the alumina ceramics industry, but suffice to say that despite its smaller tonnage, the alumina ceramics industry at the $10 billion order of magnitude is only one order of magnitude in dollar value smaller than the aluminum smelting industry. This has become the case only relatively recently.

Therefore, although the tonnages are smaller in the alumina ceramic industry compared to the aluminum metal smelting industry, the alumina ceramic industry is by no means insignificant in dollar terms, due to the high value adding involved in alumina ceramic products. While aluminum metal is worth about $2 per kg wholesale ($2000 per tonne), ceramic grade alumina powder is in the order of $1000–$2000 per tonne, depending on many factors including tonnage ordered, particle size (degree of grinding), impurity content, and GMP/ISO certification. Some very specialty high sintering grade alumina powders such as the spray-dried nanocrystalline powder can be as high as $10,000 per tonne; however, the cost of these raw materials is insignificant compared to the value adding that is possible with high-end biomedical-grade commercial alumina ceramics.

An alumina ceramic bearing used in a hip replacement is worth about $1000 and weighs about 50 g, which equates to about $20,000 per kg, 10,000 times the value of the raw alumina powder, albeit under ISO13485 biomedical quality manufacturing standards, a higher priced alumina powder is used in biomedical products, but the value adding is still in the 1000-fold range. In specialty electronic applications and laser applications, the value adding is comparable.

On a bulk industrial scale, the value adding is less, but still substantial. A mass-produced standard 100 mm × 75 mm alumina wear tile for the mining industry retails for about $5 wholesale and weighs about 350 g, which at $10–$15/kg is still 5–15 times more valuable than the raw alumina powder. A specialty, or custom-made wear tile, may be more than 10 times that price, perhaps 100 times more valuable than the raw alumina powder. Similarly, a specialty wear component such as an auger or wire drawing cone may be perhaps 100 times more valuable than the raw alumina powder. A standard 250 mm × 300 mm alumina body armor insert plate weighs about 2 kg and costs about $25–$50 wholesale, which at $12–$25/kg is 5–25 times more valuable than the raw alumina powder.

Millions of alumina wear tiles and specialty machinery components are produced per year, about a million alumina hip joints bearings, millions of alumina bionic feedthroughs, and hundreds of thousands of body armor plates, and many more armor mosaic tiles. This is why each of these relatively new global industries is in the hundreds of millions to billion dollar order of magnitude.

So while the aluminum industry is very large in both tonnage and dollar value, approximately 60 million tonnes a year and $120 billion per year, and in tonnage terms by far the world's biggest consumer of alumina powder, the alumina ceramic industry is now approaching a level only one order of magnitude lower in dollar value, the $10 billion order of magnitude, even though the tonnages are very much smaller.

This was not the case for the alumina ceramic industry in 1984 when the last monograph was written on alumina ceramics. Year 1984 was the time point midway between the beginning of the alumina ceramics industry in 1950, and now, but in dollar terms alumina ceramics in 1984 were a small percentage of what the alumina industry is today, and in 1984 alumina ceramic commercial applications were dominated by alumina refractories and low-tech electrical applications such as spark plugs. Most of the alumina ceramics in 1984 were primarily low-tech electrical applications of alumina, or refractory uses, with a small but significant wear ceramics market.

As discussed previously, while alumina as a mineral is found in nature as the mineral corundum, it is a relatively a rare mineral. Corundum is also very hard (Mohs 9), which makes it very expensive to grind, causing expensive wear and tear on grinding equipment. Bauxite combines the three key attributes of natural abundance, high recoverable aluminum content, and it is generally much softer than corundum, and hence easy to crush and grind. Therefore, the global aluminum and alumina industries depend almost entirely on mined bauxite as the feedstock for their refineries.

2.1 Bauxite, alumina refining, and aluminum smelting: A mega-industry

Because of its abundance, its high aluminum content, and its ready availability to open-cut mining, the primary commercial mineral source of aluminum for manufacturing aluminum metal is the relatively common aluminum ore bauxite. Bauxite is essentially hydrated alumina in an ore-specific (every bauxite mine is different) blend of its three common mineral forms: gibbsite $Al(OH)_3$, boehmite γ-AlO(OH), and diaspore α-AlO(OH), combined with many impurities, most commonly clay, silica, iron oxide, and titanium dioxide.

A bauxite is often defined by its weight% available alumina as Al_2O_3. Gibbsite and boehmite are the usual sources of Al_2O_3 in bauxite ores and diaspore to a lesser extent. The theoretical limit of Al_2O_3 content for bauxite is 65.4%. This would be for pure $Al(OH)_3$. In practice, an extremely high-quality bauxite will contain, on an equivalent oxides basis, over 50% Al_2O_3, with a large loss on ignition (mostly, the water of hydration, also some organic content), and a few percent each of SiO_2, Fe_2O_3, and TiO_2. In practice, bauxites containing 50% alumina are considered high quality, and some are as low as 30% Al_2O_3. Of course there are enormous variations between mines, but in terms of bauxite quality, this is a good guideline for typical production quality bauxite. Also of importance is the relative amount of the three phases: gibbsite, boehmite, and diaspore as their dissolution characteristics in the Bayer process are quite different. By far, gibbsite is the cheapest to process.

In 1873, bauxite mining began in Villeveyrac in France. Since then, bauxite mining has grown to a huge scale. By the 1960s, world bauxite production had reached 40 million tonnes. It is now around 260 million tonnes. In 2016, Australia is the world's largest bauxite producer, producing 80 million tonnes. Australia produces about 30% of the world's total bauxite. Australia exports more than 80% of its bauxite, the rest being used primarily for the Australian aluminum industry [36].

How much bauxite remains on the Earth? In 2012, Australia reportedly had 6.3 billion tonnes of minable bauxite economic demonstrated resources (EDR) and 22% of the world's known estimated 28 billion tonnes of Bauxite EDR [36,37]. This is shown in Table 2.1. While 28 billion tonnes may sound a lot, at 260 million tonnes a year, the world will run out of commercially viable bauxite reserves in about 100 years. It is sobering to note that China, currently the world's leading aluminum producer, will consume all of its relatively small bauxite reserves in 15 years at current consumption rates, representing an export opportunity for other countries with larger reserves.

It is estimated by the USA Geological Survey [37] that the world's total bauxite reserves are 55–75 billion tonnes, in Africa (33%), Oceania (24%), South America and Caribbean (22%), Asia (15%), and elsewhere (6%). The majority of these are not EDR bauxite, but as bauxite becomes scarce, it will drive up prices and increase the proportion of current non-EDR that becomes viable EDR. It is worth noting that at current consumption, the 55–75 billion tonnes stretch out our global depletion date to 210–285 years into the future. This is a more comfortable horizon, but still very finite.

Table 2.1 **World's top 15 bauxite producers (millions of tonnes per year) [37]**

Rank output	Country	Bauxite (Mt)	% Globally
1	Australia	82	31
2	China	65	25
3	Brazil	35	13
4	India	25	10
5	Guinea	20	7.5
6	Jamaica	8.5	3.2
7	Russia	5.4	2.1
8	Kazakhstan	4.6	1.8
9	Saudi Arabia	4	1.5
10	Greece	1.8	0.7
11	Guyana	1.6	0.6
12	Vietnam	1.5	0.6
13	Indonesia	1	0.4
14	Malaysia	1	0.4
15	Other	6.4	2.4
	Total	262	

There are of course other less commercially viable sources of aluminum, such as clay, and potentially some even less accessible bauxite reserves, which are unviable at current world prices. However, in the past, and for the foreseeable future, bauxite extracted by large-scale open-cut mining is the industry norm, since most bauxite deposits are large in area, close to the surface, and in the order of 10 m thick, which makes it a very easy ore to mine.

Many of these bauxite reserves have problems with purity, undesirable silica contamination, or low gibbsite proportion. Gibbsite is easy to refine, and it is very difficult to refine boehmite and diaspore. Europe and China have a big problem with low gibbsite proportion, making refining problematic. China has the largest reserve of diaspore bauxite in the world [38]. China and Europe work with major refining challenges. Some Australian bauxite deposits are outstanding; some have problems with high silica content. Nonetheless, in the northern (tropical) states of Queensland and Northern Territory, Australia has some of the world's highest alumina bauxites around the 50% level. Bauxite refining issues will be discussed in Chapter 3 of this book. World economic reserves of bauxite are summarised in Table 2.2.

Bauxite ores are generally categorized by the intended end use, the main uses being (Figs. 2.2 and 2.3):

- Metallurgical (aluminum refining)
- Alumina abrasives
- Alumina cement
- Refractories
- Chemical grade (includes alumina ceramics)

Table 2.2 **World economic reserves of bauxite [37]**

Country	Bauxite (MT)	Rank (output)	% Globally	Current usage (MT/Y)	Years at current usage
Guinea	7400	5	27	20	376
Australia	6200	1	22	82	76
Brazil	2600	3	9	35	75
Vietnam	2100	12	8	1.5	1400
Jamaica	2000	6	7	9	235
Indonesia	1000	13	4	1	1000
China	980	2	4	65	15
Guyana	850	11	3	1.6	531
India	590	4	2	25	24
Suriname	580	–	2	–	–
Saudi Arabia	210	9	1	4	53
Russia	200	7	1	5	37
Kazakhstan	160	8	1	5	35
Greece	130	10	0	1.8	72
Malaysia	110	14	0	1	110
USA	20	15	0	w	–
Other	2700		10	–	–
Total	27,830				106

Usually, the quality of alumina used to make alumina ceramics needs to be of the highest grade possible, that is, chemical grade. Purity is paramount with alumina advanced ceramics. The great majority of bauxite, currently around 85% of global production, is used to make aluminum. As the world leader in bauxite production and alumina production and world number 6 in aluminum production, Australia makes a good case study. It has been active in the aluminum industry for over half a century and has 5 bauxite mines, 6 alumina refineries, and 4 aluminum smelters. In 2011, Australia produced 1.96 million tonnes of aluminum, of which about 90% was exported, creating exports of $3.8 billion, and employing 17,600 Australians in all aspects of the nationally integrated industry: bauxite extraction, alumina, and aluminum production [32].

It is also notable that while Australia is the world's largest bauxite producer, the world's second largest bauxite exporter, and the world's largest alumina exporter (20.5 million tonnes), it exports most of its bauxite and is only the world's sixth largest aluminum producer behind China, Russia, Canada, India, and UAE [39]. Of the ~60 million tonnes of aluminum produced in the world per year, Australia produces 1.7 million tonnes while China produces more than half at around 31 million tonnes. China is undergoing massive expansion.

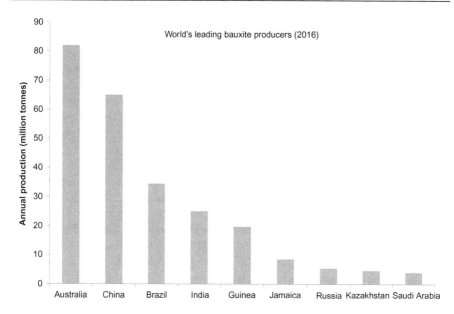

Fig. 2.2 World's top bauxite producers (millions of tonnes per year) [37].

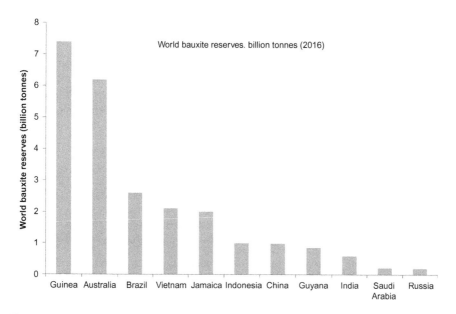

Fig. 2.3 World economic reserves of bauxite (billions of tonnes) [37].

In tonnage terms, over 85% of the alumina made in the world today (more than 85 million tonnes a year) is used as feedstock for an aluminum smelter. Aluminum is electrochemically produced from alumina using the Hall-Héroult process. This involves dissolving the alumina feedstock in molten cryolite (Na_3AlF_6, sodium hexafluoroaluminate) in a carbon-lined "pot" (cathode) and then using electrolysis via an immersion anode to produce aluminum metal by carbothermic reduction. Depending on the quality of the bauxite, on average globally, 3.6 t of bauxite is needed to produce 1 t of aluminum metal. About 263 million tonnes of bauxite mined globally, 222 million tonnes (85%) of that used to smelt the world's global output of 60 million tonnes of aluminum. However, in some cases, it can be as little as 1.9 t of bauxite per tonne of aluminum.

Aluminum smelting is a very energy-intensive process, and therefore electricity is a major input to the aluminum production process, approximately 15 MW hours per tonne, accounting for between 25% and 30% of total operating costs. For example, the Australian Tomago aluminum plant, which employs 950 people at its plant 150 km north of Sydney, produces 580,000 t of aluminum a year, a quarter of Australia's aluminum production output, and consumes 10% of the electricity production of the State of NSW [33]. Overall, aluminum smelting consumes approximately 13% of Australia's entire electricity consumption. There have been many books and handbooks written on aluminum smelting, and the reader is referred to these for more detail on aluminum smelting. The focus in this book is the ceramic applications of the alumina feedstock.

There are two main types of bauxite ores: lateritic (also known as equatorial) and karst. Lateritic bauxites account for about 90% of the world's EDR minable bauxite and karstic bauxites about 10% [40]. Lateritic bauxites are found in equatorial regions of the world and are much preferable because they are easier to digest in Bayer liquor than karstic bauxites, requiring less concentrated NaOH, lower temperatures, and holding times, as discussed in Chapter 3. Karst bauxites, 10% of the world's bauxite EDR, are found mainly in Eurasia, particularly Eastern Europe and China [41].

Bauxites contain many minerals, of which the most significant to Bayer processing are as follows [42,43]:

The hydrated alumina polymorphs are
- Gibbsite $Al(OH)_3$—common polymorph of aluminum hydroxide
- Bayerite $Al(OH)_3$—less common polymorph of aluminum hydroxide
- Nordstrandite $Al(OH)_3$—less common polymorph of aluminum hydroxide
- Boehmite AlOOH—aluminum oxy-hydroxide (resistant to Bayer digestion)
- Diaspore AlOOH—aluminum oxy-hydroxide (very resistant to Bayer digestion)

Commonly gibbsite is referred to as the trihydrate alumina (THA) component of a bauxite ore and (boehmite + diaspore) as the monohydrate alumina (MHA). Total available alumina (TAA) = THA + MHA.

The sources of the silica contaminant:
- Kaolinite clay (silica): Highly soluble at all commercial Bayer digestion temperatures.
- Quartz (silica): Soluble above 200°C in Bayer liquor.

Also minor silica contributors chamosite and illite, if present, usually only present in trace amounts.

The sources of the "red mud" waste sediment from bauxite refining are as follows:
- Goethite (iron oxide—red mud). Can be fine grained = slow sedimentation and therefore an alumina contamination risk in bauxite refining
- Hematite (iron oxide—red mud)
- Maghemite (iron oxide—red mud)
- Magnetite (iron oxide—red mud)
- Ilmenite (iron oxide and titania—red mud)
- Anatase (titania—red mud). Can deposit scale in the Bayer equipment
- Rutile (titania—red mud). Can deposit scale in the Bayer equipment
- Calcite (calcia—red mud)
- Apatite (calcia—red mud)
- Crandalite (calcia—red mud)

As mentioned earlier, lateritic bauxites are not only more common (90% prevalence) but also more easily digested. This is because they are primarily gibbsite with some boehmite. Karstic bauxites, on the other hand, are rare (10% prevalence) and much harder to digest because they contain primarily boehmite and diaspore, the MHA aluminum oxy-hydroxides. Gibbsite, the THA, is comparatively easy to digest in the Bayer liquor, whereas boehmite and diaspore, the MHA, are difficult to digest.

Refining of alumina: The Bayer process

3.1 Overview of the Bayer process for refining bauxite into alumina

The Bayer process is so industrially important to both the alumina and the aluminum industries, that it will receive a detailed focus in this chapter, given that it is the foundation upon which both the alumina and the aluminum industries are built. As discussed in Chapter 1, the Bayer process was invented by Austrian chemist Karl Josef Bayer in 1887, and he patented it in two separate patents, one filed in 1887 and granted in 1888 [15] and one filed in 1892 and granted in 1894 [16]. The Bayer process has since become the cornerstone of alumina and aluminum production worldwide. This chapter will overview the Bayer process in detail, and also look at the key alternatives to the Bayer process in terms of their economics and technological platform.

Prior to the 1887–94 invention of the Bayer process for commercially viable alumina refining, and the 1886 invention of the Hall-Héroult process for commercially viable aluminum smelting, aluminum metal was so expensive to produce it was regarded as a precious metal. Anecdotal tales from the 19th century recount that the French government had aluminum bars on show next to the crown jewels, Napoleon III used highly prized aluminum cutlery for special guests, and in the United States, the Washington monument was capped with a 6Lb aluminum pyramid in the 1880s. However, like the "Bitcoin" crash of 2018, the aluminum market crashed in the late 19th century thanks to Bayer, Hall, and Heroult. Widespread commercialization of aluminum metal, beginning in the early 20th century, and alumina ceramics, beginning in the mid to late 20th century, were predicated on the low cost efficient Bayer process.

The scale of Bayer refining globally is huge today. Building a typical Bayer refinery may involve a capital outlay of $1 billion. Such a plant might process 1 million tonnes a year of bauxite, producing up to 500,000 t a year of alumina, at a market value of up to $500 million. Globally, 260 million tonnes of bauxite are currently processed.

The energy efficiency of using the Bayer process compared to the sinter process of Le Chatelier is about threefold, which is a massive energy cost saving when processing on this scale, representing annual energy cost savings in the $10–$100 million order of magnitude for a Bayer plant. We know this threefold energy saving is valid in today's money because the sinter process is still used in some Eastern European and Chinese plants that have to process karstic bauxite (see Section 3.8). Moreover, optimization of the many aspects of the Bayer process can yield annual cost savings of several $million a year, creating many opportunities for chemical and industrial engineers to use their initiative to earn their keep.

3.1.1 Chemistry of the Bayer process

The Bayer process is based on the precipitation of aluminum hydroxide crystals, as the gibbsite mineral polymorph, from caustic sodium aluminate solutions. The chemical basis of the Bayer process can be summarized in the following reversible chemical equation:

$$Al(OH)_{3(s)} + NaOH_{(aq)} \leftrightarrow NaAl(OH)_{4(aq)}$$

The enabling chemical principles are fivefold:

(1) Hydrated alumina (gibbsite/boehmite/diaspore) is soluble in concentrated sodium hydroxide solutions (NaOH), as aqueous sodium aluminate, under autoclave conditions.
(2) $Al(OH)_3$ can be reversibly reprecipitated into the gibbsite polymorph upon cooling and seeding the aqueous sodium aluminate Bayer liquor with gibbsite fines.
(3) Silica SiO_2 is also soluble in hot NaOH but remains in solution under the Bayer $Al(OH)_3$ reprecipitation conditions, and therefore soluble silicates can be removed from the NaOH solution before or after $Al(OH)_3$ precipitation, usually early in the Bayer process before the $Al(OH)_3$ autoclave digestion stage.
(4) Other than SiO_2, clay, and $Al(OH)_3$, the other mineral content of bauxite is generally insoluble in hot NaOH, and can therefore be removed as waste sediment, known as "red mud."
(5) The NaOH Bayer liquor can be recycled, by cleansing it of impurities, and concentrating it by evaporation, thereby enhancing the commercial viability of this process.

In this way, a cost effective, large scale, chemical process can be used to convert bauxite into pure $Al(OH)_3$, in the form of a pure $Al(OH)_3$ gibbsite precipitate, which can then be converted into pure Al_2O_3 by high-temperature calcination. Pure Al_2O_3 can then be used for smelting of pure aluminum metal by the Hall–Héroult process, or as raw material for the alumina ceramics or alumina powder industries.

3.1.2 The stages of the Bayer process

There are many variations in the Bayer processing parameters, depending primarily on the bauxite feedstock, and the numerous disturbances that occur when running such a complex multistaged industrial process, with interacting processes, and significant amounts of dead time. Engineers are always dealing with the problem of silica and oxalate impurities, which are discussed in Sections 3.4 and 3.6, respectively. It is difficult to attain steady-state or optimal conditions in a Bayer refinery, and small improvements in process throughput can equate to millions of dollars in cost savings due to the scale of the process.

The following dot-point summary is a general summary of the Bayer process as it is done currently. In practice, many plants deviate significantly from this general process, depending on the specific characteristics and problems of their particular bauxite ore feedstock. The following summary (read in conjunction with Fig. 3.1) is very useful to the reader who has only a passing interest in alumina refining as it captures the essence of Bayer processing. The rest of this chapter is a detailed review of all of the aspects of Bayer process, and is useful to the reader who seeks deeper knowledge of the complex and very industrially important Bayer process.

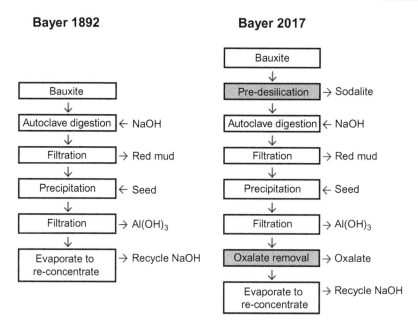

Fig. 3.1 Modern-day enhancements (gray highlight) to the original 1892 Bayer process: (A) the removal of silica, primarily precipitated out as insoluble as sodalite before the bauxite digestion stage; (B) oxalate removal, for example, by calcium oxalate precipitation before the NaOH is evaporatively concentrated for recycling.

- Bauxite is crushed to <30 mm in hammer crushers then mixed with the NaOH Bayer liquor and fine ground as a slurry in the Bayer liquor at about 60°C. Often silica dissolution from kaolinite impurities and silica removal by precipitation as sodalite and other insoluble silicates is done at this stage.
- The powdered bauxite is then sent to an autoclave and heated for digestion for 30–40 min in the hot NaOH solution (the Bayer liquor) at temperatures from 140°C (high-gibbsite bauxite) to 240°C+ (high boehmite), to as high as 280°C (high diaspore bauxite). Typical NaOH concentration is around 3 M for gibbsite, and up to 8 M for boehmite or diaspore. The majority of the aluminum hydroxides from the bauxite dissolve in the NaOH "Bayer liquor," as well as any residual reactive silica in the bauxite, which dissolves in the liquor then ideally precipitates out as sodalite. Unreactive silica is the silica component that does not dissolve in the Bayer liquor.
- The Bayer liquor slurry is then cooled to about 100°C. Flash tanks and heat exchangers are used for cooling, for efficiency and minimal energy wastage. Iron oxides and other insoluble mineral content form the so-called "red mud" sediment in the autoclave, which is removed by sedimentation and then filtration
- The red mud is washed, to minimize NaOH losses, and neutralize the alkalinity for disposal, and then pumped away to a waste storage reservoir. Red mud is a somewhat environmentally hazardous (alkaline and saline) waste product.
- With the liquor cooled to the gibbsite precipitation temperature, around 60–80°C, it is seeded with gibbsite fines, to precipitate out the aluminum hydroxide as gibbsite crystals approximately 1 μm in size, while any residual dissolved silicates, and hopefully also the sodium

oxalate soluble impurity, remain in solution. Silica and sodium contamination prevention is critical to alumina ceramics
- The purified aluminum hydroxide precipitate is removed and washed.
- The Bayer liquor is recycled for the next batch of bauxite. Often sodium oxalate impurity cleansing is done at this stage, prior to evaporative reconcentration of the Bayer liquor.
- Evaporators are used to reconcentrate the Bayer liquor for the next digestion cycle.
- The purified gibbsite is calcined in an inclined rotary kiln. Huge water vapor release occurs between 250°C and 400°C turning the kiln contents into a fluidized bed.
- At around 1000–1250°C, alpha alumina forms by an exothermic process, which increases the kiln temperature. Catalysis of the alpha alumina transformation is assisted with the use of mineralizers.
- A hold time of 1 h is required for complete transformation of gibbsite to alpha alumina as per the chemical reaction:

$$2Al(OH)_3 \rightarrow Al_2O_3 + 3H_2O$$

- The result is pure alpha alumina powder, ready for commercial use either as feedstock for an aluminum smelter, or as a raw material for the alumina ceramic industry, or abrasives or other related industries that utilize alumina powder.
- Depending on bauxite purity, it takes approximately 2 t of bauxite to produce 1 t of alumina, but it can be over 3 t.

3.2 A century of Bayer refining: Review of problems and lessons learned

The Bayer process was invented 130 years ago and remains the global method of choice for converting bauxite to alumina for aluminum and industrial alumina production. In general, the three largest problems in the Bayer process are as follows:

(a) The removal of silica (a problem with high silica bauxite).
(b) The removal of oxalate (a problem with high organic content bauxite).
(c) Disposal of the sedimented waste: red mud.

There are many other problems of lesser magnitude of which the most significant are:

- Optimizing the digestion parameters for the specific Gibbsite/Boehmite/Diaspore ratio in the ore. Gibbsite requires less aggressive digestion conditions than Boehmite and Diaspore.
- Optimization of gibbsite precipitation.
- Optimization of calcination in terms of energy efficiency.

Given the pivotal importance of the Bayer process for both alumina and aluminum production, both in terms of production cost in monetary and manpower terms (given the global scale of the industry), and the quality of the end product, particularly in terms of alumina ceramics thereby produced, it is important to review these issues here.

The following sections will discuss these issues in the following sequence:

3.3. Autoclave digestion of bauxite
3.4. Silica problems (dissolution, removal, and scale)
3.5. Precipitation of the gibbsite

3.6. The sodium oxalate problem (bauxite organic content)
3.7. The "red mud" disposal problem
3.8. Alternatives to Bayer processing

3.3 Autoclave digestion of bauxite

3.3.1 The Bayer liquor

A Bayer liquor, the concentrated sodium hydroxide solution, with or without dissolved sodium aluminate, is commonly defined by four parameters:

(1) The "Free Caustic" content, which is grams per liter of NaOH sodium hydroxide. Typically, 100–120 g/L for THA gibbsite (~3 M) and as high as 300 g/L for MHA boehmite/diaspore (~7–8 M).
(2) The "Caustic" content, also known as "Total Caustic" content, or "C" in g/L, which equates to the "Free Caustic" plus sodium aluminate, in g/L.
(3) The "Soda" content, which corresponds to "Total Caustic" plus sodium carbonate, represented as grams per liter of equivalent sodium carbonate.
(4) The total aluminum content of the Bayer liquor is defined as grams per liter Al_2O_3, the total alumina content of the Bayer liquor "A" in grams per liter.
(5) The A/C ratio that is a measure of aluminum solubility in the Bayer liquor. THA and MHA have differing NaOH concentration requirements.

3.3.2 Hydrated alumina minerals: Digestion in the Bayer liquor

The three equations of relevance to digestion (dissolution) in the Bayer liquor of hydrated alumina minerals of bauxite are as follows:

THA Gibbsite: $Al(OH)_3 + H_2O + NaOH \rightarrow Na^+ + Al(OH)_4^-$
MHA Boehmite/diaspore: $AlO(OH) + NaOH \rightarrow Na^+ + Al(OH)_4^-$
Alumina: $Al_2O_3 + 3H_2O + 2NaOH \rightarrow 2Na^+ + 2Al(OH)_4^-$

Bauxite ores vary from mine to mine in terms of the relative ratio of the three main hydrated alumina minerals: most importantly, gibbsite (THA) to (boehmite+diaspore) (MHA) ratio or THA/MHA.

- Gibbsite $Al(OH)_3$
- Boehmite γ-$AlO(OH)$
- Diaspore α-$AlO(OH)$

3.3.3 Bauxite digestion parameters

Bauxite refineries are classified as either low temperature (~150°C) for processing predominantly THA gibbsite bauxite or high temperature (>240°C) for bauxite with >6% boehmite or diaspore MHA. Boehmite requires >240°C for digestion. Diaspore requires the highest temperatures of all, 260–280°C. Digestion temperatures are due to the chemistry of the hydrated aluminas. Gibbsite is a trihydrate, and the OH bonds are relatively weak. Boehmite is an oxy-hydroxide, which is a much stronger bond. Diaspore is the same as boehmite, an oxy-hydroxide, but it has a different crystal structure to boehmite, which is more stable and therefore more difficult to breakdown [44].

The Bayer bauxite digestion stage involves heating a slurry of the bauxite in NaOH under autoclave conditions so as to dissolve the hydrated alumina in the bauxite and thereby form the sodium aluminate solution from which pure gibbsite can subsequently be precipitated. The digestion temperature depends on the type and proportions of the aluminum hydroxide THA and MHA polymorphs in the bauxite. Gibbsite (aluminum hydroxide) requires much less severe digestion conditions than do the oxy-hydroxides boehmite and diaspore. The higher the THA/MHA (gibbsite:boehmite+diaspore) ratio, the less severe the digestion conditions required.

For pure gibbsite, the digestion can be performed in the temperature range of 140–150°C, and the NaOH concentration is around 100–120 g/L (around 3 M). For boehmite, 240–260°C is required and much stronger NaOH concentrations of around 300 g/L or ($\sim$7–8 M). For diaspore, temperatures of 260°C and higher are needed as well as higher A/C, that is, stronger NaOH concentrations [44,45]. Digestion is done in an autoclave, and therefore pressures are temperature dependent and correlate with the steam saturation pressure at the particular temperature. At 150°C, this is about 10 bar, that is, 10 times atmospheric pressure, and by 260°C, it is approaching 50 bar, which requires a very strong autoclave pressure vessel, a large energy input and cost, and the associated high capital cost for the autoclave. This is why gibbsite-rich ores are preferable.

Bauxite refineries are set up to operate under specific conditions, and in broad terms there are two types: low-temperature refineries ($\sim$150°C) for ores that are 95% or higher gibbsite; high-temperature refineries for ores that have >6% MHA (boehmite+diaspore). A particular refinery works to a particular bauxite feedstock. Obviously, constant lab quality control checks are required of the bauxite feedstock for fine tuning as required.

The higher the digestion temperature, the faster and more complete is the alumina digestion, but conversely the greater the risk of other impurity oxides dissolving in the Bayer liquor. This particularly applies to quartz. Quartz needs higher temperatures than kaolinite clay for dissolution. Quartz dissolution begins around 200°C and is a problem in terms of loss of sodium from the Bayer liquor, a production loss cost as discussed in Section 3.4. It also applies to iron to a small extent. Iron contamination can end up in the gibbsite precipitate, which is very undesirable.

Digestion time is typically 30–40 min, which is much longer than necessary since the actual alumina digestion happens very quickly at the correct temperature, but the prolonged digestion time is done to allow sufficient time for the reactive silica to dissolve in the Bayer liquor and then reprecipitate as sodalite into the red mud. It is desirable to have as little silica as possible in solution in the Bayer liquor when it is cooled and seeded for gibbsite precipitation.

After digestion is complete, the slurry is cooled to around 100°C for the removal of the red mud and then gibbsite precipitation stages.

3.3.4 Digestion of lateritic versus karstic bauxites

As discussed in Chapter 2, the hydrated alumina polymorphs in bauxite ores vary significantly between ore deposits from different locations, and essentially there are two broad classes of bauxites: lateritic (90% of the world's bauxite EDR) and Karst about 10% [40]. Karst bauxites require much higher digestion temperatures, which are unfortunately high enough to dissolve quartz and thereby load up the Bayer liquor with silica. This becomes problematic if the karst bauxite has a high-quartz level.

3.3.5 Double digestion for heavily quartz-contaminated karst bauxite

One solution to the problem of low gibbsite high-quartz bauxite is double digestion. [46–48]. In this scenario, the bauxite ore is first digested in freshly recycled Bayer liquor (i.e., recycled from previous Bayer cycles as usual), at $\sim 150°C$ to digest only the gibbsite component, and thus no quartz is digested. Then the Bayer liquor is sent for $Al(OH)_3$ precipitation. The (now gibbsite depleted) bauxite is then digested in a second stage with freshly recycled Bayer liquor at 220–270°C to digest the boehmite/diaspore, and it is only in this second digestion that quartz dissolution may happen. A patent of Rodda and Shaw [49] addresses this approach as a way of reducing silica dissolution in the Bayer liquor and the associated sodium losses (see Section 3.4), although it has only a marginal benefit on reducing the silica content of the Bayer liquor.

3.3.6 Bauxite precalcination

One way of enhancing the relatively indigestible alumina oxy-hydroxides, as well as the gibbsite (not that gibbsite enhancement is necessary), is precalcining the bauxite to about 500–550°C to render the hydrated alumina phases amorphous. Only a very brief soak time, if any, is required at this temperature to achieve this [50,51]. This significantly reduces the required digestion temperature. Moreover, it removes organics, thereby eliminating the oxalate problem (Section 3.6), and the lower digestion temperature reduces quartz dissolution, which helps with the reactive silica problem (Section 3.4).

3.4 Silica problems: Dissolution, removal, and scale

Without a doubt, silica is the most commercially significant bauxite impurity in terms of the Bayer refining process. It represents a production loss cost in many ways. Before we get to those issues, it is important to define the problem of reactive silica.

3.4.1 Reactive silica

Most bauxite ores contain silica contaminants in various forms. Particularly relevant to Bayer processing are forms of silica that yield soluble silica dissolution in the Bayer liquor, which are most commonly clays and quartz. The silica content of a bauxite ore that is capable of dissolution in the Bayer liquor during autoclaving is known as "reactive silica." In terms of silica content, the principal defining parameter of a bauxite for Bayer processing is the reactive silica content as weight% SiO_2. There is also the nonreactive silica. In low-temperature digestion ($\sim$150°C), reactive silica equates to silica portion of the kaolinite content of the ore. Kaolinite is presumed to be 100% soluble in Bayer liquor, even at 150°C. In high-temperature digestion (>240°C), reactive silica equates to kaolinite-silica plus partial quartz-silica. In high-temperature digestion, weight% reactive silica needs to be determined by laboratory testing under digestion conditions.

Another common parameter is the bauxite A/S ratio, which is (aluminum as Al_2O_3)/(silicon as SiO_2). This is for quantifying the silica contamination in relative terms. Obviously, the silica problem is ore related. Ores low in silica will have little reactive silica problem. Ores high in MHA-hydrated alumina (boehmite+diaspore) with high quartz are most problematic. High silica bauxites are considered to be bauxites that contain >8wt% silica. These cannot be processed by conventional Bayer processing but require a sinter process (see Section 3.8).

A significant proportion of global bauxite ore reserves have too high a silica content to be commercially viable for Bayer refining because high silica levels are costly in terms of sodium losses from the Bayer liquor. However, since bauxite ores are a finite resource globally, ways of managing silica in a commercially viable way are a focus of intense research. In future decades, as global bauxite reserves dwindle, silica will become an increasingly acute problem.

Therefore, Bayer refineries are increasingly needing to develop strategies to manage high silica bauxites. The main strategies pursued involve premineral processing to remove silicate minerals from the bauxite ore, engineering the Bayer chemistry to minimize the amount of sodium lost to the Bayer liquor through silica precipitation, and attempts to recover this lost soda from the red mud.

Most bauxite ores are not processed before refining, in terms of beneficiation by mineral processing techniques to remove undesirable impurities such as clay and silica. However, in some cases, clay and quartz may be removed by mineral processing before refining. There is a lot of benefit in beneficiation, particularly in the future as quality bauxites dwindle, and refineries will be increasingly confronted with ores of increasingly high levels of impurity, and so beneficiation of bauxites is likely to become a lot more common in the future. It is always a cost-benefit question. With high-grade ores, there is little need for beneficiation, but with low-grade ores, the cost of beneficiation may be a lot less than the refinery penalty for no beneficiation.

The main problem with silica is its consumption of sodium and thus soda loss (and some aluminum loss) as sodalite precipitation into the red mud (a production loss cost) from the Bayer liquor. Sodium recovery from the red mud has proven to be commercially unviable. This is discussed in Section 3.7. This leaves engineering the Bayer chemistry as a key focus, which will be overviewed briefly in the following sections. Silica contamination is also most undesirable in Bayer-processed alumina used to make alumina ceramics.

3.4.2 The problem with reactive silica

It is undesirable (but usually inevitable) to have reactive silica in bauxite. There are several reasons for this.

First, silica is a detrimental contaminant in aluminum metal, and especially in industrial alumina powder. While this is not such a problem for alumina feedstock to be used for aluminum smelting, it is a very significant issue for alumina feedstock to be used for alumina ceramics, since silica contamination is such a problem for alumina ceramics in all advanced ceramic applications, with the possible exception of lower grade alumina armor (Chapter 11) and lower grade alumina wear-resistance components (Chapter 12).

Silica contaminant in high-performance alumina ceramics is harmful as it concentrates in the grain boundaries and makes the alumina susceptible to moisture-induced static fatigue (Chapter 4), a serious problem if it occurs for dental or orthopedic implants, reduces the refractoriness when alumina is used as a refractory (Chapter 15), reduces the hardness, indeed silica compromises most of the industrially important properties of alumina.

Generally silica is mostly removed from Bayer liquor by sodalite precipitation (see Section 3.4.5), and residual silica in the Bayer liquor generally does not contaminate the gibbsite precipitate significantly, but for advanced alumina ceramic applications, the importance of silica contamination, however small, cannot be overemphasized.

Second, silica is problematic in the Bayer process as in some cases silica-based scale can be deposited in the Bayer digestors or in the process heat exchangers, which causes production downtime for acid cleaning [52]. Ideally all silica scale is precipitated into the red mud and removed that way.

Silica usually precipitates out of Bayer liquor as insoluble sodium aluminosilicate compounds, such as sodalite, depleting the Bayer liquor of sodium, which must be replaced, representing a production loss cost. This is currently considered the biggest problem with silica in Bayer refineries: the high cost of sodium hydroxide loss due to silica precipitation. In summary, the five key problems with silica in bauxite are as follows:

(1) Scale contamination of the Bayer equipment (production downtime cost).
(2) Irreversible loss of sodium from the Bayer liquor (production loss cost).
(3) Irreversible loss of aluminum from the Bayer liquor, precipitated as sodalite (production loss cost).
(4) If silica level is above 8%, Bayer processing is not viable and sinter processes need to be used, and these have three times the energy cost of Bayer processing.
(5) Risk of silicon contamination of the aluminum hydroxide precipitate (primarily a concern with refined alumina feedstock for alumina advanced ceramics).

One possible solution to the silica problem is double digestion, as discussed in Section 3.3.5. Double digestion has proved beneficial in a number of ways, but it has not proved to be an effective solution to the silica problem. However, calcium additions to Bayer liquor have proved an effective solution (Section 3.4.4).

3.4.3 Silica dissolution in the Bayer process

In terms of the Bayer process, silica dissolution commonly begins right at the beginning, during the bauxite grinding in 60°C Bayer liquor, forming the Bayer slurry. Kaolinite clay ($3Al_2Si_2O_5(OH)_4$), a platy nanoparticulate aluminosilicate mineral of very high surface area, is the most soluble of the silicate contaminants in bauxite, and therefore the main cause of silica dissolution at this stage of the Bayer process. The chemical reaction can be defined as:

$$3Al_2Si_2O_5(OH)_4 + 18NaOH \rightarrow 6Na_2SiO_3 + 6NaAl(OH)_4 + 3H_2O$$

It is worth noting that kaolinite is a source of both reactive silica and soluble alumina, and for this reason, it is possible (though usually uneconomic due to its high silica content) to refine alumina from kaolinite clay using a variant of the Bayer process. In a future world depleted of bauxite reserves, this will be of growing importance.

Most Bayer refineries preface the aluminum hydroxide digestion with a desilication stage, which involves holding the Bayer slurry at around 100°C for a few hours so as to enable the following sodalite ($Na_6(AlSiO_4)_6 \cdot Na_2X$) reactive silica precipitation reaction to go as close to completion as practically possible, typically 80%–90% completion is achieved, thereby precipitating 80%–90% of reactive silica, primarily as sodalite, into the "red mud" waste sediment.

$$6Na_2SiO_3 + 6NaAl(OH)_4 + Na_2X \rightarrow Na_6(AlSiO_4)_6 \cdot Na_2X + 12NaOH + 6H_2O$$

There are a number of factors that increase the solubility of sodalite [52]:

- Increasing Bayer liquor temperature
- Increasing NaOH concentration in the Bayer liquor
- Increasing sodium aluminate concentration in the Bayer liquor

The remaining 10%–20% of reactive silica is converted during the bauxite digestion stage under the autoclave conditions used for aluminum hydroxide dissolution, and ideally the remaining reactive silica is precipitated into the red mud as sodalite (or cancrinite, hydrogarnet, or other insoluble silicates) during the bauxite digestion stage, and has all been removed by precipitation into the "red mud" bayer sediment prior to the seeding/gibbsite-precipitation stage of the Bayer process.

If the digestion conditions are severe, for example, if it is a high boehmite/diaspore bauxite (see Section 3.3 for Bayer digestion conditions), some of the less reactive silica (quartz) will dissolve into the Bayer liquor during digestion. In high-quartz bauxites, therefore, it is sometimes necessary to reduce the severity of the digestion conditions (see Section 3.3 for Bayer digestion conditions) so as to avoid excessive quartz dissolution. This problem will arise specifically with a high-quartz bauxite with low gibbsite content and high MHA, either boehmite or diaspore, or both, that is, a quartz-rich karst bauxite.

As mentioned previously, the main commercial problem with reactive silica is the loss of sodium from the Bayer liquor (sodium consumed when it forms the silica precipitates), which is a production loss cost. This is why bauxites higher than 8% reactive silica are generally considered economically unviable. It is difficult to reduce the amount of silica dissolution into the Bayer liquor, especially in bauxites that are high in reactive silica and high in boehmite or diaspore or both (severe digestion conditions required), but strategies can be considered such as mineral processing to separate the clay or quartz from the bauxite prior to Bayer processing, or chemical recovery of sodium from the sodalite in the red mud. Neither of these strategies seems to be commercially justifiable in the current situation of plentiful bauxite ore globally, although on a country-by-country basis, some countries might see these measures justified for their local aluminum industry, particularly countries high in karst bauxite with significant silica contamination.

In the distant future, in a bauxite-depleted world, the economics will make such bauxites economically viable. In 100 years, at current bauxite consumption rates, bauxite EDR reserves will be finished. Refineries will have to adapt to the (currently) uneconomic bauxite reserves. It will simply be a case of supply and demand and adaptation to reduced quality. However, already a significant proportion of the current bauxite EDR reserves are high in silica, and bauxite consumption is growing yearly, and so it is likely that this situation will become prevalent within a few decades.

3.4.4 Minimizing sodium losses by using calcium to precipitate the silica

Desilication products (DSP or scale) involve chemical combinations of soda ash, silica, alumina, and calcia, with the driving component being the silica. Silica scale in the Bayer process comes in various forms [53,54]:

Sodium aluminosilicates (sodalite):

> Consumes sodium from the Bayer liquor.
> For example, Sodalite $Na_6(AlSiO_4)_6 \cdot Na_2X$ OR $3Na_2O \cdot 3Al_2O_3 \cdot 6SiO_2 \cdot Na_2X$.
> (X includes SO_4^{2-}, CO_3^{2-} or $2Cl^-$ or $2OH^-$)

Sodium calcium aluminosilicates (cancrinite):

> Consumes sodium from the Bayer liquor and forms in the presence of both silicon and calcium, generally forms only at very high temperatures over 200°C.
> For example, $Na_6(AlSiO_4)_6 \cdot 2CaCO_3$.

Silico calcium aluminates (hydrogarnet):

> Forms in the presence of both silicon and calcium.
> $Ca_3Al_2(SiO_4)_n(OH)_{12-4n}$.
> ($0.8 > n > 0$ in Bayer refining [54]).

Iron silico calcium aluminates (iron hydrogarnet):

> $Ca_3(Al, Fe)_2(SiO_4)_{3-x}(4OH)_x$.
> (where $1 < x < 2$).

3.4.5 Sodalite and cancrinite

In standard Bayer processing, where silicon oxygen and sodium are in abundance, the usual precipitate is sodalite, and it ends up in the red mud where it is disposed of along with the red mud. The benefit of sodalite precipitation is minimizing the risk of silica contamination of the gibbsite precipitate, and enabling recycling of the Bayer liquor, which does not accumulate dissolved silica through repeating cycles. The detriment of sodalite precipitation is major sodium losses from the Bayer liquor, which is a production loss cost. Also there is some aluminum loss from the Bayer liquor as aluminum is a component of sodalite. However, in the presence of calcium, cancrinite, and hydrogarnet, precipitates can form.

All of these silicates consume not just silicon but also aluminum, and therefore represent a production loss in aluminum terms, as well as sodium loss in the case of sodalite and cancrinite. In the case of sodalite, the $Na_2O:SiO_2$ molar ratio is 2:3. For every mole of silica removed, 0.67 mol of sodium is lost. In the presence of free calcium, cancrinite can form and can reduce sodium loss by as much as 25% compared to the sodalite sodium loss in the absence of calcium, that is, from 8 mol of sodium to 6—Na_6 remains but Na_2X substitutes for $2CaCO_3$ (both have an Na_6 component in their chemical formula, cancrinite does not have the Na_2X). However, since cancrinite generally forms only at very high temperatures, over 200°C, and long digestion time periods, conditions which cause quartz dissolution, this defeats the purpose of a cancrinite strategy. Moreover, there is the cost of the lime to offset a rather marginal sodium saving. Thus, the cancrinite strategy is not an attractive option.

3.4.6 Hydrogarnet

Hydrogarnet, in contrast to sodalite and cancrinite, takes no sodium at all which is a positive. Moreover, it is very stable under the severe Bayer digestion conditions [55,56]. However, its aluminum consumption is much higher than sodalite. While sodalite has a 1 to 1 Al:Si molar ratio, hydrogarnet has a 2Al:nSi ratio, where n ranges from 0 to 0.8, so in the best-case scenario hydrogarnet has a 2.5–1 Al:Si molar ratio, and in the worst-case scenario it takes no silicon at all. So hydrogarnet has 2.5 times the aluminum losses of sodalite, in the best-case scenario.

Having said this though, the actual amount of aluminum loss through the hydrogarnet route is still a very small proportion of the total aluminum yield. In practice seeding Bayer liquor with lime ($CaCO_3$) to precipitate silicate out as hydrogarnet, with negligible sodium loss in total, and negligible aluminum loss proportionately, has been used with commercial success in some Chinese bauxite refineries [57].

Additions of up to 16% calcium oxide to the bauxite stream have worked successfully [58]. However, offsetting the saved cost in NaOH retention is a significant cost for a 16% CaO addition as lime. In China, lime is abundant and inexpensive but for other locations, depending on local availability and price of lime, the cost of lime may make the hydrogarnet strategy commercially unviable.

3.4.7 Iron hydrogarnet—An excellent solution to Bayer liquor desilication

An even better solution is the iron hydrogarnet route, where iron partially substitutes for the aluminum group [59–61]. If the iron hydrogarnet can be formed with a value of 1 for the x-parameter, this yields a 1.5:1 Ca:Si molar ratio, in comparison with of 3.75:1 or more for iron-free hydrogarnet. 3.75:1 is for $n = 0.8$, the maximum value for n, and the lower n, the higher the Ca:Si molar ratio for higher for iron-free hydrogarnet.

Moreover, the aluminum loss is less for iron hydrogarnet, due to the partial substitution of Fe for Al. Therefore, the iron hydrogarnet strategy is a win-win solution for dealing with the difficult bauxite ores of Eastern Europe and China, assuming the cost of lime is not offsetting the saved NaOH. This strategy gives maximal silicon removal, negligible sodium loss, and much less aluminum loss than the alternatives. The iron hydrogarnet strategy was patented by Medvedev [62] and has been adopted in quite a few alumina refineries in Eastern Europe [63].

3.5 Precipitation of the gibbsite

The gibbsite precipitation stage is one of the most critical steps in the Bayer process. The equation of relevance to precipitation is

$$Al(OH)_4^- + Na^+ \to Al(OH)_3 + NaOH$$

Precipitation is not intended to be spontaneous. It is managed in a controlled way by the twofold strategy of cooling the liquor via heat-exchangers and flash tanks and seeding it with gibbsite particles from previous production batches. The lower the NaOH concentration in the Bayer liquor, the greater the risk of spontaneous gibbsite precipitation, which is undesirable. Three key observations have been made for the precipitation of $Al(OH)_3$ from Bayer liquor [64]:

- There is usually a latent period, in the order of half an hour, between seeding and the observation of precipitation.
- The lower the temperature, the shorter the latency period.
- The higher the seed concentration and amount, the shorter the latency period.
- Without seeding, no precipitate forms even after a few days at room temperature.

The precipitate is not necessarily 100% gibbsite. Higher temperatures ($\sim 70°C$) favor gibbsite, but as the liquor approaches room temperature, bayerite is favored. Both gibbsite and bayerite are THA polymorphs of $Al(OH)_3$. It has been found that at temperatures of 65°C, 75°C, and 80°C, a small amount of Bayerite forms during the initial precipitation stage, but no other $Al(OH)_3$ polymorphs other than gibbsite and bayerite [65]. However, when the liquor is seeded with gibbsite, only gibbsite precipitates out [65,66].

The next stage after initial gibbsite precipitation is growth in the size of the gibbsite precipitates. Nucleation followed by growth are well-known mechanisms in many precipitation processes, both solid-state precipitation (e.g., recrystallization of glass ceramics), and the case for precipitation from a liquid, which is the relevant one in the Bayer process. However, in the case of the Bayer process, some key issues should be noted [67]:

- Growth rate increases with seed concentration up to a limiting seed concentration.
- Above this limiting seed concentration, growth rate decreases. Thus, there is maximum growth-rate/seed-concentration correlation.
- Increased stirring rate of the Bayer liquor reduces growth rate.
- Higher $Al(OH)_4^-$ concentrations in the Bayer liquor increase the growth rate.
- Organic contaminants in the Bayer liquor, not just oxalates, but the whole spectrum of organic contaminants that can come from bauxite, inhibit gibbsite precipitation.
- Some organic contaminants can even increase the sodium content in the gibbsite precipitate [68]. This is most undesirable, especially for alumina for the ceramic industry. Sodium contamination (strong flux) is even worse for alumina ceramics than silicon contamination (weak flux and strong static fatigue enhancer). Sodium is particularly detrimental to the electrical properties of alumina (Chapter 13).

The worst culprit in compromising the precipitation process is sodium oxalate. As mentioned in Section 3.6. The two most significant effects are inhibition of growth of gibbsite precipitates (i.e., precipitation of fines) and a decrease in precipitation yield [69,70]. Both result in a significant production penalty.

3.6 The sodium oxalate problem (bauxite organic content)

Sodium oxalate precipitation is a problem at the gibbsite $Al(OH)_3$ precipitation stage of the Bayer process. It is one of the biggest problems for the Bayer process, but it only occurs when using bauxite with significant organic content. It can cause two serious problems: reduced gibbsite yield and sodium contamination of the gibbsite precipitate and subsequent alumina.

So what is the cause of the oxalate problem? Bauxite ore, like many surface minerals, often contains organic matter, a residual from decomposed vegetation. Organic matter in the ore is highly problematic to the Bayer process because the pH 14 caustic NaOH solution of the Bayer liquor converts this organic matter into various organic sodium salts, of which the most problematic is sodium oxalate. Obviously, the higher the organic content of a bauxite ore, the larger the sodium oxalate problem. Because the NaOH liquor is recycled from batch to batch, the sodium oxalate concentration can build up until it reaches a critical supersaturation point, at which point sodium oxalate coprecipitates with the gibbsite during the gibbsite precipitation stage of the Bayer process. As a result, gibbsite fines are produced instead of the desirable gibbsite agglomeration scenario, and fines are problematic for the gibbsite calcining process [71].

The other significant problem with sodium oxalate is sodium contamination of the gibbsite. When sodium oxalate coprecipitates with the oxalate, it can result in sodium ash during calcination, thereby contaminating the alumina [69]. Sodium contamination is a very big problem for alumina ceramics since it is a flux and compromises its electrical properties (Chapter 13).

The solution is to engineer the Bayer liquor such that oxalate remains in solution during gibbsite precipitation, and conversely that oxalate is precipitated out of solution (for removal) at another stage in the Bayer process when it can safely be done, ideally after all the precipitated gibbsite has been separated from the Bayer liquor. This has been achieved with the use of quaternary ammonium compounds in the Bayer liquor. These have been shown to suppress the precipitation of oxalate during the gibbsite precipitation stage, and not adversely affect either the gibbsite yield or gibbsite growth. They can also be deactivated through thermal degradation so as to allow oxalate precipitation at the appropriate point in the Bayer cycle [72].

In addition to the ammonium compounds, several other methods have been reported for oxalate removal, including [73]:

- Using calcium to precipitate it out as calcium oxalate
- Ion exchange resins
- Oxidation via strong oxidizing agents
- Precipitation in a fluidized bed
- Ozone treatment
- Adsorption on activated carbon

Finally, as discussed in Section 3.3.6, bauxite precalcination can totally eliminate organics. One way of enhancing the relatively indigestible MHA alumina oxyhydroxides is by precalcining the bauxite to about 500–550°C to render the hydrated alumina phases amorphous. Only a very brief soak time, if any, is required at this temperature to achieve this [50,51]. This significantly reduces the required digestion temperature. Moreover, it removes organics, thereby eliminating the oxalate problem, and the lower digestion temperature reduces quartz dissolution, which helps with the reactive silica problem (Section 3.4). There is a significant energy cost associated with this calcination.

3.7 Red mud

Red mud is the alkaline sludge waste product of bauxite refining, essentially the sediment that forms from all the insoluble bauxite content. Red mud is typically comprised of iron oxides, sodium aluminosilicates (such as sodalite), silica, calcium carbonate, calcium aluminate, and titanium oxide. It also contains traces of rare earths. Of course its composition depends on the impurity types and concentrations in the parent bauxite ore. It represents a relatively manageable problem during Bayer refining, but a large and intractable disposal problem postrefining, as it is purely a waste product, with negligible commercial value. This section will look at red mud from two perspectives:

- Red mud management issues.
- Research into commercial utilization of red mud.

3.7.1 Red mud removal during Bayer processing

After the red mud is removed by sedimentation, the turbid Bayer liquor is then filtered to remove red mud fines so as to minimize the risk of fines of iron oxide or other impurities contaminating the gibbsite precipitate during the seeding and precipitation stage.

Red mud can be a problem during the actual Bayer processing because of iron oxide fines. The iron oxide polymorph commonly found in red mud is goethite, which has a tendency to reprecipitate as submicron fines. This will obviously have a very slow sedimentation rate and therefore frequently contaminates not only the recycled liquor but also the gibbsite precipitate. Flocculants are used in an attempt to overcome this, with varying degrees of success [74]. Iron contamination in the precipitated gibbsite is bad for both aluminum production and alumina ceramic production. It will compromise the properties of the alumina ceramic in wear, biomedical, armor, and electronic applications, and only tiny traces of iron are sufficient to give pearly white alumina a red tinge, a problem for biomedical uses.

As much NaOH as possible is reclaimed from the red mud. This is both a cost-saving imperative (recycle as much as possible of the NaOH) and also an environmental management imperative as alkaline red mud is an environmental hazard. Reclamation is usually done using washer tanks, which work on the principle of countercurrent decantation. Then the washed red mud is pumped away for disposal/storage, and the recovered NaOH is recycled.

3.7.2 Red mud disposal—The scale of the problem

For every tonne of alumina produced, between 1 and 2 t of red mud result (dry weight), depending on the bauxite ore used [75]. Given that the current global production of alumina is about 100 million tonnes, this corresponds to 100–200 million tonnes of red mud accumulating every year globally.

In China, one refinery alone, after 30 years of refining, has accumulated 22 million tonnes of red mud from which it has created a red mud hill approximately 800×700 m and 70 m high, the size of a town of 1000 people [76]. The global bauxite industry red mud output represents five or more such hills a year of this size. It is both a disposal cost and an environmental cost that has seen significant research attention. At the root of the problem is the fact that the alkalinity and salinity of red mud mean that it is difficult to revegetate red mud tailings dumps. Red mud would not be a problem if red mud tailings ponds could readily be chemically neutralized, revegetated, and reclaimed as usable land.

3.7.3 Red mud disposal—Problems with land reclamation

Red mud is a wet sediment whose liquid component is the caustic NaOH of the Bayer process, and therefore, unless neutralized, for example, with carbonic acid, it is a toxic caustic sludge. Even when neutralized, red mud is a significant environmental and waste management challenge. Partly because of the salinity and the staggering amounts accumulated over the years at a refinery. Typically, red mud has a pH of 10–12.5 and 15%–30% solids loading [77].

It has proven difficult to revegetate red mud dumps because of the alkalinity, the salinity, and the low organic and nutrient content. [78]. The term "plough your land with salt" somewhat applies to the red mud tailings field.

The biggest problems with red mud are threefold:

- *Groundwater hazard*: It is caustic. NaOH is relatively toxic. Disposal/storage must be done in such a way as to prevent the caustic NaOH from contaminating groundwater.
- *Huge quantity*: 100–200 million tonnes a year of red mud are generated currently, as mentioned above in Section 3.7.2, and it accumulates. Thus large quantities of red mud are generated every production cycle, and they accumulate at the production site.
- *Useless*: It is purely a waste product, and red mud dumps cannot be revegetated. Despite a lot of research, there have been no commercially viable uses found for it. So its disposal is a significant proportion of the production cost of Bayer alumina.

3.7.4 Red mud disposal—Management strategies

Red mud management needs to be done in such a way as to prevent contamination of the groundwater. It usually involves one of the following options [75]:

- A simple storage pond in which the red mud can gradually dry out has been the traditional approach. This solution has not proven totally adequate in protecting groundwater.
- Storage pond containing both clay lining and polymer membrane, with a drainage network incorporated into the system. This is more expensive but also more effective.
- Dry disposal with evaporative drying and dewatering systems incorporated. This is the best for the environment and can even be cheaper than wet storage in a pond.

3.7.5 Recovery of sodium and aluminum from red mud

Red mud contains about 5%–12% Na_2O, mainly as sodalite. Rather than introducing calcium during the Bayer alumina digestion process, as discussed in Section 3.4, another process for recovery of lost sodium is to simply conduct Bayer digestion in the conventional way, removing the reactive silica by sodalite precipitation into the red mud as usual, then at the very end of the Bayer process, recover the sodium from the sodalite precipitates in the red mud. This can be done by various means, involving lime sinter of the red mud (Section 3.8 for lime sinter), or carbonic acid infusion of the red mud to yield recoverable soda [79]. In practice, carbonation of red mud is routinely used to chemically neutralize it so it will not leach alkalinity after disposal, but as a sodium recovery methodology, it has not proven commercially viable.

Sodium recovery from red mud has not been adopted in Bayer refineries because, in general, the extra cost and complexity of this red mud posttreatment does not justify the recovered sodium. With high silica bauxites, clever strategies such as iron hydrogarnet are preferred.

Red mud also contains about 15%–25% Al_2O_3, mainly as sodalite and unreacted diaspore, which is almost as much as a low-grade bauxite (30%). While not really commercially viable now, in future years, as bauxite ore dwindles, there may be value in reprocessing red mud through a bauxite refinery to refine the residual Al_2O_3.

3.7.6 Recovery of calcium silicate (Portland cement) from red mud produced by the bauxite sinter process

The bauxite soda/lime sinter process, used for low-gibbsite high-silica bauxites, is discussed below in Section 3.8. One of the by-products of the sinter process is calcium silicate $2CaO \cdot SiO_2$, also known as Portland cement. Portland cement is one of the world's largest industrial production products at around 5 billion tonnes a year (an order of magnitude larger than the bauxite industry) servicing global concrete needs. Specifically Portland cement is a blend of the two aggressively hydrating compounds $3CaO \cdot SiO_2$ and $2CaO \cdot SiO_2$. The calcium:silicon ratio is important, it should be in the ratio of 2–3.

Portland cement is produced by calcining silica and lime in the correct ratio at 1450°C. Bauxite sintering is done at only 1200°C and uses both soda and lime. Portland cement uses only lime. So the calcium silicate residue from the bauxite sinter process is far from optimal Portland cement as it does not necessarily have the optimal stoichiometry or adequate calcination. The other problem is that cement standards restrict sodium content, and red mud is high in sodium.

Nonetheless, the red mud of the bauxite sinter process contains a large load of calcium silicate, along with the other iron and titanium residue. This brings some benefits in weakly cementing the mud together and therefore making it more mechanically stable in a waste dam. Bear in mind, bauxite processing by the sinter process is a very small proportion of world alumina production (Section 3.8). Almost all bauxite refining globally is done by the Bayer process.

When the alkalinity of the red mud is neutralized, it becomes quite a manageable waste product [80]. Moreover, some of the calcium silicate can be reclaimed from red mud (made by the sinter process) for use by the cement industry. The quantity is small, in the context of the vast global cement industry, and the reclamation cost is high.

An interesting case study comes from China. The Shandong Aluminum Company processed 22 million tons of red mud (accumulated over three decades of bauxite refining) from which they reclaimed 6 million tons for cement production, one thousandth of the world's annual cement needs. The remaining 16 million tons of red mud were disposed of in a hill measuring 700×800 m in area and an average height of 70 m [76]. This gives some idea of the scale of the problem and the modest benefits in reclamation of calcium silicate.

A disturbing case study comes from a red mud disaster in Hungary in 2010, when a prolonged period of rain caused the wall of a red-mud containment dam to collapse at the Akja alumina plant and release about 1 million cubic meters of alkaline liquid waste from the red mud dam, creating a local flood in the surrounding areas, killing 10 people and injuring 150 people in a nearby village, and creating a very difficult clean up task. A lesson learned from this is that red mud pond containment systems need to take into account abnormal rainfall events.

3.7.7 Recovery of rare-earth elements from red mud

Red mud contains various rare earth metals including yttrium (60–150 ppm) scandium (60–120 ppm), and gallium (60–80 ppm) and radioactive metals including uranium (50–60 ppm) and thorium (60–150 ppm), which can be recovered by acid leaching [81]. Some studies have also attempted to extract aluminum, sodium, iron, or other metals from the red mud [82–85]. For titanium, and iron, their current ore prices are not economically viable. Moreover, none of this metal recovery, other than perhaps iron, deals with the real issue, which is finding a way of using or safely disposing of the whole 100 million plus tonnes a year of red mud.

3.7.8 Other commercial uses for red mud

Finally, much research has gone into finding uses for red mud. These have all proved commercially unviable. Moreover, even if viable, the scale of these red mud commercial usage programs are too small and so most of the red mud still remains afterward.

- $CaO\text{-}SiO_2\text{-}Al_2O_3$ glass ceramics [86].
- Heavy clay ceramics [87].
- Sintered ceramic bricks [88].
- Cements [89].
- Glazes [90].
- Road construction [91]
- Wastewater heavy metal ion adsorbent [92].
- Remediation of heavy metal-contaminated soils [93].

One of the problems with using red mud to make building materials such as bricks is the higher than normal concentration of radioactive elements, due to the refining process [81]. In the mid-1990s, a sports pavilion was built of bricks made from silicate bonded red mud. The radioactivity concern was assuaged when testing showed the dose equivalent from the building was about 2 mSv/year, where 3 mSv a year is considered a safe limit. This seems like a risky endeavor. Mildly radioactive buildings will never become a mass market architectural pathway to consuming 100+ million tonnes a year of red mud a year.

Indeed a search through the literature will demonstrate that the list of red mud applications that have been prototyped, researched, or merely hypothesized is almost endless, and the only thing all these uses have in common is that none of them have been able to develop a commercially viable red-mud utilization industry on the scale that would be required to resolve this problem.

3.7.9 Red mud practical options

Red mud solutions need to be:

- Low cost
- Huge in scale
- Able to consume most or all the red mud available

The best way forward is research into land remediation using an effective affordable and safe means of leaching, draining, and neutralizing the red mud stockpile, while shielding it from groundwater, until it can safely be covered with landfill.

In the opinion of the author, another good possibility, given the scale of the iron smelting industry and the fact that red mud is high in iron content, is research into blending red mud with iron ore in blast furnace feedstock, or using pure red mud as feedstock, ideally with a colocated small iron smelting facility. Transportation of 500,000 t a year of red mud from an alumina refinery to a distant steel smelting facility would involve about 250 return truck journeys a week and haulage in 40 t loads around $3/km. Offsetting that is the returns on 4000 t a week of smelted iron, with steel billet at $300/t, it would be breakeven at an iron smelter distance of 800 km (1600 km return), if a smelter is available within that distance. Shorter distances make the process potentially profitable.

3.8 Alternatives to Bayer refining

Bauxite is not the only mineral that has been explored for alumina refining. Any mineral that is low in cost, widely available, and has a significant alumina content is a candidate, for example, clay, shale, nepheline, and fly ash. Clay has seen the most attention as a bauxite alternative, but it has been primarily research or pilot plant level activity, not significant commercial activity.

Many alumina refining processes have been proposed and even trialed. These can be summarized as:

(1) Wet alkaline processing, that is, the Bayer process with NaOH in an autoclave.
(2) Wet acid processing
   - HCl, HNO_3, H_2SO_4, H_2SO_3
(3) Carbothermic furnace processing
(4) Alkaline sinter processing
   - Soda sinter (Le Chatelier) or lime sinter or soda/lime sinter
   - Other reducing agents, such as sulfates or chlorides.

The first of these processes listed above is the Bayer process, and of the other three, two of them were technically viable but not commercially viable (wet acid; carbothermic). The fourth (alkaline sinter) was the original 1855 process of Le Chatelier, a very energy inefficient process, but it has continued to see limited use in China and Eastern Europe as a solution to processing the difficult karstic bauxites of Eurasia.

3.8.1 Wet alkaline processing

This is of course the Bayer process which has dominated global bauxite refining for over a century, because of its low cost, efficiency, and the knowledge benefits of over a century of fine tuning the process. It has already been discussed in depth.

3.8.2 Wet acid processing

This process has been trialed in the 20th century using acid to extract alumina from clays [94–96]. A couple of pilot plants were built in the United States, which has very limited bauxite reserves (see Table 3.1), one in 1945 and one in 1963. The process did not prove to be cost effective due to equipment corrosion, a limited capacity for recycling the acid, and the high pick up of iron from iron oxide dissolution. On the plus side, there was little silica dissolution.

3.8.3 Carbothermic furnace processing

This process was first patented a century ago as a means of making alumina abrasives [97,98]. The technology never gained significant traction over the subsequent decades.

3.8.4 Alkaline sinter processing

The original bauxite refining method invented by Le Chatelier in 1855 was a soda sinter process. Bayer's innovation of 1887–92 revolutionized bauxite refining because of his low-temperature (low-energy) NaOH digestion process. However, for bauxite ores containing 8%–15% silica, conventional Bayer digestion cannot be used as the sodium consumption by the reactive silica makes the process commercially unviable.

In China, and Russia/Eastern Europe, there is an abundance of bauxite ores that are either high in silica, high in MHA boehmite/diaspore, or both. Therefore, many refineries are compelled to use either a pure soda/lime sinter process for refining or a *combined* process involving a Bayer digestion followed by a soda sinter of the red mud [99]. Sintering is a much higher energy consumption process as shown in

Table 3.1 **Alternative bauxite refining processes and associated energy cost [45]**

Refinery name	Production method	Ore type	Energy intensity GJ/t Al_2O_3
Zhengzhou	Combined	Diaspore	34.15
Shanxi	Combined	Diaspore	37.28
Guizhou	Combined	Diaspore	43.31
Zhongzhou	Combined	Diaspore	52.17
Pinguo	Bayer	Gibbsite/Diaspore	15.1
Shandong	Sinter	Diaspore	40.5
(Un-named) France	Bayer	Diaspore	13.52
Pinjarra (Australia)	Bayer	Gibbsite	11.21
Shennigola (Greece)	Bayer	Diaspore	14.86
Stratford (Germany)	Bayer	Gibbsite	9.6

Table 3.1 [45]. Only China and Russia sinter, with their karstic and high silica bauxites, the rest of the world uses lateritic bauxite for which Bayer processing is ideal.

The sintering process is done in much the same way as the cement industry manufactures Portland Cement (calcium silicate), using large rotary kilns, and in fact one of the by-products of the bauxite sinter is indeed calcium silicate. The rotary kilns are operated at temperatures of $\sim 1200°C$, with blends of the bauxite (red mud if it is a combined process) with either lime alone (lime sinter) or lime plus soda ash (lime/soda sinter). This produces sodium aluminate and calcium silicate. The following chemical reactions are relevant [100,101].

$$Na_2CO_3 + Al_2O_3 \rightarrow Na_2O \cdot Al_2O_3 + CO_2$$
$$\text{(sodium aluminate − dehydrated Bayer liquor)}$$

$$2CaO + SiO_2 \rightarrow 2CaO \cdot SiO_2 \text{ (Portland cement)}$$

$$Na_2CO_3 + Fe_2O_3 \rightarrow Na_2O \cdot Fe_2O_3 + CO_2 \text{ (red mud)}$$

When sintering is complete, the sodium aluminate and calcium silicate sinter residue are digested using an alkaline solution, usually sodium carbonate, followed by gibbsite precipitation, usually with gibbsite seeding as for the Bayer process, or else the pre-Bayer (Le Chatelier) method of carbonic acid seeding [102]. As Table 3.1 shows the energy consumption of a sintering process, or a combined Bayer/Sinter 2-stage process, is at least three times greater than the standard Bayer process, and in some cases up to five times greater. This was of course the driver for bauxite smelting to switch to the Bayer process over a century ago when Bayer invented his energy-efficient bauxite refining process. However, with high silica karstic bauxites, Chinese and Eastern European/Russian refiners have no option but to use the sinter method.

The red mud of the sinter process contains a large load of calcium silicate (Portland Cement), along with the other iron and titanium residue. This brings some benefits in weakly cementing the mud together, and therefore making it more mechanically stable in a waste dam. When the alkalinity of the red mud is neutralized, it becomes quite a manageable waste product [80].

Processing, structure, and properties of alumina ceramics

This chapter provides an overview of the key ceramic engineering principles that underlie alumina ceramic technology. Processing of alumina ceramics—forming, and sintering; structure of the various alumina polymorphs from a crystallographic perspective; properties of alumina ceramics (mechanical, electrical, and thermal). Therefore this chapter provides a ceramic engineering foundation for the 10 applications chapters that follow. Specifically the three key reasons for the dominance of alumina in the global advanced ceramics industry are (1) Alumina has a set of very industrially valuable properties. (2) High-quality alumina raw material is abundant and inexpensive. (3) Alumina can be pressureless sintered to full density in air.

Aside from its impressive mechanical, electrical, and chemical inertness properties, the main reason for the great success of alumina commercially is its low production cost, for a material of such outstanding properties. This is due to two main factors:

(1) The plentiful supply of low-cost alumina raw material thanks to the Bayer process, and the huge scale of Bayer refinement globally, due to the aluminum industry. Such a low raw material cost is unique among the advanced ceramics.
(2) The ease with which alumina can be pressureless sintered to full density in air. This is a rare capability among the advanced ceramics.

The focus of this book is on biomedical and industrial applications of alumina, with emphasis placed on those with the largest commercial and humanitarian impact in the world today. Important properties for the commercial applications of alumina are therefore comprehensively summarized in their relevant chapters. Specifically:

- Biological properties in Chapter 5, but also where relevant in Chapters 6–10.
- Electrical properties in Chapters 13 and 14, and to a lesser extent in Chapters 8–10 and 14.
- Properties of relevance to ballistic performance in Chapter 11.
- Properties of relevance to industrial wear resistance in Chapter 12, and orthopedic wear resistance in Chapters 6 and 7.
- Refractory properties in Chapter 15.

The purpose of this chapter is to outline the basics of alumina the ceramic material and to give an overview of how it is fabricated, its structure and crystallography, and its key properties. Important properties that are not described elsewhere will be given special attention in this chapter. To this end, this chapter will outline the following topics:

(1) A brief history of alumina ceramics
(2) Preparation of alumina ceramic-precursor powders
(3) Forming of alumina ceramics

(4) Sintering of alumina
(5) Structure and crystallography of alumina
(6) Overview of the properties of alumina ceramics

4.1 A brief history of alumina ceramics

4.1.1 Background to ceramics

Alumina is an oxide ceramic, formed in almost all cases by conventional powder forming. In essence, form a shape from the powdered ceramic (dry or wet), dry the shape (if it was wet formed), debind if it was formed with a binder, then densify by solid-state sintering in a furnace. In special cases alumina is formed by unconventional powder forming, or nonpowder forming methods.

Oxide ceramics, formed by powder forming methods, date back to the dawn of civilization in the form of ancient pottery: clay pots made from mud (clay), and fired (sintered) in a wood-fired fireplace or brick-furnace construct (kiln). A significant aspect of archeology involves recovery of buried pottery artifacts from ancient civilizations, and historical interpretations thereof. Given the durability of pottery artifacts, and ceramics in general, and the cultural information they retain for millennia, ceramics have a key role in the ancient historical chronicle. Recent archeological research suggests that pottery arose around 20,000 years ago, possibly even earlier than advanced stone tools (Neolithic), and that pottery predates agriculture by perhaps as much as 10,000 years [103]. While the precise chronology remains the subject of active research and debate, there is no doubt that pottery is an ancient craft, and along with agriculture, and the discovery of metal smelting, pottery is one of the key innovations that drove the rise of early civilization (Fig. 4.1).

Fig. 4.1 Ancient Japanese Jomon pottery from the Hinamiyama site. This artifact is approximately 9000–13,000 years old.
Image Author PHG [104].
Reproduced from Wikipedia (Public Domain Image Library).

Thus, in its simplest terms, powder forming of oxide ceramics involves forming a shape from powdered clay, sourced as natural "mud." Clay is a natural nanoparticulate aluminosilicate mineral. The nanoparticulate size, plate-like morphology, and surface chemistry render clay inherently able to be formed into a moldable plastic mass by blending it with an optimal amount of water. This renders it suitable for plastic forming by hand, pottery wheel, or other simple methodology. It is then dried, and finally densified by solid-state sintering (firing) in a kiln. During sintering, it shrinks uniformly as the powder nanoparticles fuse and coalesce, but the shape, if well constructed, retains its original form, and the bulk item manifests the inherent ceramic properties of the clay particles, hard, brittle, chemically inert, and fire resistant. This is the craft known as pottery, or traditional ceramics.

Clay has a natural plasticity owing to its platy nanoparticulate powder morphology. The primary component of mud is clay, which gives mud the slippery soft plasticity that clay is known for. In contrast, powdered silica (sand) and other powdered nonclay minerals such as alumina are gritty, dilatant, only very weakly cohesive when wet, and when dry not cohesive at all. It was not until the 20th century that clay-free ceramics began to be made commercially from minerals and ceramic powders other than clay, for a long time ceramics had been made by blending clay with other minerals, for example, china ware, stoneware, and porcelain and all clay-silicate blends. Indeed mud is usually not pure clay but a blend of clay, sand, organics, and other trace minerals. However, the concept of ceramics that contain zero clay is a relatively recent one. More specifically, the concept of high-purity single component nonclay ceramics, for example, high-purity alumina ceramics, or precision-blended high-purity nonclay ceramics such as zirconia-toughened alumina (ZTA), is a 20th century technology and only became widespread in the late 20th century, forming the basis of the advanced ceramics industry.

In contrast to clay, all other ceramic powders, including alumina, are much like sand. Hard, brittle, and if crushed down to fine particles, these particles themselves are hard, angular, and roughly equiaxed in shape. Indeed alumina powder is used as an abrasive powder, and in fine particle sizes as a polishing powder, because it is hard, gritty, and dilatant, exactly the opposite of the plasticity of clay. Hard, angular equiaxed brittle particles do not manifest natural plasticity when moistened, unless the particle size is ground down to the submicron range, and even then the particle cohesion is much inferior to clay. A combination of fine grinding and synthetic chemicals (deflocculants, binders, and plasticizers) in the moistening water or solvent are required. For conventional oxide ceramics such as porcelain, clay is added to the ceramic body formulation to impart plasticity. However, for refractory high-purity oxide ceramics such as alumina, magnesia, and zirconia, clay cannot be added as clay constitutes a source of silica and alkali metal impurities that will compromise the properties of the advanced ceramic.

This was exactly the challenge a century ago when the first rudimentary experiments began with forming ceramics from Bayer alumina powder that had recently become available. For millennia, ceramists and potters had become familiar with the simplicity of molding ceramics from moistened clay. Just add water, and shape it. Not so with alumina.

4.1.2 The first alumina ceramics

After the world-changing inventions of Bayer in alumina refining in the 1880s, the world began to have, for the first time, a plentiful supply of high-quality low-cost alumina powder. However, the alumina ceramics industry did not begin for some time. The focus for some decades was on aluminum smelting from Bayer alumina. This is unsurprising given the industrial importance of aluminum. Two decades were to elapse before the scientific literature records the first rudimentary attempts at manufacturing a pure alumina ceramic. While there are a number of early patents exploring alumina ceramics made by using clay or flux additives, for example, the 1867 patent of Seymour (ground emery and flux) [105] and the 1902 patent of Buchner (alumina and clay) [106], it was Graf (Count) Botho Schwerin of Germany who made the great leap forward in 1910. In the year 1910, the first publication on the fabrication process of a pure alumina ceramic appears in the scientific literature in the form of a patent for making a pure alumina ceramic by slipcasting [17]. This was followed by two more patents by Schwerin, in 1912 [107] and in 1913 [108]. Experimenting with Bayer alumina, and facing the challenge of its dilatant nonplastic properties, Schwerin made three truly ground-breaking innovations:

- Fine grinding of Bayer alumina to render it colloidal.
- The use of acid as a chemical deflocculant to colloidally stabilize aqueous suspensions of finely ground alumina.
- Forming of alumina ceramics by slipcasting of the alumina colloids.

Slipcasting (Section 4.3.2.3) does not require plasticity, merely colloidal stability, as it involves the dewatering of a ceramic colloid (known as the casting slip) in a porous mold, gradually building up a deposit on the wall of the mold. At some point the casting slip is poured out of the mold, the mold and its coating are left to dry, and once dry, the natural drying shrinkage of the wet-powder-formed ceramic, combined with the use of a multicomponent mold that can be disassembled, enables the dried ceramic shape to be removed from the mold and sintered. Slipcasting is one of the many standard clay-based ceramic forming techniques, and is used for many industrial ceramics characterized by a relatively thin wall and complex shape, such as lavatories, sinks, vases, and fine tableware.

Three years after the first Schwerin patent, the first British Patent on pure alumina appeared in the literature, granted to the Thomson-Houston company in 1913 [109]. It was the first example of alumina green machining (Section 4.3.3.1). A pure alumina body with 10% gum tragacanth binder was pressed into a 6-mm 20-mm thick diameter disk, presintered to 1400°C, green machined into the end-use component, then resintered at 1800–2000°C. The proposed use was wear-resistant applications such as tools, bearings, dies, and drills.

World War I intervened at this stage. The first commercial application of alumina ceramics did not appear in the literature for another 14 years: slipcast high-purity alumina ceramic spark plug insulators, the innovation of Hans Reichman who published four patents in the early 1930s on this invention [110–113]. His original patent [110] was filed in 1927, and granted in 1931. Essentially Reichmann took Schwerin's alumina slipcasting innovation and attempted to commercialize it in the application of spark plug insulators. Exactly as per Schwerin's 1910 patent, Reichmann did this by slipcasting fine-ground Bayer alumina using acid as a deflocculant, which was successful at the forming stage, but the required sintering temperatures for pure alumina components proved problematic commercially and so Reichmann went on to explore flux additions, up to 5%, compromising the purity and electrical resistivity of the alumina, but bringing the sintering temperature down to a barely commercially viable 1620–1700°C. Unfortunately for Reichmann, the advent of alumina sintering aid research, enabling nonflux additions and well below 1% that could enable much lower sintering temperatures, and a pure alumina ceramic, was a couple of decades in the future (Section 4.4.3).

The work of Reichmann, though innovative for its time, was derivative of Schwerin, and largely unsuccessful commercially, though Reichmann certainly advanced the commercialization potential of slipcast alumina much further than Schwerin.

The next big breakthrough in alumina forming also involved spark plug insulators, and came shortly after the Reichmann patents, in the early 1930s. However, it involved a fundamentally new approach to making alumina ceramics: cold isostatic pressing (CIPing) in rubber molds, which ultimately became one of the industry standard means of making alumina spark plug insulators, up until the present day. Spark plug insulators were to become a mass-market product in the postwar period. This innovation first appeared in the literature in the 1934 patent of Homer Daubenmeyer of the Champion Spark Plug Company [18], and was to ultimately become the standard means of manufacturing spark plug insulators, and the Champion Spark Plug Company continues to thrive to this day.

It is noteworthy that the commercial prototype of isostatically pressed alumina spark plug insulators came half a century after the original Bayer innovation of 1887. Moreover, the Daubenmeyer patent was filed 2 years after the Reichmann patent was granted in 1931, so it is probable that Reichmann's 1927 work patented in 1931 inspired Daubenmeyer. Reichmann himself was most probably inspired by the 1910 patent of Schwerin, which was in itself enabled by the 1887 innovation of Bayer, whose innovation was enabled by the 1855 process of Le Chatelier. However, all told, there was not much progress for 80 years.

The landmark patent of Daubenmeyer, filed in 1933, and granted in 1934, is really the first successful commercial prototype of an alumina ceramic. If one was to nominate the year in which the first commercial alumina ceramic prototype was developed, 1933 would be the best estimate. Alumina spark plug insulators made by isostatic pressing were a relatively pure alumina ceramic that had a ready and ongoing market, and remain with us to this day, including the Champion brand. However, the first commercial prototype developed is one thing, actual mass-market commercialization of alumina ceramics did not really begin until the 1950s, and it was a very gradual process of market penetration for many decades, from the 1950s onward (Fig. 4.2).

World War II intervened at this stage, and no significant recorded alumina innovation happened after the Reichmann/Daubenmeyer spark plug innovations of the 1920s and 1930s, until the late 1940s. In 1949, the next major breakthrough happened with the innovation by Karl Schwartzwalder: powder injection molding (PIM) of alumina ceramics [19]. It is notable that Schwartzwalder was also granted a patent in

Fig. 4.2 A modern spark plug. The white alumina insulator clearly seen on the right extends all the way through the center of the engine-block metal connection sleeve, at the left-hand end of the spark plug, to the electrode tip—the small white region on the far left.
Image Author Sonett72 [114]. Reproduced from Wikipedia (Public Domain Image Library).

1943 [19] on isostatic pressing of spark plug insulators filed in 1938. The PIM of alumina and other advanced ceramics was to become a mass-market technology many years later (Section 4.3.2.1).

4.1.3 Modern industrial alumina ceramics

Although there was some rudimentary alumina development in the period between 1910 and World War II, the alumina ceramics industry did not really begin in earnest until the postwar period. Specifically, beginning in the 1950s, there was a gradual increase in publications on alumina in the scientific literature, and the emergence of a fledgling alumina ceramics industry, which reached a position of some strength by the 21st century.

In the postwar period, as the fledgling alumina industry began to evolve, the four alumina forming methods that have been invented prior to the 1950s in a rudimentary form (slipcasting, die pressing, isostatic pressing, and injection molding), as well as tapecasting and extrusion which were adapted to alumina around this time, were to form the foundation of the emerging alumina industry. They were each to evolve to a high level of sophistication in the ensuing decades. In addition, over the next few decades, a number of other advanced forming methods were to be invented and refined. The great majority of alumina ceramics made in the world today are still produced by these six methods, to which can be added green-machining and 3D printing. Thus the eight most important alumina forming methods in the world today are as follows:

(1) uniaxial die pressing
(2) cold isostatic pressing (CIPing)
(3) PIM
(4) tapecasting
(5) slipcasting
(6) extrusion
(7) green machining
(8) additive manufacturing (3D printing)

In addition to these mainstay methods, there are a large number of specialty ceramic forming methods. However, in this chapter a discussion of specialty forming methods will be confined to the following list, these being the most significant for the alumina industry:

(1) gelcasting
(2) thixotropic casting
(3) fuse casting

All the alumina forming methods described above will be overviewed in Section 4.3. However, before doing so, it is important to address the issue of powder preparation. All alumina ceramics, other than for fuse casting, are shaped from alumina powder into a powder preform, and consolidated into a dense ceramic by solid-state sintering of the powder preform. No melting takes place, unlike for metals and plastics, for which melt forming is the default process. An exception is fuse casting, the only mainstream alumina forming method that involves pouring a liquid melt (molten alumina) into a mold. In essence forming an alumina ceramic (other than in the case of fuse casting) involves the following basic steps:

- Prepare and characterize the powder.
- Form the powder preform (dry- or wet-forming powder methods).
- Densify the powder preform into a dense ceramic by solid-state sintering.

Solid-state sintering is a complex process and will be discussed in Section 4.4.

4.2 Alumina powder preparation

The Bayer process, described in detail in Chapter 3, provides a virtually limitless resource of low-cost high-quality alumina raw powder. Bayer alumina powder is in the order of $1000–$2000 per tonne ($1–$2 per kg), depending on many factors including tonnage ordered, particle size (degree of grinding), impurity content, and GMP/ISO certification. This is one of the most important factors that underpins the success of the alumina ceramic industry. Specifically the three key reasons for the dominance of alumina in the global advanced ceramics industry are as follows:

- Alumina has a set of very industrially valuable properties.
- High-quality alumina raw material is abundant and inexpensive.
- Alumina can be pressureless sintered to full density in air.

For boutique and specialist alumina ceramic applications, some very specialty high sintering grade alumina powders are available, such as the spray-dried nanocrystalline powders which can be as high as $10,000 per tonne ($10 per kg). However, costs of these raw materials are insignificant compared to the value adding that is possible with high-end commercial alumina ceramics ($100 to $1000 per kg or more). Moreover, across the alumina industry, the value adding in converting alumina powder into high-end ceramics is substantial.

For the production of high-purity high-quality alumina ceramics, a significant amount of postprocessing of Bayer alumina is often carried out. Generally, this simply involves fine grinding. However, in some cases, where extreme purity, ultrafine crystallite size, or other specialty requirements are involved, alumina powders are chemically synthesized from precursors such as salts and alkoxides. This is particularly the case for sol-gel forming of alumina ceramics, an advanced forming approach involving controlled gelation of chemically synthesized high-purity colloidal sols of alumina. Sol-gel forming is widely practiced for many advanced ceramics, not just alumina.

4.2.1 Definition of important terminology

Before discussing powder preparation, some important terms need to be defined:
- slurry
- deflocculant
- solids loading
- binder
- plasticizer
- milling
- spray drying

Slurry: This is a ceramic engineering term referring to a suspension of ceramic particles in a liquid medium, commonly water. When the ceramic particles are coarse (microns to tens of microns in size) the slurry will be colloidally unstable and prone to sedimentation in the absence of agitation. When the ceramic particles are submicron, the slurry may be colloidally stable (colloidal sol) and therefore will not sediment out readily, or the particles may form robust interconnected networks (gel).

Deflocculant: This is a chemical agent added to a slurry that prevents the particles from forming robust interconnected networks (gel). The optimal use of a deflocculant enables the slurry to be optimally colloidally stable and to have a maximum solids loading.

Solids loading: This is a measure of the slurry concentration: total volume of particles as a percentage of total slurry volume. The accurate measure is volume% (vol%). Weight% (wt%) is also commonly used, but wt%, while useful in process management, is misleading since the density of ceramic particles varies depending on composition, and ceramics are much denser than water. For example, alumina has a specific gravity of 4, silica 2.6, and zirconia 5.7. A 50 wt% slurry of each has a totally different vol% (28 vol% for silica, 20 vol% for alumina, and 15 vol% for zirconia). A typical ceramic slurry is 10–20 vol%. A very high solids loading is above 40 vol%. Most "slurries" are a plastic "gel" above 20 vol%.

Binder: This is a chemical agent added to a slurry that imparts green strength to the powder preform when dry. Commonly it is an organic compound. The principle that sandcastles collapse when they dry is well known. If sandcastles were made using water containing a binder, they would not collapse. This is the principle underlying sand-mold casting of metals, using a binder in the sand.

Plasticizer: This is a chemical agent added to a slurry that imparts plasticity and flexibility to the powder preform when dry. The plasticity of modeling clay is well known. Wet sand is not "plastic," but when a plasticizer is added, can manifest a rudimentary doughy sticky plasticity similar to "plastic" modeling clay.

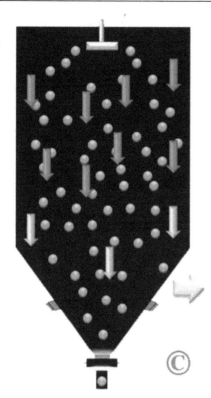

Fig. 4.3 Spray drying process. Yellow droplets of slurry sprayed at top center immediately dry into spherical "soft agglomerates" of the parent powder, and are collected at the base of the conical spray drying chamber.
Image Author U.S. Machineries [115]. Reproduced from Wikipedia (Public Domain Image Library).

Milling: Milling (comminution) is a process involving fine grinding of a powder to reduce its particle size. Commonly done using a ball mill, or for more severe milling, using an attrition mill.

Spray drying: This process produces spherical agglomerates of primary crystallites. Essentially the spray-dried spherical agglomerate can be envisaged as "snowball" (microns to tens of microns in size), the primary crystallite of the sprayed powder as the "snowflake" (tens to hundreds of nanometres in size). Spray drying enables nanopowders to be dry-handled and dry-processed with ease. Spray drying involves spraying a slurry (powder + binder + water) through an atomizer, thereby producing droplets microns to tens of microns in size. The atomized droplets are sprayed into a conical chamber containing a vortex of hot air that immediately dries the droplets into soft microparticle agglomerates, commonly spherical in shape. These spherical microparticles, which are essentially "soft agglomerates," then fall into a collecter (Fig. 4.3).

4.2.2 Powder characterization

The term "powder" conjures up the expectation that it is dry. In fact the majority of alumina powder forming methods are wet-forming methods, in which the powder is suspended, with various chemical additives, in a liquid medium, commonly water, and sometimes a polar solvent or organic binder.

A typical sintering-grade alumina powder has a primary particle size in the order of 1 μm or less. It is uncommon to sinter alumina ceramics from coarser powders. It is quite common to use much finer powders. At the fine end of the scale, nanopowders can have primary crystallite sizes down to tens of nanometres in size. However, if they are to be used for dry forming (pressing), these nanopowders will be spray dried into spherical agglomerates that are microns to tens of microns in size. At the coarse end of the scale, tabular alumina (calcined) powder can be tens or even hundreds of microns in size, but such powders are unsuitable for sintering. They are used as "grog" in refractory blends, or as abrasives.

The key characteristics for an alumina powder are as follows.

4.2.2.1 General terminology

Purity: Generally a measure of inorganic contaminants in wt% or ppm. This sets an upper limit on end product purity.

Particle shape: Usually determined by microscopy. Alumina particles produced by attrition tend to be angular, regardless of size. Particles produced by chemical synthesis tend to be nanosized and agglomerated.

Particle size: Sometimes defined by sieve size, sometimes by optically measured size (laser diffraction), and sometimes determined as stokes size (by sedimentation rate). Particle size is not necessarily primary crystallite size, as, for example, with hard agglomerates.

Specific surface area: The higher the specific surface area, the more efficient the sintering. However, conversely, the more difficult the powder is to process by wet-forming methods as the maximum solids loading for a flowable slurry will be very low—the alumina powder will have an enormous capacity for absorbing liquid. Also, the drying and firing shrinkage will be very high.

4.2.2.2 Nanopowders

Crystallite size: This term is generally used to refer to nanopowders, colloids, gels, and spray-dried agglomerates. The smaller the crystallite size, the more efficient the sintering (lower possible sintering temperature). However, the drying and firing shrinkage will be very high. Generally measured by transmission electron microscopy, or by x-ray diffraction line broadening.

Flowability: The best flowability is achieved by spray drying the raw powder. In the case of a nanopowder that is to be pressed, spray drying is virtually essential. Micropowders that are not spray dried have poor flowability. Nanopowders that are not spray dried have little or no flowability.

4.2.2.3 Coarse powders

Calcination state: Calcination involves heat treating a powder to fuse it and enlarge its primary crystallite size. It has the effect opposite to that of milling. A highly calcined powder will have a low specific surface area, a large crystallite size, will require a very high sintering temperature, and will have low drying and firing shrinkage. It is good for refractories or plasma spraying, but not for sintering into a ceramic.

Mesh size: Coarse powders are sieved into size grades, which are commonly referred to as mesh size. The larger the mesh size, the smaller the powder (mesh is a measure of wires per inch in a sieve). While mesh sizes go well above 1000 mesh, sintering-grade alumina is not commonly defined this way. Mesh size is the standard sizing method for alumina abrasive powders.

4.2.3 Common powder preparation methods

4.2.3.1 Postprocessing Bayer alumina

Comminution/milling: A pathway to particle size refinement of Bayer alumina, increasing the specific surface area by decreasing the particle size. In the low-tech to medium-tech segments of the alumina ceramic industry, alumina powders for ceramic forming are prepared from Bayer alumina by mechanical attrition methods, such as ball milling and attrition milling. Submicron sintering grade is easily achieved, but there is commonly "mill pickup" contamination from erosion of the milling media (balls or rods) and liner, unless high-purity alumina milling media and liner are used.

4.2.3.2 Chemical synthesis of ultrapure alumina nanopowders

SolGel powders: Chemical precipitation of nanocrystalline alumina as colloidal sol or gel using precursor high-purity aluminum alkoxides or aluminum salts. Provides good control of powder chemical purity, porosity, crystal size, particle size distribution, and other physical-chemical properties. Commonly spray dried after chemical precipitation.

Hydrothermal method: This method hydrolyzes anhydrous ceramic powders directly from a heated salt solution (autoclave), producing submicron powders (commonly nanopowders) with narrow particle size distribution, weak agglomeration characteristics, crystalline or amorphous structure, and very high sinterability.

Spray pyrolysis: Commonly involves an aluminum salt solution sprayed through an atomizer into a heat source such as a flame. This method allows powder processing from microparticles to nanopowders with precise morphology (spherical, hollow, porous, fibrous) and precise chemical composition (pure and doped alumina).

Chemically synthesized "SolGel powders" are orders of magnitude more expensive than Bayer alumina, but for extremely high value-added ceramics this approach can be justified.

4.3 Forming of alumina ceramics

Forming of alumina ceramics[1] can be done by a number of methods. A typical low-tech alumina ceramic such as a wear tile, slurry pipe, faucet seal, or armor tile, will be made by low-tech methods such as die pressing or slipcasting from Bayer alumina, after some ball milling or attrition milling. Alumina ceramics of complex shape will

---

[1] The author's involvement with alumina forming methods began in the 1980s, publishing his first paper on an alumina forming method in 1988 on polyelectrolyte-based micro-slipcasting and subsequently publishing on thixotropic casting, gelcasting, tapecasting, PIM, and hydrostatic shock forming of alumina. See "About the Author."

use the same milled Bayer feedstock for injection molding or isostatic pressing. Electrical substrates will commonly use the same milled Bayer feedstock for tapecasting.

More advanced and high-purity alumina ceramics such as specialty wear components, biomedical components, or high-purity electrical components will be made from spray dried powders. Bayer alumina powder can be spray dried into microsphere agglomerates if it is first heavily milled, for example, by attrition million. The highest quality alumina powders are chemically synthesized nanopowders, spray dried into microsphere agglomerates. The most advanced alumina components, such as those in specialty microelectronics, may be made from nanopowders by advanced sol-gel methods.

Green machining and 3D printing, once the domain of prototyping only, are increasingly commonly used, not just for prototyping, but even in production.

Forming of alumina ceramics can be divided into three categories:

- dry forming
- wet forming
- additive manufacturing (3D printing)

4.3.1 Dry forming

The two main dry-forming methods are uniaxial die pressing and CIPing. The main advantage of dry forming is that drying is not required, which eliminates drying shrinkage problems, which are threefold:

- risk of cracking during drying
- time involved in drying (hours to days)
- energy savings on drying costs

In addition, dry forming is a rapid and simple mass-production process. It's main disadvantage is shape complexity. CIPing is much superior to uniaxial die pressing on shape complexity, but it is inferior to PIM on precision.

4.3.1.1 Uniaxial die pressing

Uniaxial die pressing[2] is the low-tech simple production method of choice for the following main applications:

- Simple prototyping, for example, laboratory testing, and R&D.
- Mass production of simple shapes, such as tiles, using an automatic or semiautomatic press. Huge production output is possible, but tooling is expensive. Wear tiles (Chapter 12), body-armor breastplates, and armor mosaic tiles (Chapter 11) are typical examples for alumina.
- Forming blanks for subsequent green machining, for example, CeramTec use this approach for making orthopedic hip bearings (Chapter 7).

---

[2] The author spent several years assisting in the setup and optimization of a semiautomatic uniaxial-pressing mass-production process in the lightweight body armor industry, including development and implementation of quality management systems, one of which concerned radiographic examination for internal flaws and cracks. See "About the Author."

Feedstock is commonly simple milled Bayer alumina. Sometimes it is spray dried milled Bayer alumina. At the high end, it is spray-dried chemically synthesized nanopowders. Unlike ductile metals, for which more pressing pressure produces more consolidation, for ceramics, more pressure usually causes problems such as delamination and nonuniform firing shrinkage. This is why die pressing of ceramics involves only low to moderate pressures and is generally used only with thin simple shapes, such as tiles. A typical pressure for ceramic die pressing is 20–50 MPa. It is rarely higher than 100 MPa. A pressure of 200 MPa is considered an extreme upper limit for ceramic powders. In contrast, 500 or even 1000 MPa can be used for ductile metal powders.

In its standard form, uniaxial die pressing involves essentially dry powders, incorporating a minimum amount of liquid and organic binders. The powders are fed into a uniaxial hollow die, which generally comprises two opposing pistons or pressing plates. A typical die cross section is a simple shape such as a cylinder, square, or rectangle. The press applies pressure to the two opposing pistons of the die to consolidate the powder into a "green" ceramic component. For small components, tooling commonly involves a mold with multiple identical cavities, for example, in mass production of small mosaic-armor tiles.

The primary benefits of uniaxial die pressing are as follows:

- Drying is not required (no shrinkage and cracking problems, energy saving on drying costs).
- Ease of automation (automated presses, with multiple cavity molds can press thousands of ceramic components per hour).

There are a number of disadvantages also:

- simple shape capability
- nonuniformity of particle packing
- high upfront capital cost in tooling manufacture, requiring long production runs to amortize
- die degradation through abrasion from the alumina

The key disadvantage of uniaxial pressing is the nonuniform particle packing, which is a result of pressure gradients developed in the compacted-powder component during pressing. This problem is exacerbated: the thicker the component, the more complex its shape. Nonuniform particle packing, depending on its severity, can result in nonuniform firing shrinkage, distortion, and even cracking during sintering. Obvious cracks enable immediate rejection. Invisible or internal cracks can only be detected by two main QA processes:

- X-ray radiographic examination
- proof testing

Such testing adds to production cost, but the benefit is the prevention of catastrophic failure at a later stage during service life. A judgment must be taken based on product cost, and consequences of failure.

High green density by uniaxial die pressing can be problematic. Particle packing models predict that the highest particle packing density theoretically possible for monosized rigid spheres is 74% (hexagonal close-packed particle packing). In practice, milled Bayer alumina comprises angular particles of irregular shape, which generally reach a limiting green density of about 50% from die pressing. To achieve higher green densities, powder grading is required: mixing coarse medium and fine powders in optimized ratios. By this means, pressed densities approaching 70% are achievable. Other aids to enhanced pressed density include the use of a vibratory press (compressed-air driven), or optimized use of lubricating binders. Such strategies are generally only used in specialized cases. However, the vibratory press offers two key benefits:

- Significant reduction or elimination of nonuniform particle packing. This has benefits in component shape complexity and elimination of distortion and cracking.
- Involves greatly reduced pressures. Typically 1–5 MPa, making the press much more portable than 20–50 MPa hydraulic press for the same component.

In the opinion of the author, vibratory pressing is a much-under-utilized process that offers enormous benefits to the industry in product quality.

In the alumina industry uniaxial die pressing could be described as the mainstay of the low-tech end, ideal for tiles of all shapes and sizes, and other simple flat components, and ideal for anything from prototyping to large production runs.

4.3.1.2 Cold isostatic pressing (CIPing)

The main advantages of CIPing over uniaxial die pressing are twofold:

- More complex shapes are possible for CIPing compared to uniaxial die pressing.
- Distortions and cracking from nonuniform particle packing (due to pressure gradients) is much reduced in CIPing compared to uniaxial die pressing. This is critical for complex-shaped components. However, in the case of a simple-shaped component such as a thin tile, uniaxial die pressing can give an excellent result.

When complex-shaped components are required, beyond the shape capabilities of uniaxial die pressing, and a large-volume production throughput is needed, CIPing and injection molding are the two main options used. CIPing was the first high-tech method reported in the literature for manufacturing alumina ceramics, by Daubenmeyer in 1934 [18]. Indeed, spark plug insulators are probably the largest production-volume CIPed ceramic component globally. A significant proportion of the 3 billion spark plug insulators manufactured annually are CIPed. As for uniaxial die pressing, CIP feedstock is commonly simple milled Bayer alumina, milled and spray-dried Bayer alumina, or in rare cases, high-tech spray-dried SolGel nanopowders. However, in general, CIPing is not as common industrially as injection molding. CIPing is generally only used when very complex shapes are required and injection molding is not practical for whatever reason.

CIPing has superior shape capability and particle packing uniformity compared to uniaxial die pressing because it involves uniform pressing from all directions in three dimensions. It involves placing dry ceramic powder in a thin-walled soft "bag" or "mold," generally made of an elastomer such as latex, polyurethane silicone, neoprene, or nitrile. In some cases the bag or mold is shaped. In some cases, a powder preform (uniaxial die pressed, or otherwise premolded) is inserted in a sealed elastomer bag, a process known as post-CIPing.

The bagged powder or preform is then immersed in an oil chamber which is subsequently pressurized. Thus a CIP is an oil-filled chamber with the capability of being pressurized, and easily opened for loading and removal of the bagged components. Typically, CIPing pressures are ~70 MPa, although pressures of several hundred MPa, up to about 400 MPa are sometimes used. The pressurization and depressurization cycles need to be conducted at a slow controlled rate to prevent green-component damage from sudden application or reduction of stress. The process is readily automated, and can be conducted on a large-scale with many bags pressed simultaneously in one cycle, which can be automatically loaded and unloaded on racks or other fastening systems.

Some of the key advantages of CIPing for alumina ceramics are as follows.

In comparison to uniaxial die pressing:
- more complex shapes are possible
- greatly reduced pressing pressure gradients (distortion and cracking)

In comparison to all ceramic forming methods:
- Low mold cost (ideal for complex parts with small production runs).
- No mold cost for post-CIPing
- No size limitation, other than the limits of the press chamber. It is therefore ideal for very large components, especially complex-shaped ones. Components heavier than 1 t have been CIPed.
- Short processing cycle times (no drying or binder burnout required).

Some of the key disadvantages of CIP for ceramics are as follows:
- Limited dimensional control. The elastomer mold or sheath needs to be very thin and of uniform thickness but even so, high precision is problematic.
- Shape complexity is superior to uniaxial die pressing but inferior to PIM.
- Powders for CIPing require excellent flowability which generally means spray drying, or in less stringent cases, mold vibration during filling.

CIPing is commonly used for mass production of advanced ceramics of complex shape or extremely large ceramic components. Components other than spark plug insulators commonly CIPed include specialty wear components (pumps, textile industry), and electrical insulators.

In the alumina industry CIPing could be described as the mainstay of the spark plug industry, and also a relatively low-cost process ideal for shape complexity and large production runs.

4.3.2 Wet-forming methods

Wet forming is the original ceramic production process: plastic forming and pottery-wheel forming of clay-based ceramics goes back millennia. The main benefit of wet forming is shape complexity, but it has large negative effects in terms of the need for drying (if water is the liquid medium) or binder removal (injection molding and gelcasting) because of the three associated problems:

- Risk of cracking during drying (or debinding). Manifest as both visible cracks and invisible cracks (which may appear only after sintering or during operational service).
- Time involved in drying or debinding (hours to days).
- Energy costs for drying or debinding.

Drying shrinkage is the key problem, which can be up to 10% and the ceramic powder compact at this time has little cohesion, so cracking is a real risk. Some of the issues to be mindful of with drying are as follows:

- The finer the particle size, the greater the shrinkage and therefore the higher the risk of cracking. Finer particles give enhanced sintering so this is an optimization dilemma.
- The broader the particle size distribution, the lower the shrinkage and therefore the lower the risk of cracking. Broad particle size ranges give inferior sintering so this is also an optimization dilemma.
- The higher the solids loading (minimized liquid) the lower the shrinkage and therefore the lower the risk of cracking. However, less liquid makes the body less workable, less suitable to precision mold filling, and require very high forming pressures, so this is also an optimization dilemma.
- Longer drying times assist with reducing cracking.
- The larger the component, the larger is the total shrinkage and therefore the cracking risk.

Some high-tech solutions to drying include drying in ethylene glycol, supercritical CO_2 drying, and controlled humidity chambers. All add time, cost, and complexity to a production process.

So in summary, wet forming offers excellent shape complexity, at the high cost of the time energy and cracking risk in the drying/debinding process. Having said that PIM is one of the most common production methods today for advanced alumina ceramics. Tapecasting is the mainstay of the electrical substrates industry. Slipcasting and extrusion are less commonly used for alumina but each has important niche applications.

4.3.2.1 Powder injection molding

The PIM process, also known as ceramic injection molding (CIM), has long been considered the premier production process for high-precision alumina components. Like all "wet-forming" methods this comes at the cost of debinding problems: particularly cracking risk, and also the time and energy involved. This is particularly the case because thermoplastic binders are used in large amounts, up to 40 wt%. The primary competitor for PIM is CIPing, but CIPing does not have the precision capabilities of PIM. However, in the current era of low-cost 3D printing and the trend toward green machining, PIM now has three serious competitors.

The PIM process involves first preparing a feedstock by mixing the alumina powder with up to 40 wt% thermoplastic polymer binders, waxes, or other organics. After hot mixing and cooling, the feedstock is granulated into pellets a few millimeters in size.

Feedstock pellets are fed into an injection molding machine, which is similar to a plastic injection molding machine. However, the molding machine and mold need to be designed to accommodate the problem of erosion from the abrasive binder-alumina mix. The granulated feedstock is heated to around 250°C to form a flowable paste before injection into the mold under high pressure. The binder-alumina feedstock has very high viscosity, even at the molding temperature.

After cooling and removal of the component from the mold, the next step is debinding. This is a critical process because of the risk of cracking. First, the majority of the binder is dissolved or removed by chemical and/or thermal methods. The final residue is then commonly removed as a burnout stage in the sintering process. While high component shape complexity is achievable, and automation is possible, the PIM process is expensive, and binder removal is problematic. For small components, mold tooling can be designed to accommodate multiple components. There is a high mold-amortization cost with PIM that dictates long production runs.

A prominent example of PIM alumina is Cochlear bionic feedthroughs which are manufactured by PIM (Chapter 9). Another example would be specialty wear components where high-dimensional precision and shape complexity are required, such as slurry pump components (Chapter 12).

Two key issues of unique importance with PIM of ceramics are as follows:

Wall thickness: Minimization of wall thickness (ideally below 7 mm), and uniformity of wall thickness (as much as possible) is essential in order to minimize distortion during binder removal. However, if the wall is too thin this can also be problematic with the risk of distortion during debinding, due to its fragility.

Draft: Draft angle is a term referred to a slight taper required in protruding features of the molded component, for ease of mold removal. For example, a square protrusion with totally parallel walls may break when removed from the mold. If the walls are slightly tapered by 1 or 2°, this will assist in mold release. In the case of PIM alumina, draft angle is not always required as the polymer binder used in PIM of ceramics generally releases more easily from the mold than for injection molded plastics. Moreover, if PIM components are ejected from the mold before they fully cool and shrink around the mold, mold removal is easier and the need for draft may be avoided.

In the alumina industry, PIM could be described as a high-end, expensive, and complex process, with significant cracking risk, used when the imperative is for complex shapes with high precision, irrespective of cost.

4.3.2.2 Tapecasting

Tapecasting[3] is the mainstay of the electronic substrates industry (Chapter 13) and is the method of choice for manufacturing thin-film and thick-film insulator substrates for microchips, thick-film devices, and micro-feedthroughs for bionics (Chapter 10). Tapecasting involves forming thin sheets of ceramic from a powder slurry, generally a nonaqueous slurry containing various binders, plasticizers, and other additives to produce a dried "tape" that is quite robust for postprocessing. The tapecasting process involves feeding slurry through a micro-screed called a "doctor blade," which screeds the slurry into a thin layer of fixed thickness, which then rapidly dries by evaporation. This is much like concrete screeding, but on a submillimeter thickness scale.

---

[3] The author was involved for some years in a project that involved micro-screen printing, laser machining, and lamination of tapecast alumina tapes in the manufacture of micro-feedthroughs for bionic eye retinal implants. See "About the Author."

Thickness of the cast "tape" is determined by the adjustable doctor blade/substrate gap width. Screeding can be done in one of two configurations:

- Stationary doctor blade screeding slurry over a moving substrate.
- Stationary substrate over which the screeding doctor blade moves.

The dried thin sheet is called a "tape" which is commonly then processed in various ways including the following:

- Thick-film microcomponent deposition by screen printing and laser machining.
- Micro-feedthrough manufacture by screen printing, laser machining, and laminating.
- Substrate for silicon on insulator (SOI) microchip manufacture.

Tapecasting is not just used for alumina, but is in fact widely used in the electronics industry for making many electronic components including capacitors and piezoelectrics. Some of the important processing parameters of the tapecasting process include the following:

- Particle characteristics of the ceramic powder (specific surface area, particle size, particle size distribution).
- The type of suspending solvent used to form the slurry.
- The solids loading of the slurry.
- Composition and concentration of the additives which include
   - deflocculants (enhance colloidal stability during tapecasting)
   - binders (enhance green strength)
   - plasticizers (enhance flexibility of the tape)
- Doctor blade/substrate clearance gap.
- Doctor blade/substrate relative velocity.
- Drying parameters (heating rate, drying temperature, hold time).

In the alumina industry, tapecasting could be described as the mainstay of the electrical substrates industry sector.

4.3.2.3 Slipcasting

The original alumina publication, by Graf Botho Schwerin in 1910 [17], was a patent for a slipcasting process for manufacturing alumina. Thus, slipcasting is the original alumina forming process. Slipcasting[4] is a wet-forming process specifically for making hollow shapes that combines shape complexity with simplicity and is low cost. A slipcasting slurry requires as little as four components:

- alumina powder (ball milled to submicron size range)
- demineralized water
- chemical deflocculant (this helps optimize solids loading and maintain colloidal stability of the slurry during the casting process)
- organic binder (this imparts green strength to the cast component)

---

[4] The author published a series of papers on polyelectrolyte micro-slipcasting of alumina ceramics over the period 1988–96. See "About the Author."

Schwerin's unique achievement was to use acid as a deflocculant [17]. Acids are ashless as their residues are anions (sulfate, nitrate, chloride) which volatilize on sintering. Today polyelectrolytes are commonly used as deflocculants such as ammonium polyacrylate and sodium carboxymethylcellulose. Binders are optional but highly recommended. Sodium carboxymethylcellulose is ideal as it functions effectively as binder and deflocculant, however, it does impart a small amount of sodium contamination. For highest purity alumina, ashless deflocculant and ashless binder must be used.

Slipcasting is ideal for small-production run specialty components, and in the alumina industry has seen greatest use in the manufacture of specialty wear-resistant ceramics such as pipes, liners, and other hollow components or tiles for mineral processing.

Slipcasting involves the following steps, as depicted in Fig. 4.4:

- A master template is produced in the shape of the ceramic to be cast, allowing for drying and sintering shrinkage. Template is produced commonly by machining or 3D printing.
- The template is used to produce a porous casting mold (or several duplicate molds), typically made of plaster of paris. The mold usually involves several blocks of plaster with matching mating points, which can be easily assembled and dismantled.
- An alumina slurry is prepared by ball milling Bayer alumina. Deflocculants and binders are added and the solids loading is maximized to the limit of pourability.
- The plaster mold is assembled, embodying a hollow cavity of the shape of the template.
- The alumina slurry (casting slip) is poured to the brim of the mold cavity.

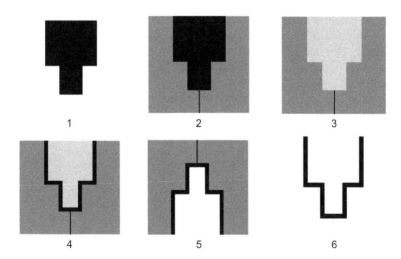

Fig. 4.4 Flowchart of the slipcasting process: (1) mold template; (2) casting the plaster mold around the template; (3) assembled plaster mold filled with alumina casting slip; (4) slipcast deposit buildup on inside wall of plaster mold; (5) mold inverted to drain residual alumina casting slip, and dry out; and (6) mold disassembled and removed to leave slipcast green alumina ceramic. Refer to the paragraph above for commentary.

- The porous mold dewaters the slurry by capillary forces, resulting in the gradual buildup of a layer of plastic alumina deposit on the wall of the mold.
- The slurry is regularly topped up until the casting time (desired wall thickness) is reached.
- The mold is drained of residual casting slip and left to dry.
- As the component dries, it shrinks away from the mold, making mold release an automatic mechanism.
- When dry, the mold is disassembled and the alumina component removed.
- The alumina component is oven-dried and sintered.

Slipcasting is an enormously versatile process. It can be used to make components from a few millimeters in size with wall thicknesses in the order of 100 μm, up to a meter or more in size with wall thicknesses of up to a centimeter or more. The range of possible molds that can be easily and cheaply fabricated is almost limitless. In the era of 3D printing, slipcasting may one day become less important for prototyping than it used to be, but it is such a simple, versatile, low-cost process that it remains a cornerstone of specialty alumina component manufacture. And is likely to remain so for the foreseeable future.

In the alumina industry slipcasting could be described as a simple low-cost and versatile prototyping process, and also an ideal low-cost process for mass production, principally for hollow components, though it is also useful for flat and curved tiles.

4.3.2.4 Extrusion

Extrusion is a well-known ceramic and metal forming process involving the extrusion of a ceramic slurry (or metal or polymer melt) through a die to produce articles of semiinfinite length and fixed cross section. It is commonly used to manufacture pipes, bricks, and aluminum tubing. In the alumina industry, extrusion is commonly used to manufacture alumina tubular items such as furnace tubes, pipes for mineral processing, thermocouple tubes, and various other such items. All that is required is to prepare a plastic mix of alumina (milled Bayer alumina) in a solvent (commonly water) with binder, deflocculant, and other additives such as plasticizers.

In the alumina industry extrusion could be described as a niche process for making tubing.

4.3.3 Direct manufacture

4.3.3.1 Green machining

Three years after the first Schwerin patent, the first British Patent on pure alumina appeared in the literature, granted to the Thomson-Houston company in 1913 [109]. It was the first example of alumina green machining. A pure alumina body with 10% gum tragacanth binder was pressed into a 6-mm thick 20-mm diameter disk, presintered to 1400°C, green machined into the end use component, then resintered at 1800–2000°C. The proposed use was wear-resistant applications such as tools, bearings, dies, and drills.

Today, green machining is increasingly the process of choice for making precision components. It is very simple conceptually. It generally involves manufacturing a blank by uniaxial die pressing, followed by precision CNC machining, then sintering. As such it eliminates all the problems and risks involved in processes capable of shape complexity such as CIPing and PIM.

Most importantly, green machining combines shape complexity with the elimination of the problem of drying and associated cracking risk of PIM, its main competitor. A million alumina hip bearings are manufactured a year by CeramTec using green machining, and CeramTec is virtually a monopoly supplier to this industry.

Green machining could be described as the mainstay of the orthopedic alumina industry sector.

4.3.3.2 Additive manufacturing (3D printing)

Like the internet, 3D printing is a 21st century disruptive technology that is so pervasive, and so universally well known, it needs no further introduction here. Suffice to say, there is almost no area of the manufacturing industry that has not been disrupted in some way by 3D printing. 3D printing has revolutionized many industries, and is making a significant impact on the alumina industry also. To date, the principal use of 3D printing in the alumina industry has been for research and for prototyping, but increasingly it is being used for short production runs, and is ideal for specialty component manufacture. Specialist biomedical applications of alumina particularly lend themselves to 3D printing given the shape complexity requirement, and involving high value adding. Dentistry is a good case in point. While 3D printing is an expensive process, it is getting cheaper by the year, and there is no tooling cost. Therefore, for short-production runs, it is very attractive.

There are three common means of 3D printing alumina ceramics:

- Inkjet 3D printing.[5] Commonly done by precision inkjet water injection into a powder bed of binder-coated powder.
- Selective laser sintering (SLS). Direct laser-sintering in a powder bed as part of the 3D printing process.
- Electron beam melting (EBM). Direct sintering by EBM in a powder bed as part of the 3D printing process.

Between green machining and 3D printing, it is possible that in a few decades the high-tech manufacturing processes CIPing and PIM will become obsolete in many cases. The lower tech methods (uniaxial die pressing, CIPing, slipcasting, tapecasting, and extrusion) are unlikely to be supplanted since they operate at a low-cost base.

---

[5] The author has extensive experience in 3D printing including inkjet printing of biomaterials. See "About the Author."

4.3.4 Specialty methods

The number of boutique ceramic forming methods is large, and beyond the scope of this book. Three are worthy of a brief mention for their relevance to the alumina industry:

(1) gelcasting
(2) thixotropic casting
(3) fuse casting

4.3.4.1 Gelcasting

Gelcasting is process that shows promise, but has failed to become a mainstream industrial ceramic forming process. It is a relatively recent technology, arising in the early 1990s. It involves the formation of a slurry in which a monomer is used as the solvent, typically 2%–5%. The slurry is cast in a mold and then polymerized. Debinding and sintering follows. Gelcasting not only shares some common elements with PIM, but also the following key differences:

- The amount of binder in gelcasting is 5–10 times less than for PIM, making the binder removal stage less problematic.
- Gelcasting involves simple pouring of a slurry into a nonporous mold. It does not require pressure injection at high temperature, and all the associated machinery and tooling of PIM.

The gelcasting concept has also been adapted to the tapecasting process, and a centrifugal-casting version of gelcasting has also been developed.

Of greatest relevance to this book is the fact that gelcasting has successfully been used to make ZTA[6] by Abbas et al. in 2015 [116].

In spite of its great promise, gelcasting has not made much traction in the ceramics industry in general, or the alumina industry specifically. This is probably because of the following limitations:

- The primary problem with gelcasting is the toxicity of the polymer component used.
- The second most important problem is that gelcasting is not easily automated.
- Strength after gelation is sometimes insufficient for handling postcasting.
- High cracking susceptibility during drying.

In summary, gelcasting shows some promise, but has not become a mainstream industrial process as yet.

4.3.4.2 Thixotropic casting

Like gelcasting, thixotropic casting is process that shows promise, but has failed to become a mainstream industrial ceramic forming process. The author published the seminal paper on thixotropic casting in 1992 [117], and many subsequent papers on the topic [118–124], and can therefore speak with the benefit of some hindsight on the potential advantages of thixotropic casting for the alumina ceramics industry.

---

[6]The author was a co-author in the Abbas ZTA Gelcasting paper, work that was conducted at SIMTech in Singapore.

Thixotropy is a rheological property whereby the viscosity of a liquid decreases when it is agitated, which is the opposite of dilatancy. Thixotropic casting involves the preparation of a highly thixotropic paste which, when vibrated, flows and can be cast into a mold, and in the absence of vibration is quite rigid. This is achieved combining the following two factors:

- Extremely high solids loading achieved by blending, coarse, medium, and fine powder grades in optimized ratios, in combination with a powerful chemical deflocculant. By this means, solids loadings over 70 vol% can be achieved, which is the highest solids loading achievable in any ceramic forming process. Most ceramic forming processes operate below 20 vol%, very few at above 40 vol%.
- Extreme thixotropy is achieved by the use of powerful chemical deflocculants.

Because of the extreme solids loading, there is little or no drying shrinkage. Published and unpublished work by the author has demonstrated the following:

- Thixotropic casting is the only method by which a 1 m^2 (or larger) tile can be wet formed without cracking, because of the near-zero drying shrinkage. Ordinarily a 1 m^2 tile can only be fabricated by uniaxial die pressing at around 4000 t, which is very fragile in the green state, and the cost of a 4000+ tonne press is prohibitive. Thus thixotropic casting is ideal for large architectural ceramics.
- Thixotropic casting is a very efficient means of casting fiber reinforced composite materials without the risk of fiber segregation.
- Thixotropic casting is ideal for making large refractory blocks, and to date, this has been its primary commercial application.
- An extensive unpublished study by the author in the 1990s on thixotropic casting of alumina demonstrated that, while ideal for refractories, fiber composites, and large architectural tiles, it is not ideal for alumina ceramics due to the requirement to have coarser powders in the mix which do not sinter as effectively as a purely submicron alumina body. Sintering by hot pressing overcomes this problem, but is only justifiable for alumina ceramics in the case of fiber composites.

Thus, while thixotropic casting shows great promise, in the alumina realm its primary application is in refractories, or fiber-reinforced alumina-matrix composites.

4.3.4.3 Fuse casting

Fuse casting of alumina is almost exclusively used for making alumina glass-contact refractories. Alumina wear tiles are occasionally also made by fuse casting. Alumina glass-contact refractories represent a $300 million industry. Fuse casting is discussed in detail in Chapter 15 under the topic of alumina refractories.

4.4 Sintering of alumina

4.4.1 Diffusion in alumina

In the mid-19th century, Adolf Fick introduced his laws of diffusion to the world, and these laws remain one of the fundamental principles of engineering today. Indeed the author spent a number of years teaching Ficks Laws of Diffusion, and associated diffusion principles. While a detailed treatise on diffusion is beyond the scope of this book, nonetheless, it is helpful to provide a brief summary on diffusion here by way of background in terminology, and key issues of relevance to alumina, to assist with the following discussion on the sintering on alumina. In this regard, some of the most important terms in diffusion are as follows:

- Diffusion coefficient (D)—proportionality factor from Fick's law of diffusion, with units length^2/time. Diffusion coefficient relates to a specific ion in a specific lattice, for example, aluminum ions in alumina.
- Activation energy of diffusion (Q)—in kJ/mol. This also relates to a specific ion in a specific lattice, for example, aluminum ions in alumina.
- Ficks first law (steady-state diffusion), for example, diffusion through a membrane.
- Ficks second law (unsteady-state diffusion), for example, cooling of a block of material.
- Schottky defect (lattice vacancy defect). Schottky defects are up to several orders of magnitude more significant in solid-state diffusion of alumina than Frenkel defects.
- Frenkel defect (interstitial defect). Less significant in alumina.

There are various types of diffusion: gas diffusion through a membrane, osmotic-pressure diffusion, liquid diffusion into porous materials, solid-state diffusion, to name but a few examples. While gas and liquid diffusion have relevance to porous alumina, for example, a tissue scaffold or a porous ceramic membrane, in the case of dense alumina ceramics, solid-state diffusion is the specific case for diffusion that is of relevance. Solid-state diffusion is not just the phenomenon underlying sintering in alumina, it is also the phenomenon underlying most of the high-temperature properties of alumina, and its ionic conductivity as an electrical ceramic. The diffusion process controls the kinetics of the mechanism involved. The key diffusion-controlled properties of relevance to alumina ceramics can be summarized as follows:

- sintering mechanisms (particle bonding, grain growth, and other microstructural changes)
- phase changes
- high-temperature creep (slow deformation under load) and high-temperature slow crack growth
- electrical conductivity (ionic conductivity)

Electrical conductivity and ionic conductivity are discussed in detail in Chapter 13. High-temperature creep is relevant to the application as a refractory (Chapter 15). Phase changes are beyond the scope of this book since this book concerns pure or near-pure monolithic alumina ceramics. However, the following discussion on sintering would be incomplete without prefacing it with a brief overview of solid-state diffusion principles in alumina as they relate to sintering.

In essence, solid-state diffusion is determined by the mobility of the diffusing ions within the crystal lattice. The key factors in diffusion are as follows:

- Temperature (higher temperature = higher diffusion rate). This creates intrinsic defects for diffusion.
- Impurity, or dopant ion, content (effect depends on the ion type and concentration). Impurities create extrinsic defects for diffusion.
- Oxygen partial pressure (oxygen dominates the alumina crystal lattice).

Extrinsic defects dominate over intrinsic defects in alumina diffusion. However, while the above parameters have different effects on diffusion, the complex interaction of the effects of these parameters can make diffusion a difficult process to model and predict. Temperature is a key driving force. Temperature creates point defects (intrinsic), which enable the ions to diffuse in the lattice, and the higher the temperature, the more point defects are created, and also the higher is the energy of the random thermally activated motion of the ions in the lattice. However, in the case of alumina, impurity-derived extrinsic defects are extremely important. There are essentially three types of ions diffusing in the alumina lattice:

- aluminum ions: small, high mobility (Section 4.5)
- oxygen ions: large, low mobility (Section 4.5)
- impurity or dopant ions: mobility depends on the ion

One of the unique factors with alumina, in relation to diffusion, is the narrow solid solubility range of aluminum in oxygen. While many ceramic compounds have a broad solid solubility range, as reflected in their associated phase diagram, the aluminum-oxygen phase diagram has a negligible solid solubility range for the two-component phase Al_2O_3, which is essentially a point compound of fixed stoichiometry. Dopant ions are of great significance in the alumina lattice. Tetravalent ions such as titanium cause aluminum vacancies to form (extrinsic defects), while divalent or monovalent ions such as magnesium or sodium cause oxygen ion vacancies to occur (extrinsic defects). The effects of these dopants on diffusion and the associated electrical properties of alumina are described in detail in Chapter 13 in relation to their effect on the electrical properties of alumina. Electrical properties are arguably the most important properties of alumina from a commercial perspective, given that uses of alumina in bionic implants and industrial electrical applications account for the majority of commercial applications in the world today, in dollar value.

Magnesium is critical in controlling the solid-state-diffusion of alumina in a microstructural sense, during the sintering process, as outlined in Section 4.4.4.

So in broad terms, diffusion in alumina can be summarized as follows:

- Defects provide the vacancies that allow aluminum, oxygen, and impurity or dopant ions to diffuse through the alumina lattice.
- Intrinsic vacancies are created by temperature.
- Extrinsic vacancies are created by dopant or impurity ions.
- The higher the temperature, the higher the ionic mobility.
- Schottky defects dominate (rather than Frenkel defects) in alumina.

While a detailed treatise of diffusion in ceramics is beyond the scope of this book, the diffusion-related issues are discussed in detail from an applications focus, where relevant. Sintering and electrical resistivity are the most important to the commercial applications of alumina, with refractoriness another important diffusion-related property. The main types of diffusion of relevance in this regard are as follows:

- lattice diffusion
- grain boundary diffusion
- surface diffusion

These issues are discussed in Section 4.4.3. However, before moving to a discussion of diffusion in sintering, we will first look at sintering from the perspective of sintering as an industrial process.

4.4.2 The key parameters of sintering

Most ceramics are formed by the following three steps:

- powder preparation
- powder forming (green ceramic)
- densification of the powder compact (green ceramic) by solid-state sintering to form a dense ceramic

Solid-state sintering involves densifying a powder into a dense solid object without melting. Sintering is the result of atomic motion enhanced by heat, and is dominated by solid-state diffusion forces. Sintering occurs via the bonding together of the particles, when heated, through mass transfer by surface transport and bulk transport (solid-state diffusion) mechanisms. The bonds formed between sintering particles are identical to the pre-existing bonding within the particles. The goal is for the green ceramic to coalesce into a fully dense solid component, although in most cases some residual porosity remains. The elimination of surface energy is the main objective and driving force (along with heat) of sintering. Thus the higher the specific surface area of the powder (finer particle size) the better it will sinter.

Alumina is capable of being sintered to full density by pressureless sintering. This is a rare capability among the advanced ceramics, many of which require hot pressing to attain full density. The excellent sintering characteristics of alumina, combined with the plentiful supply of low-cost high-quality Bayer alumina, are the primary reasons for the relatively low cost of alumina ceramics, compared with most advanced ceramics. This, combined with its excellent properties, explains why alumina has been such a successful engineering ceramic since the 1950s.

Sintering is significantly correlated with the many factors. Temperature is most important as it has an exponential effect. Particle size is also very important. However, there are many parameters of importance:

- *Sintering temperature*: Sintering rate increases exponentially with temperature, in a characteristic sigmoidal (upper and lower dual asymptote) temperature-density correlation. The solid-state diffusion rate approximately doubles for every 20°C rise in temperature. Therefore, raising sintering temperature by 20°C can theoretically halve the sintering time. Sintering temperature is usually 70%–80% of the melting point, depending on particle size, purity, and sintering aids.
- *Sintering time*: Also known as "soak time." For a given temperature, the degree of sintering correlates asymptotically with sintering time.
- *Particle size*: The surface energy per unit volume is inversely proportional to particle diameter. Therefore, smaller particles have more energy and sinter more rapidly. Thus smaller particles (higher specific surface area) have enhanced sintering efficiency. This is why cost expenditure in attrition (milling) produces a cost saving in the heat energy required for sintering. More importantly, furnace cost increases exponentially with temperature capability (the amortized furnace cost is therefore less for low-temperature capability furnaces), and

furnace life decreases with increasing operating temperature (furnace maintenance cost decreases with decreasing operating temperature).
- *Green density*: Higher green density is better because this reduces firing shrinkage. Higher green density can be achieved by higher pressing pressures (dry forming); higher solids loading (wet forming); broader particle size distribution or deliberately graded coarse/medium/fine particle blends. However, if the coarsest particle size is much above 1 μm, sintering efficiency will be reduced.
- *Agglomerates*: Hard agglomerates can compromise sintering and cause residual pores. Soft agglomerates (such as from spray drying) are broken down during the forming process.
- *Pressure*: Sintering is enhanced by the application of pressure, manifested by two processes: hot pressing and hot isostatic pressing (HIPing). For many advanced ceramics, pressure sintering is essential to attain full density. Alumina does not require pressure, but in some cases it is hot pressed or hot isostatically pressed (HIPed) so as to minimize grain size and totally eliminate any residual porosity or internal flaws, for example, alumina hip-implant bearings are HIPed (Chapter 7).
- *Sintering atmosphere*: Many advanced ceramics require a protective sintering atmosphere (argon, nitrogen, or vacuum). Alumina has the advantage that it can be sintered in air which greatly simplifies the sintering process and greatly reduces sintering cost. There is some evidence to suggest that alumina sintering can be slightly influenced by sintering atmosphere (air, hydrogen, argon, vacuum, water vapor), but in most industrial applications alumina is generally sintered in air.
- *Heating rate*: If volatilization occurs (such as binder burnout) heating rate needs to be slower. Also, large ceramic cross sections require a slower heating rate to accommodate thermal diffusion and prevent harmful thermomechanical stresses developing from thermal gradients. Moreover, very slow heating rates can influence the degree of sintering in the situation where short soak times are involved. Primarily this is a problem in the final 20°C of heating, in a furnace operating at or near its power limit. In such cases, heating rates slow down significantly near the sintering temperature.
- *Cooling rate*: As for heating, the faster the cooling rate, the higher the risk of thermal shock due to thermomechanical stresses developing from thermal gradients. This is exacerbated with larger ceramic cross sections.
- *Grain size*: Grain growth increases with increasing temperature and increasing sintering time. Grain growth is undesirable as it diminishes mechanical properties. Grain boundaries are defective regions with high atomic mobility and play a crucial role in late-stage sintering. Small grain size is contingent on a high sintering rate.
- *Porosity*: The goal of sintering is to minimize or eliminate porosity, except in the case in which a porous ceramic is the goal, such as a nanoporous filter or tissue scaffold.
- *Shrinkage*: The coalescence of the particles causes "firing shrinkage" which is typically 10%–20% linearly. Firing shrinkage is increased by finer particle size (higher specific surface area), and by lower green density.
- *Liquid-phase sintering*: The use of fluxing agents (alkali earth ions and silicates) creates a glassy phase which can enhance sintering, and then concentrates in the grain boundaries. For high-performance or ultrapure alumina this is most undesirable. For lower-purity alumina, such as spark plug insulators, AD85 body armor panels, and low-end wear tiles, liquid phase sintering is a key attribute.

There are two key advantages of using heavily milled ceramic powders, or nanopowders produced by sol-gel methods:

- Reduced cost of sintering.
- Reduced grain size (enhanced mechanical properties).

4.4.3 Sintering mechanisms in alumina

An important issue to define in sintering is closed and open porosity. Open pores are accessible to the external environment, that is, capable of absorbing water. The "apparent porosity" of a ceramic is a measure of the percentage of open porosity. Closed pores are isolated sealed pores. Total porosity = (open porosity) + (closed porosity). During sintering, porosity gradually transitions from 100% open porosity, to 100% closed porosity, to (ideally) zero porosity. The evolution of closed porosity in alumina ceramics generally begins at around 15% porosity (85% dense) and by 5% porosity (95% dense) all porosity is closed porosity.

Sintering involves mass-transport mechanisms, which are as follows:

- *Evaporation condensation*: Evaporation of the solid from the surface of a particle, which can then be reprecipitated on the surface of the same or another particle. This is not a very significant mechanism in alumina.
- *Surface diffusion*: Diffusion along the surface of a particle. This involves neck growth without a change in particle spacing, that is, it does not cause shrinkage.
- *Volume diffusion*: Also known as lattice diffusion. Solid-state diffusion through the interior crystalline lattice of a particle. Aluminum ion diffusion is much more rapid than oxygen ion diffusion in the alumina lattice. This is also shown in Chapter 13, Fig. 13.2. Volume diffusion results in shrinkage.
- *Plastic flow*: Atomic transport by plastic deformation of a particle. This is not significant in pressureless sintering of alumina, although it can occur during hot pressing and HIPing. Plastic flow results in shrinkage.
- *Grain boundary diffusion:* This is the dominant mechanism in alumina sintering. Solid-state diffusion along the grain boundary between two grains, which in the early stages of sintering is the "neck" between two particles. Grain boundary diffusion results in shrinkage.

These mechanisms are shown schematically in Fig. 4.5.

The stages of sintering are shown in Fig. 4.6. These stages are much the same for all ceramics, but the following commentary is specific to alumina:

- *Early stage 1. Initial neck formation*: Shown in image (2) in Fig. 4.6. No shrinkage occurs. Initial "neck weld" particle bonds are forming. Surface diffusion dominates.
- *Early stage 2. Neck growth*: Shown in image (3) in Fig. 4.6. Negligible shrinkage occurs. More extensive particle bonds are forming. Surface diffusion dominates. There may be a small contribution from evaporation/condensation. Early stage sintering is considered to end when neck diameter reaches 30% of particle diameter.
- *Intermediate stage. Shrinkage and coalescence*: Shown in the transition from image (3) to image (4) in Fig. 4.6. Pore structure becomes smoothed. Major shrinkage occurs in this stage. Grain boundary diffusion and volume diffusion (lattice diffusion) dominate here. This is the stage in which the transition from 100% open pores to 100% closed pores occurs, that is, from 70% to 95% dense.
- *Final stage*: Shown in the transition from image (4) to image (5) in Fig. 4.6. Small amount of shrinkage. Closed spherical pores shrink slowly by vacancy diffusion to the grain boundaries. Any gas trapped in the pores can be problematic to full densification. This stage is

Processing, structure, and properties of alumina ceramics

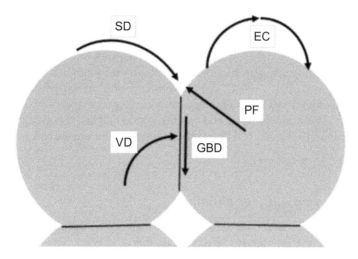

Fig. 4.5 Mass-transport mechanisms in solid-state sintering. *EC*, evaporation condensation; *SD*, surface diffusion; *VD*, volume diffusion; *PF*, plastic flow; *GBD*, grain boundary diffusion.

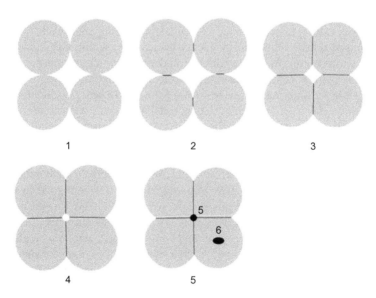

Fig. 4.6 Stages of sintering. 1. Green ceramic, particles in contact only, density ~50% theoretical. 2. Early sintering—neck formation, density ~50% theoretical. 3. Intermediate stage, pores are smoother, density is 70%–92% theoretical. 4. Final stage, above 92% theoretical, with closed porosity only. 5. Shows a fully sintered ceramic with residual porosity with two pore types shown: pore on the grain boundary (5); pore in grain interior (6).

crucial to the attainment of 100% density. In this stage, breakaway grain growth can leave pores trapped inside grains (image (5) in Fig. 4.6).

The temperature at which these stages are reached depends on the particle size of the alumina, and the impurity type and content. For high-purity alumina, the onset of neck sintering is around 1050–1100°C, and the shrinkage and coalescence stage becomes significant at around 1400°C. Final stage sintering is around 1600°C, but it can be as low as 1500°C with nanopowders and optimized sintering aid, and as high as 1700°C with Bayer alumina that has had little or no milling and inadequate or no sintering aid addition.

4.4.4 Sintering aids for alumina

In the early days of spark plugs in the 1930–50s, alkali earth/silicate fluxes (sodium/potassium silicates) were used to assist the sintering of alumina. This was a very damaging approach because alkali earths badly compromise the electrical properties of alumina (see Chapter 13) and silicates badly affect the hydrothermal stability and static fatigue resistance of alumina (see section "Purity and chemical inertness"). Indeed, alkali earth and silicates compromise almost all important properties of alumina. Moreover, the amount of flux added was in the order of several wt%, an extremely high contamination level.

The main purpose of a sintering aid is to allow complete sintering, approaching 100% density, before secondary or discontinuous recrystallization can proceed to an excessive extent. If secondary recrystallization occurs significantly, and pores become trapped, then 100% density cannot be achieved within a practical sintering time and temperature. Small particle size and good particle packing in the green ceramic also help. Small grain size is crucial to a high sintering rate.

Many alumina sintering aids were trialed in the early years, including oxides of magnesium, manganese, chromium, silicon, titanium, cobalt, iron, barium, strontium, cerium, nickel, tantalum, thorium, zirconium, calcium, copper, vanadium, beryllium, zinc, lead, bromine, tungsten, sodium, tin, antimony, and of course clay. Better ones included magnesium, chromium, and silicon. Eugene Ryshkewitch, of the Air Force Research Laboratories (Ceramic Division) at Wright Patterson Air Force Base, was the leading pioneer of alumina, who proposed MgO as a sintering aid in the 1950s [125]. In 1964, leading German pioneers in alumina Wilhelm Dawihl and Erhard Dorre recommended an MgO amount of 0.25% which has ultimately proved to be approximately the currently accepted level [126]. MgO certainly qualifies as a sintering aid on the eutectic criterion. When a sintering aid is used, the sintering temperature must be kept below the eutectic temperature of alumina/sintering aid. The MgO/Al_2O_3 eutectic temperature is 1925°C, well above normal alumina sintering temperatures.

However, it was to be some time before consensus was reached on MgO as the optimal alumina sintering aid, and the optimal amount. Gradually, over a period of some years from the 1950s to the 1980s, the realization dawned that MgO was the best sintering aid, and that somewhere around 0.15%–0.25% was the optimal amount.[7] By the 1960s MgO was known as one of the better possible sintering aids. By the 1970s MgO was seen as the preferred sintering aid, though the optimal amount was debatable. The amount of MgO explored in the early days was anywhere from as low as 0.02 wt% up to 2.5 wt%. MgO proved extremely effective in the role enabling 100% density to be attained at temperatures as low as 1500°C, with fine particles, down from 1700°C in the early alumina era, and also the amount of MgO required was extremely small (0.15%–0.25%). Finally while traces of MgO do compromise the electrical properties of alumina somewhat (Chapter 13) they do not compromise any of the other important properties of alumina significantly.

It is believed that sintering aids enhance the sintering rate if they increase the diffusion rate or if they inhibit grain boundary movement. The mechanism of sintering in alumina is believed to be secondary recrystallization, that is, discontinuous grain growth of a few large grains at the expense of adjacent smaller grains. Thus grain growth involves the swallowing up of adjacent grains by growing grains, and also secondary recrystallization engulfs the pores. MgO, which has long been the standard sintering aid in alumina, can inhibit grain growth without significantly compromising the properties of the alumina. Moreover, the amount of MgO required is very small.

It is believed that the mode of action of MgO is that surface diffusivity is enhanced by magnesia doping. MgO also works as an additive controlling breakaway of grain boundaries from pores in the final stages of sintering. Prevention of this is important in the attainment of 100% density in pressureless sintered alumina. MgO functions in this role is equally effective irrespective of whether it is a high-purity alumina with MgO the only "impurity," or it is an impure alumina, in which case the presence of the other impurities does not seem to affect MgO carrying out its function as a sintering aid.

4.4.5 Enhanced sintering methodologies

There have been many attempts to develop mathematical models to model sintering. These have had varying degrees of success. The essential problem is that other than the rare case of perfectly packed monosized perfect spheres (possible with sol-gel precipitation such as the Stober method for silica), the reality is angular particles of a range of sizes and shapes, imperfectly packed. Therefore, the best way to map sintering is to make a set of test pellets and sinter each at a different temperature, measuring density and shrinkage for each temperature, and plotting a density-temperature graph, which is usually sigmoidal.

Enhanced sintering techniques are commonly used, although in the case of alumina they are not really necessary unless a perfect product is paramount. Alumina hip-implant bearings are HIPed as discussed in Chapter 7. This is to minimize grain size and eliminate porosity. A perfect product is paramount for hip replacements, and HIPing delivers that, with a failure rate in service of below 1 in 100,000.

---

[7] Dr. Bernie Baggaley, a mentor of the author in the 1980s, spent many years researching MgO as an alumina sintering aid and believed that 0.06 mol% (0.15 wt%) was the optimal amount. The author has always found this to be optimal, through long experience since the 1980s.

The following methods are commonly used for enhanced sintering:

Pressureless methods:
- self-propagating high-temperature synthesis (SHS)
- microwave sintering

Methods involving pressure:
- hot pressing
- hot isostatic pressing (HIPing)
- spark plasma sintering (SPS)
- hydrostatic shock forming

SHS: This method can only be used in the rare case in which the precursor particles are combustible. SHS is not appropriate for alumina.

Microwave sintering: This can be an extremely energy-efficient and rapid heating process, in the case where the ceramic couples well with microwaves, which alumina does not. Alumina is microwave transparent. Microwave transparent ceramics need to be placed inside a tubular susceptor (commonly SiC) to act as a radiant heater, to heat alumina ceramic to a temperature at which it can couple with microwaves. Heating control and temperature measurement can be problematic in microwave sintering, with the concept known as "thermal runaway" (uncontrollable excessive heating) a common problem. The author spent several years researching temperature control and temperature measurement in microwave sintering, with the conclusion that on a case-by-case basis it can be an excellent method, but pure alumina is not one of those cases [127–132]. However, ZTA is one of those cases. Zirconia is a microwave absorber, and given that ZTA now dominates orthopedic and dental applications of alumina, this is of some importance.

Hot pressing: This is a costly, often labor-intensive and slow-throughput process, which is only used when essential. High-temperature dies (usually graphite) are used, and die release and ejection can be problematic, commonly requiring complex die design, graphite foil or boron nitride mold-release agents. For example, carbide ceramics such as B_4C and SiC are not easy to densify by pressureless sintering and are commonly hot pressed. This makes them very expensive. In the case of alumina, hot pressing is primarily a laboratory tool, and also useful for fiber-reinforced alumina ceramics.

HIPing: This involves heating components in a pressurized gas chamber, typically argon at around 100 MPa. It is an excellent enhanced sintering method as it is capable of sintering large number of components simultaneously and requires no dies or tooling. It is the method of choice for making alumina hip bearings. A million a year are made in this way by CeramTec (Germany) as discussed in Chapter 7. HIPing requires a surface-sealed ceramic, which means the component is either presintered to closed-porosity density (>95%) or it is vacuum encapsulated in glass or metal foil. The author has extensive experience in sinter HIPing and encapsulated HIPing and can state unequivocally that while sinter HIPing is a very commercially viable and attractive process, encapsulated HIPing is costly and complex and primarily a research tool, not a production method.

SPS (Spark Plasma Sintering): SPS is also known as PECS (pulsed electric current sintering) or FAST (field-assisted sintering technology). The SPS mechanism involves repeated application of a DC pulse voltage and current between the powder particles. The spark discharge point and joule heating point generate a mass-transport driving force for the formation of sintering neck. The SPS is sometimes used in combination with hot pressing. SPS is not necessary for alumina, other than perhaps fiber-reinforced alumina composites. It is commonly used for specialized materials such as high-performance magnets and high-wear composites.

Hydrostatic shock forming: Also known as high-rate forming, this involves densification by hydrodynamic flow, by means of a focussed and controlled explosive detonation. It is useful in densifying functionally graded ceramic-metal composites for which the metal melting point is very much less than the ceramic melting point, such as alumina-aluminum. The author has spent some time working with this method [133]. There is a high cost in machining the encasement molds, which are generally destroyed in the blast. Explosive ejection of the ceramic can also occur. The method can be effective, but it is expensive and complex. The conclusion is that the cost and complexity of this method is only justified in very specialized cases.

4.5 Structure and crystallography of alumina

The crystalline structure of alumina and its polymorphs is well known, having been comprehensively quantified over half a century ago. Alumina (Al_2O_3) is aluminum metal in its highest oxidation state. Alumina ceramics are generally polycrystalline ceramics of the α-alumina crystalline structure. α-alumina is also known as the mineral phase "corundum," a rhombohedral crystal system with space group D_{3d}^6. α-Alumina sapphire is a blue single crystal α-alumina gemstone mineral, and synthetic sapphire is a clear synthetic single-crystal α-alumina ceramic.

The crystalline structure of α-alumina consists of a hexagonal closed-packed array of large oxygen anions (140 pm ionic radius) in an A-B-A-B arrangement. The small aluminum cations (53 pm ionic radius) occupy the octahedral interstices in the oxygen HCP lattice, but only two-thirds of these octahedral interstices are occupied by an aluminum ion, so as to maintain stoichiometry and charge balance. This gives coordination numbers of 6 for the Al^{3+} cation and 4 for the O^{2-} anion. This is shown in Fig. 4.7. The α-alumina lattice parameters are as follows:

- $a_o = 0.476$ nm
- $c_o = 1.299$ nm
- density (theoretical) $= 3.99$ g cm^{-3}

Beta alumina (β-Al_2O_3) is a common industrial crystalline form of alumina, which is actually a hexagonal crystalline phase with space group D_{4h}^6 and the approximate formula:

- $Na_2O \cdot 11Al_2O_3$
- $K_2O \cdot 11Al_2O_3$
- $MgO \cdot 11Al_2O_3$
- $CaO \cdot 6Al_2O_3$
- $SrO \cdot 6Al_2O_3$
- $BaO \cdot 6Al_2O_3$

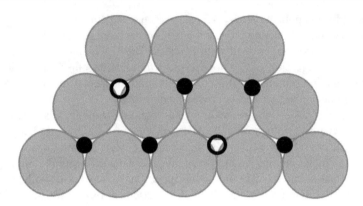

Fig. 4.7 Crystalline structure of α-Al$_2$O$_3$. The large spheres are the oxygen ions in an HCP lattice. The small black spheres are the aluminum ions in the octahedral interstices of the HCP oxygen lattice. Note that every third interstitial site is vacant.

There is also a cubic zeta alumina phase with space group O_h^7 Li$_2$O·5Al$_2$O$_3$, and a cubic (spinel) phase with space group O_h^7 AlO·Al$_2$O$_3$. Other known alumina phases include Chi (cubic), Eta (cubic spinel), Gamma (tetragonal), Delta (tetragonal), Iota (orthorhombic), Theta (monoclinic), and Kappa (orthorhombic). In addition, there are the hydrated phases AlOH$_3$, commonly found in bauxite, which include Gibbsite (α-Al$_2$O$_3$·3H$_2$O), Bayerite (β-Al$_2$O$_3$·3H$_2$O), Nordstrandite (Al$_2$O$_3$·3H$_2$O), Boehmite (α-Al$_2$O$_3$·H$_2$O), and Diaspore (β-Al$_2$O$_3$·H$_2$O).

In commercial practice, polycrystalline alumina is used as α-alumina, the main crystalline structure of relevance. β-Al$_2$O$_3$ has some applicability in the electrical industry (Chapter 13) where it is primarily a problem, as the presence of the β-Al$_2$O$_3$ phase greatly diminishes the electrical resistivity of alumina. β-Al$_2$O$_3$ is sometimes useful as an ionic conductor, and it is also useful in the refractories industry in fused-cast alumina glass-contact refractories (Chapter 15). The hydrated alumina phases are relevant to alumina refining (Chapters 2 and 3). The remainder of the alumina crystalline phases listed above are obscure and of little industrial relevance.

4.6 Overview of the properties of alumina ceramics

The key properties of alumina of industrial relevance are as follows:

Manufacturing advantages
- Low production cost—both in raw materials and in manufacture.
- Capable of being pressureless sintered in air to high density.
- With sintering aids the grain size and sintering temperature can be greatly reduced.

Manufacturing issues are discussed at length in Sections 4.2–4.4.

Mechanical properties
- hard and wear resistant (second only to diamond in the Mohs scale)
- high strength

- high stiffness
- adequate toughness, greatly enhanced as ZTA

Electrical properties
- outstanding electrical insulator
- extremely low dielectric loss tangent up to GHz frequencies

Thermal properties
- high chemical inertness and corrosion resistance
- outstanding biocompatibility
- good refractoriness
- outstanding high-temperature oxidation resistance

Important properties are discussed in detail in the relevant biomedical or industrial application as per the following list.

- Biological properties in Chapters 5–10.
- Electrical properties in Chapter 13, and to a lesser extent in Chapters 8–10 and 14.
- Properties of relevance to ballistic performance in Chapter 11.
- Properties of relevance to industrial wear resistance in Chapter 12, and orthopedic wear resistance in Chapters 6 and 7.
- Refractory properties in Chapter 15.

When one looks through reference sources on the properties of alumina, one finds a diversity of reported values for each property, in many cases without explanation of the key characteristics of that alumina specimen (particularly purity and grain size). Therefore, in order to do a comparative evaluation of the properties of alumina, a single source of alumina property data must be used that is both reliable and industrially relevant. To this end, the CoorsTek published data table on alumina, which cites purity and grain size, which is the reference source used here [134]. CoorsTek is one of the three leading alumina manufacturers in the world, as outlined in Chapter 1. The key data for alumina within the 85%–99.99% CoorsTek purity range is compiled in Tables 4.1–4.3 with an associated commentary of the relevant properties by category:

- mechanical properties
- thermal properties
- electrical properties

For context, ZTA is also included in the tables, although it must be said that the CeramTec biomedical grade ZTA is a highly evolved ceramic made by HIPing and proof tested and has superior properties to the CoorsTek industrial ZTA product discussed in this section. Chapter 7 provides a great deal of detail on the highly evolved CeramTec ZTA orthopedic bearing ceramic.

4.6.1 Mechanical properties

4.6.1.1 Young's modulus

Young's modulus of elasticity for alumina is very high for an oxide ceramic. In comparison, TZP zirconia is only 210 GPa and Porcelain is merely 104 GPa [134]. However, the carbides are higher: high-purity silicon carbide is 460 GPa, and tungsten carbide is 627 GPa [134]. The highest known Young's modulus value is that of

Table 4.1 A summary of key mechanical properties of alumina, with purity and grain size reported, with ZTA (zirconia-toughened alumina 80% alumina, 20% zirconia) included

	Test used	AD85	AD90	AD94	AD96	FG995	AD995	AD998	PP99.99[a]	ZTA
Purity (%)	N/A	85	90	94	96	99.5	99.5	99.8	99.9	80
Density (g cm^{-3})	ATSM-C20	3.42	3.6	3.7	3.72	3.8	3.9	3.92	3.92	4.01
Grain size (μm)	Thin-section	6	4	12	6	6	6	6	3	2
E (GPa)	ASTM-C848	221	276	303	303	350	370	370	386	360
Hardness (GPa)	Knoop-1kg	9.4	10.4	11.5	11.5	13.7	14.1	14.1	14.5	14.4
MOR (Mpa)	ASTM-F417	296	338	352	358	375	379	375	400	450
σT (Mpa)	ACMA Test4	155	221	193	221	248	262	248	283	290
σC (Mpa)	ASTM-C773	1930	2482	2103	2068	2500	2600	2500	2700	2900
K_{Ic}	Notched-beam	3.5	3.5	4.5	4.5	4.5	4.5	4.5	4.5	5.5
Poisson's ratio	ASTM-C848	0.22	0.22	0.21	0.21	0.22	0.22	0.22	0.22	0.23

[a]Plasma pure.
Data are from CoorsTek, Ceramic Product Specification Document, 2008.

Table 4.2 **Fracture energy relative to pure alumina, defined as the ratio of the squares of K_{Ic}, using K_{Ic}**

Ceramic	K_{Ic} (MPa m$^{1/2}$)	Fracture energy $\left(\dfrac{K_{Ic}\ \text{ceramic}}{K_{Ic}\ \text{alumina}}\right)^2$
Porcelain	2	0.25
SiC	3.5	0.75
Alumina	**4**	**1.00**
WC	6	2.25
ZTA	7	3.06
Mg-PSZ	11	7.6
Y-TZP	13	10.6

Data are from CoorsTek, Ceramic Product Specification Document, 2008.

diamond, which is both the hardest material known and has the highest elastic modulus known of ~1210 GPa [135]. From an industrial point of view the two most important issues regarding Young's Modulus of alumina are as follows:

- The fact that alumina has twice the modulus of zirconia is the secret to the resistance of ZTA to hydrothermal aging, a problem that plagues pure zirconia orthopedic bearings in vivo (Chapter 6).
- The high modulus can create stress shielding in the jawbone when alumina is used as a dental implant (Chapter 5).

Young's modulus correlates linearly with alumina purity, as shown in Fig. 4.8.

4.6.1.2 Hardness

Hardness is one of alumina's most industrially important properties. Of the key commercial uses for alumina, hardness is relevant to wear resistance for orthopedic bearings (Chapters 6 and 7), industrial wear-resistant linings (Chapter 12), and to ballistic resistance (Chapter 11). Alumina wear resistance depends particularly on hardness, with toughness also very important, and a number of other properties are also important to wear resistance as discussed in Chapter 12. Ballistic resistance only requires that alumina hardness exceeds the projectile hardness, which alumina does for all but the rare scenario of the tungsten carbide projectile.

Alumina is the hardest oxide ceramic and is defined as Mohs 9, second only to diamond on the Mohs hardness scale. However, the Mohs scale is not a linear scale, and representative only. Microhardness in GPa is the accurate measure. While the microhardness of alumina has been reported to be as high as 20 GPa or more, this is for low indenter loads. The measured microhardness does depend on the indenter load to a certain extent, the higher the indenter load the lower the measured hardness. For a 1-kg indenter load, the generally accepted microhardness value for pure alumina is around 15 GPa as presented in Table 1 and Fig. 4.9.

Table 4.3 **Thermal properties of alumina** [134]

	Test USED	AD85	AD90	AD94	AD96	FG995	AD995	AD998	PP99.99[a]	ZTA
Thermal conductivity (W/mK)	ASTM-C408	16	16.7	22.4	24.7	27.5	30	30	35	27
Coeff. thermal expansion (μ/mK)	ASTM-C372	7.2	8.1	8.2	8.2	8.2	8.2	8.2	8.1	8.3
Specific heat (J/kg/K)	ASTM-E1269	920	920	880	880	880	880	880	870	885
Thermal shock resistance (°C)	Quench test[b]	300	250	250	250	200	200	200	200	300
Maximum use temp (°C)	N/A	1400	1500	1700	1700	1700	1750	1750	1750	1500

[a]Plasma pure.
[b]Thermal shock resistance—Tests are run by quenching samples into water from various elevated temperatures. The change in temperature where a sharp decrease in flexural strength is observed is listed as ΔTc which is quantified in °C in this table.

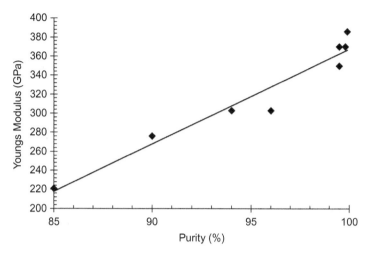

Fig. 4.8 Young's modulus of elasticity for alumina, as a function of purity level [134].

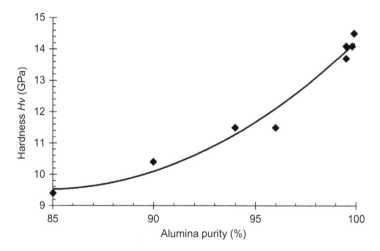

Fig. 4.9 Alumina hardness as a function of purity [134].

The reported diamond microhardness is as high as 120 GPa, and certainly over 80 GPa. Below diamond is the cubic diamond analogue ceramic cubic boron nitride which is around 40 GPa hardness. Between cubic boron nitride and alumina are the carbide ceramics, of which the hardest are silicon carbide and boron carbide, both of which are of much the same hardness. CoorsTek reports SiC microhardness at 27 GPa [134] which accords with the general consensus in the literature. Tungsten carbide, another key competitor for alumina in the industrial wear resistance realm, and a rare ballistic threat in the body armor realm, has a hardness value in the 12–16 GPa range, depending on metal content, and closely mimics the hardness range of alumina, as a function of purity, as shown in Fig. 4.9. Tungsten carbide is tougher than alumina and at least four times more expensive.

The hardness-purity correlation for alumina follows an exponentially increasing correlation up to the 100% purity point, as shown in Fig. 4.9.

Hardness is essentially a measure of the resistance of a material to penetration by another material. Thus it is a measure of resistance to surface pressure and has the SI unit GPa. Hardness in ceramics is commonly measured as microhardness, by indentation with a diamond indenter, spherical (Rockwell), square pyramid (Vickers), or rhombohedral pyramid (Knoop). Indentation is done at a precise metered load in Newtons, producing a plastic indenter impression in the test specimen surface, with micron dimensions. The geometry of the Vickers test is shown in Fig. 4.10. The significance of the red corner cracks is discussed in relation to toughness measurement in Section 4.6.1.4.

Hardness is then determined by measuring, with a microscope and scale bar, the dimensions of the plastic impression formed in the test material surface at the measured load (from 5 g to several kilograms, typically within the range 100 g to 1 kg), where hardness (pressure) is equal to (test calibration constant)x(load/area). For example, Vickers hardness is calculated in accordance with ASTM E384 [136]:

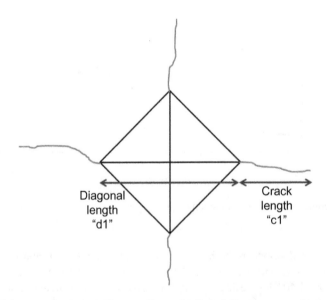

Fig. 4.10 Vickers test geometry. The two diagonals d1 and d2 are measured where "d" is the average of d1 and d2 (for simplicity only d1 is labeled here). The length of each of the four corner cracks c1, c2, c3, and c4 are measured where "c" is the average of c1, c2, c3, and c4 (for simplicity only c1 is labeled here). The higher the indentation load, the larger is the value of both "d" and "c."

$$Hv = 1854.4\left(\frac{P}{d^2}\right)$$

where Hv is the Vickers hardness in GPa, P is the load in Newtons, and d is the average of the two diagonal dimensions of the indentation in μm.

Diamond is used as the indenter since it is the hardest known material, and therefore there is no risk that the indenter will be deformed or distorted by the material it is penetrating, although there is always the risk of chipping of the indenter tip. Therefore, the indentation thus produced is a true reflection of the plastic deformation of the test material surface at the measured load. At the low loads involved, the indentations are measured in microns, and commonly microhardness indentations test the properties of a single grain of material.

The association between hardness and wear resistance is so obvious that it is common for technologists to assume that hardness is the sole determinant of wear resistance. Hardness is certainly a very important wear resistance-determining parameter in general, and for alumina specifically, but wear resistance depends on a complex interplay of many properties, of which hardness is the most important and toughness also very important, as discussed in Chapter 12.

It is important to note that the hardness of alumina can be enhanced by Cr_2O_3 additions [137,138]. Alumina with 3% Cr_2O_3 is about 10% harder than pure alumina, but hardness decreases at higher Cr_2O_3 content. A formula was proposed to correlate hardness with Cr_2O_3:

Hardness = 1945 + 15.23 (mol% Cr_2O_3) for a 500-g microindenter load [138].

4.6.1.3 Strength

Strength (σ) is a measure of the stress required to cause failure of a material. Strength is measured in MPa. Three common types of strength are commonly reported for alumina, and ceramics in general:

- modulus of rupture (MOR)
- tensile strength (σT)
- compressive strength (σC)

For ceramics, MOR is widely accepted as the standard measure of strength. All of these tests are a measure of tensile strength because ceramics always fail in tension. Even the compressive strength test, which involves crushing cylinders, involves a tensile failure in the classic double-conoidal failure configuration (Fig. 4.11).

The strength of ceramics is problematic in terms of its usefulness in characterizing a ceramic. Because ceramics are brittle (near-zero failure strain) measured strength is an unpredictable quantity, and not necessarily indicative of performance in service. Indeed a high measured strength can give a misleadingly positive representation. The reported strength value relates to the following three factors:

(1) The theoretical strength of a ceramic, based on interatomic bond strength, is approximately $E/10$ to $E/5$. In the case of alumina with an E-value of 380 GPa, this equates to about

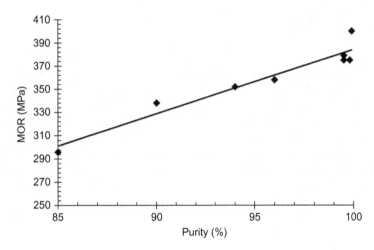

Fig. 4.11 Modulus of rupture (MOR), also known as flexural strength, of alumina as a function of purity [134].

20–40 GPa (20,000–40,000 MPa). In reality, the actual measured strength of ceramic is generally about 0.1% of alumina, that is, $E/1000$.

(2) The amount and severity of the flaws in the material. A porous ceramic is full of internal flaws (pores). The surfaces of all ceramics are covered in microscopic scratches which act as flaws. Flaws act as stress concentrators during mechanical loading.

(3) The amount of the surface of the ceramic which is exposed to the measured tensile stress at the point of failure during the test has a significant effect on the measured strength.

With regard to point 2, the quality and uniformity of the surface polish is critical to the test result.

With regard to point 3, a 3-point bend test exposes a very small proportion of the ceramic test bar surface to the measured tensile stress at the point of failure—just the face in tension, and just a small region in the center of that face, aligned with the load bar. Thus the chances that the critical flaw in the surface will receive the measured tensile stress at the point of failure is <5%, and so the strength as measured by the test is very likely, on average, to be higher than the true average strength. Consequently, the 3-point bend test, on average, gives not only the highest reported strength value, but also the largest scatter in strength values from the standard collection of about 10 duplicate test bars.

The 4-point bend test exposes a much greater proportion of the ceramic test specimen materials' surface to the measured tensile stress at the point of failure—still just the face in tension, but up to a third of the entire face. Thus the chances that the critical flaw in the surface will receive the measured tensile stress at the point of failure are around 30%–50%. So the strength as measured by the test is quite likely, on average, to be higher than the true average strength. Therefore, the 4-point bend test, on average, gives the medium reported strength value, and the medium in the scatter in strength values from the standard collection of about 10 duplicate test bars.

An internal hydrostatic loading test of a hollow cylindrical specimen exposes the entire outer surface of the cylinder to the measured tensile stress at the point of failure. Thus the chances that the critical flaw in the surface tested will receive the measured tensile stress at the point of failure are 100%. So the strength as measured by the test represents the true average strength. Consequently, the internal hydrostatic loading test, on average, gives not only the lowest reported strength value, but also the smallest scatter in strength values from the standard collection of about 10 duplicate test specimens.

What does this mean in practice? The reported strength of alumina as MOR is 400 MPa. This is approximately $E/1000$, which accords with long experience from the fracture mechanics community: in practice, the measured strength of a ceramic is approximately $E/1000$, rather than the theoretical strength based on interatomic bonding of $E/10$. The reason that measured strength falls short of theoretical strength for ceramic by a factor of $1/1000$ is explained by fracture mechanics theory.

It is well known from fracture mechanics theory that tensile failure in a ceramic originates from the sharpest deepest flaw. This is mathematically quantified, for crack initiation "mode I" conditions by the following relationship, which applies to tensile failure originating at a critical flaw. A critical flaw is most commonly a large pore or a deep surface scratch:

$$r = \frac{4\sigma^2 c}{\sigma_y^2}$$

where r is the flaw "tip" radius, σ is the macroscopic (not local) stress, c is the flaw "depth" or "crack length" (as shown in Fig. 4.13). If it is a surface flaw it is depth, if it is an internal flaw it is length, and σ_y is the yield strength of the material.

With ceramics, flaws are commonly atomically sharp, thus r approaches ionic radius. Flaw depth then becomes the critical parameter. The essential question then becomes: for that deepest flaw (deepest scratch or largest pore), what is the probability that the strength test used loaded that critical flaw to the measured stress at failure, as defined by the tensometer load cell reading and the test geometry? It ranges from <5%–100% depending on the test used, as described above and as shown in Fig. 4.12. Therefore, the results of strength testing of ceramics are problematic for two reasons:

(1) The type of test used influences the measured strength.
(2) The quality of surface polish influences the measured strength.

The problems described above are somewhat esoteric, and primarily only known and discussed within the fracture mechanics community. However, the fiberglass example demonstrates the truth of the esoteric fracture mechanics analysis. Fiberglass fibers are coated in surface polymer film (sizing) as soon as they are formed, and therefore have a genuinely pristine flaw-free surface. Young's modulus for E-glass fiberglass is around 90 GPa. The measured strength of E-glass fiberglass is typically 4500 MPa, which is $E/20$. In contrast, window glass has a typical strength of $\sim$100 MPa or $E/1000$.

In practice, the MOR test, and to a lesser extent the tensile and compressive strength test, continue to be held in good regard by engineers reading test reports and product specification documents relating to ceramics. The reality is that K_{Ic} (fracture toughness) and γi (fracture energy) are a better indicator of ceramic mechanical reliability in service.

Fig. 4.12 Common ceramic strength test configurations. The shaded zone on the underside of the beam (3-point and 4-point bend) represents the surface tensile stress: top: 3-Point bend test, note the small area (essentially a single point in the underside center) where surface tensile stress is as measured by the test, labeled "T"; middle: 4-Point bend test, note the moderately sized zone where stress is as measured by the test; labeled "T"; bottom: tube internally hydrostatically loaded, the entire outer surface of the tube uniformly experiences the tensile stress as measured.

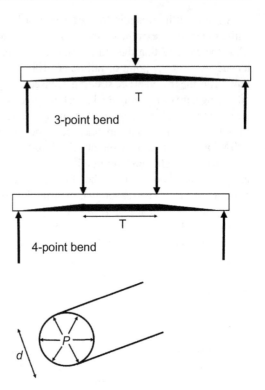

As a solution to the problem of the unpredictability and scatter in strength testing, the Weibull modulus was developed, which gives a quantitative measure of the scatter of measured strength values for a ceramic. Essentially the Weibull modulus is a statistical parameter, specifically designed to quantify ceramics. Most engineers simply rely on the standard deviation of their test result, and do not take the trouble to generate a Weibull plot.

None of the above is intended to suggest that reported strengths for alumina, and other advanced ceramics, have no value. On the contrary, they can be very useful, if they are appropriately validated. Rather it is to say that more than most other material properties, strength of ceramics is highly dependent on sample preparation and test used. The MOR is considered the standard measure of ceramic strength, but reported MOR values are dependable only if it can be verified that ASTM-F417 was rigorously complied with, and it is more meaningful if purity and grain size are also reported.

4.6.1.4 Toughness

Toughness is essentially a measure of the energy required to cause failure of a material. This is not to be confused with strength which is simply a measure of stress required to cause failure. Toughness incorporates both stress and strain, and in simple terms equates to the area under the stress-strain curve.

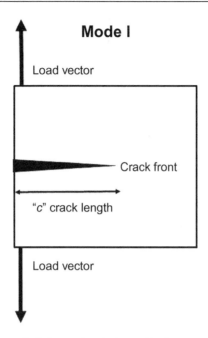

Fig. 4.13 Crack opening mode I, the mode relating to K_{Ic} fracture toughness.

Toughness is commonly reported as K_{Ic} fracture toughness (units $\text{MPa}\,\text{m}^{1/2}$) where K_{Ic} equates to the critical stress intensity for unstable crack propagation in crack opening "mode-I," which is shown in Fig. 4.13. K_{Ic} is a somewhat esoteric parameter that does not correlate linearly with material toughness. Fracture energy γi is the energy ($\text{J}\,\text{m}^{-2}$) required to create new fracture surface by the propagation of a crack. In simple terms γi fracture energy is a measure of the energy required to fracture a material, or the "impact strength" of a ceramic. The relationship between γi and K_{Ic} is as follows:

$$K_{Ic} = \left(\frac{2E \cdot \gamma i}{1 - \nu^2}\right)^{1/2}$$

where E is Young's modulus and ν is Poisson's ratio. Or to put it simply, K_{Ic} is proportional to the square of γi.

In practice, therefore, the relative "impact strength" or "damage tolerance" of one material compared to another can be determined by a simple ratio of the squares of K_{Ic} for each. Thus, while K_{Ic} is the commonly reported parameter for toughness of a ceramic, the most meaningful comparison for comparing the toughness of two ceramics is by a ratio of the squares of their reported K_{Ic} values. By this criterion, Table 4.2 illustrates how pure alumina ranks in toughness (impact strength) against its enhanced version ZTA, and against its leading ceramic competitors porcelain from the electrical realm, silicon carbide (SiC) from the armor and wear realm, and tungsten carbide (WC) from the wear realm.

Table 4.2 demonstrates just what an outstanding performer ZTA is. Superior to tungsten carbide on toughness, comparable to tungsten carbide in hardness, and very much cheaper than tungsten carbide, ZTA is also resistant to hydrothermal aging, unlike zirconia. Zirconia is also an impressive performer on toughness, but it suffers from three key problems:

- Refined zirconia powder is not an abundant low-cost raw material, unlike alumina which benefits from the huge scale of Bayer alumina refining by the aluminum industry. Quality alumina powder is available for as little as $1 per kg. This makes it difficult for zirconia to compete with alumina in the low-cost mass-market industries like wear-resistant linings in the mining industry.
- Zirconia is very dense ($5.7\,g\,cm^{-3}$) which makes it too heavy to compete with alumina in the body armor realm.
- Zirconia suffers from hydrothermal aging which precludes its use in the orthopedic industry.
- Stabilized zirconia is an ionic conductor, and so it does not compete with alumina in alumina's very industrially important role as heat-sinking insulator.

There are a number of methods and associated testing standards for measuring K_{Ic} using macroscopic test specimens, which require precision machining, polishing, and the machining of a precise notch. Alternatively and more commonly, K_{Ic} fracture toughness can simply be measured by Vickers indentation using a high indentation load (usually 1 kg or more), in accordance with the testing geometry shown in Fig. 4.10, by the formula of Anstis [139]:

$$K_{Ic} = A \left(\frac{E}{H}\right)^n \left(\frac{P}{c}\right)^{1.5}$$

where E is the Young's modulus, H is the Vickers hardness, P is the indentation load, c is the average corner crack length, and A and n are the relevant material constants.

Toughness is extremely important in wear-resistance applications, as discussed in detail in Chapter 12. This is relevant both for industrial wear resistance ceramics (Chapter 12), and in orthopedic hip bearings (Chapters 6 and 7). Toughness is also important in dental applications (Chapter 5), hence the recent rise of ZTA in the dental industry. In electrical, ballistic, and refractory applications toughness is not such an important property and the moderately high K_{Ic} value of alumina of 4.0 is more than adequate for these roles.

4.6.1.5 Inherent materials properties—Mechanical property effects

There are several inherent material properties that have a significant bearing on the mechanical properties of alumina and other advanced ceramics.

Porosity
While porosity can enhance thermal resistivity and is therefore useful in some refractories, porosity is a negative in dense advanced ceramics of the type discussed in this book. Porosity makes a ceramic friable, and pores commonly act as crack-initiating

flaws as per the toughness discussion above. Pores have the following negative effects:

- reduce ceramic wear resistance
- greatly diminish ceramic armor properties
- compromise the seal of bionic feedthroughs
- render fuse-cast alumina refractories susceptible to glass attack
- render corrosion-containment alumina more susceptible to attack
- increase the risk of fracture in alumina ceramics in general, due to the presence of internal flaws

Therefore, the lower the porosity of the alumina the better, ideally it should be 0%.

Grain size

A crack follows the cleavage plane of a grain, which corresponds to the atomic planar orientation direction of that grain. A grain boundary involves two grains in contact at which their atomic planes, aligned to different orientations, meet as a crystalline discontinuity. As such it is common to talk of "grain boundary angle." A crack traveling from one grain to the next has to change direction, by an amount corresponding to the grain boundary angle, which consumes some of its energy. This is shown in Fig. 4.14. Therefore, the more grain boundaries a crack must traverse, the more energy is required to drive it, that is, the tougher and stronger the material. This is why fine-grained ceramics have higher strength, higher wear resistance, and higher toughness. A fine-grained alumina, when pore free, can also be transparent (for dental use) as discussed in Chapter 5. Therefore, the finer the grain size of the alumina, the better.

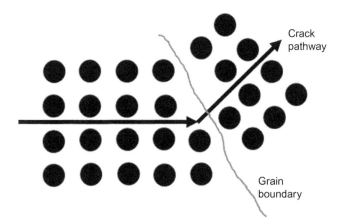

Fig. 4.14 Concept of grain boundary angle. The crack pathway, through the atomic planes of a grain of material, upon meeting a grain boundary, has to change direction by an angle corresponding to the "grain boundary angle" which is about 45° in this illustration. This directional change requires energy, and therefore consumes some of the energy driving the crack.

Purity and chemical inertness

One of alumina's key attributes is chemical inertness. This not just important for corrosion-containment applications, but also crucial in implantable biomedical applications such as bionic feedthroughs, orthopedic bearings, dental implants, and tissue scaffolds. It is also important in wet wear resistance applications. The majority of wear resistance applications are wet.

One of the biggest problems with impure alumina is moisture-enhanced static fatigue, which is exacerbated by silicates. Impurities tend to concentrate in the grain boundaries, and silicate impurities, concentrating as they do in the grain boundaries, render alumina susceptible to moisture-enhanced static fatigue. Static fatigue is a problem suffered by ceramics in which slow crack growth occurs in a moderate local stress intensity. A prolonged period of slow crack growth has the same end result as the traditional cyclic fatigue in metals. Sudden, unexpected failure after a significant time in service is catastrophic. Specifically static fatigue (slow crack growth) is defined as follows:

$$K_{Io} < K_I < K_{Ic}$$

where K_{Io} is mode I stress intensity level below which no crack growth occurs and K_{Ic} is mode I stress intensity level at and above which catastrophic unstable crack growth occurs.

This is shown schematically in Fig. 4.15.

Conclusions about the mechanical properties of alumina

In concluding this discussion about mechanical properties, the following key points should be noted:

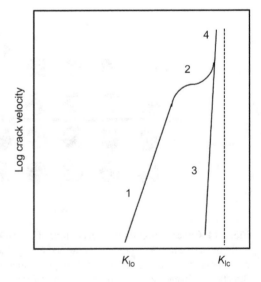

Fig. 4.15 Moisture-enhanced static fatigue. This phenomenon occurs when alumina has significant silicate content in its grain boundaries. Log (crack velocity) is plotted against K_I. For $K_I = K_{Ic}$, there is catastrophic crack propagation in all cases. For $K_{Io} < K_I < K_{Ic}$, (subcritical crack-growth load) slow crack growth is possible in the scenario that silica is in the grain boundaries, and the alumina ceramic is in a moist environment and under moderate load (pathway 1–2–4). In the absence of moisture, pathway 3–4 is followed.

- *Young's modulus*: For alumina it is extremely high, which can be good or bad, depending on the application.
- *Hardness*: Very important for determining wear resistance, but toughness is of comparable importance, as are properties of many other materials. Hardness also has some relevance to ballistic performance but it only needs to be "high enough" for the application.
- *Strength*: Due to the difficulty in measuring strength in a meaningful way, it is wise to ensure that reported strength values are only compared to one another if they were generated by the same test. Moreover, reported strength values should only be depended on if they have rigorously followed the relevant standard, and ideally the ASTM-F417 MOR test, and report purity and grain size. Following standards is always important, but for strength tests it is crucial.
- *Toughness*: This is undoubtedly the most important mechanical property for alumina as it goes to the heart of wear resistance, mechanical reliability, impact resistance, and thermal shock resistance. It is also important to be aware that K_{Ic} does not bear a linear relationship to fracture energy but is proportional to the square of fracture energy.
- *Porosity*: Should ideally be 0%.
- *Grain size*: Should be as small as possible.
- *Purity*: Should be as high as possible and silicate impurities should especially be avoided.

4.6.2 Thermal properties

4.6.2.1 Thermal conductivity

Alumina has a high thermal conductivity for a ceramic, which increases with purity as Table 4.3 illustrates. Alumina is one of the very few materials that combines extreme electrical resistivity (electrical insulator) with high thermal conductivity, and low cost, making it a superb heat sink, and the workhorse of the electrical insulation industry, as discussed in Chapter 13. The insulating heat sink, in microchips and thick-film microcircuits, is one of the largest commercial applications for alumina. Likewise spark plug insulators, another large market for alumina. Spark plugs are just a few centimeters long and have one end in the 800°C hot corrosive engine combustion zone and the other end out in the cold, exterior to the engine block.

For context, conductivity of highly conductive metals such as copper and silver are around 400 W/mK. The conductivity of a moderately conductive metal like steel is around 50 W/mK, not far above alumina. Concrete is a good insulator at around 1 W/mK, and insulating brick can be below 0.2 W/mK.

Paradoxically, alumina is also commonly used as a refractory (furnace lining). In this application, thermal insulation, heat resistance, and affordability are all important. Alumina has two out of three applications and the third (thermal insulation capability) can be engineered: commonly alumina is fibrous or in a porous brick form in this role, and the fibrous or porous form imparts thermal insulation properties to an otherwise conductive ceramic. However, in glass tanks, alumina refractories are fully dense and thermally conductive. In that scenario, chemical inertness at high temperature, for an affordable cost, is the key requirement.

The high thermal conductivity of alumina is also important in some wear resistance applications where frictional heat buildup on ceramic surfaces can cause wear, for example, in pump rotary seals there is a risk that the pump could periodically run dry.

4.6.2.2 Coefficient of thermal expansion/thermal shock resistance

The coefficient of thermal expansion of alumina is around 7–8 µ/mK, which is about average for a ceramic. It is not one of alumina's strengths. This means the thermal shock resistance is not as good as it could be, though it is helped by the high thermal conductivity of alumina. As it is, alumina is average when it comes to thermal shock resistance. Slightly better than porcelain, slightly worse than SiC.

4.6.2.3 Refractoriness (maximum use temperature)

Alumina is unique in combining a low cost with a very high refractoriness. Its melting point is around 2050°C. There is great variation in the reported melting point of alumina within the 2000–2100°C range, depending on the method used and the alumina tested, but it is approximately in the mid-2000s. The maximum use temperature of pure alumina is 1750°C which is above virtually all common industrial processes. It is significantly above the ~1540°C melting point of steel, well above the sintering temperature of all clay-based ceramics, and well above the processing temperature of most glass formulations. This combined with the chemical inertness of alumina make it one of the leading refractory materials in the world (Chapter 15). Alumina of varying purity levels is a standard liner in most high-temperature kilns. One of the more significant refractory uses of alumina is in dense fused cast alumina blocks as glass contact refractories. This is one of the harshest refractory applications in the industry, albeit fused cast zirconia is used for the most extreme glass contact refractory roles.

4.6.2.4 Specific heat

Specific heat is a measure of the energy required to heat a substance. The specific heat (heat capacity) of alumina is 900 J/kg K, which could be described as average. It is neither unusually high nor unusually low. It is not one of the standout properties of alumina.

4.6.3 Electrical properties

Alumina is one of the best electrical insulators known, both at room temperature and elevated temperature, and when its low cost and its high thermal conductivity are taken into account, alumina is the premier choice for cost effective extreme-performance insulators in general, and heat-sinking insulators in semiconductors.

Chapter 13 provides a detailed treatise on the electrical properties of alumina, and the science and engineering underlying alumina as an electrical resistor. The key applications for alumina in the electrical field are therefore as an advanced electrical insulator from microelectronics up to macro-electrical insulators, and as an electrical feedthrough insulator in bionic implants (Chapters 8–10), and in a wide range of devices in the electrical industry requiring high-performance electrical feedthroughs: both large and small devices; both high and low voltages; both high and low temperatures (Chapter 14).

Alumina also has an extremely low dielectric loss tangent, which enables a number of high-tech electrical applications, particularly in the microwave field (ovens, radar, and communications), and to a lesser extent in the RF field. This is also discussed in detail in Chapter 13.

Dental, tissue scaffold, and other specialized biomedical applications of alumina

5

Section B of this book, Biomedical Applications of Alumina, involves 7 chapters, from chapter 5 to 11. This chapter (Chapter 5) concerns dental, tissue scaffold, and other niche biomedical uses. Chapters 6 and 7 concern the use of alumina as an orthopaedic bearing material. Chapters 8, 9, and 10 concern alumina in implantable bionics. Chapter 11 gives a detailed overview of alumina in lightweight body armor, which is neither a traditional industrial application, nor a biomaterials application, but it does involve protecting the human body, combining materials science and the study of traumatology. The biomedical applications of alumina are not just important commercially, they each provide important benefits as life-saving and/or life-enhancing technologies. The bionic pacemaker, defibrillator, and pacemaker/defibrillator as well as lightweight alumina body armor are life-saving technologies, with millions of lives saved over recent decades. The orthopaedic bearing, and the other 11 bionic implants are life-enhancing technologies.

As mentioned in the beginning of Chapter 1, one of the most unusual aspects of the global alumina industry is how little cross-pollination there is between the various alumina market sectors. There are a number of large global markets for alumina, but those involved in each market sector have little or no interaction with those from the other market sectors. This is much more pronounced for those involved in the biomedical applications (orthopedic bearings, feedthroughs for implantable bionics, lightweight body armor, dental implants, and tissue scaffolds) than for those involved in the industrial applications (wear-resistant linings, refractories, microchips, insulators, and electrical feedthroughs).

Section B of this book, Biomedical Applications of Alumina, involves seven chapters, from Chapters 5 to 11. This chapter concerns dental, tissue scaffold, and other niche biomedical uses. Chapters 6 and 7 concern the use of alumina as an orthopedic-bearing material. Chapter 6 has a focus on the history, background, and technology evolution of alumina in orthopedics. Chapter 7 looks at the state of the art today and where the future is heading. The dominance of alumina in orthopedics is so well established now that the term "ceramic" in the orthopedic context is taken to mean alumina, either pure alumina or zirconia toughened alumina (ZTA).

Chapters 8–10 concern alumina in bionics. Chapter 8 concerns the invention of the alumina bionic feedthrough, which was invented for the application of the world's first bionic implant: the pacemaker. Chapter 9 looks at the world's second bionic implant, the bionic ear, and the advancements in alumina feedthrough technology that made the bionic ear possible. Chapter 10 looks at the 12 subsequent bionic implant types that have come on the market since the bionic ear, with a focus on the bionic eye and the extraordinary advancements in alumina feedthrough technology that made the bionic eye possible. Chapter 10 concludes with a look at the future for bionic implants, with a focus on the brain-computer interface via implants in the cerebral cortex, and the role of alumina feedthroughs in this futuristic technology.

Finally, Chapter 11 gives a detailed overview of the other major biomedical application of alumina ceramics: lightweight body armor. This chapter has an extensive coverage of all aspects of body armor design, optimization, and commercial applications.

Traditional industrial applications and biomaterials applications for alumina are easily categorized to Biomedical or to Industrial. Lightweight body armor uses of alumina, and its spin-offs in lightweight vehicle armor, are neither a traditional industrial application nor a biomaterials application. They do involve protecting the human body, combining materials science and the study of traumatology, and have delivered an enormous humanitarian benefit to society with millions of lives protected, and many saved. Therefore, body armor and its spin-offs have been categorized in this book as belonging to the biomedical engineering category, though obviously they are not biomaterials, unlike the orthopedic bionic, dental, and scaffold applications.

The biomedical applications of alumina are not just important commercially, they each provide important benefits as life-saving and/or life-enhancing technologies.

The bionic pacemaker, defibrillator, and pacemaker/defibrillator as well as lightweight alumina body armor are life-saving technologies, with millions of lives saved over recent decades.

The orthopedic bearing and the other 11 bionic implants are life-enhancing technologies. Alumina bearings greatly prolong the service life of a hip replacement, enabling us to move forward from the earlier era when short-lived metal-on-polyethylene bearings were the main choice, and so patients were encouraged to endure their pain and postpone surgery as long as possible given that the (non-alumina) hip implant had such a short service life. In the bionic realm, the implants that are not life savers are still life-enhancers, treating as they do many debilitating and otherwise untreatable disease states such as sensorineural deafness, drug-resistant Parkinsons disease, dystonia, obsessive compulsive disorder (OCD), depression, chronic pain, and incontinence. The same could be said of the potential for alumina (as ZTA) in dentistry and in foamed-alumina bioactive-doped bone tissue scaffolds, to be discussed in this chapter. Although these are both new areas, both show promise for a bright future.

5.1 The early history of alumina in biomedical engineering

The first documented reference to alumina as a biomaterial is in a 1933 German patent [140], but this patent did not address any specific application for alumina. It was to be three decades before alumina again appeared in the literature in a biomedical context. This was in 1965, with the publication of a British Patent on an alumina dental implant developed by Sandhaus of Switzerland [141]. In 1969, Eyring published a paper concerning alumina osteotomy spacers [142]. Thus, prior to the year 1970, the literature contained just three reports on alumina as a biomedical material. Alumina before 1970, as a biomedical material, was obscure to the point of virtually unknown. Since 1970, there has been an explosion of activity on alumina in biomedical engineering, both in research and commercialization.

The year 1970 was an extraordinary breakthrough for alumina in biomedical engineering. Two completely unrelated, but equally significant events played out simultaneously in two different countries. In Australia, David Cowdery, an Electrical Engineer with a strong research background in specialist welding technology, invented the alumina-feedthrough/titanium-casing hermetic encapsulation system for implantable bionics and managed its commercialization in 1971. This was the world's first hermetic cardiac pacemaker and the world's first hermetic implantable bionic medical device (Chapter 8). Simultaneously, in 1970, in France, Pierre Boutin, a visionary orthopedic surgeon, developed and implanted the first alumina ceramic hip implant [143,144].

These two events happened quite independently, in the same year, in two different countries: Australia and France. It is merely a coincidence that they both happened in the same year. Neither Boutin nor Cowdery was aware of the existence, or activity, of the other at the time. However, each was certainly aware of the technology race that was going on in their respective fields: pacemakers and orthopedics. Each of these paradigm shifting events was the tip of the iceberg of an active global technology race.

In the case of bionics, pacemakers had used epoxy encapsulation since the first pacemaker implantation in 1958 (Sweden) [145]. Epoxy was not hermetic and lasted barely a year in the body cavity. Finding an alternative was the global technology race. US researchers were working hard on the same problem at the same time. The Cowdery alumina-feedthrough/titanium-casing system changed the industry overnight as it was stable for decades in the body cavity (Chapter 8). This alumina/titanium bionic feedthrough system remains the key enabling platform technology of all 14 different types of implantable bionic devices on the market today (Chapter 10). Implantable bionics has grown to a $25 billion global industry in the half a century that has elapsed since the 1970 Cowdery invention [146,147].

In the case of orthopedics, Boutin was the first to develop an alumina hip replacement (Chapter 6). Pierre Boutin was a French orthopedic surgeon and one of his patients was Vice-President of Ceraver (a ceramic company). As a result, a partnership formed between the orthopedic surgeon and the ceramic company, a perfect partnership for a rapid breakthrough. Boutin's reasons for developing the alumina hip were, according to him, "its hardness which renders it perhaps indestructible for the life of a human, its perfect chemical inertia, its low coefficient of friction, and its tolerance by the organism" [148]. However, it was a technology race and a large body of German researchers rapidly caught up and soon overtook Boutin, and it was German engineering and innovation that ultimately revolutionized the world of orthopedics. Japan was also a significant participant in this technology race.

Interestingly, also in the year 1970, in the United States, Hulbert [149] published the first paper on calcium aluminate bone tissue scaffolds. Strictly speaking this was not an alumina implant, it was calcium aluminate, but it foreshadowed the concept of calcium-doped alumina tissue scaffolds, which was to become a focus of research attention in the 21st century. Hulbert followed up in 1972 with a further study of calcium aluminate as a soft tissue scaffold [150].

Further details on how the Boutin invention revolutionized the field of orthopedic hip replacements are discussed in Chapters 6 and 7. Further details about how the pacemaker and ultimately the huge implantable bionics industry evolved from the Cowdery invention are outlined in Chapters 8–10.

Orthopedics, bionic implants, and lightweight body armor, as discussed in Chapters 6–11, account for nearly 100% of the current commercial applications of alumina in the biomedical sector.

For completeness, the purpose of this chapter is to explore all other significant reported uses for alumina in the biomedical sector. In this regard, the main two uses of significance are dental technology and tissue scaffolds. Therefore, this chapter will be divided into three areas:

- Dental applications of alumina
- Alumina tissue scaffolds for bone tissue engineering
- Other biomedical uses of alumina

Dental technology is the oldest biomedical application for alumina, dating back to the 1960s. As a pearly white high-strength ceramic alumina has a lot to offer the dental field. However, alumina in dental technology languished since the 1980s due to the need for enhanced toughness. Alumina in dentistry has recently seen a new lease of life with the advent of ZTA. Alumina as ZTA appears to have a bright future in dentistry.

Alumina tissue scaffolds are the newest biomedical application for alumina, which reached the proven prototype stage just a couple of years ago in 2015. They have an unrivaled combination of properties (strength/porosity/bioactivity) that eclipse contemporary tissue scaffold technology: bioactive low-strength calcium phosphates and nonbioactive low-strength biodegradable polymers.

5.2 Alumina in dental technology

While alumina is a monopoly technology in bionic implant feedthroughs and holds 55% of the global market in hip replacement bearings, alumina languished for many decades in the dental field due to the high toughness requirement. In dentistry, alumina has occupied experimental and niche commercial roles in crowns, dental implants, abutments, and bridges, all roles which are currently dominated by titanium and dental porcelain. However, all this is beginning to change with the recent advent of ZTA. Currently, the main role for alumina in dentistry where it is approaching market dominance is orthodontic brackets.

Alumina has been used in the whole spectrum of dental devices, particularly the main ones:

- Orthodontic brackets
- Dental crowns
- Dental bridges
- Dental implants
- Dental abutments

5.2.1 Alumina orthodontic brackets

Orthodontic brackets are the brackets attached to teeth to which the bracing wires of dental braces are attached. Traditionally made of metals, alumina is increasingly commonly used in this role[1] due to the aesthetics of using a white or translucent bracket. This is the only one of the five alumina applications in dentistry discussed in this section for which alumina is currently approaching market dominance. There is an obvious cosmetic benefit in using pearly white, translucent, or transparent orthodontic brackets as shown in Fig. 5.1, and alumina is much stronger than the alternative white/clear plastic brackets.

Translucent alumina has been in existence since the 1950s, having originally been developed by Robert Coble of General Electric for sodium lamps, under the product name "Lucalox" [153]. Transparency or translucency in alumina is attained when light scattering is minimized below a certain threshold. This requires porosity to be below 0.1% and pores to be ultrafine in size [154]. Grainsize needs to be ultrafine: certainly below 2 µm, and ideally below 0.8 µm [153]. Obviously, the thinner the cross section the better the light transmission.

Single-crystal transparent alumina is known as sapphire. Polycrystalline transparent alumina is easier and cheaper to manufacture and is the commonly used option in alumina orthodontic brackets. Orthodontic brackets of polycrystalline transparent alumina are ideally made by powder-injection molding (Chapter 4).

5.2.2 Alumina in dental crowns

A crown is a veneer tooth made of pearly white ceramic and color-toned to closely resemble the existing teeth in a patient's mouth. The two main scenarios in which a crown is used are:

- A cap on a broken tooth, in the case where the broken tooth retains load-bearing capabilities.
- A cap on the abutment of a dental implant.

Crowns are traditionally made of dental porcelain, sometimes with a gold supporting underlayer. The porcelain crown was patented by in 1889 by Charles Land [155] and is now a highly optimized product aesthetically, although its mechanical integrity in the oral environment can be problematic due to moisture-enhanced static fatigue (-Chapter 4), which is why it commonly has a gold supporting underlay. A crown is used in two common scenarios. (1) When a tooth breaks, but still has load-bearing capability, it is ground down such that a crown can be fitted over the top and glued into position and (2) with a dental implant, a crown is fitted over the abutment of a dental implant. See Fig. 5.2 (left image).

While it is pearly white and much superior in strength to porcelain, alumina is not an ideal material for a dental crown because alumina is much harder ($\sim 14\,\text{GPa}$) than teeth ($\sim 4\,\text{GPa}$), and therefore an alumina crown would tend to grind down the softer opposing teeth. Nonetheless, because of the superior strength and static fatigue

---

[1] The author was involved in a translucent alumina orthodontic bracket development with SIMTech in Singapore in 2011/2012. See "About the Author".

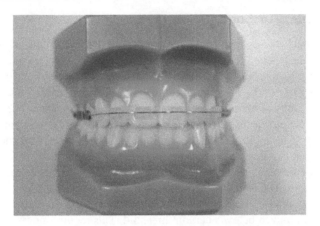

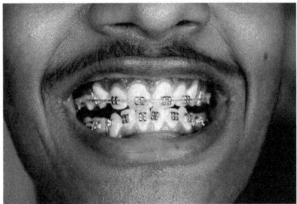

Fig. 5.1 Orthodontic brackets. Above: Translucent orthodontic brackets. Below: Metal orthodontic brackets.
Above: Image Author Hanabishi [151]. Below: Image Author Suyash Dwivedi [152]. Reproduced from Wikipedia (Public Domain Image Library).

resistance of alumina compared to porcelain, alumina crowns have been developed, for example, the Procera AllCeram ceramic core (Nobelpharma, Sweden) [158]. An alternative to the pure alumina crown is the In-Ceram concept (Vita Zahnfabric, Germany), a composite crown comprising a slipcast porous alumina crown, which is infused with glass [158]. This combines aesthetics with a surface hardness that matches the opposing teeth. There is also the Procera AllCeram alumina crown, which comprises a 99.9% alumina crown to which a feldspathic ceramic is layered. Zirconia is also commonly used in similar manifestations.

While these boutique alumina-based and zirconia-based pure and composite dental crowns have made some impact in the market, the traditional porcelain crown remains the dominant crown technology in the market. Whether this is handmade pure porcelain, handmade porcelain with gold underlay, or the modern and convenient concept

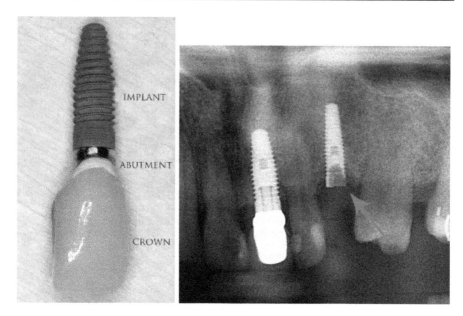

Fig. 5.2 Dental implant. Left: Dental implant, including abutment and dental crown. Right: Implanted implant abutment and crown (left X-ray image), and failed implant (right X-ray image).
Left: Image Author Coronation Dental Specialty Group [156]. Right: Author: Coronation Dental Specialty Group [157]. Reproduced from Wikipedia (Public Domain Image Library).

of the direct machined crown, manufactured on the spot in the dental surgery in a mini CNC machining unit from a porcelain blank.

5.2.3 Alumina in dental bridges

When one or more adjacent teeth are missing, there are two options for rectifying it: an implant or a bridge. The implant, in spite of being invented in the 1950s, did not become a mainstream technology until the 21st century. The dental bridge is a much older, cheaper, and long-established technology. A bridge, as the name implies, bridges the gap of a missing tooth. Where a tooth is missing in the jaw, the tooth on each side of the gap is ground down (abutment) in the same way as is done for the tooth onto which a crown is to be fitted. The two "abutment" teeth for a bridge are ground down sufficiently to slide a dental bridge over the top of each abutment tooth (the bridge support teeth). The bridge itself therefore comprises a single prosthesis, with an abutment tooth at each end, and artificial teeth (pontic) spanning the gap, as shown in Fig. 5.3.

The bridges shown in Fig. 5.3 both comprise two abutment prosthetic teeth and one pontic prosthetic tooth. The most common type of bridge is that utilizing two normal teeth as its two abutments (Fig. 5.3, left image). In rare cases, where there are three missing teeth in a row, a bridge may be mounted on two dental implants, as shown in

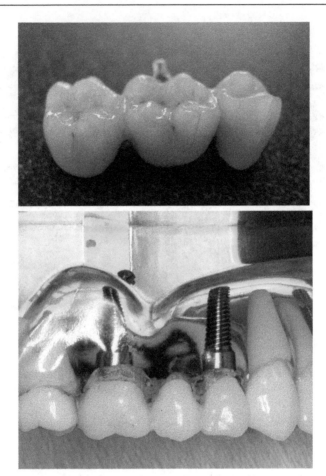

Fig. 5.3 Dental bridge. Above: Dental bridge of the conventional porcelain type, that is, type fitted to two existing teeth, in the case where the middle tooth is missing. Below: Dental bridge supported by two dental implants (one on each side). The middle tooth is the "pontic" tooth supported by the adjacent implants, which function as abutments.
Above: Image Author Wagonerj [159]. Below: Image Author Coronation Dental Specialty Group [160]. Reproduced from Wikipedia (Public Domain Image Library).

Fig. 5.3 (right image). A bridge is commonly made of dental porcelain, or metal with porcelain veneer. The problem with the bridge is that the bone of the jaw, where the tooth is missing, can atrophy due to it not receiving any load from a tooth. The other problem is breakage of the bridge due to overloading of the "pontic" teeth. Breakage is a problem with a porcelain bridge, especially because of the phenomenon of moisture-enhanced static fatigue in silicates (Chapter 4). It is not a problem with a metal bridge. A metal bridge commonly uses porcelain veneer so as to resemble real teeth.

Alumina has occasionally been used for dental bridges, both as pure alumina and as the InCeram concept discussed in Section 5.2.2, but this is a boutique market. Metal porcelain and pure porcelain remain the market dominant bridge systems in contemporary dental practice.

5.2.4 Alumina in dental implants

Single tooth replacement using an implant-supported crown has revolutionized the field of dentistry. It is the superior and much more surgically complex and expensive (~$5000) alternative to the dental bridge (~$1000), for replacing a missing tooth. The number of dental implants implanted globally is approaching 4.5 million. It should be clarified at this point that the implant concept involves three components:

- The implant
- The abutment (Section 5.2.5)
- The crown (Section 5.2.2)

The implant itself comprises a barbed or threaded screw-like component that is fitted into the reamed-out hole in the jaw bone, prepared from the cavity left by a missing tooth. An abutment is fitted to the top of the implant and protrudes through the gum. It functions as a synthetic version of the ground-down tooth "stump" onto which a crown is fitted. An implant loads the jaw normally (unlike a bridge) preventing bone atrophy in the jaw. Bone remodels itself based on its strain history and undergoes atrophy in the absence of mechanical load, a principle that is well known with astronauts who spend prolonged time in zero gravity and lose bone density as a result. Moreover, a dental implant does not suffer the problem of tensile failure through bending loads that plague ceramic bridges (though not metal bridges).

The dental implant itself is therefore a load-bearing prosthetic dental root that is fitted into a hole in the jaw as shown in Fig. 5.2. Commonly the hole left in the jaw by the lost tooth is appropriately reamed out, and the implant is usually installed with bone graft material (natural or synthetic) to enable osseointegration with the jaw so that it can develop stable long-term load-bearing capabilities. After a period of osseointegration, the crown is fitted and the patient can begin to adapt to their new tooth.

Dental implants are most commonly made of titanium, the invention of pioneering Swedish inventor Per-Ingvar Branemark. The first osseointegration studies of titanium in animals were done by Branemark in the 1950s [161,162], followed by human clinical trials with titanium dental implants beginning in the mid-1960s. It was not until the 1980s that this invention became a recognized dental procedure by the conservative and somewhat skeptical dental community, who were content with the traditional solution to tooth replacement: the dental bridge. Today, nearly 4.5 million dental implants are implanted a year, well ahead of the hip prosthesis in popularity. Therefore, the world owes a great debt of gratitude to Branemark, who did receive many accolades many decades after his original invention, once the world came to embrace this life-enhancing biomedical technology several decades after it was invented.

Not all dental implants are made of titanium. Vitallium (cobalt-chrome alloy) was trialed in the 1930s and has seen limited clinical usage, but cobalt-chrome alloy is not as biocompatible as titanium, nor does it have the same capacity for osseointegration as titanium. There have also been a number of experimental ceramic dental implants, including alumina and zirconia. A small number of experimental alumina dental implants have also been developed and clinically trialed. The first reported usage of alumina in a medical device predated the 1970 innovations of alumina in orthopedics and alumina in bionics. It was an alumina dental implant patented in 1965 [163]. This was a British Patent on a screw-type alumina dental implant developed by Sami Sandhaus of Switzerland. Positive clinical outcomes for this device were reported by Sandhaus in 1997 [163].

Other alumina dental implants followed. In 1977, Klawitter published the first paper on alumina dental implants, reporting on animal trials of porous rooted alumina dental implants [164]. The Tubigen implant, one of the best-known alumina dental implants, was patented by Heimke and Schulte in 1980 [165] and shown in Fig. 5.4. The Tubingen implant had a complex design to allow fixation by bone ingrowth and was much like the modern dental implant in many ways. Another clinically successful alumina dental implant, known as the Synthodont, was developed by Driskell in the United States [166]. In Japan, Kyocera introduced an alumina (synthetic sapphire) dental implant in 1972, which was introduced to the United States in 1980 [167,168]. A clinical follow-up many years later by in 2008 by Takahashi [169] of the sapphire implants suggested that porous sapphire was better than standard sapphire, and some good long-term outcomes were reported.

Alumina dental implants were a niche research area, and never seriously competed with titanium. ZTA—see Chapter 7) has made an appearance in the dental implants field in recent years. ZTA was first proposed for dental implants in 2008 by Ban et al. [170]. In 2010, Kohal reported on ATZ dental implants, comparing them with zirconia favorably [171]. The 2015 study by Palmero [172] demonstrates that ZTA is gaining momentum in the dental implants field. However, the stellar success of titanium as a dental implant suggests that ZTA is unlikely to supplant titanium any time soon, though ZTA could certainly compete to the point of a successful duopoly like the alumina/CoCr duopoly that currently exists with femoral heads in orthopedics. Albeit, the future in orthopedics is that alumina (as ZTA) will totally supplant CoCr in the medium term future. It is hard to predict if ZTA may supplant titanium in dental implants in the distant future. Certainly, ZTA has profound aesthetic advantages over titanium, and so ZTA could have a bright future and put alumina dental implants back at the forefront as they were in the 1960s to 1980s.

Another problem with alumina as a dental implant is stress shielding. Its extremely high modulus of 380 GPa is approximately 20 times higher than cortical bone, higher still in comparison to trabecular bone. Thus, alumina was not as effective as low modulus titanium (110 GPa) at load transfer to the supporting bone of the jaw. Moreover, in comparison to titanium dental implants, the alumina implants were quite large in diameter in order to have sufficient strength, and in some cases too large for the bone of the jaw to accommodate. ZTA can resolve the size issue but not the modulus issue. However, aesthetics is a powerful driver in the dental industry, probably more important than the modulus issue. On modulus, zirconia has the advantage being around 200 GPa, but inferior to alumina on wettability and for the zirconia hydrothermal aging problem (Chapter 6).

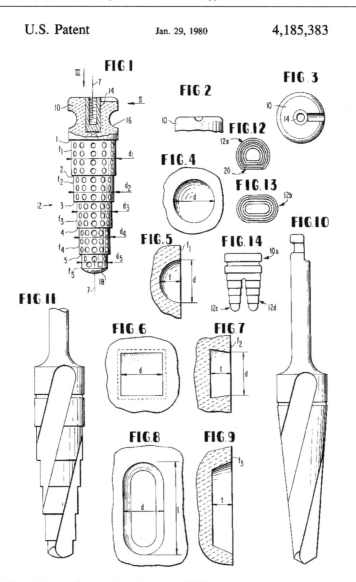

Fig. 5.4 The Tubigen alumina dental implant [165].

Alumina is attractive as a dental implant and was one of the earliest dental implants trialed, due to its high bioinertness (superior to all metals), its proven capacity for osseointegration, and its high wettability (superior to zirconia). However, it was ultimately supplanted in the market by titanium (primarily) and by zirconia (secondarily) primarily due to the superior toughness of titanium and zirconia. Zirconia has the advantage that, like alumina, it is pearly white, but zirconia has much higher toughness than alumina (Chapter 6). Hydrothermal aging (Chapter 6), which precludes the use of zirconia in orthopedics, has not yet proven to be an issue for zirconia in dental implants. A dental implant is subject to high and repetitive impact loads and requires a material of high toughness. Zirconia is the toughest biocompatible ceramic known. ZTA, as discussed in Chapter 7, has many advantages over zirconia and is now making inroads into the dental implants field.

In summary, titanium currently dominates the metal dental implants market, with zirconia the ceramic of choice when a ceramic dental implant is required. ZTA is beginning to make a significant impact and could have a bright future in the dental implants field.

5.2.5 Alumina in dental abutments

As discussed in Section 5.2.4, and shown in Fig. 5.1, the dental implant involves three components:

- The implant (Section 5.2.4)
- The abutment
- The crown (Section 5.2.2)

The word "abutment" derives from the field of civil engineering, for which an abutment is a load-bearing structure built on each opposing river bank for supporting a road or rail bridge. In a dental bridge, the abutment is a natural one: a healthy tooth that is ground down so as to function as the structural support for a dental bridge. With a dental implant, the abutment is synthetic. It is a stump-shaped biomaterial component that is connected to both the dental implant (implanted in the jaw bone and concealed beneath the gum) and to the visible crown. The abutment passes through the gum and is attached to the implant (beneath the gum) and protrudes through the gum as the "stump" onto which the crown is fitted, as shown in Fig. 5.2.

Commonly abutments are made of metals: titanium or stainless steel, even gold in some cases. Ceramic abutments also exist, including alumina abutments [173] as well as ZTA and zirconia abutments. The main advantage of a ceramic abutment is aesthetics, because in some cases with a dental implant, the gum can deteriorate and expose the abutment. A pearly white exposed alumina abutment in this situation is barely noticeable, whereas a shiny metal abutment is obvious and cosmetically undesirable.

5.3 Alumina bone tissue scaffolds

Bone tissue scaffolds have four key requirements

- Bioactivity
- Very high porosity (preferably >80%, ideally >90%)
- Large pore size of at least 100 µm to allow bone ingrowth
- High strength at the high porosity

Currently, the bone tissue scaffolds field is dominated by calcium phosphate ceramics (hydroxyapatite, tricalcium phosphate, bioglass), biodegradable polymers (PLA,

PGA, PGLA, and PCL), and calcium phosphate-doped biodegradable polymers. However, these suffer some limitations:

- Biodegradable polymers are not bioactive and are very low in strength.
- The in vivo dissolution of biodegradable polymers releases acidic monomers in the tissues, which can cause inflammation.
- Calcium phosphate doping can impart bioactivity to biodegradable polymers, but strength and acidic-monomer-inflammation are two fundamental flaws in the biodegradable polymer tissue scaffold application.
- Calcium phosphates are highly bioactive and do not cause inflammation.
- The strength of calcium phosphates is low, especially at low porosity.
- It is not possible to produce calcium phosphate ceramics with porosities anywhere near 90%.

Alumina is a very strong ceramic, far superior to calcium phosphates, and has a high strength even at porosities of over 90%, as shown by a series of papers published between 2005 and 2015 on foamed alumina tissue scaffolds published by the author[2] and Edwin Soh [174–177]. This research ultimately achieved the following: at the extremely high porosity of 94.4%, a high compressive strength of 384 MPa, with an average pore size of 300 μm, more than large enough for bone ingrowth [176].

Of course it is well known that alumina is bioinert, and therefore while bone ingrowth into an alumina tissue scaffold is possible, it is going to be a slow process because, just like biodegradable polymers, alumina has no inherent bioactivity. While biodegradable polymers can be calcium phosphate doped to impart bioactivity, mixing calcium phosphate into an alumina ceramic will turn it into an alumina-calcia-phosphate compound, and thereby take away its main advantage of strength.

In 2004, a ground-breaking paper was published by Pabbruwe, Standard, Howlett, and Sorrell [178] demonstrating that surface doping of alumina tubes (1.3 mm outer diameter, 0.6 mm inner diameter, 15 mm length) with calcium, magnesium, or chromium ions could impart bioactivity to alumina and stimulate bone ingrowth to an alumina pore.

Combining the innovation of Pabbruwe et al. [178] with the Parakala innovation of the high-strength foamed alumina scaffold [174], Soh and the author developed a high strength (380 MPa), high porosity (94.4%), large pore (300 μm), bioactive alumina tissue scaffold by surface doping the foamed alumina scaffolds with calcium, phosphorous, magnesium, or silicon. The in vitro response of bone cells (MG63 osteosarcoma cells) to this scaffold was then studied [177]. Bioactivity was demonstrated with the silicon and phosphorus doping proving the most effective at stimulating bone cell response [177]. The manufacturing process was as follows:

- $(NH_4)_2 \cdot Al_2(SO_4)_3 \cdot 24H_2O$: Salt solution is placed in an alumina crucible and foamed at 100–200°C
- $(NH_4)_2 \cdot Al_2(SO_4)_3$: Dehydroxylation at 300–500°C
- $Al_2(SO_4)_3$. + NH_3: Ammonia volatilizes at 300–500°C
- $Al_2O_3 + SO_3$: Sulfate volatilizes at 500–1000°C
- βAl_2O_3: Foamed porous beta-alumina intermediate stage

---

[2]The concept and technology of foamed alumina was originally conceived of by Dr. Padmaja Parakala and successfully executed by Dr Edwin Soh and the author (see About the Author).

- αAl_2O_3: Foamed porous alpha-alumina after calcining at 1000–1600°C
- Scaffold cooled and soaked for 24 h in bioactive dopant-ion solution
- Soaked scaffold calcined to 900°C to bond bioactive dopant ions to pore surface

This innovation has significant potential in the bone tissue engineering field. Never before has a bioactive bone scaffold been seen with such a favorable high-porosity/high-strength/large-pore combination. It is early days yet, and it remains to be seen how this 2015 innovation fares in the bone tissue engineering realm (Fig. 5.5).

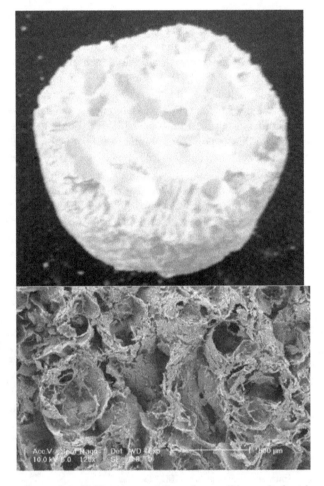

Fig. 5.5 Above: Foamed scaffold as removed from the crucible, ready for trimming. Below: Scanning electron microscopy image of a 94% porous foamed scaffold (note the 1-mm scale bar) with average pore size of 300 µm.
Images supplied courtesy of Edwin Soh.

5.4 Other biomedical uses for alumina

While data mining of the obscure reported biomedical applications of alumina will turn up some odd curiosities, there is little of significance beyond orthopedics, bionic feedthroughs, lightweight body armor, dental technology, and tissue scaffolds. However, two other minor applications of significance for alumina are worth mentioning here:

- Orbital implants (prosthetic eyeball) [179]. Orbital implants must combine the three attributes of high porosity, hardness, and biocompatibility, three properties for which alumina is an ideal match.
- Osteotomy spacers, for example, in the tibia, for the knee [142,180]. In 1969, this was the second ever reported biomedical application for alumina [142]. It remains significant in the 21st century with a clinical trial of 50 cases reported in 2002 [180].

Alumina bearings in orthopedics: Origin and evolution

6.1 Introduction to alumina in hip replacements

The global orthopedics medical devices market is currently around $40 billion PA. The global market in hip replacements is currently around $7 billion PA, and it is forecast to reach $9 billion by 2024 [181]. In 2010, the global number of hip replacements reached 1 million PA [182]. Using data from various sources [182–184] the forecast shown below was generated, showing that in the current year (2018) there were approximately 1.3 million hip replacements globally. By the early 2020s, the global number of hip replacements will reach about 1.4 million, including over 420,000 in the United States alone, and it will surpass 1.5 million by 2024. This chapter will provide a holistic overview of the evolution of the hip replacement, from its invention in Germany in the 1880s (ivory), to its evolution in the mid 20th century into the metal-on-metal hip, to its evolution in the 1960s and 1970s into the metal-on-polyethylene hip, and its ultimate evolution into the "ceramic" (alumina) hip, which began in the 1970s and achieved market dominance in about 2015.

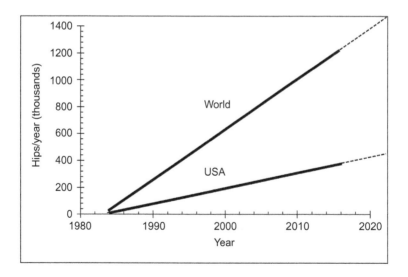

In 2010, about 13% of hip replacements in United States were surgical revisions [182]. By 2015, the revision rate had dropped to 10% [183]. This dramatic drop in hip revision rate from 2010 to 2015 mirrors the dramatic rise in the use of hip replacements with alumina ceramic femoral heads in the United States from 36% in 2012 to 49% in 2015.

Over 50% of all hip replacements implanted across the key global markets use alumina ceramic bearings, with Europe leading the world at around 55% and the US slightly behind at about 50%. Alumina bearings only entered the US market in 2003 (1994 in Europe). The market share of alumina bearings is growing rapidly, giving alumina ceramic bearings a key role in underpinning the $7 billion hip replacement market. At current trends, the CobaltChrome (CoCr) hip bearing, specifically CoCr-on-polyethylene, the market dominant hip bearing from the 1970s until the 2000s, could be completely supplanted by alumina hip bearings by the mid to late 2020s.

Numerous conference proceedings and journal papers have been published over recent decades in which alumina hip bearings are discussed, but they generally do not give a holistic overview. The bioceramics conference series (the 30th Bioceramics conference takes place this year) and the Biolox symposium series are particularly pertinent, but there are various other relevant one-off conferences, and bioceramics symposia at generalist Biomaterials, Materials Science, and Ceramics conferences around the world also. The author has been participating in such events since 1984. They generally focus very deeply on specific research projects. Thus it is very difficult for the nonexpert to be able to get the big picture: a 1970–2020 half century overview of the underlying alumina technology, the major challenges, the major successes, past trends, lessons learned, and the future directions.

The following two chapters will provide that overview, targeted to both clinicians and biomedical engineers. This chapter will provide the background to the development of the alumina hip replacement, or as it is commonly known, the "ceramic hip," and the demise of its competitor the zirconia ceramic hip. The following Chapter 7 will look at the state of the art in the alumina hip today.

Hip bearings are one of the larger applications for alumina ceramics globally with around 900,000 alumina hip-bearing components implanted per annum currently, and forecast to pass the 1 million per annum mark by 2020. The great majority of them are zirconia-toughened-alumina (ZTA): alumina reinforced with ~20% tetragonal zirconia microparticles. To date over 14 million alumina orthopedic-bearing components have been implanted globally [29]. This is a very recent development. At the turn of the century, the number of alumina bearings implanted globally was relatively small.

The big driver for the success of alumina hip bearings is the problem of polyethylene wear particles for the current market dominant hip bearing: CoCr-on-polyethylene. A normal functional CoCr-on-polyethylene hip bearing sheds about 500,000 polyethylene particles per step [185]. Given that the average person walks 5 million steps a year, this corresponds to about 2 trillion polyethylene wear particles a year. While polyethylene is not toxic, merely an irritant, this situation is generally tolerated for a few years but commonly a pathological condition called osteolysis of the bone results, followed by implant loosening. This is a leading cause of implant failure today.

The three main materials used in hip bearings are CoCr alloy (also known as CoCr alloy or as vitallium), polyethylene, and alumina (pure or zirconia toughened). Of the approximately 1.3 million hip replacements implanted annually across the key jurisdictions of the United States, European Union, United Kingdom, Australia, Canada, and the leading developed nations in Asia (Japan, Korea, Singapore), ~45% were CoCr/polyethylene bearings (no alumina), and CoCr-on-CoCr was <2% of the market. In the alumina realm, ~55% of the 1.3 million hip replacements utilized alumina of which ~40% were alumina/polyethylene and ~15% alumina/alumina. This represents, very approximately, 900,000 alumina-bearing components implanted a year of which over 80% were femoral heads due to the dominance of the alumina/polyethylene bearing.

In short, in the last 15 years, alumina has dramatically revolutionized the global orthopedic hip replacement industry. This is the most recent of the mass-market commercial applications of alumina. It has been a fast-growing market for a new alumina technology, second only to the meteoric rise of alumina in the implantable bionics market since the 1970s (alumina feedthroughs—Chapters 8–10).

The ceramic hip is one of the most extraordinary biomaterials innovations in the field of medicine in the 21st century. Like all the major alumina technology applications, this development has taken place with very little cross-pollination from the other alumina technologies (electrical, refractories, wear resistance, armor, and bionics). This remarkable story has its roots in a 1970 innovation of French surgeon Pierre Boutin, and many German researchers who immediately followed Boutin, followed by the pioneering work of a leading global ceramic company from Germany: CeramTec (now UK owned). This world-changing outcome has been the result of a successful partnership of quality German ceramic engineering and global medical innovation.

Indeed, the story of the hip replacement is a very European story, specifically a British and German story, plus one very important French pioneer. The hip replacement was originally invented in Germany in the 1880s by Themistocles Gluck. The modern metal-on-metal hip (CoCr) was invented in the United Kingdom by Philip Wiles and George McKee between the 1930s and mid-1950s. The modern metal-on-polyethylene hip was invented in the United Kingdom by John Charnley in the 1960s, and dominated the hip replacement market from the 1970s until the 2000s, although today it has fallen into second place behind alumina, and it represents a shrinking market. The metal-on-polyethylene-bearing system also remains the global standard for all other joint prostheses: hip, knee, elbow, shoulder, ankle, wrist, finger, spinal disk.

The first successful hip resurfacing implant (metal-on-metal) was designed by Derek McMinn of Birmingham, UK, and its success is reflected in the fact that the metal-on-metal resurfacing implant is known as the "Birmingham Hip." The alumina resurfacing hip is now under development, with key innovators in the United Kingdom, Australia, and Germany.

The hip replacement was the first orthopedic joint replacement to be commercialized. First developed in its modern manifestation by John Charnley in the 1960s, half a century ago [186], today hip replacement is the most common orthopedic procedure. Prior to 1994, all commercially available hip replacements used either CoCr-on-polyethylene, or CoCr-on-CoCr bearings. There has been a revolution since the end of the 20th century, and particularly in the last decade. Today, over half of all hip joints implanted use alumina bearings, on average 55% globally, with the number growing rapidly every year. This is an extraordinary revolution in an industry characterized by conservatism and risk minimization, and this begs the question why the dramatic shift? The modern hip replacement has been with us for about 60 years (CoCr-on-CoCr), or 50 years in the case of CoCr-on-polyethylene, without any significant changes in its basic CoCr-on-polyethylene, or CoCr-on-CoCr bearings until the early 21st century. Then suddenly in the last decade, a dramatic turnaround has happened. This has been driven by three main factors:

(1) Polyethylene wear particles from metal-on-polyethylene joints, deposited in the trillions inside the joint cavity. The number is estimated to be 500,000 per step [185]. This is tens of trillions over the implant lifetime. Polyethylene wear particles correlate with inflammation osteolysis (see Chapter 7, Section 7.3.2), and aseptic loosening, requiring revision surgery in 10% or more of cases. In contrast, the wear rate of alumina ceramic bearings is in the order of two to three orders of magnitude lower. Secondly, alumina is highly bioinert and biocompatible and therefore, like polyethylene wear particles, alumina wear particles are only problematic if present in huge numbers. This is the default situation with polyethylene and a rarity with alumina. Only in the case of extreme edge wear can alumina begin to approach polyethylene in wear rate. Thus inflammation osteolysis can be overcome by the use of alumina bearings.

(2) Galvanic corrosion caused by the galvanic couple of cobalt chrome/titanium taper junctions, which is standard in the titanium shaft, CoCr-ball system used in modern metal-on-polyethylene hip joints. The titanium stem is the cathode and the CoCr ball the anode, and therefore the corroded metal. CoCr/titanium galvanic couples are a "battery" within the human body cavity and both cobalt and chromium corrosion products are toxic ions and therefore problematic. Equally problematic is CoCr fretting wear-induced corrosion in the taper junction, which constantly exposes fresh CoCr metal surface to feed the corrosion.

(3) Chromium metallosis and systemic cobalt poisoning caused by cobalt-chrome wear particles from metal-on-metal bearings, viz., the global ASR CoCr-on-CoCr hip recall of 2010 and the multiple $Million order of magnitude impact this had on the hip replacement market, and the thousands of patients affected. The result is that metal-on-metal bearings had fallen to <2% of hip replacements in 2017, with most of their market share taken by ceramic bearings.

The general perception is that the greatly enhanced wear resistance and the benign nature of alumina wear particles are the main drivers for the success of alumina bearings. In fact alumina-on-polyethylene is more common (40% of the global market) than alumina-on-alumina (15% of the global market), and in alumina-on-polyethylene bearings, essentially all the wear particles are polyethylene. It is true that the wear rate of polyethylene is significantly less when articulating against alumina, compared to CoCr. However, even more important than this is the galvanic corrosion and fretting corrosion problem that alumina heads resolve. Taper galvanic corrosion, and taper-fretting-wear corrosion of CoCr can cause just as much of a cobalt and chrome toxicity problem as CoCr-on-CoCr bearings. So, while the metal-on-metal problem has resolved itself by the virtual disappearance of metal-on-metal implants from the market since 2010, the CoCr toxicity risk remains given that 45% of the market still involves CoCr heads on titanium stems in CoCr-on-polyethylene implants.

Implementation of alumina bearings in the hip prosthesis has resolved these issues, in a revolutionary way, and this is all based on the unique properties of alumina ceramics, both in the pure form and in the zirconia-toughened form (ZTA).

6.2 Introduction to the anatomy of the hip and hip joint pathologies

6.2.1 Anatomy of the hip joint

The human hip is a ball-and-socket joint in which the ball (femoral head) at the proximal end of the thigh bone (femur) articulates in a socket (the acetabulum) in the pelvis. The pelvis is a load-bearing bony member at the base of the spine, which has the biomechanical role of load transfer from the single central spine to the two legs, that is, transferring the central load vector from the spine to the dual load vectors at the right and left hip joint. The pelvis has an acetabulum on each side, one for the left femur and one for the right femur. This is shown in Fig. 6.1. The articulating surface of the femoral head is coated in a smooth layer of articular cartilage, which functions as the low-friction surface for articulation. In Fig. 6.1, the femoral head is shown bare of articular cartilage for illustration clarity. The articular cartilage can be seen in Fig. 6.2.

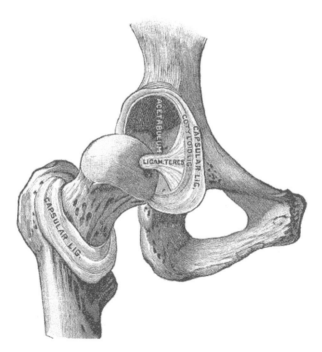

Fig. 6.1 The human hip joint shown with much of the soft tissue removed for clarity. The femur (thigh bone) is on the left. The proximal femur (top end of the thigh bone) contains a ball-like bony surface (femoral head) which is covered in smooth articular cartilage (articular cartilage not shown, for clarity). The acetabulum is the socket on the right. The acetabulum is part of the pelvis (the partial left side of the pelvis is shown here). For clarity, the femoral head is shown adjacent to the acetabulum rather than articulating inside it. Image Author Henry Vandyke Carter [187].

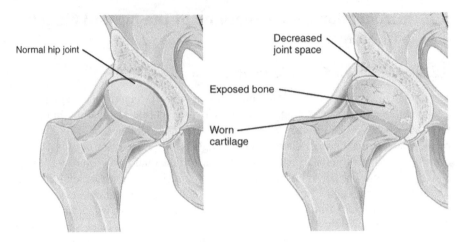

Fig. 6.2 Arthritic hip, the primary driver for hip replacement. Left: healthy hip joint; Right: hip joint suffering arthritic deterioration.
Image Author CFCF Carl Fredrik [188,189]. Reproduced from Wikipedia (Public Domain Image Library).

6.2.2 Pathology of the hip joint

The primary driver for the 1 million+ hip replacements a year is deterioration of the hip joint through osteoarthritis or rheumatoid arthritis, a pathology of the joint that results in deterioration of the articular cartilage and the underlying bone. In the United States, 70% of all hip replacements are due to osteoarthritis, and <1% are due to rheumatoid arthritis [183]. Femoral fracture accounts for 10% of hip replacements. The remainder is due to various pathologies, including avascular necrosis and osteonecrosis [183]. Osteoarthritis can be caused by traumatic joint injury, or by repetitive joint injury such as regular jogging on paved roads. Rheumatoid arthritis is a pathological condition, a slowly progressing inflammatory process that is commonly inherited. Whatever be the cause, arthritis generally develops slowly and causes pain during movement, and in more advanced cases, constant pain at rest.

Commonly arthritis worsens with age. The average age for a hip replacement in Australia is 67 years. Hip fractures are also sometimes a driver for hip replacement, for example separation of the femoral head from the proximal femur. However, often such fractures can be repaired by various medical devices involving threaded bone screw technology.

6.2.3 Anatomy of hip replacement

A partial hip replacement (hemiarthroplasty) replaces only the femoral head. A total hip replacement (THA—total hip arthroplasty) involves replacing both the femoral head and the acetabular-bearing surface. X-ray images of implanted THA hip replacement devices are shown in Fig. 6.3. The essential components of the basic THA

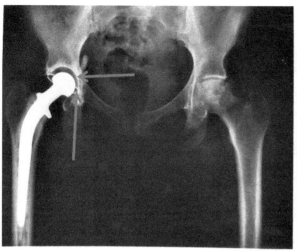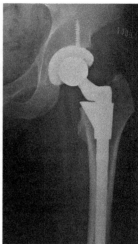

Fig. 6.3 X-ray of a total hip replacement. In both images the metal stem can clearly be seen inside the medullary canal of the femur and the metal femoral head can clearly be seen inside the acetabulum. Left: The total hip replacement is on the left-hand side, the right-hand side is the natural hip joint. In this radiograph of a hip replacement, the acetabular cup inside the acetabulum is not so clear but it can be seen as the fine white line highlighted by the red arrows, with the gap between the acetabular cup and femoral head filled with X-ray transparent polyethylene. Right: In the hip replacement in the right image the acetabular cup is more X-ray opaque and visible, and a metal fixation screw can clearly be seen.
Left: Image Courtesy of National Institutes of Health, US Government [190]. Right: Image Author. Right: Image courtesy of author KimvdLinde [191]. Reproduced from Wikipedia (Public Domain Image Library).

procedure are as follows. Hemiarthroplasty, which is a less common procedure (approximately 10% of hip implants [183]), excludes steps 3 and 5:

1. The hip is dislocated and the femoral head is cut off with a bone saw.
2. The medullary canal of the femur is reamed out to a bore sufficient for insertion of the stem of the hip prosthesis. The metal hip stem (commonly titanium) is inserted into the reamed medullary canal and bonded by bone cement, mechanical fixation, or osteogenesis.
3. The acetabulum is reamed out sufficiently for the metal (usually titanium) acetabular cup to be inserted and bonded by bone cement, mechanical fixation, or osteogenesis.
4. A femoral head "ball" (CoCr or alumina) is fitted to the taper socket of the stem.
5. An acetabular liner "socket" (polyethylene or alumina) is fitted to the tapered cavity of the metal acetabular cup.
6. The joint is reassembled surgically with the femoral head (CoCr or alumina) articulating inside the acetabular liner "socket" (polyethylene or alumina).

The hip prosthesis as shown in Fig. 6.3, is a ball-and-socket joint which comprises four basic components:

- *Stem*: Load-bearing stem, generally made of dense titanium, with surface features for enhanced bone bonding to the medullary canal. This is firmly inserted into the medullary canal of the femur enabling stable biomechanical fixation and load bearing.

- *Femoral head (ball)*: A ball, affixed to the stem generally by a tapered modular junction, traditionally made of CoCr alloy, now commonly made of alumina ceramic.
- *Acetabular cup*: A metal cup, generally made of porous titanium, firmly affixed (bone cement, bone screws, mechanical interlock, or bioactive) to the acetabulum of the pelvis, enabling stable biomechanical fixation and load bearing.
- *Acetabular liner (socket)*: An almost 180° (generally 150°–160°) socket, mechanically supported by the acetabular cup, and articulating against the femoral head (ball). Traditionally made of ultrahigh-molecular-weight-polyethylene (UHMWPE), now commonly made of alumina ceramic.

In traditional hip replacements, it has always been necessary to have both a metal acetabular cup, and a separate UHMWPE liner socket, because soft low-modulus UHMWPE is unsuitable for mechanical fixation to the acetabulum. However, in some cases, particularly in metal-on-metal resurfacing, the acetabular cup can be a single component (Figs. 6.4 and 6.5).

6.3 A brief history of the hip replacement

6.3.1 Themistocles Gluck (1880s): The inventor of the prosthetic hip

The late 19th century was the beginning of the era of aseptic surgery, a fertile ground for surgical innovation. The first hip replacement was developed in this era, in the 1880s in Germany, by Moldavian Surgeon Themistocles Gluck, who was appointed Head of Surgery at the Emperor and Empress Friederich Paediatric Hospital, Berlin, in 1890 [192], following in the footsteps of his physician father, who had been the attending physician for the Moldavian Royal Family [193].

Gluck served as a wartime surgeon in the Balkans from 1877 to 1885, which was a fertile learning ground in which his natural genius and capacity for innovation flourished. During this time Gluck successfully used steel plates to repair a broken femur, and a mandible, and he innovated with bone cements including plaster of Paris, a pumice or gypsum-based putty, and copper amalgam. Gluck was a pioneer in medical device innovation [194]. His bone cement innovations predated by 70 years the bone cement pioneers of the 1950s and 1960s: Haboush [195], Wiltse [196], and Charnley [186,197]. His intramedullary fracture fixation innovation predated by 60 years the intramedullary process of Kuntscher [198]. Gluck also innovated other medical procedures unrelated to joint prosthesis.

Gluck is famous as the inventor of the hip replacement in the mid-1880s [199], 80 years before John Charnley launched the metal-on-polyethylene hip replacement, which became the first mass-market hip replacement technology by the 1980s. Indeed Gluck also designed and implanted prosthetic knees, ankles, shoulders, elbows, and wrists, using his hinged ivory technology innovation.

Until the late 19th century, the only surgical options for joint infection (injury or tuberculosis) were amputation or joint resection. Death by sepsis happened in 30%–80% of cases as antibiotics were not discovered until 1928, and not

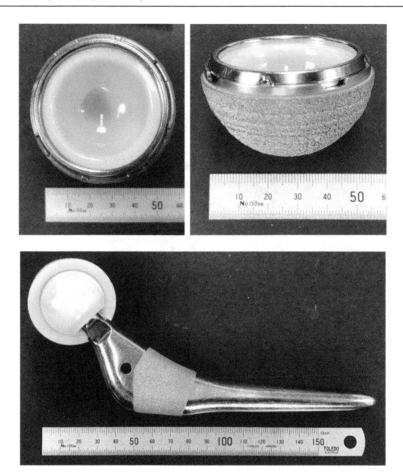

Fig. 6.4 The modern hip replacement. Top: Alumina acetabular insert (socket) inside the titanium acetabular cup, with porous titanium surface for optimal bone-titanium bonding. Bottom: Titanium stem, with alumina femoral head (ball) affixed to the stem at the taper junction, and articulating against the alumina acetabular liner (socket), shown without the titanium acetabular cup.
Exhibit provided courtesy of Professor Bill Walter. Specialist Orthopedic Group. Wollstonecraft NSW.

commercialized until the mid-20th century. Gluck wanted to change contemporary surgery from the destructive art that it had been until that time, into a reconstructive art. He did 14 joint replacements in the 1880s using ivory and pioneered all aspects of joint replacement:

> *Animal testing* prior to implanting in humans: Gluck tested his devices in animals before human implantation.
> *Stabile fixation* of the artificial joint (Gluck used "bone cement" or used cementless fixation) and taught how vital stable biomechanical fixation to the skeleton was.

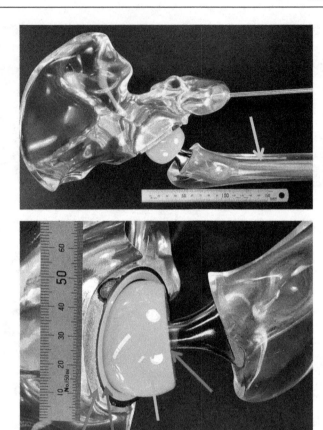

Fig. 6.5 A surgically implanted alumina hip in a human tissue model made of perspex. Above is a wide-field view, and below is a zoom-in view of the alumina ball-and-socket junction. The alumina is pink in color as it is CeramTec BIOLOX*delta* ZTA which is colored pink. Note the following features: *Red arrow*: Titanium acetabular cup (gray with porous surface texture) affixed to the acetabulum of the pelvis. *Purple arrow*: Alumina acetabular liner "socket" (pink) inside the acetabular cup, and articulating against the (pink) alumina femoral head (ball). *Blue arrow*: Alumina (pink) femoral head (ball) articulating against the acetabular liner (socket). *Pink arrow*: Taper junction that joins the alumina (pink) femoral head with the titanium stem. *Green arrow*: Titanium stem is inserted inside the femoral medullary canal.
Exhibit provided courtesy of Professor Bill Walter. Specialist Orthopedic Group. Wollstonecraft NSW.

Intramedullary fixation (with ivory cylinders). Gluck demonstrated in animal experiments that the medullary cavity will accept the shaft of the artificial joint if it is stably anchored within it—one of the keystones of the modern total joint surgery, and used this system of fixation in his human prostheses.

Modular construction of artificial joints: Gluck proposed that total joints should be assembled by the surgeon directly at the operation board from modular parts of different sizes. This is the dominant paradigm today.

Allografts. Gluck proposed that joints taken from corpses and amputated limbs may be used, although he himself did not use them. Allografts have a small but significant role in surgery today.

Biocompatibility. Gluck was the first to develop the idea that foreign materials for total joint replacements must be well tolerated by the patient's body. Indeed, Gluck pioneered many of the concepts that underpin biocompatibility.

Stress shielding. Gluck declared that the skeleton behaves according to the principle to use the minimum of bone mass just able to bear the load put on it. Therefore, he thought that ivory was the best material for total joints; it is equally light and equally strong as the bone tissue itself and has perfect elastic modulus match with human bone. Ideally the artificial joint should match the flexural properties of bone to prevent stress shielding. Modern surgeons and engineers call this principle "Wolff's law." Today titanium stem implants (modulus six times higher than bone) and CoCr stem implants (modulus 12 times higher than bone) shield the femur from the strain commensurate with the load. Bone remodels itself according to its strain history, as is well known for astronauts in space for prolonged periods (bone atrophy). Thus proximal femoral atrophy, through stress shielding, is common in contemporary metal-stem hip implants.

Gluck's superior, von Bergmann, did not share Gluck's vision and ordered Gluck to desist, writing in relation to Gluck's forthcoming presentation at the 10th International Medical Conference (Germany 1891): "as the leader of German surgery I cannot allow you to discredit German science in front of a platform of international specialists" [192]. After the cessation of Gluck's joint prosthesis development, there followed a 40-year hiatus in hip replacement development until the 1930s, and the modern hip replacement did not come into surgical use until the 1970s, and widespread use from the 1980s, a century after Gluck. This was following the innovations of British pioneer John Charnley. However, Gluck who died in 1942, was ultimately honored for his accomplishments by a listing on the honor roll of the German Surgical Society [192], and is seen today, along with Charnley, as one of the key innovators of the orthopedic joint replacement.

From the 1890s to the 1920s, the field was dormant. The only significant activity in this era was surgical experimentation with interposing various natural tissue products, such as submucosa, pig bladders, fascia lata, and skin, between the femoral head and acetabulum of arthritic hips as an arthroplasty strategy [200]. From the 1930s to the 1950s, a series of unsuccessful trials was conducted on hip replacement devices, with a number of problems hampering progress:

- poor understanding of bone necrosis and bone nonunion joints,
- mechanical failure of biometals,
- corrosion of biometals, and
- poor understanding of fracture healing biomechanics

On the positive side, a series of biomaterials innovations in the 1910–40s laid the foundations for the future.

6.3.2 1910–40s: Biomaterials innovations

Before overviewing the history of hip replacement device development from the 1930s to the 1950s, it is important to summarize the key biomaterials innovations that underpinned the medical device design and development of the hip replacement.

Glass: In 1923, US surgeon Marius Smith-Peterson discovered bioactive arthroplasty when a shard of glass found in a patient's back had developed a synovium and established that a glass acetabular cup develops a synovial cartilage layer in vivo [201]. He adapted this finding, surgically fitting a hemispherical hollow glass hemisphere over the arthritic femoral head and thereby providing a smooth articulation surface for the acetabulum. This was a standalone femoral head "resurfacing" concept made of glass, arguably the world's first resurfacing trial (see Section 6.3.5). On the plus side the glass was biocompatible and smooth. On the negative side, its brittleness was inadequate for the biomechanics of the application, resulting in fracture of the glass.

Stainless Steel 1912: Vanadium steel was first developed for medical use by Sherman in 1912 (Sherman plate), and then 18-8-Mo stainless steel was introduced for fracture fixation plates in 1926 by Lange [202]. A 316-L stainless steel supplanted 18-8 in the 1950s and ultimately was to see small but significant use in hip replacements although largely obsolete now due to crevice corrosion problems [203].

CoCr (Cobalt-chrome alloy also known as vitallium) 1929: Charles Scott Venable, an early US pioneer in hip implant technology, was the first to establish that electrolysis was the main cause of failure of biometals in vivo. He studied a range of alloys and found only one that was completely electrically inert (passive) in vivo: CoCr—cobalt-chrome-molybdenum alloy [204]. Commonly called CoCr, this alloy is also sometimes called "vitallium" or "cobalt-chrome alloy." This book will refer to it hereafter as CoCr. This discovery, made in 1929, was to revolutionize the field of hip replacement and orthopedic medical devices from the 1930s onward. Initially used in dentistry, CoCr rapidly became the mainstay of stem technology until the late 20th century when it was supplanted by titanium. It remains a common femoral head biomaterial.

CoCr is a hard metal and performs well in articulating wear both in metal-on-metal joints and metal-on-polyethylene joints, and was therefore the mainstay of the femoral head (ball) until the current decade, when it began to be supplanted by alumina. The hip is the only joint prosthesis that currently has a mass-market alumina-bearing system. For this reason, CoCr remains dominant in all other joint replacements: knee, ankle, shoulder, elbow, etc. However, only the hip and the knee are in the million plus implants per year market size, other joints are in the thousands a year range. So the hip replacement represents roughly half the joint replacement market, knees the other half, and the rest of the joints are a very small market size in comparison.

Tantalum was first trialed in 1936 by Burch and Carney [202]. Ultimately it came to occupy a niche application as trabecular metal in acetabular cups, pioneered by leading multinational orthopedics company Zimmer Biomet.

Titanium was first trialed as an implantable orthopedic biometal in 1947 by Cotton [202]. Its much lower modulus (lower stress shielding) than CoCr, its greater biocompatibility than CoCr, and its capacity for osseointegration (unlike CoCr) led titanium to dominate stem and acetabular cup technology in hip replacements by the late 20th century, a role it continues to dominate today. Titanium has also become dominant in dental implants, since the innovation of Branemark in the 1950s [161,162], in some cases mated with alumina dental crowns. Moreover, titanium casings with alumina feedthroughs dominate the global $25 billion bionic implant industry (Chapters 8–10). *The titanium-alumina combination is therefore the most important biometal-bioceramic partnership in the global medical device industry.* Titanium is a relatively soft metal and cannot be used in articulating wear. This role is serviced by a duopoly of CoCr and alumina.

6.3.3   1930–50s: The era of hemiarthroplasty

Between the 1880s and 1930s, no further progress took place on Gluck's intramedullary fixation. However, beginning in the 1930s, half a century after Gluck, gradual progress over the 1930s to 1950s slowly brought the research direction back to intramedullary fixation, and toward a modern hip prosthesis.

CoCr metal implants using intramedullary fixation evolved gradually in this period, with something of a hiatus during World War II, however, from 1937 to 1952, primarily implants for hemiarthroplasty were evolving: that is, no acetabular cup, the metal femoral head articulates into the tissues of the acetabulum itself. This can be problematic for the acetabulum and most modern hip replacements are THA, with only about 10% of current hip implants a hemiarthroplasty [183]. There were a number of pioneers in the United States in the early stem/ball hemiarthroplasty developmental period from 1937 to 1952, including Harold Bohlman, Austin Moore, and Fredrick Thompson, as well as Jean and Robert Judet of France. The two most significant developments in this period were as follows:

1939—Harold Bohlman (USA) rediscovered the Gluck intramedullary fixation concept and developed a rudimentary CoCr hip replacement comprising a short femoral shaft and femoral head (ball). This was the first hip replacement innovation of the 20th century to resemble the Gluck concept, and the modern hip hemiarthroplasty device [204]. It was implanted for the first time in 1940 by Bohlman.

1950—Austin Moore (USA) and Harold Bohlman (USA) jointly developed and implanted the first long-stem femoral shaft [205]. This was the first device to resemble the modern hip replacement, however it still articulated directly into the acetabulum as a hemiarthroplasty. These Moore/Bohlman devices were the first hip replacements that were widely distributed, through Austenal Laboratories (which later merged with Stryker), even though they were only for hemiarthroplasty.

6.3.4   George McKee (UK 1956). Metal-on-metal: The first total hip replacement

There are unconfirmed reports in the literature that the first implanted THA, a CoCr metal-on-metal device, was designed and implanted by Philip Wiles of the Middelsex Hospital, UK, in the 1930s. There are no details in the literature as to the dates, patient numbers, or device specifications, as records were lost during World War II, but it is reported that one patient still had a functioning Wiles implant 35 years later [206].

Other than this unconfirmed report for Philip Wiles, until 1956, all confirmed hip replacement trials until the 1950s involved hemiarthroplasty, with CoCr stem/ball implants. The first confirmed report of THA came from George McKee in the United Kingdom. In 1953, he began using the stem developed by Fredrick Thompson, but with the addition of a new one-piece CoCr socket as the acetabular implant [201]. This was the first documented use of an acetabular implant, and therefore represented the world's first documented THA, by the modern definition. However, it is probable that it was a follow-on from the work of Philip Wiles. Indeed George McKee did his clinical training with Wiles some years earlier, which makes the validity of the Wiles report all the more probable, albeit McKee used an implant developed by US pioneer Fredrick Thompson.

The McKee implant was also the world's first documented metal-on-metal (CoCr) hip replacement. McKee implanted this device in 26 patients between 1956 and 1960 with a 57% success rate [207], and seven other surgeons also used the McKee implant with success between 1966 and 1987. These metal-on-metal implants had good long-term performance, with a 28-year survival rate of 74% [208]. However, by the mid-1970s, metal-on-polyethylene implants developed by Charnley (Section 6.3.6) began to supplant metal-on-metal due to the effects of metal particles on the joint cavity [209]. Metal-on-metal THA continued to have a niche role up until the 2010 De Puy ASR recall, after which it all but vanished from the market. However, the McKee metal-on-metal innovation was very important for two main reasons:

(1) It was the first successful modern total hip replacement.
(2) It was a stepping stone to the very important hip-resurfacing technology which evolved several decades later.

6.3.5 Hip resurfacing: A very gradual evolution

Before turning our attention to the Charnley metal-on-polyethylene revolution of the 1970s, it is important to address the other key innovation in hip replacement technology that slowly evolved from the McKee metal-on-metal hip in the late 20th century: hip resurfacing. Resurfacing DOES NOT use intramedullary fixation. It is a maximal femoral-preservation procedure that involves placing a thin-walled large-diameter CoCr metal cap over the natural femoral head, to articulate against a large-diameter CoCr metal socket in the acetabulum. Resurfacing offers the following benefits:

- minimal bone loss
- normal femoral loading, that is, no intramedullary metal stem and therefore avoidance of stress shielding
- maximum proprioceptive feedback
- minimal risk of dislocation
- ease of revision
- restores natural hip anatomy

Resurfacing is commonly used for younger patients. While the median age for hip replacements in Australia is 67 years, males below the age of 65 have been shown to do well with a resurfacing hip implant [210]. Given that THA causes atrophy of proximal femoral bone, there are a limited number of times revision surgery can be done before the patient runs out of viable load-bearing femoral bone stock.

Thus the resurfacing implant has greatly lengthened the length of time possible for living with a hip replacement in accordance with the following progression:

(1) *Hip resurfacing.*

Then, in the event of revision surgery being required...

(2) *Traditional THA.*

Then, in the event of revision surgery being required, and given the likelihood of proximal femoral atrophy from the THA...

(3) *Long-stem THA.*

Stage 3 is potentially the end of the road, on a case-by-case basis. At some point after a series of revisions the patient inevitably will run out of load-bearing femoral bone stock. However, this resurfacing strategy means that it is now viable for younger patients to receive a hip replacement. The traditional view, which arose from Charnley, is that given the problem of osteolysis and aseptic loosening from THA, it is wise for surgeons to counsel patients to delay hip replacement surgery for as long as they can possibly endure the pain of arthritis. This is neither a workable nor a humane strategy in practice (Fig. 6.6).

Early unsuccessful resurfacing trials were conducted by US surgeon Marius Smith-Peterson in the 1920s [hemispherical hollow glass hemisphere over the arthritic femoral head articulating against the native acetabulum (Section 6.3.2)]. The glass trial was unsuccessful because the glass shattered under the load. The next resurfacing trial was a Teflon-on-Teflon resurfacing trial conducted by Charnley in the early 1950s (Section 6.3.6). The Teflon trial was unsuccessful due to the failure of Teflon as a bearing material. The next resurfacing trials, again unsuccessful, were done in the 1970s with polyethylene femoral component and CoCr acetabular cup resurfacing trials. This strategy was conducted independently by both Freeman [213] and by Furuya [214].

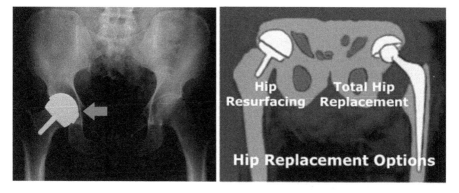

Fig. 6.6 The hip resurfacing concept. Left: X-ray of an implanted hip resurfacing system. Right: Schematic of resurfacing versus total hip arthroplasty.
Left: Image Author James Heilman MD [211]. Right: Image Author Patricia Walker [212]. Reproduced from Wikipedia (Public Domain Image Library).

The general consensus from these resurfacing trials of the 1970s was that the failure rates were high due to the collapse of the femoral head. While disappointing at the time, this is now seen as being due to a materials failure not indicative that the resurfacing concept is a flawed idea. On the contrary, resurfacing is a very good idea, it has just taken a long time to perfect it, a process that is ongoing today.

The first successful resurfacing implant was designed by British orthopedic surgeon Derek McMinn, of the BMI Edgbaston Hospital (Birmingham UK), and first implanted in 1991. It was a metal-on-metal CoCr device, and its success is reflected in the fact that the metal-on-metal resurfacing implant is commonly known as the "Birmingham hip." Successful clinical usage has now established that polymethyl methacrylate (PMMA) bone cement is the best fixation system for the mushroom-shaped femoral component (Fig. 6.7, top right image), and (HA)-coated/macroporous cementless fixation is best for the acetabular cup [215].

In August 2010, DePuy Orthopedics, issued a voluntary recall of two metal-on-metal orthopedic implants [216]:

- DePuy ASR™ XL Acetabular Hip System (metal-on-metal total hip replacement)
- DePuy ASR Hip Resurfacing System (metal-on-metal hip resurfacing implant)

A total of 93,000 patients globally had received an ASR implant. The lawsuits followed continued for some years, and awards to patients collectively ran into the $100 million order of magnitude. De Puy identified reasons for the recall as follows [216]:

According to 2010 data from an independent national registry in the UK that tracks implant performance and outcomes, while most patients with the ASR Hip system had not required additional hip surgery, the rate of revision was not in line with data previously reported in that registry. In a revision surgery, the existing hip implant is removed and is replaced with a new hip implant.

The 2010 data indicated that within 5 years of having an ASR resurfacing device implanted, approximately 12% of patients had revision surgery, and that within 5 years of having an ASR total hip replacement, approximately 13% of patients had revision surgery, which was not in line with data previously reported in that registry.

Complications arising were related to elevated levels of metal wear particles due to a design feature which generated greatly increased wear. The ASR acetabular cup had a 3-mm recessed area right at the edge of the cup where it was not the spherical surface. It was a recessed area with a lip at the edge of the spherical area, a slightly wider diameter (larger radius) there than the spherical diameter would be. This is shown in Fig. 6.7 (most clearly in the bottom image). This lip was for engaging with the surgical insertion tool for the acetabular cup. This lip caused problems at high angles of motion of the hip, in which scenario, the femoral head could grind on the lip edge which could cause exacerbated wear [210].

The systemic effects of high wear-particle loads were primarily elevated cobalt levels although cobalt is generally excreted by the renal system. The local effects on the joint cavity were more problematic, for example chromium metallosis. With well-designed metal-on-metal implants, wear rates are sufficiently low that these

Fig. 6.7 The De Puy ASR resurfacing hip replacement that was recalled in 2010. It comprises three components (seen in top image from left to right): acetabular cup, acetabular liner (fits into the cup), and mushroom-shaped femoral head sleeve. The lip at the edge of the spherical cavity in the acetabular liner is clearly visible in the zoom-in of the bottom image. Exhibit provided courtesy of Professor Bill Walter. Specialist Orthopedic Group. Wollstonecraft NSW.

issues are not usually problematic, as experience with metal-on-metal implants from the previous half century demonstrated.

This recall was most unfortunate given that the metal-on-metal implant concept had been just as successful as the metal-on-polyethylene implant for half a century since it was first implemented by McKee in 1956, and more importantly, the metal-on-metal resurfacing implant was deservedly growing in popularity at that time. As a result, the use of metal-on-metal implants dropped dramatically across the industry. In Australia, usage went from 12% in 2008 to 2% shortly after the DePuy recall [217]. Resurfacing is now experiencing a recovery, but metal-on-metal total hip replacements are virtually obsolete now in some jurisdictions, for example, they are <1% of the US market [183].

The second outcome of the ASR recall was a positive one. Coming as it did just 7 years after FDA approval of alumina bearings for hip replacements, and given that it raised in the public mind a question mark over the safety of CoCr, it had the effect of accelerating the market penetration of alumina bearings into the hip replacement industry, supplanting CoCr femoral heads. 85% of all hip replacements on the market today use a polyethylene liner, 40% are alumina-on-polyethylene, and 45% are CoCr-on-polyethylene. 15% are alumina-on-alumina. Currently 55% of the 1.3 million hip implants implanted PA use an alumina femoral head, this percentage was significantly lower in 2010, down around the 30% level.

6.3.6 John Charnley (UK): The 1970–90s metal-on-polyethylene revolution

Sir John Charnley, a British orthopedic surgeon, is seen as the father of the modern hip replacement. He believed that metal-on-metal was the wrong paradigm. His reasoning was that the high frictional torque would eventually loosen the implant, regardless of the lubrication effect of synovial fluid, and he devoted a decade to finding a self-lubricating bearing concept. As it turned out, he was right for the wrong reason. The problem with metal-on-metal is the metal wear particles generated, and their undesirable local and systemic effects.

Charnley devoted a decade to developing a metal-on-polymer total hip replacement [218]. In the early 1950s Charnley trialed Teflon-on-Teflon, but found this wore out in 2 years. From 1958 to 1962 he trialed CoCr-on-Teflon, implanting this device in several hundred patients. Unfortunately, the high wear of the teflon and resulting large quantity of Teflon wear particles caused severe osteolysis (Chapter 7, Section 7.3.2) and loosening, requiring many surgical removals [218]. He also observed that the volumetric wear rate was highest for the largest femoral head diameter, due to the increased contact area, and surmised that a smaller diameter head would reduce wear. While this proved to be the case, it also resulted in linear penetration into the Teflon socket, shortening its service life.

In 1962, Charnley trialed CoCr-on-polyethylene (ultrahigh-molecular-weight-polyethylene UHMWPE) the most wear resistant polymer known. This proved successful, but after previous disappointments, Charnley waited until 1967 to announce his success. It is now half a century since 1967, when CoCr-on-polyethylene became public domain knowledge as a technology. It became the dominant paradigm in orthopedics from the 1980s until the 2000s. In summary, Charnley innovated four main concepts:

- the concept of the metal-on-polymer bearing for joint replacements
- the concept of CoCr in combination with the polymer

- ultrahigh-molecular-weight-polyethylene (UHMWPE) as the polyethylene component, well known today as the most wear resistant polymer known, with excellent natural lubricity
- fixation by PMMA bone cement

By 1967, the McKee metal-on-metal hip had become quite well established in the fledgling field of total hip replacements; however, metal-on-metal never gained much traction in the market. While it is true that the first total hip replacement was developed in 1953 and first implanted in 1956 by George McKee (UK), the metal-on-metal total hip replacement has always remained a niche product. Its most important manifestation is for the resurfacing hip implant (Section 6.3.5).

In the 1970s, McKee metal-on-metal hips and Charnley metal-on-polyethylene hips coexisted, both being regarded in this era as new and successful to date, but not yet proven. However, Charnley set up a demonstrator at the biomechanics laboratory he established at Wrightington Hospital (Lancashire UK) which involved a McKee and a Charnley hip articulating while connected to a pendulum. The Charnley metal-on-polyethylene swung freely while the McKee metal-on-metal ground to a halt. Numerous visiting surgeons witnessed this and were convinced of the superiority of metal-on-polyethylene. Charnley died in 1982 believing that he had brought to the world the last word on hip replacements. McKee died in 1991 believing that his metal-on-metal hip was a failure. With the benefit of hindsight, we can now see that both were wrong in these perceptions.

The Charnley concept of metal-on-polyethylene (CoCr/UHMWPE) became the industry gold standard from the 1970s to the 2000s. Although it is now a shrinking market, it is still 45% of the global hip replacement market today. Around the early 1990s, when the author first got involved in orthopedic ceramics, a perception prevailed across the industry that the Charnley concept was the final word on the hip prosthesis. Nobody at that point foresaw the revolution that lay just ahead. The consensus at this time was that the Charnley concept, while not without its problems, was the best system for the past, present, and future of hip replacement.

The key problem with the Charnley concept is the huge amount of polyethylene wear particles generated, which generally remain in the joint cavity, causing irritation, and a very high revision rate due to osteolysis (Chapter 7, Section 7.3.2). Unlike metal wear particles which are toxic and can cause metallosis or bone necrosis, polyethylene wear particles are merely an irritant, a "sawdust in the joint" kind of irritant. This has been the cause of an epidemic in polyethylene wear particle-induced osteolysis, implant loosening, and revision surgery. The consensus today is that the Charnley concept is quite satisfactory for elderly inactive patients, and not so suitable for younger or more active patients [219–222].

The Charnley concept is today a shrinking market, in the process of being supplanted by alumina hips, because of its very high associated osteolysis rates. Disenfranchisement with the Charnley concept is a recent development. It was only in the last decade that the metal-on-polyethylene hip saw a serious challenger: alumina bearings. Moreover, this has so far only been in hip replacement, not yet the other joints. The trajectory is clear. CoCr-on-polyethylene is on the way out and alumina is on the way in, the crossover point 50:50 CoCr:Alumina was traversed just a few years ago, around the middle of the current decade (2010s). The reasons for the benefits of alumina bearings are discussed in the second half of this chapter.

Before moving to the evolution of the alumina hip bearing or "ceramic hip" as it is commonly known, it is important to summarize the market trends up to the present day:

(1) In the late 20th century, CoCr/UHMWPE became the standard for all prosthetic bearings in all joints.
(2) In the late 20th century, CoCr stems were almost totally supplanted by titanium stems.
(3) PMMA bone cement remains a significant fixation method, with two alternative "cementless" fixation methods becoming equally significant: bioactive coatings on stem and acetabular cup (plasma-sprayed HA); porous surface features (sandblasting, mesh, macrospheres) on stem and acetabular cup to enhance biological fixation.
(4) In the last decade, alumina has taken over from CoCr as the dominant femoral head.
(5) In the last decade, alumina has significantly competed with UHMWPE as the acetabular liner (socket).
(6) Metal-on-metal total hip replacements (the McKee concept) have become virtually obsolete
(7) Metal-on-metal resurfacing implants became a significant option for younger patients.

Although polyethylene dominates the acetabular liner market, there are two other acetabular liners commonly used: alumina (15% of all implants on the market are alumina femoral head and alumina acetabular liner) and CoCr (primarily in resurfacing implants, which are <2% of the market). Some experimental materials have also been used, such as the biocompatible wear-resistant polymer poly-ether-ether-ketone (PEEK). Though uncommon, there are a few instances of the use of PEEK liners, as for example the one shown in Fig. 6.8. The PEEK (black polymer) is articulating against alumina in this case. PEEK dominates the spinal cage market, but it is relatively uncommon in hip replacements.

6.3.7 Generic enhancements to the hip replacement

By the late 20th century, the standard commercial hip replacement had evolved to the following general features:

- titanium stem
- titanium acetabular cup
- CoCr femoral head (ball)
- UHMWPE polyethylene acetabular liner (socket)
- cementless implants and cemented implants were both commonplace

Two of these enhancements have not been significantly discussed yet and need to be briefly overviewed as these concepts arise in the alumina hip section of this chapter (Section 6.4).

6.3.7.1 UHMWPE polyethylene acetabular liner

UHMWPE was the key contribution of Charnley to the modern hip replacement, and indeed virtually all joint replacements in the late 20th century, before the alumina-bearing era, and just under half the 1.3 million or so commercial hip joints implanted annually in the present, involve the CoCr/polyethylene paradigm.

Alumina bearings in orthopedics: Origin and evolution 159

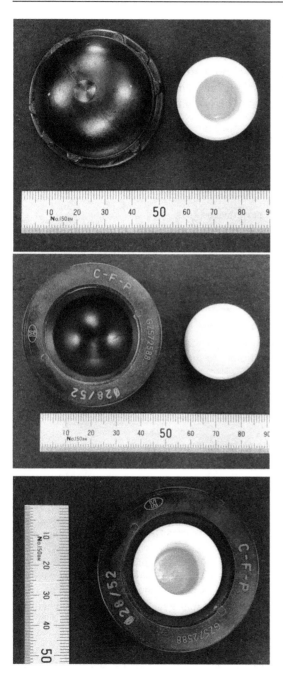

Fig. 6.8 Alumina-on-PEEK—poly-ether-ether-ketone. Exhibit provided courtesy of Professor Bill Walter. Specialist Orthopedic Group. Wollstonecraft NSW.

Polyethylene is the world's most common commercial polymer (engineering plastic) with annual production in the order of 80 million tonnes, exceeding aluminum production. It is comprised of long, noncross-linked chains of the simple ethylene monomer CH_2. Polyethylene comes in several forms, based on average chain length. Usually the chains are linear with very little branching, typically <2%. Polyethylene is defined by molecular weight in atomic mass units (AMU):

Low-density polyethylene
- chain length 10–30 thousand AMU
- the softest weakest polyethylene
- used, for example, in freezer bags

High-density polyethylene
- chain length 50 thousand to 1.5 million AMU
- strong tough and widely used polyethylene grade
- widely used in bottles and plastic packaging, for example, in the thin strong plastic nonbiodegradable grocery bags that were predominant in the 2000s

Ultra-high-molecular-weight polyethylene
- chain length 3.5–7.5 million AMU
- the most wear resistant and toughest polymer known, surpassing polycarbonate for toughness and polyurethane for wear resistance

UHMWPE is the grade used in acetabular liners.

There are also a number of intermediate grades between these general classifications, but these three main classifications are the basic definitions of commercial polyethylene.

6.3.7.2 Cementless implants

The Charnley paradigm was PMMA bone cement for bonding the stem and acetabular cup to the bone. PMMA gives instant high strength, enabling the patient to walk soon after surgery, but the strength gradually diminishes over time as the PMMA deteriorates in the body cavity [223].

Cementless implants utilize two main strategies for enhanced bone bonding:

Calcium phosphate coatings, generally as plasma-sprayed HA. HA is a synthetic version of bone mineral. Theoretically pure and optimized HA, just like bone, does not dissolve in the body cavity, even over a period of many years. HA is capable of bonding osteogenesis, that is, bone forms a direct chemical bond with HA in the body cavity. This plasma-sprayed HA technology was developed and commercialized during the 1990s.[1] The key issue is to optimize the plasma spraying conditions so as to maximize the HA-metal bond strength and maximize the durability of the plasma-sprayed HA. When done correctly, the thick-film coating of HA bonds strongly to the metal implant and the bone for many years. HA coatings work equally well for titanium and CoCr. When done incorrectly, the HA is not HA at all but a dehydroxylated calcium phosphate which is biodegradable in vivo. Poor HA coating adhesion is another problem that can arise from incorrect coating parameters.

---

[1] The author has been involved in hydroxyapatite bioceramic research since 1986, including plasma spraying and other hydroxyapatite coatings technology.

Titanium osseointegration, long known since the pioneering work of Branemark [161,162], titanium with a porous surface allows bone ingrowth (osseointegration). Sintered metal surface spheres in the order of a millimeter in size, coarse metal surface mesh, and sandblasting are common strategies to enhance osseointegration. The osseointegration mechanism is what sets titanium apart from CoCr, and the reason why titanium is now dominant in stem technology, while CoCr stems are almost obsolete.

Both these cementless methods form a strong bond gradually, over weeks to months, but it is enhanced over time, rather than diminished as occurs with PMMA. Thus, short-term pain, long-term gain. In the first few weeks post-surgery the implant bond is not very strong. Commonly, mechanical fixation methods are used in combination with cementless implants, such as pressing the metal implants into position, for immediate stability post-surgery.

6.4 Alumina bearings in hip replacements: The alumina hip

6.4.1 Evolution of the alumina hip

The early developmental period for alumina hip bearings took place from 1970 to 1980. Alumina from this rudimentary era is known as "First Generation" alumina.

Around the year 1980, the period of the "Second Generation" CeramTec alumina hip bearings began, defined by the introduction of high-purity alumina powders and alumina process control improvements.

In 1995, around the time of the commencement of commercialization of alumina hip bearings by CeramTec (CE-Mark 1994) the period of the "Third Generation" alumina began, defined by the introduction of hot isostatic pressing, laser marking, and proof testing. "Fourth generation" was introduced in 2003, this was the ZTA bearing known as BIOLOX*delta*. Today we enjoy the benefits of "Fourth Generation" alumina. ZTA manufactured by CeramTec to a very high standard. CeramTec held a global monopoly until 2011 when Kyocera entered the market, but CeramTec still maintain almost total global market dominance. Kyocera bearings are currently primarily only used in the Japanese market.

CeramTec call their pure alumina BIOLOX*forte*, and it is bone colored. In 2003 CeramTec introduced ZTA, colored pink, and called BIOLOX*delta*, which has about five times the wear resistance of the pure alumina product, and even higher mechanical reliability. The ZTA product is rapidly supplanting the pure alumina product in the market. ZTA is the focus of Chapter 7.

6.4.1.1 Pierre Boutin (1970): The first alumina hip

In 1970, Pierre Boutin, an orthopedic surgeon, developed and implanted the first alumina hip implant in France [143,144]. While Boutin was the first to develop an alumina hip replacement, a large body of German researchers rapidly caught up and soon overtook him (Section 6.4.1.2).

Pierre Boutin was a French orthopedic surgeon and one of his patients was vice-president of Ceraver (a ceramic company). As a result, a partnership formed between the orthopedic surgeon and the ceramic company, a perfect partnership for a rapid breakthrough. Boutin's reasons for developing the alumina hip were, according to him, "its hardness which renders it perhaps indestructible for the life of a human, its perfect chemical inertia, its low coefficient of friction, and its tolerance by the organism" [148].

The Boutin hip used an alumina-on-alumina-bearing system, with stem and acetabular cup bonded to the bone using bone cement. It was known as the Ceraver Osteal ceramic hip from France. A Ceraver Osteal alumina hip is shown in Fig. 6.9. The stem was cemented and made of Ti6Al4V alloy, the first documented Ti6Al4V stem in orthopedics. The alumina femoral head was glued to the metal stem (not a Morse taper). Boutin had impressive early success encountering no ceramic fractures in his first 200 implants.

Clinical trials on this Boutin implant ran from 1970 to 1980 and showed promise [224–227]. For example, orthopedic surgeon Sedel implanted 86 of the Ceraver Osteal alumina-on-alumina implants from 1977 to 1986 [227] with a 98% life expectancy of the prosthesis at 8 years.

However, it was a technology race in the early 1970s, and while Boutin was developing and trialing his innovative alumina hip, a large effort from a number of German researchers rapidly began to lead the field of alumina hip technology. The German innovations included a number of strategies:

(1) Mittelmeier hip: Alumina femoral head; alumina acetabular implant as a single piece
(2) Griss hip
(3) Wagner: Alumina resurfacing
(4) Salzer: Entire alumina implant, including stem
(5) Feldmuhle Germany (which became CeramTec) developed BIOLOX orthopedic alumina bearings from the mid-1970s, a product which now has over 40 years clinical history.

6.4.1.2 The Mittelmeier hip

The most widely used of the early alumina hip devices was the Mittelmeier Autophor alumina-on-alumina hip shown in Fig. 6.10. It was innovative in being one of the earliest alumina hips, the first to have a Morse taper for the stem-head join, and also in the fact that the acetabular cup was a dual-function single implant, that is, both liner and load-bearing acetabular cup. It had a macro-scale thread for cementless fixation. The stem was smooth CoCr with grooves, which caused many early failures that were not due to the alumina, but rather problematic stem design [228].

The Mittelmeier hip was first implanted by Prof Mittelmeier in Germany in 1974 and widely used in early clinical trials, not just in Germany but in various countries including the United States [229] and Australia [210]. Over 100 were implanted in the United States at Johns Hopkins University School of Medicine [229], and a large number also in Japan, and also in Australia [210]. The early Mittelmeier hip had a high mechanical failure rate of 21.5% for the cup and 22.6% for the stem in a trial of 112 implants in 1983/1984 [229].

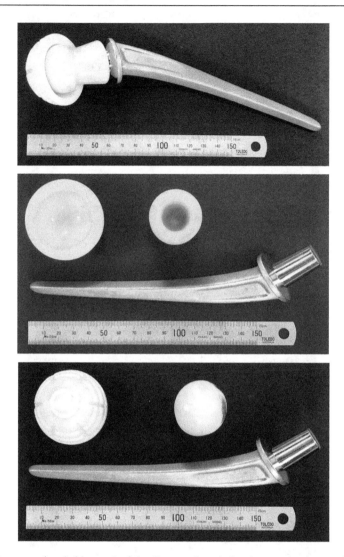

Fig. 6.9 Later version (with taper) of the Ceraver Osteal alumina-on-alumina total hip replacement, from France, from the Boutin era. The acetabular cup is also the socket that articulates against the alumina femoral head (ball). This device was implanted by many surgeons around the world in the 1970s and 1980s.
Exhibit provided courtesy of Professor Bill Walter. Specialist Orthopedic Group. Wollstonecraft NSW.

The concept of a metal acetabular cup with alumina liner (socket) is a very reliable alternative that is exclusively used today.

The Griss hip was independently developed in Germany at the same time as the Mittelmeier hip [230] but it saw limited use.

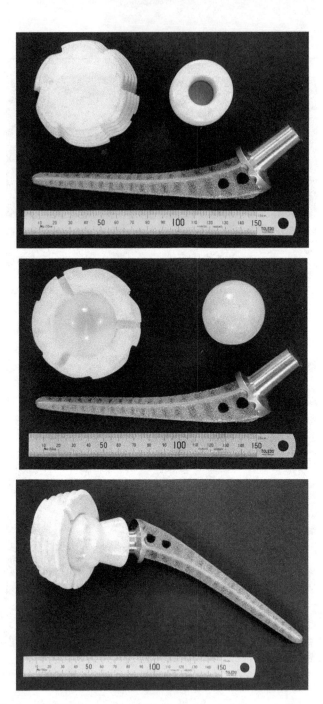

Fig. 6.10 Mittelmeier alumina-on-alumina hip replacement. The acetabular cup is also the socket that articulates against the alumina femoral head (ball). Note the coarse crew thread on the alumina acetabular cup, for mechanical fixation into the acetabulum of the pelvis. This was the Mittelmeier (Autophor) total hip replacement. This device was implanted by many surgeons in the 1970s and 1980s.
Exhibit provided courtesy of Professor Bill Walter. Specialist Orthopedic Group. Wollstonecraft NSW.

6.4.1.3 The Wagner alumina resurfacing implant

The Wagner alumina resurfacing implant was the world's first alumina resurfacing implant. It was made by Rosenthal Technik (Germany). It comprised an alumina femoral sleeve articulating against a polyethylene acetabular liner [231 – 233]. The original Wagner alumina resurfacing implants are shown in Fig. 6.11 [210].

At least two alumina resurfacing implants are currently undergoing development in the world today. The first known as Recerf is a ZTA-based resurfacing implant through an Australia/British/Canadian/Belgian surgeon team in collaboration with CeramTec and a UK-based manufacturer Matortho [210]. The second is a British program under development at the University of Southampton (UK) in collaboration with various organizations, including CeramTec, known as the EnDuRE [234]. Both devices are well advanced in their development. The challenge is to design an implant that has sufficient strength with the thin-cross-sections required, which is close to the limit of what ZTA is capable of mechanically.

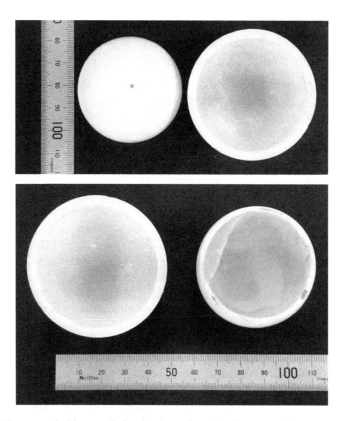

Fig. 6.11 Wagner early hip resurfacing implants from the 1970s to 1980s. Images are for a retrieved Wagner alumina/polyethylene resurfacing implant.
Exhibit provided courtesy of Professor Bill Walter. Specialist Orthopedic Group. Wollstonecraft NSW.

6.4.1.4 The novel all-alumina implant

One of the most extraordinary innovations of the German alumina hip pioneers of the 1970s was an entire hip replacement, including the stem, made of alumina [235]. This was novel but not viable as a commercial product. It was called the "femoral cap endoprosthesis;" it was clinically trialed by Salzer in the early 1970s [236]. Japanese researchers also developed a single-crystal all-alumina stem [237,238]. Novel as these developments were, they did not prove practical. Alumina's key strengths in orthopedics are chemical inertness and wear resistance. Titanium and other metals with their toughness under bending and impact loads are far more suitable for the stem.

6.4.1.5 The Shikati alumina-polyethylene hip

Alumina-on-polyethylene is today the dominant manifestation of the alumina hip, representing 25% of the global hip replacement market where alumina-alumina is 15%. The first alumina-on-polyethylene hip was developed in Japan by Shikati et al. [239]. The alumina-on-polyethylene concept rapidly gained momentum thereafter.

6.4.1.6 Origin of the ceramic Morse taper

Boutin had a problem attaching his 32-mm alumina head to the titanium stem. Initially he tried gluing it. Later he tried brazing. Both systems caused failures and fractures. This problem was resolved by the German researchers who innovated the Morse taper. The Mittelmeier hip was the first to use a Morse taper [228]. In 1977, German researchers published the original paper documenting the fact that a Morse taper was the only satisfactory means of attaching a ceramic femoral head to a metal stem, with a 6° taper as the chosen optimal taper angle [240]. From then on, Morse tapers became the standard for alumina hips. These metal-alumina taper junctions were successfully tested in dynamic fatigue for 20 million cycles: 10 Hz, 6–50 kN cycle range, immersed in Ringers solution [240]. They were also successfully tested for static fatigue, under static load for 1 year in Ringer's solution or sheep tissue [241]. Today Morse tapers are standard for all alumina femoral heads (female taper) and alumina acetabular inserts (male taper). Burst strength, due to taper hoop-stress exerted on the female taper socket of the alumina head, was a problem in the early days, until the importance of precision matching in metal and alumina taper came to the forefront.

6.4.1.7 Summary of the evolution of the alumina hip

The original driver for the invention of the alumina hip was to avoid the osteolysis (Chapter 7, Section 7.3.2) caused by polyethylene wear particle debris [242]. Thus alumina-on-alumina was dominant in the early days. A second driver was to develop a safer, low-wear rate and therefore more durable long-life-span hip replacement, so as to improve the long-term reliability of hip replacements in younger and more active patients. This was because of the short life span of the CoCr-on-polyethylene hip and its associated limits on patient activity. In general the consensus was reached by the early 20th century that alumina-on-alumina implants were ideal for young active patients and that CoCr-on-polyethylene was satisfactory for older less active patients.

Interestingly, the drivers for alumina bearings are different today. Alumina-on-polyethylene is now dominant and generally found satisfactory because the two main drivers for alumina hips today are:

- elimination of galvanic corrosion between CoCr head and titanium stem
- reduction in polyethylene wear rate

With the more wear-resistant high-quality cross-linked polyethylenes of today, polyethylene osteolysis is not such a problem as it once was and it is substantially less of a problem with alumina-on-polyethylene than CoCr-on-polyethylene. Thus ~40% of the global market is now alumina-on-polyethylene, with ~15% alumina-on-alumina.

By the turn of the century, at the 30-year point, the number of alumina hips that had been implanted globally was over a million and the technology was rapidly being refined and optimized. The key issue for longevity of an alumina-on-alumina bearing is high wear resistance and high mechanical reliability. These two factors depend on the following key parameters:

- high fracture toughness
- low porosity
- fine grain size, ideally below 2 μm

The detailed analysis of the tribology of alumina in Chapter 12 (Section 12.1) outlines the significance of these parameters on alumina wear resistance in sliding abrasion.

By the late 1990s, Europe already had Regulatory Approval (CE-Mark 1994) for alumina hips, and USA FDA approval was imminent (FDA 2003). At this 30-year point, the alumina hip had essentially "come of age." Some of the general statistics on alumina-on-alumina hips at the 30-year point, around the turn of the century and soon after its 1994 commercialization, were [243]:

- cemented 10-year survival rate: 83%
- cemented 15-year survival rate: 70%
- cemented 15-year survival rate for patients under 50 years: 86%
- osteolysis (Chapter 7, Section 7.3.2) was found in <1% of patients
- for younger patients stem loosening was very rare
- aseptic loosening of the cemented acetabular cup was the main driver for revision

Clinical trials of alumina-on-alumina hips in the United States began in 1997, under Osteonics and Wright Medical Technology [244,245].

Within a couple of decades of the invention of the alumina hip, leading German ceramics company CeramTec had perfected and commercialized the alumina hip bearing, attaining European Union medical approval (CE Mark) in 1994. Thus commercialization began in 1994 in Europe. FDA approval in the United States came 9 years later in 2003.

The commercialization issues are discussed in Chapter 7 (Section 7.4). However, before concluding this chapter, it is pertinent to look at the key ceramic engineering issues of the modern hip. In the late 1990s, the most significant ceramic issues that lay ahead were the zirconia disaster of 2001/2002 and the ZTA revolution of the early 21st century.

6.5 Other ceramic orthopedic bearings

The following three main ceramic bearing alternatives to pure alumina are all zirconium based:

(1) Oxidized zirconium (OXINIUM$^\lozenge$)—zirconium bearing with thin zirconia coating
(2) Zirconia—pure tetragonal zirconia ceramic
(3) ZTA—alumina reinforced with ~20% tetragonal zirconia microparticles

Oxidized zirconium is a niche product, and zirconia an obsolete product. They are both relevant to this chapter in the context of lessons learned from the past.

ZTA, on the other hand, has changed the world of orthopedics. ZTA is essentially toughened alumina. In the last decade, ZTA rose to become the dominant ceramic bearing of the 21st century, and will be the focus of Chapter 7, because ZTA is the present and the future for orthopedic bearings.

6.5.1 Oxidized zirconium (OXINIUM$^\lozenge$)

OXINIUM$^\lozenge$ is the brand name of the current commercial orthopedic manifestation of oxidized zirconium, a smith&nephew biomaterial used in hip and knee bearings. Oxidized zirconium is not a ceramic bearing, but a ceramic-coated metal bearing. It comprises wrought zirconium alloy (zirconium with 2.5% niobium) which is oxidized by thermal diffusion to create a 5-μm thick oxidized zirconia (monoclinic zirconia) ceramic surface film. Smith&nephew is a British-based multinational company, renowned as one of the world's leading medical device companies in the orthopedics realm. OXINIUM$^\lozenge$ was first introduced by smith&nephew, two decades ago in 1997, for both hips and knees. Thus two decades of clinical experience can be called upon to evaluate it.

In fact oxidized zirconium, as a means of enhancing zirconium wear resistance, dates back to the 1961 patent of Watson [246] for wear-resistant zirconium valve parts in nuclear reactors. Zirconium is a long-established metal in nuclear reactors. In contrast, zirconium is a recent, and boutique material in orthopedics, exclusively found in mainstream commercial use as the smith&nephew OXINIUM$^\lozenge$ biomaterial. Titanium and CoCr are the dominant biometals in orthopedics. Zirconium, stainless steel, and tantalum are the boutique biometals in orthopedics.

Oxidized zirconium biomaterials can be produced by heating zirconium in air at about 540°C for 3 h in accordance with the Davidson patent of 1991 [247], or by immersing zirconium in a molten salt bath that contains an oxidizer (normally Na_2CO_3) at 700°C for 4 h, in accordance with the Haygarth patent of 1987 [248].

A rival patent for the smith&nephew manifestation is that of Zimmer, another leading medical device company in the orthopedics realm. Filed in 2006 the Zimmer oxidized zirconium patent involves a zirconium film deposited onto a metal implant, which is then surface oxidized [249]. The patent claims that this oxidized zirconium manifestation is less susceptible to delamination because it is done by ion-beam-assisted deposition and engineered to have a 100nm "intermix zone" between the zirconia and the α-phase [249]. The patent implies that oxidizing zirconium metal in air or a salt bath results in a much sharper zirconium:zirconia interface.

Oxidized zirconium provides the high wear resistance and low friction of a ceramic surface, with the toughness, fatigue strength and manufacturing flexibility of a metal-bearing component [250]. Traditional ceramic-coatings technology involves surface deposition of the coating onto the metal, with the associated adhesion and interfacial bonding and delamination problems. In contrast, oxidized zirconium is a surface controlled-oxidation process that provides a ceramic coating grown from the surface of the metal itself. As a result, oxidized zirconium has proven to be less susceptible to the problems of poor adhesion that plague some traditional ceramic coatings. Moreover, the oxidized zirconium surface has approximately double the hardness of CoCr, which has the potential to significantly reduce wear rates. It should be noted that the zirconia film of oxidized zirconium is the stable monoclinic phase, and therefore not susceptible to aging (see Section 6.4.1.5). However, oxidized zirconium metal still has a sharp zirconium:zirconia interface which makes the zirconia coating susceptible to delamination under high applied loads [249].

Commercially, oxidized zirconium is exclusively used articulating against polyethylene, as an alternative articulation system to CoCr/polyethylene, in hip and knee prostheses. Simulator tests have reported a number of impressive facts for oxidized zirconium in comparison to CoCr, for polyethylene articulation:

- 4900 times less volumetric wear and 160 times less roughness than CoCr [250]
- 45% less wear than smooth CoCr heads in hip implants [251] and 60% less wear when roughened [252]
- 78% decrease in UHMWPE wear than CoCr [253]
- enhanced wettability and lower friction than CoCr [254]
- 85% less polyethylene wear than for CoCr for a 90% walking gait/10% stair climbing knee simulation [255].

From a biocompatibility perspective, oxidized zirconium reportedly has enhanced biocompatibility compared to CoCr in terms of the release of metal ions, associated with metal sensitivity [247,256]. From a mechanical perspective, the bulk zirconium metal of oxidized zirconium has similar fatigue strength to CoCr: both of these metal alloys, CoCr and the zirconium alloy of oxidized zirconium exceed 450 MPa for 10 million cycles [257]. These various reported advantages of oxidized zirconium over CoCr, from simulator and laboratory testing, suggest that oxidized zirconium implants have the potential for an enhanced implant survival time. Therefore, like alumina hips, oxidized zirconium hips and knees offer the potential to be well suited to younger patients.

Given that alumina hips are long established and performing well, but alumina knees are purely experimental, the real area of benefit for oxidized zirconium is in the knee prosthesis realm. The two largest joint replacement markets in terms of annual replacement numbers are the hip and the knee, both having a market of over 1 million a year. Other joint replacements such as shoulder, elbow, and ankle are only in the 10,000 PA order of magnitude.

One of the great dangers of a coated metal articulating against polyethylene is the risk of third-body wear (Chapter 12, Section 12.1) resulting from particles of delaminated coating becoming lodged between the polyethylene and coated-metal articulating surfaces. This is all the more problematic with coated metals in that polyethylene is soft, and delaminated ceramic coating particles are hard and likely to be jagged, not well-rounded particles. Oxidized zirconium is possibly a superior coating concept to the traditional ceramic coating methodologies, but this does not guarantee the elimination of third-body wear risk, or other unforeseen problems, over long-term in vivo articulation conditions. Only clinical experience can validate this.

To date, the superior properties from simulator tests and other laboratory tests of oxidized zirconium have not been reflected in in vivo studies. Clinical results have been mixed. Some studies have shown a higher wear rate for oxidized zirconium/polyethylene, than CoCr/polyethylene [258–260] and some studies have shown no clinical benefit of oxidized zirconium/polyethylene over CoCr/polyethylene in terms of clinical, subjective, and radiographic outcomes [261,262].

A 2011 study [261] of subjective outcomes in involved patients who received both implants (CoCr and oxidized zirconium) found that 38% of patients preferred the Co-Cr implant, 18% of patients preferred the oxinium, and 44% of patients had no preference. Clinical outcomes have shown scratches, cracking, gouging, and delamination in oxidized zirconium femoral heads from clinically retrieved hip replacements [250,263,264]. Surface damage such as this can have several results, including [264]:

- loss of oxidized zirconia coating
- exposure of soft metal subsurface to wear
- third body wear from coating fragments

This is not only problematic for the polyethylene. A 2009 study [264] of damaged oxidized zirconium heads showed very low wear rates for the oxidized zirconium compared to a fresh oxidized zirconium new head [124]. This is probably because the parent zirconium alloy is a softer alloy (2.85 GPa) than CoCr (4.2 GPa) [253]. This was verified by hardness testing of retrieved OXINIUM$^\diamond$ heads compared to CoCr [250].

A problem with fracture of oxidized zirconium led to the recall in 2009 of 38,750 units worldwide of the Journey II Uni tibial baseplates (knee) [265]. Up until April 2013 the FDA received >340 individual adverse event reports for the device [265]. These problems were all knee-prosthesis related. Nonetheless, one should not be too hasty to judge oxidized zirconium. It has only been in clinical use for 20 years, and recalls and failures are routine and commonplace across the orthopedics industry. Probably the two most important points about oxidized zirconium in hips are galvanic corrosion and ceramic/ceramic bearings.

Galvanic corrosion, discussed in Chapter 7 (Section 7.3.1), occurs when mixing metals at taper junctions. Alumina heads mated to titanium stems eliminate galvanic corrosion and thus chromium and cobalt release into the joint cavity. Oxidized zirconium heads involve a titanium/zirconium taper junction, which is potentially problematic.

Alumina can be used in extremely low wear rate ceramic-on-ceramic bearings. Oxidized zirconium with its very thin coating is only used with polyethylene, never as a ceramic-on-ceramic bearing, due to the risk of coating delamination.

In conclusion, it would be fair to say that to date the theoretical promise of oxidized zirconium from simulator and laboratory tests has not translated into a paradigm shift in clinical success. Oxidized zirconium remains a significant market player today in hips and knees, with the knee industry the one most in need of oxidized zirconium. The hip industry has the long-established option of alumina bearings, while alumina bearings, or indeed any ceramic bearings, are not used in commercial knee implants (other than OXINIUM$^{\diamond}$) for reasons discussed in Chapter 7 (Section 7.4.4).

6.5.1.1 Invention of zirconia toughening

ZTA is rapidly displacing pure alumina from the ceramic orthopedic bearings market. Zirconia toughening is one of the most significant advances in ceramic technology of the late 20th century.

Zirconia (ZrO_2) has three crystalline polymorphs:

- *Monoclinic*: Thermodynamically stable at room temperature, largest specific volume.
- *Tetragonal*: At around 1170°C, the monoclinic phase transforms to the tetragonal phase, which has a smaller specific volume than the monoclinic phase.
- *Cubic*: At around 2370°C, the tetragonal phase transforms to the cubic phase, which has a smaller specific volume than the monoclinic phase, the smallest specific volume of the three phases.

Ordinarily these volumetric changes at the transition temperatures cause cracking of zirconia making it an impractical engineering ceramic. The critical issue is that on cooling tetragonal zirconia below its thermodynamic stability limiting temperature (about 950°C on cooling), it transforms back to the monoclinic phase, which is accompanied by a 3%–5% volume expansion. Therefore, if a way could be found to produce tetragonal zirconia that is metastable at room temperature, energy (e.g., the hydrostatic energy created by an impinging crack), could be sufficient to transform the tetragonal crystal into the thermodynamically stable monoclinic crystal. In doing so, the crystal will expand by 3%–5%, creating localized compressive forces that will pinch off the crack, thereby arresting crack propagation by two mechanisms:

- crack-energy consumed by the tetragonal-monoclinic transition
- local compressive forces from the tetragonal-monoclinic transition pinching off the crack

This is shown schematically in Fig. 6.12. This is now a well-known phenomenon called "transformation toughening" or "zirconia toughening".

Zirconia toughening would not exist were it not for visionary British physicist Ron Garvie, who devoted the 1960s to finding a way of stabilizing tetragonal zirconia at room temperature, and the 1970s to perfecting this principle into a practical transformation toughening technology. In the early 1960s, Garvie first began investigating the high-temperature properties of zirconia at the Metallurgical Research Laboratory, US Bureau of Mines [266]. Garvie was not the first to produce metastable tetragonal zirconia. Garvie's 1965 paper [267] references a study by Ruff, Ebert, and Stefan who in

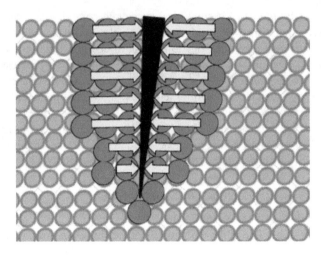

Fig. 6.12 Transformation toughening (crack pinching) in a tetragonal zirconia (TZP) ceramic. The smaller blue circles are the metastable tetragonal crystals. As the crack impinges on the tetragonal crystals the crack energy transforms them to thermodynamically stable crystals of monoclinic zirconia. Monoclinic has a larger specific volume than tetragonal, and thus is represented as the larger red circles. This volume increase of the crystals creates a localized expansion, which results in localized compression (yellow vectors) onto the impinging crack, thereby pinching it off.

1929 first reported metastable tetragonal zirconia produced by igniting zirconium salts [268]. However, Garvie was the first to see the engineering and commercial potential of this metastability, and to turn it into a world-changing technology. Garvie used X-ray diffraction to show that a zirconia powder contained the tetragonal structure at room temperature. At this time it was believed that the tetragonal structure could not be retained at room temperature. Garvie had the view that since the tetragonal structure had a lower surface energy than monoclinic, it should be possible to retain the tetragonal phase at room temperature provided that the particle size was smaller than a critical dimension.

In the mid-1960s, Garvie investigated partial stabilization of tetragonal zirconia by alloying zirconia with calcia (CaO). Calcia lowers the phase transformation temperature to produce a multiphase zirconia: a blend of monoclinic, tetragonal, and cubic phases—"partially stabilized zirconia" (PSZ) [269]. Earl Parker had recently invented transformation-induced plasticity (TRIP) steels which undergo a plastic-deformation-induced martensitic transformation, a toughening mechanism. Garvie was inspired by the TRIP innovation and adapted it to develop his concept of transformation based on his theory that a similar transformation could be possible in a calcia-PSZ. In 1969, Garvie presented this theory at an American Ceramic Society meeting, and was met with some scepticism, but also some support for what were seen as radical ideas at the time [269].

In 1972, Garvie joined the CSIRO (Division of Materials Science) in Melbourne Australia, and in that year proved his theory, achieving an extraordinarily high strength of 600 MPa in (PSZ). Garvie's original PSZ patent was filed in 1974 with co-inventors Richard Hannink and Richard Pascoe, followed by their ground breaking "Ceramic Steel" paper in the journal "Nature" in 1975 [270]. This was a revolution in ceramic technology and precipitated a global burst of intense research activity in the global ceramics community from the late 1970s until the late 1990s during which time the following general findings were made:

Dopants: Various stabilizing dopant oxides can be used for stabilizing zirconia, of which yttria and magnesia were the most common, with ceria and calcia also sometimes used.

Critical crystallite size: For a tetragonal zirconia crystal to be metastable, it needs to be between upper and lower critical crystallite size limits. The upper limit is set by the size above which spontaneous monoclinic transformation occurs during cooling after sintering. The lower limit is the size below which monoclinic transformation is impossible, even under stress. The optimal size is generally in the order of a few hundred nanometers.

Full stabilization: A fully stabilized zirconia is produced by adding sufficient dopant oxides to stabilize the highest temperature polymorph (cubic) to room temperature. This has uses as a refractory (stable on heating and cooling all the way to its melting point of $\sim 2700°C$) as discussed in Chapter 15, as an ionic conductor, for example zirconia oxygen sensors, as also discussed in Chapter 15, and as a synthetic single-crystal gemstone "cubic zirconia" which exploits the high refractive index of zirconia.

Partial stabilization: A PSZ has less dopant added, sufficient to stabilize the tetragonal phase to room temperature. PSZ ceramics containing metastable tetragonal micron-sized crystals can be produced as

EITHER

  (1) PSZ which comprises tetragonal microparticles in a matrix of cubic and monoclinic zirconia ceramic. Commonly done with magnesia, but it is not as tough as Y-TZP, as shown in Table 9.3.

OR

  (2) A zirconia ceramic comprised of pure tetragonal phase (tetragonal zirconia precipitate: TZP). This is the highest toughness zirconia ceramic and commonly produced as Y-TZP, containing about 3% yttria (3Y-TZP), as. shown in Table 9.3.

OR

  (3) Zirconia as a toughening additive: The intensive burst of zirconia toughening research activity in the 1980s and 1990s also involved the addition of tetragonal zirconia microparticles to other oxide ceramics to enhance their toughness. Most notably this was done with alumina to produce a unique material that combined the hardness of alumina with the toughness of zirconia—ZTA. A typical ZTA contains about 80% alumina and 20% tetragonal zirconia.

High toughness: Toughened zirconia was found to be characterized by both very high strength and greatly enhanced toughness compared to a regular oxide ceramic such as alumina. Toughness is commonly reported as K_{Ic} (units $MPa\, m^{1/2}$) fracture toughness where K_{Ic} equates to the critical stress intensity for unstable crack propagation in crack opening mode-I (plane strain). K_{Ic} is a somewhat esoteric parameter that does not correlate linearly with material toughness. Fracture energy γi is the energy ($J\, m^{-2}$) required to create new fracture surface by the propagation of a crack. In simple terms, γi fracture energy is a measure of the energy required to fracture a material, or the "impact strength" of a ceramic. The relationship between γi and K_{Ic} is as follows:

$$K_{Ic} = \left(\frac{2E.\gamma i}{1-\nu^2} \right)^{1/2}$$

where E is Young's modulus and ν is Poissons ratio. Or to put it simply, K_{Ic} is proportional to the square of γi. In practice, therefore, the relative impact strength of a material compared to another can be determined by a simple ratio of the squares of K_{Ic} for each. Thus the data in Table 9.10 illustrate the impact strength improvement of zirconia toughening for a number of key ceramics. Y-TZP is one of the best performing of the transformation-toughened zirconia ceramics. It has a K_{Ic} fracture toughness value in the order of 13 MPa m$^{1/2}$, whereas Mg-PSZ is lower at around 11 MPa m$^{1/2}$. ZTA is around 5–7 MPa m$^{1/2}$ while pure alumina is in the order of 4 MPa m$^{1/2}$. In short, Y-TZP has 10 times the impact strength of alumina, while ZTA has three times.

> *High Wear Resistance*: As discussed in the detailed tribology overview in Chapter 12 (Section 12.1), the two most important parameters for high wear resistance are high hardness and high toughness. Also very important are fine grain size, low or zero porosity, and chemical inertness. Zirconia is not as hard as alumina but it has comparable wear resistance owing to its greatly enhanced toughness.

6.5.1.2 Zirconia hip bearings: The 2001 catastrophic failure

In the 1980s, and indeed ever since, alumina was performing extremely well in the role of orthopedic bearings. In the 1980s, the alumina-bearing breakage rate was 0.014% (1 in 7000) and by the mid-1990s, 0.004% (1 in 25,000) [273]. Therefore, there was no real need to find an alternative ceramic-bearing material. It was more a case of the 1980s being an era of great enthusiasm and optimism for the new miracle material zirconia: "ceramic steel." I well remember attending a number of ceramics conferences in the 1980s and 1990s where zirconia was the foremost theme. In the 1980s, the zirconia hip-bearing development was motivated by a desire to further reduce the already very low ceramic breakage rate associated with alumina.

Based on the spectacular strength and toughness values in Table 6.1, it is very understandable that newly invented toughened zirconia was trialed as a ceramic

Table 6.1 Properties of alumina compared to three common zirconia-toughened ceramics

Ceramic	MOR[a] CoorsTek (MPa)	MOR CeramTec (MPa)	H-Vickers CoorsTek (GPa)	K_{Ic} CoorsTek (MPa m$^{1/2}$)	Fracture energy $\left(\dfrac{K_{Ic}\text{ ceramic}}{K_{Ic}\text{ alumina}}\right)^2$
Alumina	380	500	14.1–14.5	4	1.0
ZTA	450	1000	14.75	7	3.1
Mg-PSZ	760		13	11	7.6
Y-TZP	1240		13	13	10.6

"Fracture Energy" gives an indication of the relative fracture energy (impact strength) of each ceramic, compared with alumina [134,271,272].
[a]MOR (modulus of rupture) also known as flexural strength.

hip bearing. Moreover, given that British Physicist Ron Garvie invented toughened zirconia in Australia, it is not surprising that zirconia was first trialed in orthopedics in Australia, beginning with Mg-PSZ in 1984. Zirconia trials followed soon after in the United States. Clinical testing was very limited with Mg-PSZ. In the late 1980s, Y-TZP was trialed due to its superior mechanical properties. Thereafter Y-TZP dominated the zirconia hip bearings field.

As described in Section 6.4.1.4, the toughening mechanism of zirconia is a metastable tetragonal phase which can be transformed into the thermodynamically stable (and volumetrically larger) monoclinic phase by the hydrostatic energy created by an impinging crack. In doing so, the crystal expands by 3%–5%, creating localized compressive forces that will pinch off the crack. While this proved to be an extremely effective strengthening/toughening mechanism, in 1981 Kobayashi was the first to discover that the tetragonal-monoclinic transition could also occur spontaneously on the surface of zirconia ceramics in moist warm conditions [274]. Now known as hydrothermal aging, this spontaneous process can be catastrophic in the warm moist in vivo physiological environment of the human body. In 1985, it was further discovered that aging was reversible by high-temperature annealing [275]. However, this finding was of no practical use to orthopedic bearings: explanation for high-temperature annealing to reverse the aging is obviously an impractical option.

Obviously the ideal scenario for zirconia ceramics is tetragonal zirconia particles that are retained to room temperature after sintering, and ONLY transform to monoclinic zirconia at or near room temperature/body temperature when exposed to stress. Hydrothermal aging involves transformation at body temperature in the absence of stress. In 1988, some of the key issues for the aging phenomenon were reported by Yoshimura, including the following issues of great relevance to the orthopedic-bearing application [276]:

- Degradation is caused by the tetragonal-monoclinic transition and is accompanied by microcracking and macro-cracking.
- Degradation is time dependent.
- The higher the temperature the faster the aging, up to an optimum of 200–300°C.
- Water or water-vapor enhances the aging transformation.
- A decrease in grain size retards the aging transformation.
- An increase in stabilizing oxide content retards the transformation.
- The transformation process progresses from the surface of the zirconia ceramic to the interior.

Transformation is something of a domino effect in that once one tetragonal crystal has hydrothermally transformed, it places stress on its neighboring grains, increasing the risk of their hydrothermal transformation.

Thus autoclave steam sterilization was potentially catastrophic for the zirconia, and long-term exposure to the warm moist hydrothermal in vivo environment also problematic.

Had the proponents of zirconia in the 1980s and 1990s been aware of the deeper implications of the above zirconia science they would not have proceeded.

However, the Yoshimura paper was interpreted as implying that temperatures below 200°C were safe. This assumption turned out to be incorrect. Therefore, a zirconia disaster unfolded around the turn of the century that, in retrospect, could have been forecast and prevented based on 1980s science.

Over 600,000 Y-TZP zirconia femoral heads were implanted from the late 1980s until the early 2000s [277]. These were manufactured by various companies including St-Gobain (France), Metoxit (Switzerland), Morgans (UK), and Kyocera (Japan). Zirconia heads were only ever coupled with polyethylene in commercial use. There were no commercial zirconia-on-zirconia hip bearings. In 1997, the US FDA issued a report on the risk of steam sterilization of zirconia bearings. In 2001, the TGA (Australia) issued a hazard alert on a batch of zirconia heads (1998 batch) for which a 30% failure rate by disintegration had been reported within 3 years of implantation. In 2001, 400 zirconia Prozyr femoral heads failed in a very short-time period. It later transpired that these 400 heads had accelerated aging due to insufficient densification as a result of a production problem. Orthopedic surgeons were advised to inform patients with a Saint-Gobain Desmarquest Prozyr head prosthesis to seek urgent medical attention, and Saint-Gobain Desmarquest issued a global recall of selected batches.

Production of zirconia femoral heads declined by 90% between 2001 and 2002 and never recovered [277]. This was seen as one of the worst disasters in the history of bioceramics and could be one of the reasons for the slow uptake in alumina hip bearings in the United States early this century. In the orthopedic industry, the word "ceramic" or "metal" is used for the femoral head. While "metal" always means CoCr, "ceramic" can mean zirconia, alumina, ZTA, or even OXINIUM$^\diamond$. This conflation of all ceramics in the one basket unfairly misrepresented the reputation of alumina. However, to put it in perspective [273]:

- CoCr-on-polyethylene (2001): 10% (1 in 10) failure by osteolysis
- Zirconia heads (2001): 0.66% (1 in 150) failure by breakage
- Alumina heads (2001): 0.004% (1 in 25,000) failure by breakage
- ZTA heads Today: <0.001% (1 in 100,000) failure by breakage

Clearly its about perceptions, and the zirconia disaster of 2001 created a bad perception. The surgical community is surprisingly forgiving of the extraordinarily high 1 in 10 osteolysis failure rate of CoCr-on-polyethylene, but was initially extremely cautious of ceramic heads, although their 1 in 100,000 breakage rates rapidly overcome that perception in the 21st century.

Essentially there were two problems with the zirconia heads:

- Catastrophic breakage caused by cracking associated with the aging-induced monoclinic transition.
- Surface roughening caused by the monoclinic transition which begins at the surface and works its way in. The surface roughening causes accelerated wear of the polyethylene (compared to smooth alumina) but it is much the same as the accelerated wear that CoCr causes on polyethylene.

Zirconia, however, did go on to become a successful dental ceramic. It has other engineering applications also.

Fig. 6.13A–F is of a 1990s zirconia-on-polyethylene bearing implanted in Australia in the 1990s. Some of these have now been retrieved, including the one shown here. Significant monoclinic transition was seen but most seemed to be satisfactory [210]. Fig. 6.13A and B is of the femoral head of an explanted implant after ∼20 years. The monoclinic transition zone on the surface of the zirconia ball is clearly visible as a brown stripe. Fig. 6.13C and D shows the femoral head with the polyethylene acetabular liner showing the significant wear: the polyethylene acetabular liner is significantly larger than the zirconia femoral head ball diameter. Fig. 6.13E and F is of a pristine (unused) implant for comparison.

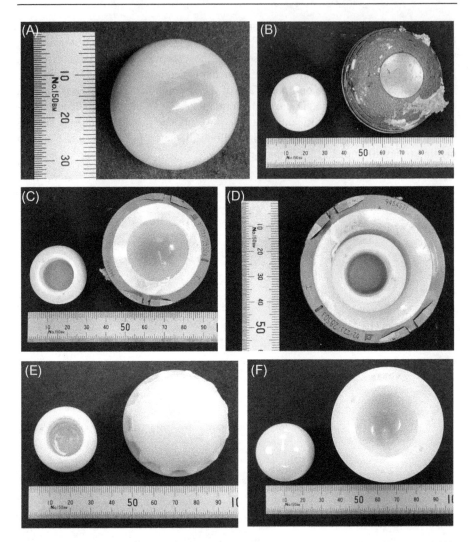

Fig. 6.13 (A) 1990s zirconia-on-polyethylene bearing. Note the monoclinic aging zone, visible as a brown stripe. (B) Explanted 1990s zirconia-on-polyethylene bearing, with acetabular cup. Note the monoclinic aging zone on the zirconia head visible as a brown stripe. (C) Explanted 1990s zirconia-on-polyethylene bearing, with acetabular cup and polyethylene liner.
(D) Explanted 1990s zirconia-on-polyethylene bearing, with zirconia femoral head inside the polyethylene liner. The liner is quite worn as shown by the looseness of fit, as is normal for polyethylene liners. Essentially all is well with this implant other than the wear stripe on the head. (E) Pristine zirconia femoral head and polyethylene acetabular liner. (F) Pristine zirconia femoral head and polyethylene acetabular liner. Note the unspoiled appearance of the zirconia head compared to (A) and (B) of the explanted zirconia head.
Exhibit provided courtesy of Professor Bill Walter. Specialist Orthopedic Group. Wollstonecraft NSW.

Alumina bearings in orthopedics: Present and future

7

There have been four main technology drivers underpinning the success of the alumina orthopedic bearing:

- Zirconia toughening (the technology underlying ZTA) is greatly enhanced the strength and toughness of alumina bearings.
- Greatly reduced rates of wear compared to non-alumina bearings: two-to-three orders of magnitude less alumina wear particles generated (alumina-alumina bearings) compared to polyethylene wear particles (CobaltChrome (CoCr)-polyethylene bearings).
- The benign nature of alumina wear particles compared to CoCr particles.
- The elimination of CoCr galvanic corrosion and CoCr fretting wear at taper joints, a problem of potential cobalt and chromium toxicity.

Chapter 6 outlined the long process of evolution of the hip replacement over the last 130 years. This began with the invention of the first hip replacement, made of hinged ivory, by Themistocles Gluck (Germany) in the 1880s. This was followed by the first metal-on-metal CoCr total hip replacements of the 1930–50s (UK), which were supplanted by the metal-on-polyethylene hip which evolved in the 1960s (UK), and the resurfacing hip in the 1990s (UK). This story culminated in the gradual development of pure-alumina bearings over the period 1970 to the turn of the century (Germany), which have now significantly replaced the metal-on-polyethylene alumina hip with currently 45% of the market metal-on-polyethylene and 55% involving alumina bearings (40% alumina-on-polyethylene; 15% alumina-on-alumina). Metal-on-polyethylene is a shrinking market while alumina bearings are a growing market.

While the pure-alumina-bearing technology evolved substantially over this period, it was but the precursor to the ultimate alumina-bearing technology, ZTA (zirconia-toughened alumina). ZTA (Germany) was introduced into the market in 2003 and is now the dominant ceramic bearing on the market.

The primary focus of this chapter is ZTA which has changed the world of orthopedic bearings, and today it has almost completely supplanted pure-alumina bearings from the market. The story of the hip replacement in the 21st century is the story of ZTA.

7.1 Zirconia-toughened alumina

7.1.1 Evolution of ZTA

ZTA is the equal most important new commercial development in alumina technology in the 21st century, along with the 1000+ channel-number alumina/platinum bionic micro-feedthrough featured in Chapter 10.

While pure alumina was steadily improving and performing well in the orthopedics-bearing market in the late 20th century, much of the excitement in the ceramic orthopedic realm in this era was focused on the exciting "ceramic steel" zirconia concept. In relation to zirconia toughening and orthopedics, the 1970s was the era of zirconia discovery, the 1980s the era of intensive global zirconia science, and the 1990s the era of intensive global zirconia commercialization. The 2000s began with the zirconia disaster of 2001 and then became the era of global alumina orthopedic bearing commercialization. This book was written in 2018, near the end of the 2010s, a decade which has been the era of ZTA market dominance over alumina in the orthopedic bearings market, and also the decade in which alumina/ZTA bearings overtook Metal-on-Polyethylene bearings as the market dominant hip replacement bearings, in about the year 2015. They now represent 55% of the market.

The failure of zirconia bearings in orthopedics was something of a setback for bioceramics researchers at the turn of the century (Chapter 6, Section 6.4.1.5). However, as the saying goes, every cloud has a silver lining, and in this cloud there were two silver linings:

1. the extremely low breakage rate of contemporary third-generation CeramTec BIOLOX*forte* pure-alumina hip bearings of 0.004% (1 in 25,000);
2. BIOLOX*delta* ZTA, a derivative of zirconia, that revolutionized alumina hip bearings in the last decade just as dramatically as did pure alumina in the 1970–90s. ZTA breakage rates are now <0.001% (below 1 in 100,000).

Therefore, from the ashes of the zirconia disaster, rose the Phoenix: ZTA as the pink BIOLOX*delta* femoral heads and acetabular inserts manufactured by CeramTec (Fig. 7.1).

ZTA seemed to appear out of nowhere in the mid-2000s after CeramTec launched their pink ZTA hip bearings on the market as BIOLOX*delta* in 2003, just 2 years after the Saint-Gobain zirconia hip bearings disaster. However, while BIOLOX*delta* ZTA *seemed* to appear out of nowhere, this was only a perception, not the reality. The reality was that ever since 1976 and up until the turn of the century, just prior to the 2001 zirconia disaster, ZTA research had been slowly evolving although ZTA did not evolve to its contemporary manifestation of 20% tetragonal-zirconia microparticles in alumina until around the turn of the century. ZTA prior to 2003 was seen as a mere scientific curiosity by most researchers and was not a particularly active research field. Fortunately ZTA was not seen this way by CeramTec (Germany). They were engaged in intense commercial ZTA development in this era. In retrospect, CeramTec are to be commended for such foresight.

The author was involved in ZTA research in this era [278–281] and presented conference papers on ZTA in Munich (Germany) in 2001 [279], in Osaka and Tokyo (Japan) in 2001 [280], and Italy in 2002 [281] and continued to be involved in the field up until recently with a 2015 paper [116] on gelcasting of ZTA. The perception of ZTA in 2001 was that pure alumina was considered the functional option, with the main cloud over it being memories of high alumina breakage rates in the 1970s still pervasive. Zirconia was seen as the far superior option that was fast rising to dominate the market, but had something of a cloud over it with an emerging body of evidence of

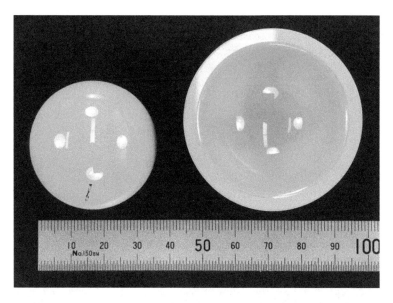

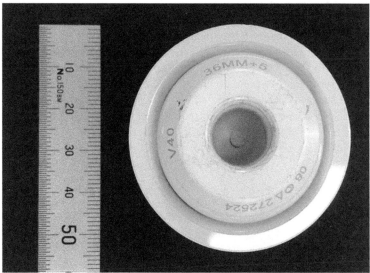

Fig. 7.1 CeramTec BIOLOX*delta*.ZTA hip bearings
Exhibit provided courtesy of Professor Bill Walter. Specialist Orthopedic Group. Wollstonecraft NSW.

problematic aging in vivo. ZTA was seen as an interesting, and somewhat novel, pure research topic: in essence, a promising idea with no obvious commercial demand driving its development. After 2003, the status of alumina did not change but ZTA then took the mantle lost by zirconia in 2001/2002: the new potentially superior (to pure alumina) commercial ceramic hip bearing. As of today, 15 years on, ZTA has indeed proven to be superior to alumina and it has come to dominate the market. Reportedly, well in excess of 14 million of these CeramTec pure-alumina and ZTA components (heads/liners) have now been implanted [29], the majority of these being ZTA.

The concept of zirconia as an additive to alumina dates back to the 1950s, albeit the original purpose was as a sintering aid [282]. However, this was zirconia in solid solution in an alumina matrix. ZTA as the ceramic matrix composite that we know it as today involves discrete zirconia particles as a distinct toughening phase in an alumina matrix. This concept did not arise until 1976.

Nils Claussen is the originator of the ZTA concept, with his pioneering 1976 paper [283]. However, the original Claussen ZTA paper of 1976 was not ZTA as we know it today. Claussen's 1976 paper involved monoclinic zirconia as the toughening phase. It is significant to note that Claussen's 1976 paper was published just a year after the "Ceramic Steel" zirconia paper of Garvie [270]. In 1978 Claussen followed up with the first paper to produce ZTA with PSZ additions to the alumina [284]. This was the next step toward the 20% tetragonal zirconia microparticle in alumina concept that we know today as ZTA. In the 1980s ZTA as a tetragonal-ZTA began to emerge as a research concept with the early review of Wang in 1989 [285] summarizing the gradually emerging science of ZTA in the 1980s.

The 1990s was a period of pure research and development in alumina-zirconia composites [286–289]. In the 1990s, there was research activity in both ZTA, as well as ATZ. ATZ is alumina toughened zirconia, typically comprising approximately 20% alumina and 80% YTZP [286,290]. One of the first documented ZTA hip-bearing development research reports was by Salomoni in 1994 [288], who produced prototype femoral heads of TZP, AZT, and ZTA (20% 3Y-TZP, 80% alumina) by pressure slip casting, with mechanical characterization. In 1999, one of the first tribology studies of ATZ was published by Affatato [289] but ZTA in the classic 20% zirconia configuration was not even considered in this study.

7.1.2 Is ZTA susceptible to hydrothermal aging?

Given the dramatic demise of pure zirconia hip bearings due to hydrothermal aging of tetragonal zirconia, the critical question that needs to be answered, in retrospect, is why are ZTA hip bearings not susceptible to aging when they contain tetragonal zirconia in a hydrothermal environment. There are three key reasons:

- higher matrix modulus
- zirconia encasement/isolation
- percolation threshold

Higher matrix modulus: The Young's modulus of zirconia (200 GPa) is about half that of alumina (380 GPa) which means that the hydrostatic load for crack pinching, associated with the 3%–5% volume expansion of tetragonal-monoclinic transition, will theoretically be twice as high for a tetragonal zirconia particle in an alumina matrix than in a zirconia ceramic. Moreover, the 3%–5% volume expansion of the tetragonal-monoclinic transformation also results in ~7% shear strain [291]. This enhanced matrix modulus factor significantly offsets the fact that only ~20% of the ZTA composite ceramic is tetragonal zirconia, unlike tetragonal zirconia precipitate (TZP) where the ceramic is nominally 100% tetragonal zirconia. In effect, the hydrostatic crack-pinching force produced by the tetragonal-monoclinic transition is leveraged twofold in an alumina matrix. Nonetheless, as shown in Table 6.1, ZTA has at best three times higher fracture energy than alumina, while Y-TZP has over 10 times higher fracture energy than alumina. Thus fourfold to fivefold dilution in tetragonal zirconia content translates to only threefold reduction in fracture energy due to the higher Young's modulus effect. The higher matrix modulus also represents a significant barrier to monoclinic transition by hydrothermal aging. Essentially ZTA has twice the mechanical resistance to the tetragonal-monoclinic transition by hydrothermal aging than TZP pure tetragonal zirconia.

Moreover, as discussed in Chapter 6 (Section 6.4.1.4) even in TZP pure zirconia, aging was generally a manageable problem. Over 600,000 Y-TZP zirconia femoral heads were implanted from the late 1980s until the early 2000s [277], TZP zirconia on the whole had sufficient aging hydrothermal resistance in vivo for long-term use. The 1998 faulty Saint-Gobain batch of Prozyr TZP zirconia femoral heads, which produced 400 zirconia femoral head failures in a very short time period in 2001, had accelerated aging due to insufficient densification as a result of a production problem. Other than that specific disaster, TZP zirconia was generally a viable femoral head material. However, the Prozyr disaster of 2001 destroyed the market credibility of zirconia as a hip bearing.

Encasement/isolation of the tetragonal zirconia crystals: As discussed in Chapter 6 (Section 6.4.1.5), tetragonal-monoclinic transformation is something of a domino effect in that once one tetragonal crystal has hydrothermally transformed, it places stress on its neighboring grains, increasing the risk of their hydrothermal transformation. This phenomenon is not possible when each tetragonal zirconia grain is encased in a matrix of alumina and has little or no direct zirconia-zirconia grain boundaries. This phenomenon was discovered in the late 1990s and was essentially the enabling technology for ZTA as a hip-bearing biomaterial. When the tetragonal zirconia particles are isolated in a nonzirconia matrix, they do not share grain boundaries. Thus a chain effect of transformation of the sort experienced by TZP zirconia ceramics (hydrothermal aging) is prevented due to the compressive forces exerted by the alumina matrix on the isolated tetragonal zirconia grains. This is the key reason why ZTA is not susceptible to hydrothermal aging and also the reason why AZT (20% alumina 80% zirconia) is not a viable hip-bearing material, as this comprises alumina particles in a zirconia matrix.

Moreover, the isolation/encasement factor also prevents water radicals penetrating the lattice and hydrothermally transforming other zirconia grains, as occurs with pure zirconia in vivo, but cannot easily occur with ZTA [292].

Critical zirconia percolation threshold: As mentioned before, tetragonal-monoclinic transformation spreads from zirconia grain to grain, through direct contact, as a domino effect. This domino effect is related to a percolation threshold, which has been established to be 16 vol%, which corresponds to 21.3 wt% zirconia in the alumina matrix [293,294]. The zirconia content of both CeramTec and Kyocera orthopedic ZTA products is aligned with this percolation threshold. This encasement theory, and percolation threshold concept, was validated by Pezzotti [295] who conducted comparative studies of hydrothermal aging of ZTA and zirconia. These demonstrated that surface hydrothermal aging occurs in ZTA, indeed it was found to occur just as readily in ZTA as in pure zirconia. However, unlike for zirconia, hydrothermal aging did not penetrate below the surface of ZTA even after >50 h in hydrothermal conditions. Therefore, because of the inability for the aging to penetrate below the surface of the ZTA, it had no effect on surface roughness, unlike for pure zirconia for which penetration of hydrothermal aging deep below the surface creates severe surface roughening.

Some traces of aging in ZTA had been reported in ZTA studies in the early 2000s, but they were limited or negligible [293,296,297]. ZTA is not immune to surface aging, it is just that the aging cannot penetrate into the bulk of the ZTA ceramic. It remains a surface phenomenon on a $\sim$1 μm scale, which is insignificant in practice.

7.1.3 Alumina-toughened zirconia

While ZTA is typically 20% zirconia/80% alumina, ATZ is the reverse: 80% zirconia/ 20% alumina. Thus it is much more susceptible to hydrothermal aging than ZTA. ATZ products have been developed, including Bio-Hips (Switzerland) and Ceramys (Switzerland) [277,294]. However, neither product has made a significant commercial impact. ATZ is in much the same category as zirconia: a technology under a cloud with no real future.

By the early 2000s, with the problems of zirconia now well-known, ZTA came to be seen as the superior alternative to AZT, because ZTA was seen as potentially immune to zirconia-aging failure due to the fact that the alumina matrix could constrain the zirconia particles, and thereby retain them in the tetragonal state under hydrothermal conditions. In this regard, the fundamental 2002 ZTA study of de Aza et al. [291] was seen as an important step forward for ZTA development.

7.1.4 Biomedical-grade ZTA

CeramTec ZTA: The CeramTec ZTA orthopedic bearing known as BIOLOX*delta* was first commercialized by CeramTec in 2003, 7 years after CE-Mark approval (European Union) for BIOLOX*forte* pure alumina hip bearings and the same year as for FDA approval (USA) for BIOLOX*forte* pure alumina hip bearings. CeramTec BIOLOX *delta* ZTA commercial hip-bearing components are pink in color (due to chromia doping). Chromia is a well-known dopant in alumina for enhancing its hardness [298]. However, chromia also reportedly has a hydrothermal-aging suppression effect in ZTA [299]. In its early manifestation CeramTec ZTA comprised about 75% alumina, 24% tetragonal zirconia, and $\sim$1% mixed oxides to create the pink color [300]. Today it comprises 76 wt% alumina (>2 μm grain size), and 22.5 wt% zirconia ($\sim$300 nm grain size). In addition to tetragonal zirconia nanoparticle toughening, CeramTec ZTA also has a second toughening phase comprising about 3 vol% strontium aluminate platelets (up to 3 μm in length) nucleated in situ, which provide toughening by crack bridging and crack deflection [301]. Thus BIOLOX® *delta* is essentially an enhanced ZTA composite. The yttria content of the zirconia is also lower than the usual 3 mol% of 3Y-TZP. It is around 1.3%, because of the stabilizing effect of the alumina matrix phase in which the zirconia crystals are embedded [292].

Kyocera ZTA: Kyocera (Japan) launched a rival ZTA alumina-bearing material 8 years later in 2011, called AZ209. It has 19 wt% zirconia and 79 wt% alumina [294]. It also has a similar content of strontium aluminate platelets to BIOLOX *delta*. It would appear to be an attempt at duplicating the CeramTec product. It is yet to make a significant global impact.

Rationale of the microstructure: The zirconia content is as high as possible without exceeding the percolation threshold. Both ZTA products are right on the percolation threshold, within the constraints of accurate determination of the threshold and precise zirconia content. The Kyocera formula is very slightly lower in zirconia than the CeramTec ZTA. Both the CeramTec ZTA and the Kyocera ZTA have about 3 vol% strontium aluminate platelets. Obviously the toughening effect of the generic technology of platelet toughening by crack-bridging and crack-deflection are well known and have been for decades. The question remains though, why is 3% strontium aluminate platelets a helpful addition? Why not rely on the zirconia alone? Kuntz [302] has suggested that incorporating multiple reinforcing mechanisms makes the ZTA more reliable because it is more effective at deflecting cracks close to the surface. There is little in the way of in-depth studies in the literature on what the microstructural and toughening benefits may be from these platelets in the specific case of CeramTec or Kyocera ZTA. So, whether or not this is the reason CeramTec came up with this unusual innovation, and the reason why Kyocera duplicated it in their rival ZTA, there is no doubt that this complex ZTA innovation derives from several decades of German ceramic engineering expertise in the alumina orthopedic field through Feldmuhle/CeramTec. It clearly embodies a great deal of well-honed ceramic engineering knowhow.

CeramTec marketing documents suggest that the platelets enhance the toughening mechanism of the zirconia by preventing microcracks in the material from propagating by dissipating the crack energy [29]. Given the 40-year-track record of Feldmuhle/CeramTec in alumina hip bearings, the 15-year-track record of excellence for BIOLOX*delta*, and the continued global dominance of the orthopedic ceramic bearings market, it would seem that CeramTec have developed a winning formula based on their pre-eminent knowledge of the technology. In time the literature may contain independent micromechanical studies elucidating the mechanisms in comprehensive detail.

In addition, ZTA is characterized by a finer grain size than pure alumina, because of the presence of the zirconia crystals [291]. As discussed in Chapter 12 (Section 12.1), finer grain size gives enhanced wear resistance.

7.1.5 Properties of biomedical ZTA

Fig. 7.2 demonstrates the iterative improvements in CeramTec alumina, from generation 1 up to the ZTA product of the 21st century. This figure makes it very clear what a substantial improvement ZTA was in strength over its predecessors.

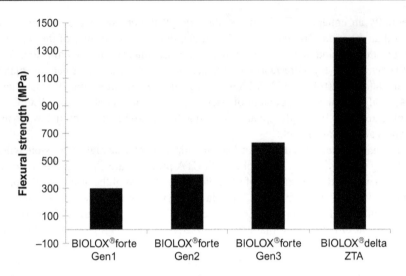

Fig. 7.2 Flexural strength of the CeramTec alumina evolutionary developments.

In general, ZTA ceramic composites need to be optimized with respect to the following parameters:

- tetragonal zirconia content
- alumina grain size

The hardness of zirconia is slightly less than pure alumina and biomedical ZTA is very slightly less hard than pure alumina. CeramTec reported strength values for their alumina and ZTA products are significantly higher than CoorsTek data of Table 6.1. This is shown in Table 7.1. [306]. Given that hardness and toughness are equally important for wear resistance (Chapter 12, Section 12.1) the small loss in hardness for ZTA is more than compensated for by its threefold fracture energy advantage over alumina (Chapter 6, Table 6.1) with the result that wear resistance for ZTA is significantly superior to alumina. However, the big advantage of ZTA is its very high strength and toughness advantage, which translates into a fivefold lower breakage rate in vivo than third-generation CeramTec pure alumina (BIOLOXforte Gen3).

ZTA in its contemporary manifestation typically contains around 20% tetragonal zirconia microparticles in a matrix of around 80% alumina. As Table 7.1 shows, a key difference between alumina and ZTA is flexural strength. The other is toughness as shown in Table 6.1. The improvement is substantial. *Both toughness (as fracture energy) and flexural strength are approximately three times higher for ZTA than pure alumina.*

This results in two key benefits for ZTA:

- Even lower failure rate in vivo, down to below 1 in 100,000, compared to 1 in 25,000 for third-generation CeramTec pure alumina.

Table 7.1 CeramTec data for biomedical alumina and ZTA [306]

CeramTec grade	Al_2O_3 %	Flexural strength (MPa)	Youngs Modulus (GPa)	Hardness Hv (GPa)
BIOLOX*forte* Gen1	99.1–99.6	>300	380	18
BIOLOX*forte* Gen2	99.7	400	410	19
BIOLOX*forte* Gen3	>99.8	630	407	20
BIOLOX*delta* ZTA	80	1390	358	17.6

- The capability to make alumina bearings with thinner cross sections, which offers many advantages in design flexibility to minimize impingement and dislocation, and new product capabilities, such as potentially the ceramic knee and ceramic hip resurfacing implants.

As discussed in Chapter 12 (Section 12.1), toughness is just as important to wear resistance as hardness. This is why ZTA has significantly better wear resistance than pure alumina. It also has a substantially (threefold) higher flexural strength. The strength and toughness improvement is seen as its primary advantage over alumina, as this enables ZTA to be used for head diameters down to 22 mm, which is not a viable size for pure alumina. The enhanced wear resistance is also a plus, but given the extremely high wear resistance of alumina compared to polyethylene, wear resistance enhancement is not such an important advantage.

A common test for ceramic femoral heads is the burst test, which is a measure of the load required to fracture a ceramic head via an axial load on the head while it is fitted to a tapered stem. A 2001 study [272] demonstrated that the burst strength of autoclaved CeramTec ZTA was double that of the pure alumina, as shown in Fig. 7.3. This demonstrates that, even 2 years before commercialization, the CeramTec ZTA had a very high mechanical strength, even after autoclaving. If ever hydrothermal aging is going to be induced, autoclaving will induce it, given the harsh hydrothermal conditions involved.

7.1.6 ZTA conclusion

It is now 15 years since ZTA commercialization, which is still not long enough for a full picture to have emerged from clinical studies and especially retrieval studies. Some recent clinical reviews have been published such as that of Kurtz in 2014 [294], but it will likely be into the 2020s before a full picture emerges. In general, ZTA has been very successful and has seen huge market uptake. Many millions of

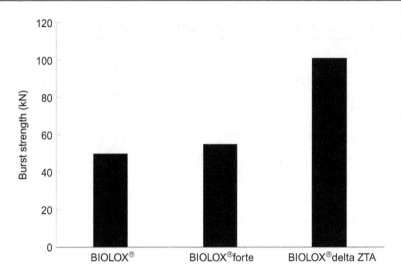

Fig. 7.3 Autoclave ZTA and burst strength. Note the forte is pure alumina and the delta is ZTA [272].

ZTA implantations later, we now know with a high degree of certainty that tetragonal zirconia embedded in an alumina matrix, at around 16 vol% zirconia, with 3 vol% strontium alumina platelets, and chromia doping, does not suffer an aging problem. The strength of CeramTec BIOLOX*delta* does not decrease in vivo, even when repeatedly steam sterilized. In addition to this, it is now known that CeO-stabilized tetragonal zirconia can exhibit toughness up to $20\,\mathrm{MPa\,m}^{1/2}$ and no significant aging in the lifetime of an implant [277]. The ceria discovery is yet to be commercially exploited.

Ever since commercialization of the CeramTec BIOLOX*delta* ZTA hip bearings in 2003, the CeramTec ZTA product has been growing in market share at the expense of the lower strength pure-alumina BIOLOX*delta* bearings, reaching the 50:50 point around the year 2009, and today ZTA has almost totally supplanted pure alumina from the market. The strength advantage is clearly conveyed in Figs. 7.2 and 7.3. Moreover, as Table 6.1 shows, the toughness of the ZTA hip bearings is three times higher in terms of relative fracture energy than the pure alumina CeramTec hip bearings, with a much lower failure rate.

7.2 Wear of orthopedic bearings

7.2.1 Tribological principles for orthopedic bearings

Tribology is the study of friction, wear, and lubrication. A discussion of the tribology of materials, and specifically the tribology of alumina ceramics, is essential for an understanding of the reasons underlying alumina's growing role since the 1960s as the world's leading wear-resistant industrial ceramic (Chapter 12), and its more recent role since 1994 as a commercial wear-resistant orthopedic-bearing material.

Wear testing in the laboratory is conducted in a number of ways, including:

- sliding abrasion simulation: pin-on-disk, pin-on wheel
- slurry erosion: rotating pin in a slurry
- impingement: grit blasting

In the case of orthopedic bearings, wear testing is conducted in joint simulators, which typically subject the bearing to 5 million cycles, or so, while immersed in a simulated body fluid, and subjected to the appropriate biomechanical loads. On average a human hip is subjected to 5 million steps a year, with loads that commonly exceed five times body weight (descending stairs, brisk walk downhill). In the case of running it can exceed 10 times body weight.

There are many different modes of wear that can be encountered in real-world scenarios. The following nonexhaustive list gives some indication of the range of wear scenarios that can be encountered:

Sliding wear:
- sliding abrasion between surfaces
- fixed abrasive particles, for example, grinding using a grinding wheel
- rolling abrasive particles, for example, rounded grit particles between bearing surfaces

Impact wear:
- micro-impact (Impingement)—impact by small hard objects, for example, ore particles in a moving slurry
- macro-impact, that is, impact by large objects, such as large chunks of ore

Other types of wear:
- adhesive wear, that is, when the "abraded" material is removed by adhering to the "abrading" surface. This occurs in the case where the adhesive force between two articulating surfaces exceeds the inherent material properties of either material
- chemical wear, either as pure corrosion, or as one or more of the above mechanisms corrosion enhanced
- third-body wear, this is when a hard particle becomes embedded between two articulating surfaces. This is a common scenario when a ceramic is articulating against another surface, and a fragment of the ceramic spalls off its surface and becomes lodged between the articulating metal and ceramic surfaces and plows the surfaces of the ceramic and its articulation partner

With regard to hip replacements, the mechanisms of greatest importance are

Number 1 importance: abrasive wear—CoCr-on-polyethylene
Number 2 importance: adhesive wear—alumina-on-polyethylene; CoCr-on-polyethylene
Number 3 importance: sliding Wear—Substantial for CoCr-on-CoCr, extremely low for alumina-on-alumina
Number 4 importance: third-body wear (fragments of bone cement or fractured bearing).

Wear of hip replacements is inevitable, and this has two effects: widening of the radius of the acetabular liner, and deposition of wear particles into the joint cavity. The much more serious problem is wear particles. Here, there are two key issues of importance: rate of wear and tissue response to the wear particles.

Rate of wear for orthopedic joints in joint simulators is generally measured in mm/yr. The following wear couples are currently on the market:

- ~45% Market share—CoCr-on-polyethylene (shrinking market)
- <2% Market share—CoCr-on-CoCr (almost obsolete)
- ~40% Market share—alumina-on-polyethylene (growing market)
- ~15% Market share—alumina-on-alumina (growing market)

Note that when we say "alumina" in this context we mean either pure alumina or ZTA.

The wear rate of alumina-on-alumina is approximately two to three orders of magnitude lower than CoCr-on-polyethylene, and one or two orders of magnitude lower than CoCr-on-CoCr [303–305]. Alumina-on-polyethylene is from 2 to 10 times lower than alumina-on-polyethylene [307]. Or to put it another way, *1 day of CoCr-on-polyethylene equates to more than a year of alumina-on-alumina*. Moreover, the Alumina wear particles are benign. Polyethylene wear particles are also benign, but they are produced in such large quantities (500,000 a day [185]) that they act as an irritant. Only in rare cases of extreme edge loading do alumina wear particles reach levels required to be an irritant. CoCr particles are potentially toxic since the tissue response to wear particles follows the progression:

- Metal: toxic above a certain level
- Polyethylene: benign (irritant in the standard case of high levels)
- Alumina: benign (irritant in the rare case of high levels)

Finally, when alumina or CoCr is articulating against polyethylene, the great majority of wear particles are polyethylene.

Given the above data, the following conclusions can be drawn:

(1) Total obsolescence of metal-on-metal total hip replacements is appropriate. Indeed they are almost totally obsolete today.

(2) Given the importance of resurfacing, the risks of metal wear particles, and the fact that all resurfacing implants on the market today are metal-on-metal, there is a strong imperative to develop an alumina-based resurfacing implant [210]. None currently exists although in the 1970s and 1980s an experimental one (Wagner see Chapter 6, Section 6.4.1.3) was trialed, and some are under development currently such as the Recerf (Australia/UK) and the EnDuRE (UK). The challenge is to develop a reliable alumina acetabular component which needs to be very thin, and therefore represents a higher fracture risk than current thicker acetabular components.

(3) Polyethylene wear rates are about 2–10 times lower with alumina femoral heads than CoCr, but they are still significant. Polyethylene particles are an irritant, implicated as a cause of osteolysis (see Section 7.3.2), a leading driver for revision surgery.

(4) With the use of alumina-on-alumina bearings, the wear particle problem is virtually eliminated.

Wear rate with polyethylene liners is, to a large extent, dependent on the quality of the polyethylene. Cross-linked polyethylene technology was greatly improved by around the year 2000 due to the development of cross-linking technology (Section 6.3.7.1). Today, noncross-linked polyethylene is obsolete although some patients still carry it in their joints due to surgery pre-2000. Nonetheless, polyethylene wear rate, though significantly improved since 2000, is still substantial.

Secondly, while metal-on-metal has a lower wear rate than metal-on-polyethylene, the metal wear particles tend to be smaller than the polyethylene particles, and therefore the metal particle number tends to be similarly high for metal-on-metal systems compared to metal-on-polyethylene systems. This is a problem given that CoCr is the dominant metal used in articulation, and both cobalt and chrome are potentially toxic when released into the joint cavity or systemically.

While it is intuitively obvious that alumina-on-alumina wear rates are greatly superior to metal or polyethylene wear rates, and that ZTA-on-ZTA is superior to alumina-on-alumina, perhaps not so obvious is the fact that for a given polyethylene, an alumina femoral head causes a lower polyethylene wear rate than a CoCr femoral head. Early studies found that an order of magnitude is lower [307]. Today it is considered to be in the range of 2–10 times better, and closer to 2 with cross-linked polyethylene. The reasons for this are discussed in Section 7.2.7.

Alumina tribology is discussed in great detail in Section 12.1. However, the discussion in Chapter 12 is pertinent to the mining industry and other industrial applications with extreme abrasive wear conditions, including alumina-alumina industrial bearings, sliding abrasion of mineral particles on alumina, and impact (impingement) of mineral particles on alumina. These wear scenarios do not exist in orthopedic joints other than the simple alumina-alumina sliding wear.

Wear tests attempt to simulate the complex reality of the real-world wear environment, to varying degrees of success. These tests are more meaningful in giving relative wear performance than quantitative absolute wear performance. Therefore, the general principle followed is that a control wear couple of standard formulation (such as CoCr-polyethylene) should be tested alongside the test material so that the relative wear performance of a test material can be meaningfully evaluated.

7.2.2 Wear testing of alumina hips in the developmental era

In the early developmental period of alumina hip bearings, German researchers rapidly became the global authorities in alumina-bearing technology for orthopedics. Moreover, Germany (population 80 million) currently implants the most alumina hips, as a percentage market share, in the world, and has the second highest (after Switzerland) rate of hip replacements per head of population.

Consequently, most of the key biomaterials paradigms and theories in alumina hip bearings were developed by German researchers such as Erhard Dorre of Feldmuhle (the precursor company to CeramTec), beginning in the late 1970s. Indeed the 1970s and 1980s were a fertile period for the science of alumina hip bearings. In 1975, one of the key paradigms in alumina hip bearings was published as part of the first documented joint simulator tests for alumina-alumina joints, with metal-polyethylene as the reference couple [308]. Test conditions were as follows:

- Load: 5 kN
- Wear cycle rate: 1 Hz
- Environment: ringers solution
- Cycles: 10 million

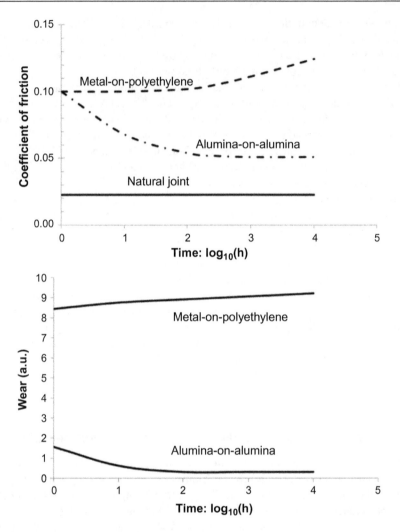

Fig. 7.4 Long-term tribology trends for alumina-on-alumina bearings compared to metal-on-polyethylene [308].

From this test was published the classic alumina wear testing chart, and the associated (now well-known) principles of alumina tribology in orthopedics, as shown in Fig. 7.4. Specifically:

- Initially the alumina-alumina wear rate is about 1/20 of metal-polyethylene.
- Over time the alumina-alumina wear rate decreases and approaches the zero asymptote, while the metal-polyethylene wear rate gradually rises.
- Friction is initially equal for alumina-alumina and metal-polyethylene.
- Over time alumina-alumina friction decreases to an asymptote slightly above the friction of a natural synovial joint, while metal-polyethylene friction gradually rises.

The explanation for these findings was that increasing wear cycles enhance the alumina surface through polishing, but conversely, increasing wear cycles cause the polyethylene surface to deteriorate and become rougher [308]. This principle was widely reported by other researchers, for example, McKellop [309]. This classic explanation remains standard in biomaterials textbooks today. Other well-known principles of alumina-on-alumina orthopedic bearings established by German researchers in the developmental are

(1) The linear correlation between load and frictional torque [310] (Fig. 7.5).
(2) The step function correlation between sphericity deviation and frictional torque, in which sphericity deviations below 15 µm have no effect on torque. Sphericity deviations above 15 µm dramatically increase frictional torque, almost as a step function, regardless of femoral head diameter in the 32–38 mm diameter range [310] (Fig. 7.5).

As a result, improvements in sphericity were introduced in the late 1970s by producing matched alumina ball/socket pairs.

7.2.3 Alumina-on-alumina hip wear rates

In the early days, wear debris was difficult to retrieve using hip simulators, and precise study of wear debris did not develop until the 1990s [311]. The wear rate of alumina-on-alumina bearings began to be measured with precision by the 1990s, and it was generally found that they were around 0.05–0.10 mm/yr, and at best 0.03 mm/yr. The latest generation ZTA-ZTA is below 0.001 mm/yr. In the late 20th century, in the era before cross-linked polyethylene, these wear rates were approximately 1000 times lower than for metal-on-polyethylene and 40 times lower than for metal-on-metal [303–305].

For normal concentric loading, the wear rate of alumina-on-alumina bearings are therefore so low as to be insignificant. Given that the wear particles are benign, and the amount is so small, wear-particle-induced osteolysis (see Section 7.3.2) is eliminated, except in rare cases of extreme edge loading. Alumina wear rates become significant in edge loading, which happens in the dysfunctional scenario in which the alumina femoral head articulates with the edge of the alumina acetabular liner. There are two ways of identifying this after explanation.

(1) Wear-particle-induced osteolysis. This should not be observed in alumina-on-alumina implants, and when observed, it is commonly an indication of edge wear.
(2) Stripe wear on the alumina femoral head, that is, when a long narrow area of damage, resembling a stripe, can be seen macroscopically on the retrieved alumina head.

Hip simulator studies have shown that BIOLOX *delta* (ZTA) has better wear resistance under edge wear than BIOLOX *forte* pure alumina [312].

A retrieval study of stripe wear in 22 alumina heads has shown that stripe depth, relative to the spherical head surface, was as high as 67 µm, but was typically <10 µm [313].

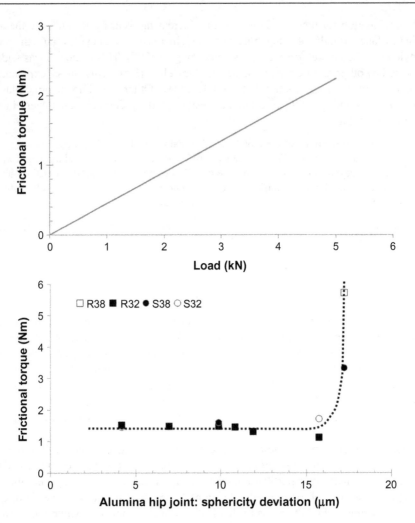

Fig. 7.5 Top: Frictional torque as a function of normal load for alumina-on-alumina bearings. Bottom: frictional torque as a function of deviation from joint sphericity for alumina-on-alumina bearings of two diameters, and immersed in one of two fluid types [310]. R38—38 mm diameter in Ringer's solution; R32—32 mm diameter in Ringer's solution; S38—38 mm diameter in synovial fluid; S32—32 mm diameter in synovial fluid.

7.2.4 Wear resistance of ZTA

As discussed in the detailed tribology overview in Chapter 12 (Section 12.1), the two most important parameters for high wear resistance are high hardness and high toughness. Also very important are fine grain size, low or zero porosity, and chemical inertness. Zirconia is not as hard as alumina but it has comparable wear resistance owing to its greatly enhanced toughness. The ultimate wear resistant oxide ceramic is one that combines the hardness of alumina with the toughness of zirconia—ZTA.

A 2009 study of ZTA wear compared three wear couples for the standard 5 million cycles, which is equivalent to 1 year of normal human activity. The relative wear rates were [314] as follows:

Wear couple	Relative wear resistance[a]
Alumina:alumina	1
ZTA:alumina	3
ZTA:ZTA	6–12

[a]Relative to alumina:alumina.

While these were simulator studies, and not necessarily a guarantee of relative in vivo performance, the data certainly accord with the tribological principle that the material with the combination of the highest toughness and highest hardness (ZTA) will have the best wear resistance (Chapter 12, Section 12.1).

7.2.5 Polyethylene wear mechanisms

Polyethylene wear mechanisms are important to alumina given that the majority of alumina hips today involve alumina-on-polyethylene, which has a wear rate significantly lower than CoCr-on-polyethylene. There are three main modes of wear for polyethylene:

- Adhesive wear
- Abrasive wear
- Fatigue wear

Adhesive wear occurs when the forces of attraction between two opposing wear surfaces exceed the inherent material properties of either surface. In orthopedics, adhesive wear primarily occurs with polyethylene. It occurs when small portions of the polyethylene surface adhere to the opposing metal surface. Adhesive wear can also happen with alumina-polyethylene, but alumina surfaces tend to be much smoother than CoCr surfaces. The adhesive wear of polyethylene results in pits and voids on the surface that are commonly very small. The adhesive wear relates to the plastic flow behavior of polyethylene. The mechanism involves local accumulation of plastic strain until a critical or limiting strain is reached, followed by the loss of surface material via small adhesive wear particles [315–317]. A plasticity-induced damage layer develops at the polyethylene surface which results in the permanent reorientation of the crystal lamellae in the polyethylene surface [318,319]. A wide range of studies are reported in the literature on adhesive wear of polyethylene, beyond the scope of this book.

Abrasive wear also occurs with polyethylene, but it is primarily caused by CoCr, not alumina. This is generally caused by surface plowing from sharp surface protrusions on the opposing surface. CoCr surfaces tend to have hard particle protrusions, such as carbide precipitates. Alumina surfaces tend to be very smooth down to the submicron level. This is one of the key reasons why alumina-polyethylene has a much lower wear rate than CoCr-polyethylene.

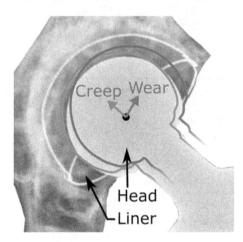

Fig. 7.6 Creep zone and wear zone in a polyethylene acetabular liner.
Image Author Mikael Haggstrom [323]. Reproduced from Wikipedia (Public Domain Image Library).

Fatigue wear also commonly occurs with polyethylene. It occurs when surface and subsurface cyclic shear stresses and strains exceed the fatigue limit for polyethylene, leading to subsurface delamination and cracking, and ultimately spallation of polyethylene wear particles. Fatigue wear relates to the plastic flow properties of polyethylene such as yield stress and ultimate stress [320] and the presence of microscopic voids in polyethylene [254,321,322].

Adhesive and abrasive wear of polyethylene are common in hip bearings, but fatigue wear is less common in hips and more common in knees. Also, human implants experience prolonged periods of inactivity, for example during sleep, which can result in creep and relaxation, which can influence wear behavior, in ways that are not encountered in simulators. The differences between creep and wear on the polyethylene liner are shown in Fig. 7.6. While a detailed discussion of polyethylene wear is beyond the scope of this book, the basic principle is that all three of these wear mechanisms are less severe with alumina femoral heads (rather than CoCr femoral heads) because alumina surfaces, although harder than CoCr, tend to be much smoother, down to the submicron level. The hardness issue is not relevant given that both alumina and CoCr are several orders of magnitude harder than polyethylene. The key differentiating factor is therefore surface roughness on the micron scale.

The hip replacement subjects the polyethylene acetabular liner to severe wear conditions. On average a patient subjects the joint to 5 million steps a year, with loads that commonly exceed 5 times body weight (descending stairs, brisk walk downhill). In the case of running it can exceed 10 times body weight. Moreover the smaller the diameter of the femoral head, the more the load is concentrated on the polyethylene.

In the 1990s the commercial polyethylene acetabular liners were very brittle and wear rates went down significantly because the industry made the switch from ethylene oxide sterilization to gamma sterilization. Gamma sterilization in air was oxidizing polyethylene and making it very brittle, which substantially reduced the wear properties of polyethylene. During the 1990s, it was gradually discovered that if polyethylene was gamma irradiated in a vacuum or inert atmosphere (nitrogen), this had the beneficial effect of cross-linking the polyethylene instead of oxidizing it, which substantially increased the wear rate of the polyethylene. Experienced surgeons have observed that the wear rate of polyethylene acetabular liners has improved approximately 10-fold since the introduction of cross-linking [210].

7.2.6 Alumina-on-polyethylene wear rates

A normal functional CoCr-on-polyethylene hip-bearing sheds about 500,000 polyethylene particles per step [185]. Given that the average person walks 5 million steps a year, this corresponds to about 2 trillion polyethylene wear particles a year. Polyethylene is an irritant, not a toxin, and therefore this situation can be tolerated for a few years. However, frequently the polyethylene wear debris initiates osteolysis of the bone, followed by implant loosening. This is a leading cause of implant failure. Thus polyethylene wear rate is a critical issue. The size, shape, and quantity of polyethylene wear debris have an effect, and the prevailing view is that osteolysis only occurs above a threshold level of polyethylene wear debris. Hence the switch to alumina-on-polyethylene bearings with their greatly reduced wear rate. So the question is: how much lower is the wear rate of alumina-on-polyethylene than CoCr-on-polyethylene?

Early alumina-on-polyethylene wear tests by the German researchers in the late 1970s showed that for a given polyethylene surface, the wear rate was 10-fold lower for alumina-on-polyethylene than for CoCr-on-polyethylene [307], and another later study by the same researchers reported 20-fold lower [324] In the early 1990s, a hip simulator study (water lubricant) of alumina heads compared to CoCr and stainless steel femoral heads found the alumina to be 32- to 85-fold lower [325]. Another later study in the year 2000 by different researchers reported wear rates of only 2–4 times lower for alumina-on-polyethylene, with a theory about the role of the lubricant, that is, that wear rates in simulators differed, depending on whether the lubricant was water or blood serum [326].

A wide range of values is reported in the literature for the relative wear rate of advantage of alumina-on-polyethylene compared to CoCr-on-polyethylene. In the current era of cross-linked polyethylene, the advantage is not as marked as it once was. The reported advantage ranges from about double to about 10-fold, but generally it is close to double. It obviously depends largely on the quality of the polyethylene, and the surface roughness of the alumina or CoCr head.

7.2.7 Fretting wear

Fretting wear[1] occurs as a result of micromotion at joints, commonly the taper joint between femoral head and stem, and the acetabular liner/acetabular cup joint [327]. Generally it is only problematic in metal-metal joins, and particularly with CoCr, as titanium particles are much less problematic than CoCr. While alumina femoral

---

[1] The author supervised a PhD project in fretting wear from 2011 to 2015 which provides a detailed review on the subject [327].

heads use taper joints, they are titanium-titanium or titanium-alumina joints and therefore only involve titanium wear particles, which are benign. The same applies to alumina acetabular liners which are generally alumina-titanium taper joins.

Therefore, alumina offers the advantage of eliminating the CoCr fretting wear problem encountered in traditional titanium-stem/CoCr-head metal-on-polyethylene implants, which currently represent 45% of the world market, a market share that continues to shrink at the expense of alumina bearings.

7.2.8 Biocompatibility of alumina wear particles

Alumina is renowned today as being the most biocompatible and bioinert biomaterial in clinical use. This is because alumina is aluminum metal in its highest oxidation state. It is therefore highly corrosion resistant and highly chemically inactive.

The highest level of biocompatibility is required for biomaterials in direct blood contact. Monolithic alumina has proven to be biocompatible in the blood-contact role in blood pumps [328]. Therefore monolithic alumina can be considered biocompatible in any in vivo applications of lower criticality, such as the lower risk applications for which it is commonly used: orthopedic bearings, electrical feedthroughs for bionic implants, and dental technology. In 1998, Takami et al. [328] conducted in vitro testing of alumina, motivated by validating the biocompatibility of the alumina/polyethylene pivot bearings in a Gyro centrifugal blood pump. They concluded that the use of alumina in blood-contacting implants can be recommended due to the finding of limited protein adsorption leading to platelet adhesion. In other words, alumina is biocompatible in blood contact.

It is well known that biomaterials that are safe in the monolithic form can be harmful in the particulate form. This is for a number of reasons:

(1) Particles are mobile and can travel through the body.
(2) Particles are of a size and scale comparable to the cells of the immune system. The macrophage, in particular, whose role is to engulf foreign objects by phagocytosis, is particularly stimulated by particulate synthetic materials. This is an evolutionary response to millions of years of humans and their ancestors suffering from dirt and grit in wounds, and the role of the macrophage in managing such problems. The macrophage is also capable of forming multicellular "giant cells," which occurs in chronic inflammation situations, for example, in dacron fabric in vascular grafts.
(3) A microparticle or nanoparticle has a much higher surface area than a monolithic object, making dissolution or leaching of the small particle in the body fluids a higher risk than for the parent monolithic material. This is mainly only a problem with metal wear particles and corrosion residue from orthopedic joints. It can also be a problem with polymer particles: thermoplastic polymers can leach additives, such as plasticizers; thermosetting polymers can leach unreacted monomer and oligomer. Chemical dissolution or leaching is not a problem with alumina ceramics. Because of its extreme chemical inertness, alumina is very bioinert in both the monolithic and particulate form.
(4) A nanoparticle is the most dangerous form of a biomaterial of all as it may be sufficiently small and mobile to be able to get inside the cell itself, for example by passing through the transmembrane channels of the cell membrane.

The general in vivo response to alumina wear particles is benign, that is, fibrocystic (fibroblasts), with very few macrophages and no multiple-macrophage giant cells [329–331]. Giant cells are the red flag for osteolysis. In contrast, in situations where metallic debris is generated in alumina-alumina implants, for example, due to socket-stem impingement or implant looseness, massive macrophage responses are found, often including foreign body granuloma [226,332,333]. Numerous studies have confirmed that the rate of osteolysis (see Section 9.5.2) with alumina-alumina bearings is very low, and generally only seen with long-term loosening or joint malfunction [229,242,334–338].

Catelas [339] studied in vitro response of mouse fibroblasts to alumina, zirconia, and PE wear particles and found that cell apoptosis was dependent on the concentration of the wear particles. Kranz [340] found that macrophage response to alumina is particle size dependent, with a stronger macrophage response to nanoparticles than to microparticles.

In contrast to polyethylene, for which wear particle number is orders of magnitude higher than alumina, alumina wear particles represent a very much lower risk in clinical use than polyethylene wear particles.

The final issue regarding biocompatibility is carcinogenicity. While the carcinogenicity risk is a real risk with polymers, and especially metals, it is almost unknown with alumina. In fact there is just one paper in the scientific literature which ever raised the question of possible carcinogenicity with alumina. It was published in 1987, in the early days when alumina was not yet well understood [341]. The clinical report by Ryu reported an aggressive soft-tissue sarcoma 15 months after a patient had an uncemented ceramic total hip arthroplasty (THA). This report also noted that nothing like this had ever been seen before, and that it was unclear if a causal relationship existed. Even if there was a causal relationship, alumina was not necessarily the explanation as this hip implant had a cobalt-chrome alloy stem, with alumina femoral head and alumina acetabular cup. Thus a causal relationship may have been due to the cobalt-chrome fretting wear particles, or to the alumina wear particles, or some unrelated cause. In short, the possibility of alumina being the causal factor existed, but it was only a possibility.

No report has since been documented in the literature linking alumina to cancer. Thus 1987 is the ONLY possible case of cancer development after joint replacement involving alumina. Since that time, extensive in vivo studies have been conducted on alumina particles, both microparticles and the less biocompatible nanoparticles, with no carcinogenic effects ever observed, and certainly none confirmed. One of the most notable studies was that of Griss [230] of long-term in vivo responses to subcutaneously injected alumina powders in rats, in which no carcinogenesis was observed. In 2009, Maccauro et al. [342] reported a carcinogenicity study of ZTA and concluded that ZTA does not display any long-term carcinogenic effect.

To date, well over 10 million hip replacements with alumina bearings have been implanted, all of them releasing wear particles to some extent, and some malfunctioning implants releasing large numbers of alumina wear particles. Thousands of clinical trials have been reported. Carcinogenicity has never been raised since the 1987 paper of Ryu [341]. The evidence now seems clear that alumina implants are not associated with cancer.

7.2.9 Conclusion on wear rates

In concluding Section 7.2 on wear, it is appropriate to summarize all the findings discussed above into the key issues [303–305].

- CoCr-on-polyethylene: up to three orders of magnitude higher wear rate than alumina-on-alumina
- CoCr-on-CoCr: one order of magnitude higher wear rate than alumina-on-alumina
- CoCr-on-polyethylene: 2–10 times the wear rate of alumina-on-polyethylene, due to the greatly superior surface smoothness of alumina than CoCr. Closer to 2 with cross-linked polyethylene
- Alumina-on-alumina: approximately 5 times the wear rate of ZTA-on-ZTA

The introduction of cross-linked polyethylene has improved things significantly for polyethylene joints, as summarized in a recent review [343]:

CoCr-on-polyethylene: 0.137 mm/yr (20 years)
Alumina-on-polyethylene: 0.072 mm/yr (20 years)
CoCr-on-cross-linked polyethylene: 0.042 mm/yr (10 years)
Alumina-on-cross-linked polyethylene: 0.031 mm/yr (5 years)

7.3 Other alumina-relevant issues arising from hip replacement

There are numerous complications arising from hip replacement, indeed too many to outline in this chapter. However, four complications have specific relevance to alumina hip bearings and warrant brief discussion here:

- galvanic and fretting corrosion
- polyethylene-induced aseptic loosening and osteolysis
- dislocation and chipping
- squeak

7.3.1 Galvanic and fretting corrosion

The original purpose of developing alumina hip bearings in the 1970s by Boutin, Mittelmeier, and successors was reduced wear, low friction, enhanced biocompatibility, and the elimination of osteolysis (see Section 7.3.2) induced by polyethylene wear particles. While alumina–alumina bearings have been very successful in achieving this, in 2017 we do not see a market dominated by alumina-on-alumina bearings. Germany leads the world in ceramic hip uptake (45% CoCr-PE; 40% alumina-PE; 15% alumina-alumina), and yet even Germany favors alumina-on-PE over alumina-on-alumina.

In the face of the extraordinarily low wear rate of alumina-on-alumina, the benign nature of alumina wear particles, and the extremely low breakage rate of CeramTec alumina bearings (see Section 7.4.1), logic would dictate that by now, 100% of the market should be alumina-on alumina. The reasons why this is not the case are threefold:

- cross-linked PE wear rate is 10 times better than in the 1990s
- alumina-PE wear rate is significantly better than CoCr-PE
- surgeons are slow to adopt new technology

Looking at the broader issues, the converse is the case—the uptake is surprisingly rapid. Surgeons are diligent and take their responsibility to their patients very seriously, and also are mindful of their reputation, and the expectations of their medical liability insurance providers. Many surgeons are very difficult to convince about adopting new technology, and this is understandable. They may have very long experience with a particular implant device and a comprehensive understanding of the complexities involved in its surgical implantation and its long-term behavior in vivo. A new device, however promising it may sound, involves an element of risk simply by stepping from the known to the unknown.

So taking this perspective, the question becomes instead "why is alumina-PE more popular than alumina-alumina and why has the uptake been so rapid for alumina-PE?" There are many reasons, but the three most important are:

- alumina-PE wear rate is significantly better than CoCr-PE
- alumina femoral heads eliminate galvanic corrosion
- a surgeon who has spent decades implanting CoCr-on-polyethylene would obviously see the step to alumina-on-polyethylene as a less radical step than to alumina-on-alumina

The traditional, and previously globally dominant (pre 2015), implant is the Charnley concept in its late 20th century evolution:

Femoral section taper junction
  · titanium stem
  · CoCr femoral head
Acetabular section junction
  · titanium acetabular cup
  · PE acetabular insert

It is not uncommon with modular implants for them to not only have a neck/head taper junction, but also a detachable neck with a stem/neck taper junction. Taper junctions at the join between the femoral head and the neck of the stem (head/neck join) enable modularity which provides a number of benefits including the ability to adjust the leg length offset, and the ability to replace the bearing in the event of wear [344]. However, each of these taper junctions is a site for fretting wear.

Titanium is the optimal stem material. While alumina is the optimal femoral head material, CoCr is the less optimal (and increasingly less commonly used) femoral head material. It is for reasons of long-established use and familiarity that inferior CoCr is still widely used as the femoral head.

Thus 45% of the market in hip implant technology of today involves a CoCr alloy intimately attached to a titanium stem, through a taper junction. This is a galvanic corrosion cell: when two dissimilar metals are in electrical contact in an electrolyte environment, they form a galvanic cell, or "battery" [345,346]. The galvanic principle was discovered in 1780 by Luigi Galvani and is very well known and used in many industrial technologies, most notably batteries and electrolytic coating of metals.

Galvanic corrosion is an electrochemical process which causes dissolution of the metal [345]. The titanium stem is the cathode and the CoCr ball the anode, and therefore the corroded metal. Mechanically assisted chemical corrosion is the main cause of corrosion at these taper junctions, that is, a combination of crevice corrosion and fretting. Fretting micro-motion continuously erodes the passivated stable oxide layer, exposing bare metal to electrochemical corrosion [347–349].

Thus taper junctions, whether alumina-titanium (neck/head), titanium-titanium (neck/taper-sleeve OR neck/stem), or CoCr-titanium (neck/head OR neck/stem) always involve fretting wear of the metal taper surface, and the associated erosion of the passivation layer, exposing it to corrosion. In the case of alumina-titanium or titanium-titanium, there is still fretting-induced corrosion, even in the case of a titanium/alumina taper junction, as shown in Figs. 7.7–7.12, but titanium corrosion detritus and fretting-wear particles are benign.

In CoCr/titanium taper junctions, the corrosion is enhanced by a galvanic-cell effect, thereby releasing cobalt and chromium into the joint cavity. Both cobalt and chromium are potentially problematic metals in vivo. The cobalt is not such a big problem as it tends to become systemic and is excreted via the kidneys, whereas the chromium stays around in the joint cavity and causes metallosis, discolouration, and sometimes even necrosis of the bone [210].

The above figures illustrating these principles were produced by Biomedical Engineer Selin Munir Kulaga and orthopedic surgeon Professor Bill Walter (Specialist Orthopedic Group. Wollstonecraft NSW).

7.3.2 Aseptic loosening and osteolysis

This is a problem which alumina bearings greatly assist with. Osteolysis is mentioned frequently in this chapter and needs defining. It is pathological resorption of bone in vivo by osteoclasts (bone resorbing cells), that is, pathologically induced bone

Fig. 7.7 Clinically retrieved titanium alloy (Ti6Al4V) taper that was mated to the taper socket in a CoCr femoral head. Note the extensive galvanic corrosion surface residue.
Image courtesy of Selin Munir Kulaga and Professor Bill Walter. Specialist Orthopedic Group. Wollstonecraft NSW.

Fig. 7.8 Titanium alloy (Ti6Al4V) taper sleeve, used as a spacer between the alumina femoral head taper socket and the titanium stem male taper in situations where the tapers do not match. This is a CeramTec innovation that provides surgeons with multiple offset options. It is also used to provide a better stress distribution across the taper surface when utilized with an alumina head. Note the absence of corrosion as it is in contact with alumina (outer surface) and titanium alloy (inner wall).
Image courtesy of Selin Munir Kulaga and Professor Bill Walter. Specialist Orthopedic Group. Wollstonecraft NSW.

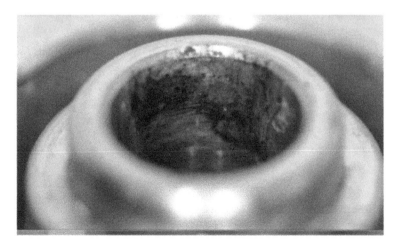

Fig. 7.9 Retrieved large metal CoCr head with female taper that was mated to a titanium neck in an in vivo taper joint, a scenario where a lot of taper corrosion would be inevitable and is clearly visible in this image.
Image courtesy of Selin Munir Kulaga and Professor Bill Walter. Specialist Orthopedic Group. Wollstonecraft NSW.

Fig. 7.10 Retrieved and corroded CoCr male taper that was mated with a titanium stem as a neck/stem taper connection in a dual-modular implant.
Image courtesy of Selin Munir Kulaga and Professor Bill Walter. Specialist Orthopedic Group. Wollstonecraft NSW.

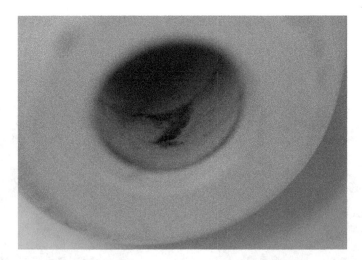

Fig. 7.11 Retrieved alumina femoral head that was mated with a titanium stem. Small amounts of residue of the titanium fretting-induced corrosion is evident on the surface of the alumina female taper.
Image courtesy of Selin Munir Kulaga and Professor Bill Walter. Specialist Orthopedic Group. Wollstonecraft NSW.

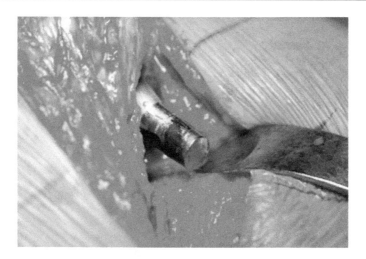

Fig. 7.12 Exposed titanium male neck taper, after removal of CoCr femoral head, with corrosion residue clearly visible.
Image courtesy of Selin Munir Kulaga and Bill Walter. Specialist Orthopedic Group. Wollstonecraft NSW.

atrophy. It commonly occurs in the presence of a prosthesis and can result in aseptic loosening. It can be caused by many mechanisms, some known and some unknown. The known ones that are relevant to alumina bearings include huge polyethylene wear particle concentrations for metal-on-polyethylene, metal wear particles from CoCr-Ti taper junctions, and metal taper-joint corrosion. The use of alumina/titanium tapers, and alumina-on-alumina bearings virtually eliminates inflammation-inducing irritants from the joint cavity, except in the case of implant dysfunction causing edge wear of the ceramic (Section 7.2.3). This just leaves stress shielding.

Hip replacement pioneer John Charnley used to discourage patients from having a hip replacement until they simply could not bear it any longer, due to the short service life of the implanted CoCr-on-polyethylene hip. The traditional view, which arose from Charnley, is that given the problem of osteolysis and aseptic loosening, significantly due to the half a million polyethylene wear particles released every step (over 2 trillion a year) implicated in causing osteolysis, it is wise for surgeons to counsel patients to delay hip replacement surgery for as long as they can possibly endure the pain of arthritis. This is neither a workable nor a humane strategy in practice. Today, thanks to the alumina hip, patients with hip implants can benefit from an implant with a very long service life, such as the discipline of orthopedics has never seen before, and need not delay surgery as long as possible and endure pain. They can have the procedure as soon as they need it, confident in the long service life of the implant.

In the early days, aseptic loosening was reportedly 45% lower for alumina-on-alumina hips than CoCr-on-polyethylene hips [350]. More recently, osteolysis was found in only 1.4% of patients with alumina-on-alumina implants compared with 14.0% of patients with CoCr-on-polyethylene [351] (Fig. 7.13).

Fig. 7.13 Preoperative x-ray image of aseptic loosening caused by osteolysis. The arrows identify the radio-opaque bone cement surrounding the stem.
Image Authors: Zhenxin Shen, Tania N. Crotti, Kevin P. McHugh, Kenishiro Matsuzaki, Ellen M. Gravallese, Benjamin E. Bierbaum, Stephen R. Goldring [352]. Reproduced from Wikipedia (Public Domain Image Library).

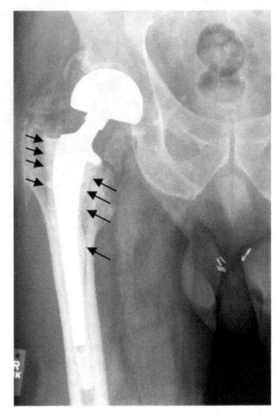

7.3.3 Dislocation and chipping

Dislocation is one of the most common complications arising from a THA. It is shown in Fig. 7.14. It can be a result of abnormal relaxation of soft tissues in the months after surgery or surgical misalignment. The geometry of the implant also has an effect. The larger the head diameter, the lower the dislocation risk, as the head diameter determines the amount of displacement required for the head to pop out of the socket. Increasing the head diameter also increases the range of motion, which reduces the risk of impingement of the cup rim and the neck of the stem, and the associated risk of chipping. The range of motion in the device, in terms of the degrees of rotation possible, obviously cannot exceed 180°, as the socket has a theoretical limit of a hemisphere (180°) [210]. It is much less than 180° in practice. In geometrical terms, impingement angle depends on head diameter and neck diameter. For example, for a 10-mm diameter neck, impingement angle relates to head diameter as follows:

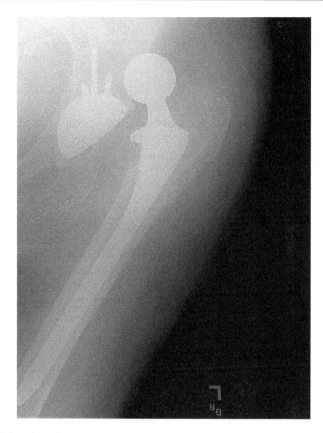

Fig. 7.14 Dislocation of a THA.
Image Author Bill Rhodes [353]. Reproduced from Wikipedia (Public Domain Image Library).

Head diameter (mm)	Impingement angle (°)
28	139
32	144
36	148
40	151
44	154

There is obviously a limit to how large the head diameter can be, determined by patient anatomy. 36–40 mm is now a common range. Alumina femoral heads are available in a range of standard sizes defined by two parameters;

- *Head diameter*: 28, 32, 36, 40, and 44 mm
- *Neck length*: (as defined by depth of female taper)
  - XL (+5 mm)—extra long
  - L (+3.5 mm)—long

- M (0 mm)—medium
- S (−3.5 mm)—short

28, 32, 36, and 40 mm are all equally commonly implanted in the United States in roughly equal proportions [183].

Smaller ceramic femoral heads are more prone to fracture although this problem has been largely resolved with the introduction of ZTA.

Chipping is a problem that is unique to alumina-on-alumina bearings, owing to the brittleness of alumina. It is shown in Fig. 7.15. Great care needs to be taken in implant design to ensure that there is no risk of impingement of moving metal edges against the edge of an alumina acetabular insert liner. This is a design issue that is pertinent only to alumina-on-alumina bearings.

7.3.4 Squeak

The alumina-on-alumina hip bearing, especially in its highly evolved ZTA manifestation would seem to be almost the perfect solution to hip replacement requirements, combining as it does the following benefits:

- extremely low wear rate, approximately 1000 times less than CoCr-on-polyethylene
- benign wear particles (unlike polyethylene which is an irritant and CoCr which is toxic)
- almost total elimination of osteolysis and aseptic loosening, which occurs 10% of the time with CoCr-on-polyethylene
- elimination of CoCr poisoning from galvanic corrosion and fretting wear in CoCr-titanium taper junctions for CoCr-on-polyethylene hips
- breakage rate in vivo below 1 in 100,000

However, it is human nature to overlook the positives and focus on the one negative that rises to prominence. This is certainly the approach taken by the mass media in reporting news. As the saying goes, "it is the squeaky wheel that receives the oil," and it is certainly true that the squeaky hip got the media attention in the early 21st century.

In the case of the alumina-on-alumina hip bearing, the one flaw is "squeak," occasionally manifest as an audible squeaking sound during specific types of leg movement by the patient. Squeak is unique to hard-on-hard bearings and has a self-reported incidence of only 1.2% [354]. It rarely causes pain, and is primarily only a nuisance. It is currently a focus of significant research attention and is yet to be fully understood. The issue of squeaking in alumina-on-alumina bearings rose to prominence in the literature and the media (particularly in the United States) around the time of the 2003 FDA approval of alumina bearings. This had the effect of harming the market entry of alumina into the US market, with focus turning instead to large diameter metal-on-metal bearings as an alternative to polyethylene in the United States [294]. The catastrophic ASR metal-on-metal recall of 2010, 7 years after the FDA approval for alumina bearings, ended the metal-on-metal concept globally, by which time the issue of squeak had fallen from prominence. Thus, it was only after 2010 in effect, that the US market ran out of reasons for tolerating the high osteolysis rate for metal-on-polyethylene, tolerating cobalt and chrome release into the joint cavity from metal-on-metal bearings, and CoCr galvanic/fretting corrosion. The US market, in the last decade has embraced alumina bearings and now almost caught up to Europe

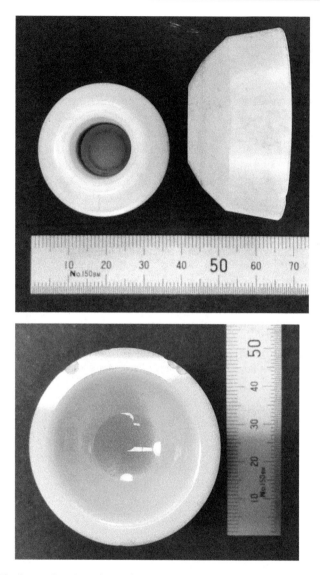

Fig. 7.15 Chipping at the edge of a retrieved alumina acetabular liner.
Exhibit provided courtesy of Professor Bill Walter. Specialist Orthopedic Group.
Wollstonecraft NSW.

(55%) in alumina-bearing usage (50% in the United States). However, thanks to media attention on squeak, things did not get off to a good start in the US market in the 2000s.

Squeaking is a well-recognized phenomenon for both alumina-on-alumina bearings [355] and metal-on-metal bearings [356], hard-on-hard, in other words. It is caused by high friction generated in hard-on-hard bearings from a loss of fluid film lubrication [355]. Polyethylene liners are not associated with squeak. This could be another reason for the current dominance of alumina-on-polyethylene over alumina-on-alumina implants. However, to put it in context, patient self-reported squeaking is 1.2% and this rises to 4.2% based on studies that evaluate squeaking via a questionnaire [354].

Squeaking does not always correlate with specific movements. It can occur in extreme hip positions, or during normal gait [355]. However, in general, squeaking almost exclusively occurs during deep hip flexion [355]. Importantly though, squeak is rarely associated with pain. It is purely an annoyance, much like the clunking sound that patients heard with the first-generation mechanical heart valves in the late 20th century.

While squeak is not yet well understood, it is now known that multiple factors are involved, all relating to mechanisms that damage the fluid film lubrication [357,358]. Alumina-on-alumina-bearings function optimally with fluid film lubrication. These factors were itemized in a recent review on squeak by Levy in 2015 [355]:

- edge loading: implant positioning can facilitate edge loading [359]
- stripe wear [360]
- impingement: implant positioning can facilitate impingement [361]
- third-body particle-based wear, for example from impingement-caused liner-rim chipping [362]
- ceramic fracture [363]. Breakage is easily diagnosed by a CT scan
- metallurgical composition: resonance of metallic parts can produce audible noise [364]
- large ceramic heads (>36mm) may accentuate noise-generating factors [365]. Another prevailing view is that large diameter ceramic heads diminish squeaking. These heads are shown in Fig. 7.16

Some of the ceramic acetabular insert liner designs were deliberately engineered to have a lower polish level at the high-angle areas up near the cup edge and the squeaking mainly happens at the high-angle areas. It is believed that there might be some correlation between this low polish edge design and squeaking [210].

In summary, squeaking is a forced vibration comprised of a frictional driving force and a dynamic response. Frictional driving force occurs when there is a breakdown in fluid film lubrication and an increase in the coefficient of friction from 0.03 to 0.3. This occurs most commonly with edge loading of the bearing and can also occur with a damaged bearing due to repeated edge loading, ceramic fragments in the joint, or a broken bearing. Dynamic response depends on the natural frequency of the stem/head construct. An implant is more likely to squeak with a flexible stem in the anteroposterior plane (lower resonant frequency) [210].

7.4 Commercialization of the alumina hip: The transition from research to commercialization

7.4.1 CeramTec: Pioneer manufacturer of alumina orthopedic bearings

By the year 1982, 60,000 trial implantations had been conducted for German-made alumina hip implants. Unfortunately, a relatively high early breakage rate of 0.026% (1 in 3800) in first-generation alumina implants, and some catastrophic

Alumina bearings in orthopedics: Present and future 211

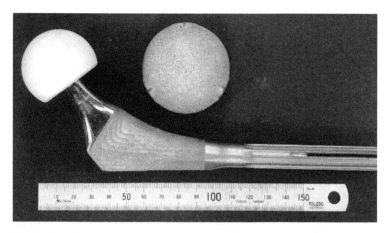

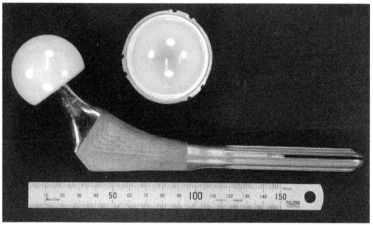

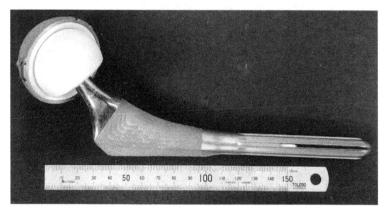

Fig. 7.16 Large diameter ceramic heads made of CeramTec ZTA BIOLOX™*delta*. Exhibit provided courtesy of Professor Bill Walter. Specialist Orthopedic Group. Wollstonecraft NSW.

failures that caused concern in the orthopedics community gave alumina hip bearings a bad reputation. Ironically 0.026% is a failure rate, 380 times lower than osteolysis failure caused by polyethylene wear particles, but the nature of a ceramic breakage in vivo is catastrophic, and distressing for an orthopedic surgeon and patient alike.

However, these breakage problems were ultimately resolved by German ceramics company Feldmuhle, who developed the BIOLOX® alumina ceramic beginning with the first-generation BIOLOX® in the 1970s, second generation from 1980 to 1995, and third generation post 1995. Feldmuhle was ultimately, through a series of mergers and acquisitions, to become CeramTec in 1996, and under the CeramTec name, pioneered the quality manufacturing of alumina orthopedic bearings, with third generation (1995 onward) with breakage rates of (0.004%) (1 in 25,000), which was 2000 times lower than polyethylene-induced osteolysis. This is summarized in Table 7.2.

The name CeramTec is today synonymous with ceramic hip joints. Indeed, the story of the commercialization of the alumina hip, just like the story of its development, is a very German story. CeramTec is a global ceramics company headquartered in Plochingen, Germany, now UK owned, and profiled in Chapter 1 (Section 1.4.2.3). A short timeline of the history of CeramTec and the alumina hip commercialization process is shown in Table 7.3.

Feldmuhle, which became Ceramtec, took the early German prototype ceramic hip concept of the 1970s/1980s and developed a highly evolved quality-manufacturing process with a core technology, known as "third-generation" alumina, that involves green machining of dry-pressed blanks, with densification using sinter hot isostatic pressing (sinter-HIPing). HIP enables minimization of grain size and flaw content, and maximization of density.

CeramTec component design is done in direct consultation with customers— orthopedic medical device companies. Designs for the green machining incorporate sintering shrinkage. Manufacturing is done using a quality manufacturing system with full traceability, right down to the production records of each (individually laser marked) component. CeramTec has a range of exacting quality standards for every aspect

Table 7.2 **General causes of revision surgery for total hip replacements, and specific causes of revision surgery for alumina total hip replacements [273,345]**

General causes of failure—All hip prostheses	Percent
Aseptic loosening	10
Fracture of stems	2
Septic loosening	1
Specific causes of revision surgery—Alumina hip prostheses	
Fracture of BIOLOX head (first generation)	0.026
Failure of BIOLOX head (second generation)	0.014
Fracture of BIOLOX forte heads (third generation)	0.004
Zirconia-toughened alumina (current era)	<0.001

Table 7.3 Timeline of CeramTec

1903	Foundation of the Thomas factory in Marktredwitz
1921	Phillip Rosenthal & Co AG cooperates with AEG on industrial porcelains.
1951	Foundation of Sudplastic und Keramik (SPK) Company in Plochingen
1954	Feldmuhle AG acquires Sudplastik and Keramik (SPK) Company
1971	Foundation of Rosenthal Technology AG for manufacturing industrial ceramics
1974	First BIOLOX alumina hip implanted by Prof Mittelmeier—known as the Mittelmeier hip
1985	Hoechst AG Acquires Rosenthal Technology AG and forms Hoechst CeramTec AG
1991	Feldmuhle Spins off its ceramics activities into an independent company. Cerasiv GMBH
1992	Feldmuhle sells Cerasiv to Metallgesellschaft AG, Frankfurt
1994	CE-Mark European Union Commercial Approval for BIOLOX alumina in Hip Implants. *BIOLOX commercialisation begins*
1996	Merger of Cerasiv GMBH (Plochingen) and Hoechst CeramTec AG (Marktredwitz) forms CeramTec AG as part of the MG technologies AG
2003	*First commercial use of ZTA BIOLOXDelta (European Union)*
2003	FDA approval of BIOLOX. *US Commercialisation begins—9 years later than European Union*
2004	*US Acquisition of CeramTec*. Purchase of CeramTec Group by Rockwood Specialties Group Inc. with Headquarters in Princeown NJ, USA Headquarters remain at Plochingen, Germany
2013	*British Acquisition of CeramTec*. Cinven, a British private equity firm acquires CeramTec from Rockwood, USA

Pioneer of alumina orthopedic bearings.

of the manufacturing process, as outlined below for the fourth-generation BIOLOX*delta* [366].

7.4.1.1 Quality manufacturing process

- powder batching and mixing (alumina, zirconia, binders, sintering aids, and other additives)
- milling
- spray drying
- manufacturing of pressed cylinders (dry pressing)
- green machining of the cylinders into femoral heads (ball with female taper) and acetabular liners (socket with male taper).
- sintering (5 day process 1300–1500°C)
- hot isostatic pressing (Sinter HIP: 1 day >120 MPa >1400°C)
- tempering (removes residual stresses from HIPing and restores correct color)
- diamond grinding
- diamond polishing
- Laser marking (avoids formation of stress concentrators)
- Crack inspection: cleaning and degreasing followed by Zyglo Crack inspection (optical test with Zyglo penetrating die)

- Cleaning to remove Zyglo die
- visual inspection
- final cleaning
- packaging and despatch

7.4.1.2 Quality assurance

- QA tests during powder preparation: LOI, surface flow, chemical composition
- green density testing
- final density testing
- mechanical strength testing
- grain size testing
- proof testing of saleable components (overload higher than normal physiological load)—Ceramtec was the first company in the industry to introduce this QA test
- comprehensive dimensional testing including sphericity testing

All CeramTec alumina-bearing components have a taper for attachment: femoral heads use a female taper for the insertion of the tapered socket on the titanium stem; acetabular inserts have a male taper for insertion into the matching female taper of the acetabular cup. The early developmental alumina acetabular inserts had 18° taper and were 8 mm thick. The more recent ones have 12° taper and are 5 mm thick. Surgeon experience has found that a 12° taper is much harder to work with because it can potentially get misaligned when you slip it in, causing it to jam. Clinical retrieval studies have found that a misaligned ZTA acetabular insert is likely to fail in vivo, and if misalignment is suspected during surgery, the liner should be replaced. Surgeon experience has found that at 18° taper, misalignment will never happen [210]. A useful CeramTec ceramic innovation involves manufacturing alumina femoral heads with a standard large taper diameter, with a range of titanium taper sleeves provided to be used as an adapter so that a single ceramic taper size can be fitted to a wide range of stem taper sizes. This is shown in Fig. 7.8.

7.4.2 Commentary on the CeramTec process

Third-generation (post 1995) CeramTec orthopedic alumina bearings are characterized by a number of key quality innovations. *Hot isostatic pressing* is probably the most important factor in making "third generation" alumina such a reliable high-quality product. Sinter HIPing is used, and this enables two key benefits:

- minimization of grain size
- elimination of porosity and internal flaws

Equally important was the introduction of *proof testing*, a well-known test that involves preloading a component to a threshold service stress level, with the aim of discarding all components that fail the test [367,368]. CeramTec was the first company in the industry to introduce this QA test, and it has greatly reduced the failure rate of their third-generation product. The science underlying the proof test of alumina orthopedic bearings dates back to the end of the 1980s [367,368]. In brief, for the femoral head, the highest tensile stress levels in service are at the dome at the deepest point of the taper socket, and the tensile hoop stress exerted on the taper socket by the titanium neck taper. The proof test is conducted in such a way as to avoid damaging the component, either by overloading or by damaging the polished surface finish, but also to ensure that any flaws above a critical threshold are detected. The impressive failure rate of below 1 in 100,000 can probably be attributed primarily to the introduction of the proof test in the CeramTec production process, and of course the high-quality ceramic that HIPing produces.

Although *laser marking* may seem a trivial innovation, this is far from the case. Traditional marking by engraving introduces surface flaws which can be the cause of catastrophic failure in accordance with fracture mechanics theory (Chapter 4). Laser marking has been a significant factor in product reliability.

The *Zyglo crack inspection* and personal visual final inspection complete the circle of exemplary quality control. A failure rate of 0.001% (1 in 100,000) is difficult for competitors to match and gives surgeons great confidence. Moreover, CeramTec do not supply their components to just anybody. The reputation of their product also depends on the exacting standards that customers (orthopedic medical device manufacturers) for CeramTec alumina bearings follow, in the design and quality control of, hip replacements that utilize alumina bearings.

HIPing and proof testing in particular, and the commitment to quality and the long experience with alumina orthopedic bearing manufacture in general, has enabled CeramTec to evolve the wear resistance and reliability of their alumina components to a very high standard indeed. The third-generation BIOLOX had a fracture rate of only 0.004% per implantation, and this has been the case for over 20 years. Current ZTA technology is below 0.001% or 1 in 100,000. This is an enormous improvement on the first- and second-generation as shown in Table 7.2. Up until a decade ago, CeramTec had a market monopoly on alumina hip bearings. Kyocera (Japan) entered the market in 2011 with a ZTA product (AZ209) but are yet to make a significant impact. CeramTec have such a strong track record for quality and reliability that this represents a significant barrier to market entry for competitors. CeramTec and Kyocera are profiled in Chapter 1 (Section 1.4.2). CeramTec pure alumina and ZTA hip bearings are shown comparatively in Figs. 7.17 and 7.18.

In the early years, a number of ceramic manufacturers were making alumina heads, all by pressureless sintering, none of them by HIPing. There is wide variation in the published fracture rates of the early experimental alumina heads, with reports ranging from near 0% for CeramTec post-1990 era heads, to 13.4% for heads manufactured before 1990 [273]. The early Mittelmeier acetabular cups had a 21.5% fracture rate 1984 [229].

Suffice to say, it has taken some time for the legacy of the early developmental period of the 1970s and 1980s to be forgotten, as well as the zirconia disaster of the early 2000s.

The statistics for reliability today are in fact extraordinarily good, as shown in Table 7.2. The same cannot be said for the high failure rate of metal-on-polyethylene hips through aseptic loosening: ~10%. This is a 10,000 times higher failure rate than the 0.001% rate of fracture of the current generation ZTA hip bearings. However, aseptic loosening also occurs with alumina-on-alumina implants in a small number

Fig. 7.17 High-purity-alumina (BIOLOX *forte* pure-alumina—CeramTec) alumina-on-alumina bearings. Note the female taper in the back of the femoral heads, and the male taper on the unpolished outside of the acetabular inserts. Exhibit provided courtesy of Professor Bill Walter. Specialist Orthopedic Group. Wollstonecraft NSW.

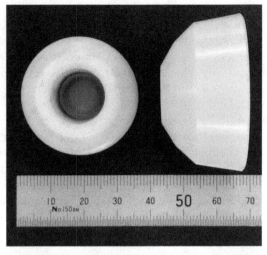

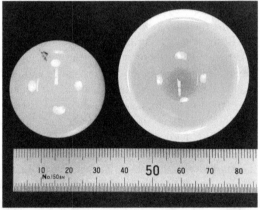

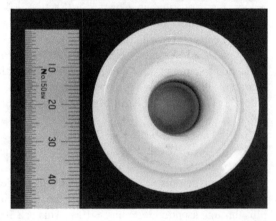

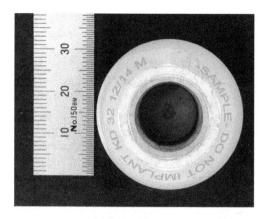
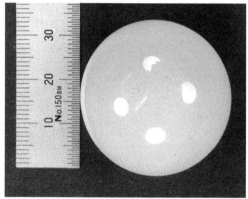
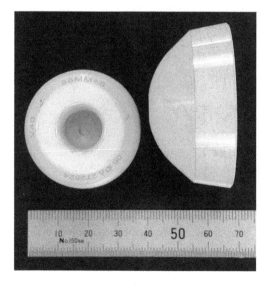

Fig. 7.18 Zirconia-toughened-alumina (ZTA) (BIOLOX *delta*—CeramTec) alumina-on-alumina bearings (sometimes called ZTA-on-ZTA). Note the female taper in the back of the femoral heads, and the male taper on the unpolished outside of the acetabular inserts.
Exhibit provided courtesy of Professor Bill Walter. Specialist Orthopedic Group. Wollstonecraft NSW.

of cases where severe edge wear or other aberrant conditions applied. The big difference between alumina-on-alumina hips and polyethylene-based hips is osteolysis. In a 2005 study, osteolysis was found in 1.4% of patients with alumina-on-alumina implants compared with 14% of patients with CoCr-on-polyethylene [351]. Logic and patient benefit is on the side of making the switch alumina bearings, and this is driving a generational change in regions of the world outside Europe. By 2030, CoCr hip bearings may be obsolete.

Over 14 million CeramTec alumina-bearing components have been implanted worldwide [29], both pure alumina and ZTA, but the majority is ZTA. CeramTec calls its orthopedic alumina BIOLOX. The pure alumina grade is bone-colored and called BIOLOX *forte*, the ZTA is pink colored and called BIOLOX*delta*. Though both ceramics have outstanding wear resistance and mechanical reliability, the ZTA product (BIOLOX*delta*) is better in wear and reliability than the pure alumina product (BIOLOX*forte*), and therefore the ZTA product has now almost totally supplanted the pure alumina product. Coloring the ZTA pink was an astute marketing strategy. Perceptions are important.

A number of standards apply to surgical biomedical-grade alumina:

- DIN58835 (Germany 1979). The original standard
- ASTM F603 (2000, updated 2016)
- ISO6474-1 (1980, updated 1994, 2010)
- ISO6474-2 (2009) relates to ZTA

Note: ISO6474 and ASTM F603 were largely derivative of DIN 58835.

7.4.3 Current market impact: Alumina hip

Other than French surgeon Boutin and his collaborating French ceramics company Ceraver being the first to produce and implant an alumina hip, the story of the ceramic hip has been an almost entirely German story. The story began with Themistocles Gluck in the 1880s whose ivory hip was the world's first "ceramic hip," albeit made of a natural bioceramic (ivory) rather than the synthetic bioinert ceramic (alumina) that was to come later. Then beginning in the early 1970s, shortly after Boutin, Germany has pioneered much of the development of the alumina hip from the 1970s onward. The developmental era of the 1970s and 1980s is discussed in Chapter 6 (Section 6.4.1).

More importantly, Germany also pioneered the commercialization of the alumina hip, through German company CeramTec, a company name that has become synonymous with the ceramic hip (Section 7.4.1).

In recent years, the US market has also embraced alumina hip bearings. In 2010, about 13% of US hip replacements were surgical revisions [182]. By 2015, the revision rate had dropped to 10% [183]. This dramatic drop in hip revision rate from 2010 to 2015 mirrors the dramatic rise in the use of alumina femoral heads in the USA, from 36% in 2012 to 49% in 2015.

As per the data presented in Figs. 7.19–7.21 show, Europe leads the world currently in ceramic hip implantation rates. 299 hips per 100,000 people are implanted in

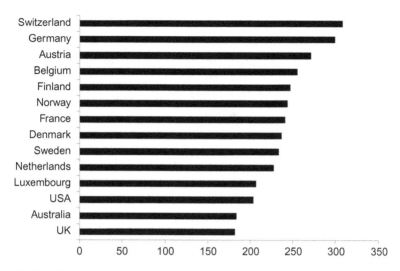

Fig. 7.19 World's leading nations in hip replacements per 100,000 population [369].

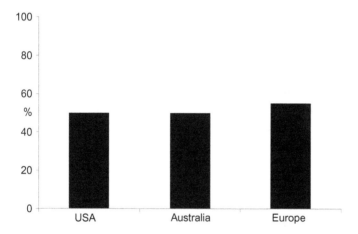

Fig. 7.20 Percentage of hip implants that contain alumina bearings (whether alumina-on-polyethylene or alumina-on-alumina). Europe leads the world and Australia essentially reflects the US trend.

Germany (population 80 million), second only to Switzerland which is world number 1 (308 per 100,000) but Switzerland has a population of only 8.5 million. With 55% of EU hip implants utilizing alumina bearings, this corresponds to 130,000 alumina hips implanted in Germany in 2015. In the United States (population 325 million), which stands at 12th ranking globally at 203 hips per 100,000, this corresponds to about 660,000 hips a year in the United States. ~50% of hip implants in the United States are alumina hips, which is about 330,000, two and a half times the number implanted in Germany.

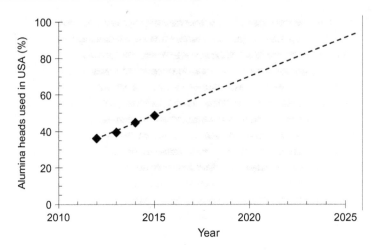

Fig. 7.21 Growth in use of alumina femoral heads in the United States from 2012 to 2015. Extrapolation into the 2020s is of course well beyond the point of statistical significance, but this is just done for illustration purposes. The upward trend is clear [183].

The process of development of a new medical device technology is a slow one. The metal-on-metal hip had its origins in the United Kingdom in the 1930s, but it did not reach mainstream use, until the 1970s. The metal-on-polyethylene hip was developed by the early 1960s, but did not reach mainstream use, until the 1970s. The alumina hip had its origins in a 1970 innovation and did not reach mainstream use until the mid-1990s. Today we have essentially four types of implant on the market. All of them arose in Europe, although the United States had a significant role in early CoCr-on-CoCr development.

- >2% metal-on-metal bearings. These have now dwindled to an insignificant market share although they are making a small comeback in the resurfacing area
- 45% metal-on-polyethylene
- 40% ceramic-on-polyethylene
- 15% alumina-on-alumina

Where, in almost all cases

- "metal" means CoCr femoral head
- "polyethylene" means cross-linked ultrahigh-molecular-weight-polyethylene (UHMWPE)
- ceramic means either pure alumina or ZTA (today ZTA is almost exclusively used rather that pure alumina).
- stem is titanium with Morse taper fittings.

As shown in Fig. 7.22, approximately 1.3 million hip replacements implanted annually, about 45% were metal-on-polyethylene, 40% were alumina-on-polyethylene, 15% were alumina-on-alumina, and <2% were metal-on-metal. This represents, very approximately, 700,000 alumina femoral heads (ball) and 200,000 alumina acetabular liners (socket). This represents significant growth as total consumption of BIOLOX in 2009

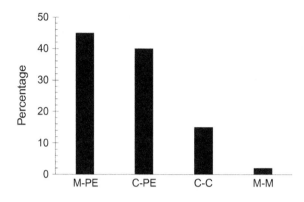

Fig. 7.22 Approximate 2015 global usage of the four main types of hip bearings. M-PE (metal-on-polyethylene), C-P (alumina-on-polyethylene), C-C (alumina-on-alumina). In the orthopedic industry, alumina is the only type of ceramic used in hip bearings (pure or zirconia toughened) but in the industry, alumina bearings are commonly referred to by orthopedic surgeons as "ceramic," hence the "C."

was 500,000 heads and 140,000 inserts, of which about 35% was pure alumina BIOLOX*forte* and 65% was ZTA BIOLOX*delta*. The annual global consumption of alumina hip-bearing components (heads and liners combined) in the current year (2018) is in the order 900,000, and it is probable that by the year 2020 this will have reached 1 million, of which the proportion that is BIOLOX*delta* ZTA would be close to 100%.

Australia is closely linked with Europe in its orthopedic trends because the Australian medical regulatory body (TGA) is linked with the EU regulatory system (CE-Mark) through the GHTF. Europe is very slightly ahead of the United States. However, it should be noted that:

- FDA (US) approval of alumina hip bearings was only granted in 2003, 9 years later than in Europe.
- Clinical trials of alumina hips began two decades later in the United States compared to Europe.
- The zirconia hip disaster which played out in 2001/2002, 1 year before FDA approval of alumina hips, damaged confidence in "ceramic" hips in the United States, unlike in Europe, where the market was familiar with the difference between alumina and zirconia, due to the long pedigree of alumina in Europe.
- The concept of squeaking, and its association with alumina-on-alumina hips found its way into the literature and the US mass media around this time, and added to the perception in the United States of ceramic bearings being problematic, a controversy later dissipated (see Section 7.3.4) [294].
- The US market made the decision to focus on large-diameter metal-on-metal bearings as a hard-on-hard alternative to polyethylene-based bearings [294]. The DePuy recall of 2010 reversed this trend dramatically and was a significant driver for the later penetration of alumina hips into the US market, a current trend.

Thus Europe had a head start of some decades over the United States. However, it is apparent that the United States has almost caught up to Europe now. Moreover, Fig. 7.21 shows a steady rise in alumina femoral head use from 2012 to 2015. If this growth rate continued linearly, ceramic femoral heads would approach the 100% level by 2025. More likely CoCr heads will dwindle in an asymptotic trend. For some years there will be surgeons who prefer to stick to the technology they know best, and CoCr has been in use for a long time. Of course it is not possible to predict the long-term trends at this stage with data to date, other than that alumina is on a steady rise, and that alumina-on-polyethylene is the preferred choice.

It would be fair to say that by the mid-2020s, alumina-on-polyethylene is likely to be market dominant, alumina-on-alumina is likely to be in number 2 position, and metal-on-polyethylene is likely to be in third place and a small proportion of the market. As an educated guess, it may be something like:

- alumina-on-polyethylene: 60%–65%
- alumina-on-alumina: 20%–25%
- metal-on-polyethylene: 10%–20%

Over the years a number of ceramic companies have been involved in the manufacture of alumina hip bearings, beginning with Ceraver Osteal in France, closely followed by Feldmuhle/CeramTec in Germany. Others have included CoorsTech (USA), IFGL (India), and Kyocera (Japan), but the CeramTec output has been pre-eminent since the 1970s, and remains so today.

7.4.4 Alumina in other joints

The two largest joint replacement markets in terms of annual replacement numbers are the hip and the knee, both over 1 million a year. A typical conventional CoCr-on-polyethylene knee (with titanium tibial component) is shown in Fig. 7.23. There have

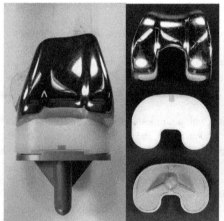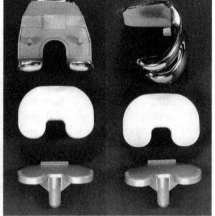

Fig. 7.23 The CoCr-PE-Ti Knee is shown assembled and fitted to a sawbone femur, and disassembled. The femoral component (shiny complex-shaped top piece of the knee) has been prototyped in alumina, but not commercialized. All commercial knees use a CoCr or Oxinium femoral component. Note the complex geometry, and complex associated tensile and point loading, are unsuited to alumina or ZTA ceramics under the >5 kN impact load conditions encountered in practice. Thus a ZTA-based ceramic knee would certainly be ZTA-on-polyethylene. Polyethylene liner is interposed between articulating CoCr femoral component, and titanium tibial plate.

been some early prototyping trials with alumina knee inserts, comprising an alumina or ZTA femoral component (top section as per Fig. 7.22) articulating against polyethylene. However, while the hip is a simple ball-and-socket joint with uniform loading of the alumina, the complex geometry of the dual-condyle femoral component of the knee, and the complex loading of the polyethylene tibial liner make the knee a much more challenging environment for alumina, pushing it to its limits and beyond in terms of tensile stresses, toughness, and impact resistance. Therefore, the CoCr-on-polyethylene-bearing system currently is virtually the sole technology in the knee market, with a very small presence of oxidized zirconium, a Smith&Nephew technology marketed since 1997 under the trade name Oxinium$^{\diamond}$, discussed in Chapter 6 (Section 6.5.1).

The unicompartmental knee can be done with a spherical joint which is more viable. This was trialed with alumina between 1972 and 1980 (73 implantations) in Germany by German pioneer in orthopedic alumina Langer [370]. The program ended in the 1980s when the CoCr/polyethylene total knee replacement (TKR) became widely available. In 1990, Oonisi pioneered a ceramic knee (TKR) in Japan known as the KOM-1 [238]. Initially it involved an alumina femoral component and tibial tray with polyethylene-bearing surface. Later versions used a titanium tibial plate. This KOM-1 program continued over some decades with over 100 implantations trialed, but has not translated into a commercial device. Unlike the simple ball-and-socket joint of the hip, with its spherical surfaces and uniform surface loading, the biomechanics and stress distribution in the knee are more problematic, and at the limit of what is mechanically possible with ZTA. Commonly the alumina or ZTA femoral component suffers a midline fracture in vivo [371].

Other joints, such as shoulder, elbow, and ankle, have very small market sizes and the CoCr-on-polyethylene system currently is virtually the sole technology in these markets.

Alumina in bionic feedthroughs: The pacemaker 8

8.1 Alumina: A key enabling technology for implantable bionic medical devices

Alumina is an essential component of the hermetic encapsulation and electrical feedthrough system used in implantable bionics. This has been the case since 1970 when the transition was made from the rudimentary epoxy encapsulation systems used in the developmental era of implantable bionics, to the alumina/titanium hermetic feedthrough system used for implantable bionics since then. Implantable bionics is a global industry currently worth $25 billion per annum, and growing rapidly. Thus implantable bionics is one of the most important commercial applications for alumina advanced ceramics in the world today. The alumina/titanium hermetic feedthrough system for implantable bionics was invented in 1970 by David Cowdery [146,147]. Cowdery did not just invent the alumina feedthrough, his invention was also the first bionic implant to use a titanium casing. The alumina feedthrough hermetically sealed in a titanium casing is the fundamental platform technology for the bionic implant. Most importantly, because of the hermeticity and durability of the alumina/titanium combination, this was the first bionic implant with a service life that could be measured in years, rather than months as was the case with the contemporary alternatives, which relied on epoxy resin cladding. This chapter overviews the evolution of the pacemaker and the invention of the alumina bionic feedthrough for the pacemaker.

When a bionic implant is described as "hermetic," the assumption is that the casing is sealed against water, ion, and gas diffusion for the lifetime of the device. Hermeticity is generally measured by the helium leak test, a very stringent test. Alumina and titanium are the two essential components of the hermetic encapsulation and electrical feedthrough system used in implantable bionics. Platinum has replaced titanium for the metal conductors used in the feedthrough and the sealing systems have seen some variations, but the alumina/titanium combination is ubiquitous to modern implantable bionic technology.

The invention of the alumina/titanium feedthrough was the key enabling technology of the world's first modern (hermetic and durable for years in vivo) implantable bionic medical device: the Telectronics P10 cardiac pacemaker. All subsequent bionic implants globally have used adaptations of the alumina/titanium feedthrough concept, which has evolved substantially in terms of the electrode density per square millimeter of alumina, and the associated device miniaturization. However the basic platform technology of the alumina/titanium feedthrough still dominates the bionic implant industry today, which has now grown to $25 billion per annum.

The invention of the alumina bionic feedthrough was a world-changing technology, and a unique chapter in the history of both alumina ceramics and bionic medical devices. It took place in Australia, with a significant British connection. David Cowdery, an electrical engineer, developed unique specialist welding expertise during his 5 years at CIG and various British welding research institutes including the British Welding Research Association in Cambridge, and the British Oxygen Company Research Laboratory in London. Thus in 1970, Cowdery was a leading expert in advanced welding technology with a strong electrical engineering background [372,373]. With his unique combination of electrical engineering and 5 years of advanced welding research, David Cowdery was the ideal person for the job when he was recruited by pacemaker company Telectronics, to develop a durable hermetic encapsulation system for pacemakers. This he succeeding in doing within a relatively short time in the year 1970.

Cowdery's invention of the alumina feedthrough triggered a global boom in the bionics industry. In retrospect, this may seem surprising given that the invention occurred in secrecy, in a relatively small company in global terms, in a sparsely populated country far from the European/United States world hub. The answer is threefold:

- *It was paradigm-shifting technology*: implantable pacemakers could now have a service life measured in years rather than months.
- *Huge global demand for pacemakers*: an abundance of desperate cardiac arrhythmia patients globally.
- *The invention was not patented*: this enabled rapid global dissemination of the technology by a global biotechnology community eager to commercialize pacemakers.

The cost to Telectronics for choosing not to patent this invention certainly would have run into the multiple million dollar range, in lost license fees and/or royalties, given the $billion order of magnitude in global pacemaker sales over the next two decades, and as a case study, this clearly demonstrates the value of a patent, and the unsuitability of industrial secrecy in the case of novel inventions with high market demand. However, in retrospect, the lack of a patent for this invention was a wonderful decision for the global community of cardiac arrhythmia patents as it had the result of rapidly disseminating this world-changing innovation globally. Pacemaker researchers were free to use the alumina/titanium feedthrough innovation, and adaptations thereof, as a platform to develop and optimize ever more advanced adaptations of the alumina/titanium feedthrough system, unfettered by licensing agreements and the 20-year patent embargo.

The result was that millions of arrhythmia patients globally have received a pacemaker in the last few decades. Today, over 1.5 million pacemakers are implanted a year, a global market exceeding $10 billion, and it is forecast to reach $12 billion and 2 million pacemakers by 2021 [374]. Moreover there are now 14 bionic implant technologies on the market today treating 25 disease states, encompassing a global $25 billion market in implantable bionics, all using the alumina/titanium feedthrough technology invented by Cowdery. This rapid explosion in bionics technology globally would have been greatly impeded had the Cowdery invention been patented.

This bionic alumina feedthrough innovation story had two subsequent episodes. The next episode (discussed in detail in Chapter 9) was the development, in Australia, by a sister company of Telectronics (Cochlear Ltd.) of the world's first bionic ear capable of speech recognition. This was the 22 channel Cochlear implant developed and commercialized in the period 1970–85 by a team led by otologist Graeme Clark, the founder of Cochlear Limited, a company within the same family of Australian companies (Nucleus) that owned Telectronics. This Cochlear implant bionic ear device depended on a more highly evolved version of the Telectronics alumina/titanium feedthrough, which was developed by Janusz Kuzma [146], who was himself recruited to Cochlear by alumina feedthrough inventor David Cowdery. This highly evolved alumina feedthrough for the bionic ear would not have been possible at the time had sister company Telectronics not paved the way a few years earlier, and the in-house expertise of Cowdery and other Telectronics engineers not been on hand at the time. The Cochlear bionic ear product continues to evolve and still dominates the global bionic ear market today.

The final episode, in which I was an active participant (see Chapter 10), was the development of a bionic eye device under a $42 million Australian Government Initiative (2010–16 ARC Special Research Initiative in Bionic Vision Science and Technology, and formerly known as Bionic Vision Australia). This was the next evolution in feedthrough sophistication, based on the invention of Gregg Suaning a biomedical engineering professor and former Cochlear engineer. This device set the world record in 2014 for the highest number of channels in an alumina feedthrough (retinal implant) of 1145 [375]. In four decades, alumina bionic feedthrough technology had evolved from a single titanium channel in a 9.6 mm diameter alumina disk (Telectronics pacemaker, 1971), to 22 platinum channels in a 14.5 mm diameter alumina disk (Cochlear implant, 1985), to 1145 platinum channels in a 7-mm diameter alumina disk (Suaning retinal implant, 2012).

The author comes from the same community from which these three chapters of bionic innovation arose: the Australian biomedical and advanced ceramics community and was therefore privileged to have known many of the innovators, both then and now, and lived through the innovation and evolution of this world-changing alumina ceramic technology, both as an observer and as a participant.

Australia is a prosperous British Commonwealth Nation of relatively small population ($\sim$25 million), similar to Canada in terms of population, prosperity, strength in high-technology and innovation, wealth of mineral resources, and the fact that both countries have the British Monarch as their Head of State. Australia is best known for being, behind China, the second largest mining nation in the world. However, Australia is more than just a prosperous mining nation, it is also a nation with a strong track record for high technology innovation. Unfortunately, it has a poor track record for commercializing innovation, most of which goes offshore.

A number of significant technology inventions originated in Australia including the Black Box flight recorder, WiFi, Google Maps (which went immediately offshore), the Frazier lens, plastic bank notes, and the power drill. However, of greater relevance to this chapter, Australia is credited with several ground-breaking biomedical inventions: spray on skin (for burns treatment), the ultrasound biomedical scanner, plastic spectacle lenses, the CPAP device for sleep apnea, the world's first hermetic pacemaker (the subject of this chapter) and the first bionic ear capable of speech recognition (Chapter 9).

This book is about alumina ceramics. Alumina, in combination with titanium, was the key enabling technology for the implantable pacemaker, the bionic ear, and indeed all implantable bionics in use today. This alumina/titanium bionic innovation, which originated in Australia, was manifest in three technological breakthroughs:

(1) The invention of the world's first hermetic implantable bionic device. The Telectronics P10 pacemaker in 1971 (David Cowdery). This technology has long since gone offshore.
(2) The invention of the world's first bionic ear capable of speech recognition (1977) by Graeme Clark, which would not have been possible at the time had the inventor of the hermetic bionic device and his team not been involved in the bionic ear development. Australia retains 70% of the global bionic ear market today.
(3) Bionic eye retinal implant developed in Australia set the world record for the largest number of channels in an alumina feedthrough of 1145 in the year 2012. This technology has been prototyped with a quality manufacturing system established, but it is yet to be commercialized.

Alumina bionics forms the basis of this chapter and of Chapters 9 and 10. This is because bionic applications of alumina are one of the largest commercial uses for alumina ceramics in the world today in dollar value. Implantable bionics, an industry worth $25 billion dollars a year, in which alumina is a critical component, surpasses alumina ceramic applications in orthopedics, body armor, and wear-resistant alumina ceramics. Only the electronics industry is comparable to the dollar value impact of alumina ceramics (Chapters 13 and 14).

This chapter outlines the invention of the alumina feedthrough, in the context of the world's first bionic device, the pacemaker, for which the alumina feedthrough was originally developed. Chapter 9 looks at the development and underlying technology of the bionic ear, the world's second implantable bionic device to be commercialized after the pacemaker. Chapter 10 looks at the 12 bionic implant technologies that evolved subsequently, with a focus on the most advanced one of them all: the bionic eye. Together, Chapters 8–10 explore the evolution of the alumina bionic feedthrough to an extraordinary level of sophistication in the era since the invention of the pacemaker. This includes not only the bionic ear and the bionic eye, but also all 14 bionic devices on the market today, and others in advanced development.

8.2 A brief history of early pacemaker technology

8.2.1 Definition of bionic implants

The term bionic refers to an electrical or electromechanical device which is functionally incorporated into the patient's body, interfacing with nerves, stimulating the nerves, and commonly receiving electrical feedback from them. Most contemporary bionics are battery powered and many have the capacity for battery charging, and system programming, in situ, by induction. Some run off external inductive power couplings.

In its broadest definition, bionics extends from the heart pacemaker and all its derivatives such as the deep brain stimulator and spinal cord stimulator, to the complex high-channel-number devices, such bionic ear and bionic eye through to external robotic devices, such as myoelectric-controlled functional artificial limbs (hard-wired into the patient's neural tissue) and thought-controlled robotic limbs coupled via a brain-computer-interface implant. All of them share many common features and can be seen to have evolved in a clear lineage, beginning with the heart pacemaker. The great majority of commercial bionic devices on the market today are implantable bionics, or to be more precise, hermetic-bionic-implant devices.

This book is about alumina, and as such the focus in this chapter is on alumina in bionics. Alumina is one of the two key enabling technologies for all implantable bionic medical devices. The following two key enabling technologies of implantable bionics were:

(1) The invention of the transistor in 1947. This was a broad enabling platform technology, applicable not just to bionics, but to the whole field of microelectronics.
(2) The invention of the alumina/titanium hermetic encapsulation system in 1970. This was an enabling platform technology specific to all subsequent implantable bionics.

Transistor and semiconductor technology is beyond the scope of this chapter. Moreover, due to the fact that alumina is an essential component of *implantable* bionics only, in this book, the focus is confined to implantable bionics.

In the early history of the pacemaker, the world's first bionic implantable medical device, there are three landmark events which laid the foundation for the modern bionics industry:

Event 1 (1928). British Doctor Mark Lidwill invented the first cardiac pacemaker—a desktop triode-valve device. It inspired decades of pacemaker research globally. This innovation did not involve ceramics, and the innovator did not become famous. He deliberately avoided public recognition.

Event 2 (1957). USA. Earl Bakken, invented the world's first portable (wearable) pacemaker. This innovation did not involve ceramics, but it kick-started the bionics industry. Earl Bakken was the cofounder of Medtronic and became globally famous.

Event 3 (1970). Australia. David Cowdery invented the alumina/titanium bionic feedthrough concept, the template for all subsequent bionic implants. David Cowdery was a Director of Telectronics from 1979 to 1988, at that time the world's third largest pacemaker company. While he was known in Australia and the global pacemaker engineering community in the late 20th century, he did not receive the global fame of Earl Bakken.

Obviously the history of bionics is much more complex than just these three landmark events. Before exploring the ceramic engineering of the alumina/titanium feedthrough, it is appropriate to look at a brief history of the pacemaker, for technological context.

8.2.2 Pre-1957: Desktop pacemaker

The history of bionic medical devices begins with the cardiac pacemaker. It is interesting to note that not only did the first hermetic implantable pacemaker have a British/Australian origin, was unpatented, and invented in Australia (see Section 8.1), but the original concept of the cardiac pacemaker also had a British/Australian origin and was unpatented. It was invented by a British doctor in 1928: Mark Lidwill, a British anesthesiologist and cardiologist, working at the time at Crown Street Women's Hospital in Sydney [145]. The original pacemaker was a large desktop external device, with electrical leads connecting it to the patient. It was initially used for pediatric resuscitation.

In spite of his ground breaking innovation, Lidwill did not patent his invention, and he deliberately avoided public recognition for his achievement, because in that era, artificially extending human life by interfering with the brain or heart was considered ethically controversial, a relic of the 19th century "Frankenstein" viewpoint on bionics. In the 21st century, bionics is seen as a positive futuristic technology, in the early 20th century, bionics had a stigma associated with it.

Over the subsequent three decades, ably assisted by the lack of a patent for Lidwill's invention, much developmental work on desktop pacemakers globally followed the innovation of Lidwill. Albert Hyman, Wilfred Bigelow and John Callaghan, John Hopps, Paul Zoll, Aubrey Leatham, and Geoffrey Davies all contributed in laying the groundwork for an understanding of cardiac pacing technology [145]. Paul Zoll, in particular, was one of the key players in this evolving technology.

8.2.3   Earl Bakken (Medtronic) 1957: The first portable pacemaker

Earl Bakken, cofounder of leading global medical device company Medtronic in the United States invented the world's first portable (wearable) pacemaker in 1957 [145]. This was a world-changing event. The invention of the transistor in 1947 (by American physicists John Bardeen, Walter Brattain, and William Shockley, who shared the 1956 Nobel Prize), and its availability to researchers from the late 1950s onward, was the key enabling event that enabled the Bakken invention and spawned the rapid proliferation of portable bionic devices in the subsequent decades. Beginning with the first wearable pacemaker in the United States in 1957, and immediately followed by the first implantable pacemaker in Sweden in 1958.

Earl Bakken is rightly acclaimed as the father of the pacemaker, and it is all the more impressive that his invention came so soon after the public availability of the transistor.

This great leap forward in pacemaker technology came three decades after Mark Lidwill's 1928 innovation, with the invention of the battery-powered wearable pacemaker. The Earl Bakken invention was the world's first portable bionic medical device. It was not an implant, but a wearable bionic device, worn around the neck and connected to the heart transcutaneously. That is, the device was external to the body, portable, and connected to the internal body cavity via transcutaneous conduit (electrical leads in the case of the Bakken pacemaker). It was therefore not just the forerunner of the implantable bionic pacemaker but also the forerunner of the modern wearable bionic devices, such as the bionic wearable pancreas (Medtronic), the bionic wearable kidney (AWAK), and the LVAD (left ventricular assist device for managing heart failure) with their transcutaneous conduit: transcutaneous catheters for injecting insulin in real time in the case of the bionic pancreas; transcutaneous catheters for transporting peritoneal dialysis fluids in the case of the bionic AWAK kidney; wearable battery pack with transcutaneous power leads in the case of the LVAD.

Bakken was a cofounder of Medtronic in 1949, and his 1957 invention was the innovation that led to Medtronic growing to become the world's largest stand-alone medical device company, capitalized today at about $100 billion. Thus the field of wearable and implantable bionics began in the post-war period almost immediately upon the commercial availability of the transistor. The transistor enabled portable electronic circuitry to be produced for the first time. Triode vacuum valve technology, which preceded the transistor era, was not portable.

A year later, in October 1958 in Sweden, Rune Elmqvist produced the first implantable pacemaker which Ake Senning successfully implanted in a human patient making it the world's first implanted pacemaker [376]. It was encapsulated in (recently invented) epoxy resin to temporarily seal the battery and circuitry from the body. While the epoxy resin was biocompatible, and an innovative solution to encapsulation, it was not stable in the body, gradually swelling due to moisture ingress, and ultimately dissolving [377]. Implantable epoxy-encapsulated pacemakers were first developed in the United States by pioneer Wilson Greatbatch in 1958 and first implanted in 1960. Paul Zoll, one of the pioneers of the external pacemaker in 1951, became a significant pioneer in implantable pacemakers in 1961. However, the early pacemakers were durable in the body cavity only for a matter of months, until Cowdery's invention of the alumina/titanium feedthrough in 1970, which gave durability measured in years.

8.2.4 David Cowdery (telectronics) 1970: The first hermetic implantable pacemaker

Telectronics was founded in 1963 by Geoffrey Wickham and Noel Gray. In 1964, electronics specialist Wickham had begun developing pacemaker prototypes using epoxy-encapsulation, like the Americans, and encountered the same problems with durability and hermetic sealing. In 1970, David Cowdery, an electrical engineer and leading expert in advanced welding technology, was recruited to Telectronics and tasked with solving the hermetic encapsulation problem. Within a year Cowdery invented the alumina/titanium hermetic feedthrough system [146,147]. It was commercialized in 1971, and it was a global technology within a few years, living proof that a game-changing technology, especially if unpatented, can change the world very rapidly.

Telectronics grew to be Australia's largest biotech company in the 1980s and 1990s, and the world's third largest pacemaker company. It was closed down in 1996 and its technology sold to leading US-based pacemaker company St Jude Medical, which itself was acquired by Abbott in 2016, a multinational biomedical company with a ~$100 billion market capitalization.

Earl Bakken was the father of the pacemaker and indeed modern bionics in general. David Cowdery was the father of the hermetic feedthrough system (titanium/alumina) which made *implantable* bionics viable and remains the industry standard today. The hermetic feedthrough has taken bionics way beyond the pacemaker in subsequent decades, with 14 bionic implant technologies on the market today, from bionic ear to deep brain stimulator, to spinal cord stimulator back-pain management devices, to the bionic eye. This comprehensive range of bionic technologies involved, in essence, "spin-offs" from the original bionic technology (pacemaker) all continue to be based on the hermetic alumina/titanium hermetic-bionic-implant concept of the pacemaker. This will be discussed in detail in Chapter 10.

8.3 The alumina/titanium hermetic feedthrough: Technology overview

8.3.1 The hermetic-bionic-implant concept

The alumina-ceramic-feedthrough/titanium-canister concept combines hermetic sealing with electrical signal feedthrough from the internal electronics to the exterior of the implant, enabling the connection of external leads which terminate in neural tissue stimulating electrodes. The titanium/alumina implant itself (also known as the generator or the implantable pulse generator—IPG) enables isolation of the internal battery and microcircuitry from the hostile fluids, ions, and gases of the body cavity. The assumption is that an hermetic bionic implant that utilizes the alumina/titanium feedthrough system is hermetic for the lifetime of the device. Hermeticity is generally measured by the helium leak test according to the military standard MIL-STD-883.1014 (A1–A4) [378] or MIL-STD-750.1071 (H1 and H2) [379], which are based on the Howl-Mann equation. [380]. Hermeticity requires a seal with a helium leakage rate of less than 10^{-9} cm^3 s^{-1}. Helium is the smallest molecule known and therefore its diffusion rate for permeating through a hermetic seal will exceed any in vitro moisture, gas, or ion.

The alumina feedthrough concept was originally invented for the bionic cardiac pacemaker, commonly known as simply the "pacemaker." This formed the template for all subsequent hermetic bionic implants. These implants have undergone significant evolution since the original invention. However, the basic concept of the titanium canister/alumina ceramic feedthrough remains the standard for hermetic-bionic-implant devices today. Hermetic-bionic-implant medical devices share several common features:

- The device requires a power supply. Usually this is an internal battery, which may be of the long-life nonrechargeable type, or may be a rechargeable battery with the capacity for transcutaneous inductive recharging. In some cases the hermetic bionic implant has no battery and is powered by transcutaneous induction.
- Hermeticity is essential, that is, the casing of the hermetic-bionic-implant medical device must be sealed against water, ion, and gas diffusion. Hermetic sealing is essential to prevent both inward and outward diffusion:
  - Inward diffusion: hermetic sealing is essential to shield the sensitive internal microelectronics, and the battery, from the harmful effects of humidity and body fluids. Common modes of failure from moisture exposure include corrosion, dendrite growth, surface, and current leakage.

- Outward diffusion: hermetic sealing is essential to shield the body from the potential harmful effects of the internal battery, and to a lesser extent the risk of toxic leaching from the internal microelectronic components.
- The hermetic-bionic-implant device requires external electrodes for interfacing with the neural tissue of the body. These electrodes must be electrically connected to the internal electronics of the device.
- A hermetic electrical feedthrough system is essential in order to communicate electronically between the internal microelectronics of the device and the external electrodes that interact with the neural tissues of the body, without permitting inward or outward diffusion of water, ions, and gases.

The battery-powered implantable cardiac bionic pacemaker, or simply the "pacemaker" was the world's first implantable bionic medical device. It remains the world's largest-selling bionic device, with over 1.5 million implanted annually at a retail price in the $5000–$10,000 per pacemaker, making a global pacemaker market size of over $10 billion [374]. The bionic pacemaker in its modern form involves five main features, known as the hermetic-bionic-implant system:

(1) Titanium case containing microelectronics and battery.
(2) Titanium case is simultaneously hermetically sealed and electrically connected to the outside environment utilizing an *alumina ceramic* electrical/hermetic feedthrough system, commonly known as the "feedthrough". The alumina performs the dual function of electrical insulator and hermetic seal.
(3) External electrode connection terminals.
(4) Hermetic bionic implant is implanted just beneath the skin in a convenient location for ease of access. Removal of the device does not necessarily involve removal of the electrode leads. A new device can be plugged into the existing leads. In the era before inductive battery recharging, this was a common procedure.
(5) The electrode leads travel often long distances to the neural stimulation site, for example, deep within the heart, deep within the brain, deep within the spinal column, a remote peripheral nerve, etc. (Fig. 8.1).

The hermetic-bionic-implant system, designed as a programmable nerve stimulator (neuromodulation implant) dates back to the late 1950s in its first manifestation as a rudimentary experimental short-service-life epoxy-clad bionic cardiac pacemaker, and in its modern titanium-alumina manifestation, invented in 1970, became the template for the hermetic-bionic-implant concept for the numerous subsequent bionic implant medical devices in use today, none of which individually has the market size of the pacemaker, but collectively they exceed it. For example, the Cochlear implant (bionic ear), the world's second ever hermetic-bionic-implant device, first commercialized in 1985, now has a global market size of about $2 billion.

8.3.2 The problem with epoxy encapsulation

The late 1950s and the 1960s was a time of intense pacemaker development and human clinical trials, but it was all constrained by the limitations of epoxy encapsulation, the primitive, unstable, and very temporary seal adopted in the early pacemakers. It takes about a year for body fluids to diffuse through the epoxy capsule,

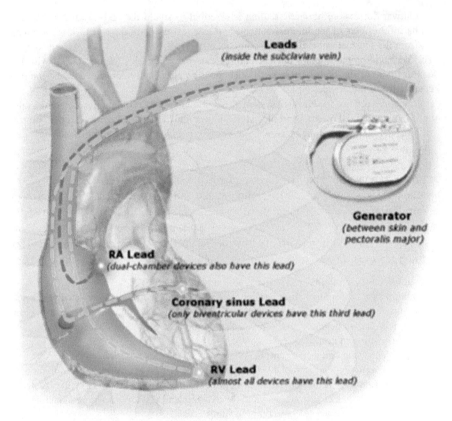

Fig. 8.1 The hermetic-bionic-implant concept, as in the case of the pacemaker. The titanium/alumina generator device (commonly known as the implantable pulse generator—IPG) is located beneath the skin in a convenient location for ease of access and removal. The leads connect to remote electrodes, in this case deep within the heart.
Image Author Nicholas Patchett [381]. Reproduced from Wikipedia (Public Domain Image Library).

resulting in failure of the implant, but the process of diffusion and the mode of failure is random and unpredictable. Diffusion through the capsule results in hazardous interaction between the battery and the body cavity, and between the body fluids and the sensitive microelectronics. In short, epoxy-encapsulated pacemakers were not a safe or viable technology for anything other than short-term experimental use.

What was needed, in order to make implantable bionics a viable technology, was the development of a biocompatible and hermetic feedthrough system. Hermeticity is usually measured using the helium leak test (MIL-STD-883.1014.A4 [378]) and very high standards of hermeticity are essential as some hermetic-bionic-implant implants, such as the bionic ear, can remain in the body cavity for decades, and for most implants residence time is many years.

8.3.3 The concept of the hermetic feedthrough

A hermetic feedthrough is a system that enables a number of electrical conductors (channels), anywhere from 1 to 1000+ channels, to pass through the wall of a sealed hermetic-bionic-implant device while retaining a perfect hermetic seal, and to simultaneously prevent the feedthrough metal conductor from electrically shorting out on the container wall. The common scenario is metal conductors (channels) passing through an insulating ceramic, with a hermetic seal at the metal-conductor/alumina interface and a hermetic seal at the titanium-casing/alumina interface. Essentially it combines an electrically insulating hermetic seal with electrical connectivity to the external environment. The feedthrough needs to be watertight and gastight, biocompatible, and hermetic over many years in the hostile biological environment of the body. It is well known that the human body is one of the most hostile corrosive environments for synthetic materials to endure, outside of a chemical manufacturing plant. This is why epoxy resin, while biocompatible, was not a stable pacemaker hermetically sealable containment system.

The feedthrough is therefore a system-critical technology for a bionic implant as it is required to enable the electrical channels for the electrodes to pass through the wall of the device while protecting, for many years in most cases, the internal microelectronics of the device from the penetration of body fluids and tissue. Equally, it is to protect the body from the batteries which can potentially represent a biohazard if they chemically interact with the body. Long-term stable hermetic biocompatible hermetic sealing technology for the feedthrough electrical conductor channels at the point where they pass through the wall of the "feedthrough" (strictly speaking epoxy is not a feedthrough) was the major challenge in early pacemaker development.

Feedthroughs are typically cylindrical, bonded at the perimeter utilizing a ceramic or glass that is bonded to the wall of a sealed containment system. It can be bonded chemically, or via a compression bond, or via a brazed bond, or combinations of these three bonding systems. Simultaneously, one or more electrical conductors usually in a rod-like form is bonded to the ceramic or glass chemically, or via a compression bond, or via a brazed bond, or combinations of these three bonding systems.

In the case of bionics, these electrical conductors are almost always platinum or its alloys, and can be in the form of a thick wire, thin wire, metal tube, or printed platinum track, and they are each individually bonded hermetically to an alumina ceramic insulator by a diffusion bond or by brazing.

Ceramic feedthroughs utilizing glass are a long-established concept and were an enabling technology for the invention of the incandescent light globe in the 19th century, later adopted in the triode vacuum valve (the basis of the pre-1950s electronics industry), and for numerous modern electrical devices ever since. The glass bulb chamber of the incandescent lightglobe is either under vacuum or filled with inert gas, to protect the hot incandescent tungsten or carbon filament from oxidation. The electrical conductors pass through the glass bulb wall and are fused to the glass at the point of passage, thereby combining both a hermetic seal and electrical conduction through the wall.

A similar principle is used for the feedthrough in Triode vacuum valves (vacuum tubes) cathode ray tubes and many electrical components. The common scenario is the use of solder glass for a Kovar wire feedthrough through a glass chamber wall or a metal chamber wall. The solder glass has a dual functionality:

- electrically insulate the electrical conductors from the metal chamber wall
- hermetically seal the metal chamber wall at the holes through which the electrical conductors pass

This is a standard electronics industry solution to the problem of hermetic electrical feedthroughs, that is, getting electric current through a glass or metal wall or shield in such a way as to not break a hermetic seal, nor electrically short out on the metal wall or shield. In a common manifestation, a solder glass feedthrough involves the solder glass acting as insulating tube seals where the Kovar metal conductors pass through the hole in the metal chamber wall, as shown schematically in Fig. 8.2.

Kovar, a Ni/Co/Fe alloy, is essentially iron containing approximately 29% nickel and 17% cobalt. It is the industry standard for glass metal feedthrough seals as it is an alloy that is engineered to have a perfect thermal expansion match with borosilicate glass across the required operating temperature range [382]. Kovar/borosilicate glass is the basis of the incandescent light bulb, the triode vacuum valve, and many other later spin-off electrical technologies of the 20th century (Chapter 14). Without a perfect thermal expansion match across the entire service temperature range (up to the softening point of the glass), thermal expansion will crack the glass-metal seal during heating or cooling. Obviously, thermal expansion matching is required only up to the softening temperature of the glass. Kovar-borosilicate glass seals are formed by heating above the softening temperature of the glass, and in service cannot be allowed to reach the softening temperature of the glass, or else the seal would be lost due to creep of the softened glass.

Kovar and borosilicate glass are not biocompatible in vivo and therefore cannot be used in bionic feedthroughs. Indeed, there are very few metals that are biocompatible in vivo. Almost all biomedical implants rely on just six bioinert biometals: titanium, cobalt chrome, tantalum, 316L stainless steel, zirconium, and platinum. To this list can be added the experimental biodegradable metals, pure iron and magnesium-calcium alloys. In implantable bionics, titanium, and platinum account for almost all commercial uses today. Kovar is certainly not on the list. Moreover, while there are numerous specialty solder glasses in existence [383], none of them are established as biocompatible, and all of them would be greatly inferior in hermetic stability in vivo in comparison to alumina, which is the most bioinert biomaterial in clinical use today.

8.3.4 Alumina-metal hermetic sealing in the early electronics industry

Alumina/metal hermetic seals, and indeed oxide ceramic/metal hermetic seals in general are a long-established technology, commonly used for vacuum electronic devices for which the hermeticity requirement is recognized as a helium leak rate of less than 10^{-9} cm^3 s^{-1}. Prior to the mid-20th century, the oxide ceramic used for such

Glass chamber hermetic feedthrough schematic

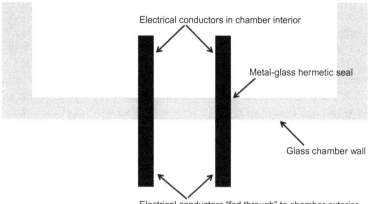

Solder glass hermetic feedthrough schematic

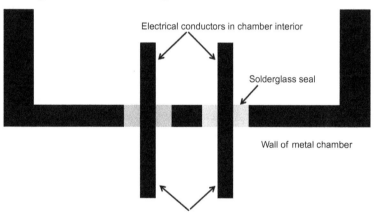

Fig. 8.2 Above: Glass-metal hermetic feedthrough of the type used in the light globe and other spin-off technologies. A metal (usually Kovar) with a perfect thermal expansion match to the glass (usually borosilicate) is fused into the wall of the glass chamber, which may be done while the chamber is under vacuum or filled with inert gas, for example, Kovar electrical feedthrough into a glass lightbulb to the (oxidation susceptible) incandescent tungsten filament. Below: Solder glass seal as used in providing both a hermetic seal and electrical insulation at the hole in the metal chamber wall through which the conductor (usually Kovar) passes. The common scenario is borosilicate solder glass and Kovar (conductor) wires passing through the solder glass seal. The solder glass (gray) fuses to both the Kovar (conductor) wires and chamber wall and provides both a hermetic seal and electrical insulation. Commonly used in electrical component manufacture.

applications was glass, bonded to Kovar [384]. The 1939 patent of Pulfrich [385] was the first step toward using oxide ceramics rather than glass for this purpose. However, te Pulfrich did not explore alumina, but rather pyrolusite, BaO, CaO, ZrO_2, and sodium tungstate.

The first documented alumina-metal hermetic seal came in the mid-1950s, with the patents of Nolte [386,387], originally filed in 1947, but not granted until 1954. The Nolte patents are primarily about the molybdenum-manganese ceramic-metal bonding method. The ceramics involved are not claimed, although the patent notes "I prefer to use the zirconium silicate or magnesium silicate bodies; however, due to the nature of the reaction it may be assumed that other materials are satisfactory, providing that they are refractory enough to withstand the prescribed firing temperature. Silicate, titanium dioxide, beryllium oxide, alumina, and others" [386].

This Nolte molybdenum-manganese ceramic-metal bonding method process was subsequently perfected and evolved into an alumina-dominated approach through the 1950s and 1960s [388–390]. By the 1970s The molybdenum-manganese ceramic-metal bonding method, bonding alumina to Kovar, had become the industry standard for ceramic-metal hermetic seals in the electrical industry, with alumina the dominant ceramic, though not the sole ceramic, used in this role.

The molybdenum-manganese ceramic-metal bonding method cannot be used for bionic implants, or indeed in any biomedical implant. This is because it involves non-biocompatible metals in the brazing (copper) and bonding to metal alloys that are both non-biocompatible and insufficiently corrosion resistant for the body cavity, that is, Kovar (iron-nickel-cobalt metal alloys). This is not to say it hasn't been tried. In the 1970s, the Mo-Mn process for alumina-metal seals, brazed with AgCu alloy, was trialed in pacemakers. The AgCu braze proved to have very poor corrosion resistance which led to failures in service.

Therefore, alumina-metal hermetic bonding techniques for electrical applications of alumina (Chapter 14) arose in the same era as the alumina bionic feedthrough innovation of Cowdery, but they evolved independently as they had very little in common with the bionic feedthrough innovation, for the following reasons:

- Alumina was not the only ceramic used in the electrical industry feedthrough systems. Alumina was, and remains the only ceramic used in bionic feedthroughs.
- The metal that the alumina (or alternative ceramic) was bonded to was Kovar, and sometimes copper, both non-biocompatible metals.
- The brazing alloys used were non-biocompatible.

This chapter has a focus on the case study of the invention, design, and development of the world's first hermetic bionic implant. For further background on the evolution of the electrical feedthrough, and the much more diverse electrical (nonbiomedical) industry applications of this technology, see Chapter 14.

8.3.5 A 21st-century biomaterials perspective

It is an interesting intellectual exercise to examine the challenge of developing an implantable bionic encapsulation and feedthrough system from scratch from a 2018 perspective. I take the perspective of a professor of biomaterials, with three decades of biomaterials and medical device research behind me, with the benefit of the contemporary highly evolved knowledge we now have about biomaterials.

Without a doubt alumina would be the only choice for the feedthrough. No other bioceramic combines the proven pedigree in biocompatibility and bioinertness, with its high electrical resistivity, high strength, and proven mechanical reliability. Moreover, solder glass the established (nonbiomedical) feedthrough seal is certainly not suitable on bioinertness, biocompatibility, or chemical durability in vivo criteria.

The question then becomes which biometal for the casing? The list of biometals that are both suitable from a biomaterials perspective and also have the credibility of regulatory approval and long-established proven implantable usage, gives us a short list of five: 316L stainless steel, cobalt chrome, tantalum, zirconium, titanium, and platinum. I would rule out 316L stainless steel due to crevice corrosion risk. Small but significant metallosis concerns would have me ruling out cobalt chrome and zirconium. Platinum is ruled out on obvious cost and weight grounds. The early pacemakers used about 200 g of titanium (SG 4.5) which equates to about 1 kg of platinum (SG 21.5) or 32 oz, at $1000/oz, this is a 1 kg pacemaker using $32,000 worth of metal. It should also be noted that in 1970 platinum was obscure as a biomaterial. It rapidly rose to prominence thereafter, but only in electrical uses, it has never been used for monolithic implants such as solid casings.

This leaves tantalum and titanium both of which have an excellent track record of long-term biocompatibility in vivo. Either could be suitable on biocompatibility criteria. The final decision would be made on thermal expansion matching criteria [391]:

Alumina: 8.1 μ/mK
Tantalum: 6.5 μ/mK
Titanium: 8.5 μ/mK

The thermal expansion of the metal needs to be higher than alumina, but not too much higher, so as to ensure that the metal-ceramic seal is enhanced, not diminished, upon cooling. The thermal expansion coefficient of tantalum is significantly lower than alumina, whereas titanium is slightly higher than alumina. Thus titanium emerges as the most suitable biometal. Of course the coefficient of thermal expansion is temperature dependent so the final decision would need to be made based on dilatometry data for the entire temperature range of the alumina-casing hermetic bonding process.[1]

None of this would have been known in 1970. Alumina ceramic was a completely new concept as a biomaterial in the 1970s although it was known at that time to a select group of specialist researchers in the electrical industry as a new and experimental electrical feedthrough ceramic, with alumina-Kovar the evolving alternative to glass Kovar in that era. The world's first experimental alumina hip was implanted in the same year. The first alumina hip was developed and implanted in France in 1970 by Pierre Boutin [144,225] (see Chapter 9). Only a small group of people in France would have been aware of this in 1970. So the use of alumina as the bioceramic in a bionic feedthrough was ground-breaking in 1970. The only documented alumina biomaterial usage in the literature prior to 1970 was an experimental alumina dental

---

[1] Fig. 14.14 contains relevant dilatometry data which verify this decision, based as it is here on documented thermal expansion coefficients.

implant that was patented in 1965 in the United Kingdom [141], and osteotomy spacers reported by Eyring in 1969 [142].

Today, titanium is the premier biometal in orthopedics and bionics due to its outstanding biocompatibility, capacity for osseointegration, excellent strength and fatigue resistance, outstanding corrosion resistance, and being nonferromagnetic, it is MRI safe. The 2010 global De Puy recall of the ASR cobalt chrome hip prosthesis further boosted titanium's preeminence given that it creates no toxic wear or corrosion residue.

However, in 1970, titanium was an obscure biomaterial. Cobalt chrome dominated orthopedics, titanium dental implants was a new and highly experimental field, known only to a small select group of specialist researchers in the dental field. While the first osseointegration studies of titanium were done by Branemark in the 1950s [161,162], this was in the small field of prosthetic dentistry, and it was not until the late 1960s that titanium dental implant research began to draw attention in the broader global dental community. Titanium was therefore known to some specialist dental biomaterials researchers in 1970, but was a novel innovation in bionics at that time. In 1970, bionics researchers exclusively used epoxy encapsulation.

8.4 The concept of the alumina feedthrough

The fundamental principles underlying the 1970 alumina feedthrough invention are essentially the "ship in a bottle" concept, where the "ship" was the electronics and battery, the "bottle" was a titanium canister, and the "cork" was the alumina feedthrough. Key issues were:

- biocompatibility of "bottle," "cork," and electrical terminal
- appropriate thermal expansion matching for "cork" and "bottle"
- appropriate sealing (brazing) technology for the cork-bottle sealing interface
- a system for passing the electrical terminal through the "cork" to the outside world to enable the electronics encapsulated inside to electrically communicate with the outside world while remaining hermetically sealed from them

Fundamentally an alumina/titanium combination, the hermetic-bionic-implant system has four essential components: the titanium casing "bottle," the alumina "cork," the titanium brazing seal, and the titanium electrical terminal that passes hermetically through the center of the alumina "cork". This is shown schematically in Fig. 8.3.

8.5 The specifics of the alumina/titanium feedthrough innovation

This section is based on an extensive interviews conducted by the author with inventor David Cowdery on the specifics of the feedthrough invention process [392].

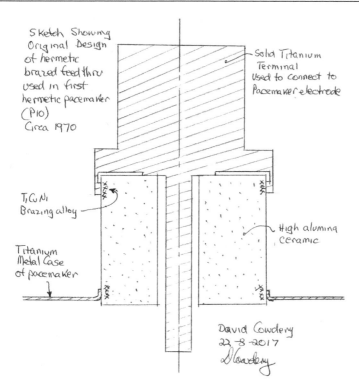

Fig. 8.3 A schematic representation of the alumina/titanium feedthrough system of the world's first hermetic pacemaker, the Telectronics P10. The alumina cylinder is 9.6 mm in diameter, with a 2.3 mm hole, through which is passed a 1.5-mm titanium terminal. Note the flanged hole in the titanium metal case of the pacemaker where it mates with the bottom of the alumina cylinder, and the flange where the terminal mates with the top of the alumina cylinder. Note the brazing at each flange. All brazing is on the periphery of the alumina cylinder so as to utilize the residual thermal contraction of the titanium flange onto the alumina cylinder due to the slightly higher thermal expansion coefficient of the titanium. This terminal feedthrough system can be seen on the top of the P10 pacemaker shown in Fig. 8.4.
Image courtesy of inventor David Cowdery.

8.5.1 Alumina ceramic

The first key innovation of the alumina/titanium hermetic feedthrough was the use of *alumina ceramic* as the "cork" in the titanium "bottle," that is, the exit seal. The alumina ceramic was brazed to both the opening in the titanium canister, and the titanium electrical terminal, as shown schematically in Fig. 8.3. This seals the opening in the titanium canister, while allowing the titanium electrical terminal to pass through the alumina ceramic "cork," passing through the center of the alumina ceramic seal, for the single channel electrical feedthrough.

The perfect suitability of the alumina feedthrough to its function is because of the outstanding physical properties of alumina outlined in Chapter 4, and the fact that alumina ceramic is well known as the most biocompatible and bioinert biomaterial in clinical use today. This principle was not well known in 1970. Also very important is the fact that alumina is an excellent electrical insulator and capable of being sintered to full density to enable a perfect hermetic seal.

Another key advantage of alumina for this role is its high strength, one of the specific reasons for its selection by Cowdery in 1970, along with its biocompatibility and good thermal expansion match to titanium. High strength was necessary because the ceramic feedthrough disk represented potentially a mechanical "weak link" joining the pacemaker casing to the external titanium terminal and was potentially subject to significant force when attaching the pacemaker lead to the titanium terminal. This can be seen by reference to Figs. 8.3 and 8.4.

In comparison to electrical porcelain, alumina is superior in every regard: greatly superior electrical resistivity, superior mechanical strength, superior durability in the body cavity, and far superior biocompatibility. Fortuitously, as was discovered a decade later, alumina offers the capacity to be co-sintered with platinum forming a solid-state diffusion bond, which was essential for the evolution of modern high channel-number feedthroughs, with the world record now standing at 1145 channels in a 5-mm diameter feedthrough [375] (see Chapter 10). However, in 1970, a key reason for the choice of alumina by Cowdery was its perfect thermal expansion match with titanium: alumina is 7.1 µ/mK and titanium is 8.5 µ/mK [391], sufficiently higher

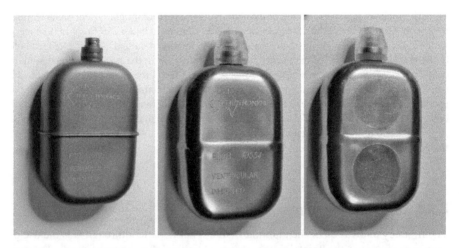

Fig. 8.4 Left: The Telectronics P10, the world's first hermetic pacemaker (1971). This is a single terminal unit, the case is the reference electrode. The terminal with alumina ceramic feedthrough is on the top of the titanium canister and is fitted with a silicon elastomer electrical sealing ring. Center: Later model of Telectronics P10 (c.1975) with flush weld and silicone elastomer coating. Plastic sealing cap is fitted over the terminal; Right: Right: 1975 version of P10 showing silicone elastomer coating.
Images courtesy of David Cowdery.

to enable the flanges of the titanium casing and terminal to shrink onto the brazed alumina/titanium joint during cooling, with sufficient thermomechanical force for sealing, but not so much thermomechanical force as to generate undue residual mechanical stress when cooled.

The specific alumina used in the 1970 innovation was 95% pure high alumina. This was necessary at the time because the equipment available at the time required a much lower sintering temperature than that required by pure alumina. The first prototype alumina ceramics for the 1970 feedthrough prototype were made by Mr. Kay of the Ceramics Corporation, a company that specialized in porcelain.

The early alumina ceramics for the feedthrough were made by die pressing, which was problematic in that die wear life was poor, and the process produced unwanted die flashing. Telectronics then approached a company called Gallard and Robinson who made extruded ceramics, and production thereafter was by extrusion (Chapter 4). The extruded alumina was also a high alumina 95% formulation. It contained a glassy phase, along with some TiO_2 to promote bonding with the titanium brazing alloy. The alumina ceramic composition (known as A14 grade) was as follows:

- Al_2O_3 95%
- Mg_2CO_3 0.5%
- TiO_2 1.66%
- MnO_2 2.84%
- Grinding size 3 μm

This formula contained sufficient glassy phase and other components to lower the sintering temperature into the practical range for the available production equipment without materially reducing its strength. With the closure of Gallard and Robinson some years later, Telectronics switched to in house production.

As can be seen schematically in Fig. 8.3, the alumina ceramic was not brazed to the titanium terminal in the central hole in the alumina ceramic through which the terminal passed through the center of the ceramic. Rather, the hole in the ceramic was slightly oversize, providing clearance between the alumina and the titanium terminal. All brazing was on the perimeter of the alumina ceramic, thereby enabling thermal expansion to do its work and shrink-fit the titanium around the alumina ceramic disk to which it was brazed at each end using the low-temperature TiCuNi brazing alloy. Thus all the brazed bonding was on the outside of the ceramic and was designed to be a compression bond due to the appropriately matched thermal expansion coefficients of alumina and titanium.

Alumina as the feedthrough biomaterial remains the standard for hermetic bionic implants today.

8.5.2 Titanium casing

The second key innovation of the alumina/titanium hermetic feedthrough was the use of titanium as the biocompatible sealable casing "bottle" for housing the battery and electronics (rather than epoxy as per the early implantable pacemakers). Titanium is highly biocompatible, corrosion resistant, and hermetic in the body for a lifetime.

It is also a perfect thermal expansion match for alumina. In 1970, Telectronics was the first company in Australia to have titanium cases deep-drawn from titanium sheeting. This specialist titanium fabrication work, now routine in the manufacture of bionic implants today, was a significant breakthrough at the time, enabled by the services of GA and L Harrington at Padstow who produced two identical deep drawn half cases from titanium sheeting. The deep drawing process radically changes the shape of the sheet metal while maintaining approximately the original wall thickness of the case over its entire surface.

Titanium casings remain the standard for hermetic bionic implants today.

8.5.3 The titanium electrical terminal feedthrough system

The third key innovation of the alumina/titanium hermetic feedthrough was the use of a titanium terminal (this first hermetic pacemaker had a single feedthrough channel) that was fed through the center of the ceramic. The titanium terminal was brazed to the perimeter of the ceramic disk, thereby providing a hermetic seal for the cylindrical terminal segment that passed through the hole in the alumina ceramic, as shown schematically in Fig. 8.3. All the Telectronics pacemakers until 1996 used titanium conductors for the feedthrough channels. Niobium and noble metal conductors for the feedthrough channels gradually appeared in hermetic pacemaker feedthroughs globally from the late 1970s and platinum was introduced in the 1980s for the first bionic ear, with its 22-μm-thickness electrode leads. Platinum, and its alloys has become the standard for bionic implants today. Moreover, platinum can be sintered directly into an alumina ceramic, opening up a range of possibilities, which will be discussed in Chapter 11.

However, in 1970, in the leap from epoxy encapsulation to alumina-titanium feedthrough technology, the use of platinum was not necessary for the pacemaker as only one feedthrough terminal, of large cross-section, was used for the pacemaker (the titanium canister itself was the reference electrode), unlike the 22 channels required for a Cochlear implant. Since it was necessary for only a single terminal to be fed through in the pacemaker feedthrough, and it could be quite large in diameter, the metal used for the terminal therefore did not require such a high electrical conductivity as platinum, given its high cross-sectional area.

In fact, neither platinum (9.4×10^6 S/m) nor titanium (2.5×10^6 S/m) has particularly higher electrical conductivity values than copper (59×10^6 S/m) or silver (62×10^6 S/m) [393]. Platinum is 3.75 times more conductive than titanium, but 6.6 times less conductive than silver. So while platinum is an improvement over titanium in conductivity, it is by no means one of the best electrical conductors, but rather the best combination of biocompatibility, corrosion resistance, and electrical conductivity of the proven biomaterials.

In respect of this combination of important criteria, titanium is certainly comparable to platinum. The electrical conductivity of platinum and titanium could be described as moderately high, but their biocompatibility and bioinertness is outstanding. However, when it comes to the ultrafine platinum conductor tracks used in a bionic eye retinal implant (Chapter 10), the higher electrical conductivity of platinum does become important.

8.5.4 Biocompatible brazing technology

The fourth key innovation of the alumina/titanium hermetic feedthrough was brazing the alumina to the titanium, that is, to braze the alumina "cork" to the titanium "bottle". This was a very challenging task. The "ship" inside the bottle is sensitive electrical circuitry and battery, and the alumina "cork" needs to be brazed to the titanium "bottle," a process involving heat, which must not damage the sensitive electrical circuitry or battery inside. It is obvious that Cowdery's unique background as an electrical engineer and leading expert in specialist welding research was an essential criterion for solving this challenging problem, but a capacity for innovation and inventiveness would have also been vital.

Specifically Cowdery innovated the use of a TiCuNi brazing because it was a high titanium alloy which had the lowest practical melting point of suitable titanium alloys. Both TiNi and TiCuNi were tested but the flow properties of TiCuNi were superior and the melting point of the TiCuNi alloy was slightly lower. Brazing temperature and time need to be minimized when brazing alumina to titanium so as to prevent recrystallization of the titanium canister, and prevent titanium penetration of the alumina ceramic which can compromise the strength of the alumina. It was also necessary for the brazing alloy to be high in titanium for maximum corrosion resistance and biocompatibility. This is an "active braze," a concept which is discussed in detail in Chapter 14.

Gold brazing could have been used in this role, but Cowdery specifically avoided gold brazing due to the concern that prevailed at that time about the risk of gold sensitivity in patients. Today, gold brazing is commonly used with platinum and other noble metal feedthrough wires in alumina bionic feedthroughs. The alumina ceramic is presputtered with gold, and then gold brazed to the electrical feedthrough wire. Gold sensitivity can be prevented through designs which prevent gold exposure to the body fluids. Gold brazing is commonly done with platinum feedthrough wires, but it would be equally viable for the original Telectronics titanium terminal with its 1.5-mm wire feedthrough. In many cases these days, platinum feedthrough conductor wires are directly sintered into the alumina ceramic feedthrough (see Chapters 11 and 12).

The 1970 development involved some prototyping brazing trials by David Miller of the Amalgamated Wireless Valve company, who had a vacuum induction furnace used to make triode vacuum valves, obviously from non-biocompatible materials. The titanium and alumina ceramic components (titanium casing, alumina disk, and titanium external terminal) were manufactured to tight tolerances. Brazing powder covered the outside diameter surface of the ceramic. The very small clearance between the titanium components and alumina disk was filled with molten braze and allowed to cool.

During cooling, the small thermal expansion mismatch of titanium and alumina meant that the titanium components would shrink more than the alumina on cooling and thereby compress onto the alumina, enhancing the seal. The system was designed so that the compression was sufficient for a hermetic seal, and not so much as to risk failure of the ceramic.

The final stage, after cooling, was the removal by sandblasting of excess brazing on the ceramic, so as to eliminate a potential short-circuit between the exterior titanium terminal, and the titanium casing which was the reference electrode. This geometry can be seen in Fig. 8.3. In later evolutions of the technology, the surface of the exposed alumina disk was coated in silicone.

In the first commercial manifestation of this innovation, the Telectronics P10, the following dimensions applied (refer also to Fig. 8.3):

- diameter of the hole in the alumina: 2.3 mm
- diameter of the titanium terminal shaft passing through the alumina hole: 1.5 mm
- peripheral clearance of the titanium terminal in the alumina hole: 0.4 mm
- outside diameter of the alumina cylinder: 9.6 mm
- length of the alumina cylinder: 7 mm

The first prototype-brazed alumina pacemaker feedthroughs were tested by the same thermal shock testing systems used for standard electrical components: boiling water and liquid nitrogen.

Subsequent to this prototyping work at the Amalgamated Wireless Valve company, Cowdery developed for Telectronics a novel alumina-ceramic/titanium-casing brazing system, and a controlled atmosphere welding machine that could produce clean welds in titanium without transmitting any damaging heat to the enclosed electronic circuits. This was one of the major challenges in the development of the alumina bionic feedthrough, and this developmental work has formed the platform technology for subsequent bionic implants to this day.

8.5.5 A brief summary of the manufacturing process

This process summary is specifically for the P10 pacemaker shown schematically in Fig. 8.3, and in the photograph in Fig. 8.4 [392]. The manufacturing principles are applicable to alumina/titanium feedthrough technology in general.

8.5.5.1 Stage 1: Assembling the feedthrough

- Two identical half-cases were deep drawn from titanium sheeting
- One half-case had a hole punched through at the top with the edge of the hole curled up at 90° forming a flange. This flange can be seen in Fig. 8.3 where the hole in the casing mates with the bottom of the alumina cylinder.
- A ceramic feedthrough cylinder was fitted into the flanged hole.
- A titanium terminal was placed on top of the ceramic.
- Nitrocellulose lacquer was applied to the exposed outside surface of the ceramic and finely divided braze alloy powder blown so as to cover the ceramic surface.
- This assembly was then placed into a molybdenum heat radiating bucket and heated to brazing temperature in an RF induction furnace.
- The outside surface of the ceramic was sandblasted to clean the alumina ceramic of any surface brazing metal. This was to ensure that the wall of the alumina ceramic was free of metal braze and therefore no conductivity pathway existed that could enable a short circuit to the titanium casing. After sandblasting, this was tested electrically.

8.5.5.2 Stage 2: Assembling and sealing the casing

- The battery and electronic circuit board assembly was placed into the two half cases.
- The two output electrical leads were joined to the terminal and case, respectively: lead 1 joined to the terminal; lead 2 joined to the inside of the top half case. The case was the reference electrode in vivo.
- The two half cases were then welded together.
- Each half case had a raised flange at the point of contact so that the final weld was in the form of a raised ridge at the join of the cases formed by partly melting the small flanges together.
- The welding atmosphere was a very highly purified mixture of Argon and Helium*. The welding process was optimized to suit this gas mixture.
- A silicone elastomer O-ring was glued over the exposed outer surface of the ceramic.
- A sealing cap was screwed over the terminal and O-ring to create a fluid seal so that the stimulating pulses were not short circuited by body fluids.
- Electrical testing was carried out on the sensing and pacing functions of the device.
- The pacemaker was cleaned and sterilized using ethylene oxide, packed in sealed packaging, and despatched.

In order to carry out leak testing using a helium-tuned mass spectrometer, it is necessary, for maximum leak testing sensitivity, to have a significant percentage of helium already present inside the implant. If no leak was detectable, the implant was considered hermetic. The leak tester was calibrated with respect to standard helium leak testing using MIL STD 202F [394].

8.6 Feedthrough evolution post-1970

The alumina/titanium hermetic feedthrough was commercialized in 1971. Production and sales commenced 15 months after the recruitment of Cowdery to Telectronics. The invention was not only unique, it was also done very rapidly. The world's first hermetic (many years in vivo) implantable bionic medical device was the 1971 Telectronics Pacemaker Model P10 [146]. This was also the world's first hermetic bionic implant: a titanium capsule containing microelectronics and battery, with a hermetic alumina ceramic feedthrough. Essentially Cowdery invented the alumina/titanium hermetic-sealing system from which all subsequent bionic hermetic-bionic-implant feedthroughs have been derived. This technology was so successful that it not only changed the world of bionics in the 1970s, it is the standard electrical feedthrough that underpins the global $25 billion bionic implants industry today.

Telectronics continued to use titanium feedthrough channels right up to 1996 when the company was shut down and its technology sold to St Jude Medical. Feedthrough channel wires of platinum and other noble metal alloys gradually became an industry norm for the 14 bionic implants on the market in the world today.

As mentioned before, Cowdery's alumina/titanium feedthrough innovation was never patented. Cowdery has patents on other inventions. His first patent was filed in 1974, 3 years after his feedthrough innovation, and granted in 1976 [395], concerning another ground-breaking innovation of his: medical-grade silicone elastomer coating the titanium casing on an hermetic bionic implant. This was for two reasons:

(1) given that the entire titanium casing was the reference electrode and thus electrically live, it was desirable that only a selected zone of the titanium casing was electrically active so as to reduce unwanted muscle stimulation; (2) silicone coating improved the anchorage of the titanium casing in the soft tissue, preventing the risk of the device migrating through the body tissues, that is, for "permitting the body tissue to act on the elastomer itself to provide a form of adhesion thereby assisting in stabilizing the pacemaker position within the body" [395]. This innovation continues to be used to this day with Cochlear implants, optimizing case electrical activity and enhancing fixation stability in the soft tissues of the implantation site. Silicone-coated Cochlear implants now number over 250,000 worldwide. Silicone coating is also commonly used for other bionic implants (Figs. 8.5–8.7).

Over the next two decades, Telectronics stuck with the titanium conductor in their feedthroughs, they did not make the switch to platinum conductors. Telectronics continued to miniaturize and evolve the technology, including further miniaturizing the alumina ceramic feedthrough and increasing the number of feedthroughs. Alumina ceramic innovators and pioneers David Taylor and Julie Taylor, founders of Taylor Ceramic Engineering (TCE: a leading alumina manufacturing company founded in 1967—see Chapter 1, Section 1.4.2.5), worked closely with Telectronics in the 1980s and 1990s on continuous improvements to the alumina ceramic feedthrough technology. TCE was active in research and development of the alumina feedthrough

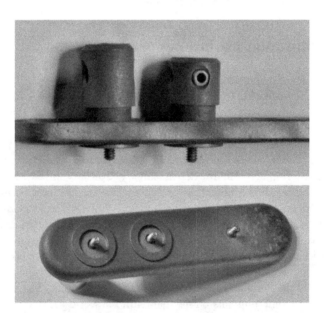

Fig. 8.5 Above: Later model of the Telectronics pacemaker terminal assembly as brazed. This is for a smaller thinner pacemaker than the P10. It is about 9 mm thick. Below: The underside of the same terminal assembly. The two electrical terminals can clearly be seen exiting through the alumina ceramic.
Images courtesy of David Cowdery.

Alumina in bionic feedthroughs: The pacemaker 249

Fig. 8.6 A later model Telectronics pacemaker (c.1976) with a bare terminal assembly identical to that used on the first P10.
Images courtesy of David Cowdery.

Fig. 8.7 Left: A more modern Telectronics pacemaker prototype (c.1995) that is only 6 mm thick and similar to pacemakers available today. Right: The multiple terminal assembly for an advanced Telectronics defibrillator (c.1996) that never went into manufacture.
Images courtesy of David Cowdery.

system and associated specialty alumina componentry, including providing significant technical support to aid Telectronics in their understanding of technical/engineering ceramics in order to determine failure mechanisms for some of the components of their technology, to advise on the most suitable Alumina components for the specified requirements, and also in developing the following precision high-purity alumina components:

- alumina insulating fixtures for terminal welding
- alumina insulator cassettes
- alumina ceramic bushes to micron tolerances (5–10 μm precision)

Fig. 8.8 Telectronics six-channel terminal including alumina componentry developed by Taylor Ceramic Engineering.
Image courtesy of Taylor Ceramic Engineering.

- alumina measurement gauges to micron tolerances (5–10 μm precision)
- manufacture of an annular porous substrate of ultrafine dimensions: 2.56 mm OD; 0.58 mm ID; 0.7 mm length (part of the electrode assembly) (Figs. 8.8 and 8.9).

Telectronics continued to use titanium conductors in their alumina right up to 1996 when the company was shut down and its technology sold to St Jude Medical. However, elsewhere, conductors of platinum and other noble metals, gradually became an industry norm for the more than 14 bionic implants in the world today.

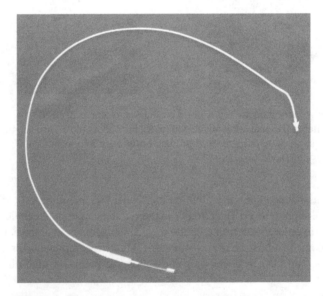

Fig. 8.9 Telectronics pacemaker electrode assembly incorporating ultrafine alumina componentry developed by Taylor Ceramic Engineering.
Image courtesy of Taylor Ceramic Engineering.

The most significant evolution in the Pacemaker since 1970 has been miniaturization. The battery microelectronics and feedthrough can now be miniaturized to the point that the pacemaker is small enough to be implanted directly at the stimulation site, with stub electrodes, rather than leads traveling to remote electrodes. This is called the lead-less pacemaker. This is a significant advance because failure in the leads (from the multiple fatigue cycles of the beating heart) is a common failure mode for pacemakers. This is not practical technology for other bionic applications, such as deep brain stimulation, or spinal cord stimulation, but for the heart and major nerves it is. The Medtronic Micra transcatheter pacing system is a pacemaker the size of a large vitamin capsule, and is implanted directly into the heart, via its stub electrodes [396]. First implanted in 2013, it is a very recent technology.

A titanium or niobium alloy provides the best thermal expansion match with alumina (see Fig. 14.7). A process for bonding tantalum and niobium to alumina was developed by Elssner in 1974 [397] from within the electrical industry. This ultimately saw no significant commercial application in the electrical industry. However, the alumina/niobium bonding technology of Elssner [397] was adapted by the newly established US pacemaker industry in the late 1970s, from which the world's second bionic pacemaker technology was launched in the late 1970s: alumina/niobium. A competitor for the alumina/titanium pacemaker feedthrough technology developed in Australia in 1970.

While niobium is biocompatible, its mechanical property limitations are its Achilles heel. This was a problem in the 1970s when it was first trialed in bionic implants, and it remains a problem today [398]. Moreover, the thermal expansion of niobium (Fig. 14.14) is so close to alumina that niobium is incapable of placing the alumina in significant stable residual compression on cooling, in accordance with the principles of seal design shown in Fig. 8.3. In the 1970s, US pacemaker manufacturers experimented with gold brazing of niobium to alumina. The softness and ductility of the gold, combined with the very close thermal expansion characteristics of alumina and niobium, resulted in minimal interfacial stresses. However, the design needs to be made in such a way that body fluids are not exposed to gold due to the hazard of gold sensitivity.

Thus while niobium can be made to work, it is not well suited to the role and never proved to be as successful as titanium in alumina-metal seals. Ultimately the alumina-niobium technology was supplanted globally by the alumina-titanium technology. Titanium on the other hand has an outstanding combination of excellent mechanical properties, and excellent biocompatibility, with over half a century track record of proven performance in orthopedics, bionics, dentistry, and other implantable medical devices.

The most significant pacemaker evolutions since 1970 have been

(1) The gradual trend toward using platinum and other noble metals rather than titanium for the channel conductors through the alumina, and co-firing platinum feedthroughs into alumina
(2) Miniaturization
(3) Increasing channel number

Less universal has been an increasing prevalence of gold brazing, and the use of titanium nitride-coated tissue-interfacing electrodes, which provide low tip-to-tissue polarization.

Alumina/stainless steel, alumina/platinum, and alumina/niobium bonding systems have all been explored. However, the only lasting materials innovation since 1980 has been the introduction of platinum conductors, originally by cochlear Ltd. for the bionic ear in the early 1980s. Platinum is now dominant in this role, as outlined in Chapters 11 and 12.

Appendix 8.1: Feedthrough Terminology

In the context of this book, and particularly Chapters 8–10, the following terms will be used to describe the components of the implantable bionic hermetic packaging system. While nonbiomedical feedthroughs commonly use hermetic insulators of solder glass or other ceramics, and even polymers, bionic feedthroughs exclusively use alumina as the hermetic insulator. Some of this terminology does not arise until Chapter 11, some not until Chapter 12.

Hermetic: Sealed and impermeable to leaks of fluid, ion, or gas, for the service life of the device. Certified by a helium leak test.

Channel: A feedthrough has from 1 to over 1000 channels passing through it, each comprised of an individual metal conductor pathway of platinum or platinum/iridium. The original pacemakers had a single channel and it was titanium. The alumina is hermetically sealed to the titanium casing, and to the platinum of the channel feedthrough wires.

Alumina: Alumina is the biocompatible electrical insulator material of which the feedthrough is comprised. It has a number of electrical channels passing through it in the form of metal conductors (titanium or platinum) that are hermetically sealed to it and insulated by it. In the context of bionics, alumina has a dual role: electrical insulator and hermetic seal through which the electrical channels (metal conductors) pass. Thus the word "alumina" in Chapters 8–10, means the feedthrough.

Hermetic bionic implant: A package of electronics with or without a battery in a hermetic titanium casing with an alumina feedthrough to provide a hermetic seal for the titanium casing and the electrical channels (metal conductors) that pass through it. There are three hermetic-bionic-implant systems used, referred to with the following terminology:

- *Titanium/alumina*: titanium casing, alumina feedthrough, metal conductor of which the feedthrough channels are comprised, is not specified.
- *Titanium/alumina-titanium*: titanium casing, alumina hermetic seal, feedthrough channel wires comprised of titanium.
- *Titanium/alumina-platinum*: titanium casing, alumina hermetic seal, feedthrough channel wires comprised of platinum.

Feedthrough: In essence a bunch of "wires" passing independently through a hermetic insulator (alumina) and bonded to it hermetically. Each channel is individually hermetically bonded to the alumina, and thereby electrically isolated from the others. Channels take many forms: thick wire, thin wire, metal tube, and printed platinum track. Usually a feedthrough has many channels and they are comprised of platinum or platinum/iridium. The original pacemakers had a single channel and it was titanium. The alumina is hermetically sealed to the titanium casing, and to the platinum of the feedthrough channel wires.

Chip side: The inside of the hermetic bionic implant, that is, the side of the feedthrough that interfaces with the internal electronics.

Electrode side: The outside of the hermetic bionic implant, that is, the side of the feedthrough that interfaces with the body cavity, either via electrical terminals connected to the feedthrough channel wires to which the electrode leads are connected (e.g., pacemaker), or directly when the channels are flush platinum "electrode pads" and the feedthrough with its flush embedded electrode pads acts as a live electrode surface interfacing directly with neural tissue (e.g., bionic eye retinal implant).

Terminal: The attachment point on either the "electrode side" or the "chip side" through which the feedthrough channel wires connect to the internal electronics (chip side) or external electrode leads (electrode side).

Electrode: Exposed live electrical conductor that injects electrical current into neural tissue in the form of AC electrical pulses. It is made of platinum or platinum/iridium alloy, in some cases coated with a biocompatible electrical conductor such as titanium nitride.

Electrode lead: The cabling that connects the terminals at the "electrode side" of an hermetic-bionic-implant feedthrough with the remotely located electrodes. The electrodes can be tens of centimeters away from the hermetic bionic implant, for example, deep in the heart or deep in the brain, while the implant is just beneath the skin in a convenient location. Most commonly, leads are silicone or polyurethane-insulated noble metal wires of platinum/iridium or titanium. An example of a miniaturized system is 25 µm platinum wires coated with a molecular-thick coating of parylene insulator (bionic ear). An example of an MRI-safe system is medical-grade nickel alloy, with silicone insulation (pacemaker).

Alumina in bionic feedthroughs: The bionic ear

9.1 Bionic ear: A global race in the 1960s, 1970s, and 1980s

Alumina had a central role in the hermetic encapsulation/feedthrough development of the world's first bionic ear (Cochlear Implant) capable of speech recognition, which was in fact the world's second implantable bionic technology to be commercialized (1985) after the worlds first implantable hermetic bionic device—the pacemaker (1971). The 1960s, 1970s, and 1980s were a time of intense and competitive bionic ear research from a large body of competing bionic ear researchers and research teams globally. It was a race to be the first to develop and commercialize a bionic ear capable of speech recognition, and be the first to market. During this time, there was much debate globally as to whether speech recognition was technically possible, and if so whether a single channel implant could enable speech recognition, or if it would require multiple channels, and if so, how many channels. Ultimately this race was won with a 22 channel hermetic titanium/alumina-platinum Cochlear implant, the world's first bionic ear capable of speech recognition. This was achieved by an Australian team led by otologist and inventor Graeme Clark, the founder of Cochlear Ltd. In the year 1985, Clark's bionic ear was FDA approved and commercialized globally thereafter by Cochlear Ltd. 250,000 bionic ear devices (Cochlear Implants) have now been implanted, the company now has a market capitalization of $8 billion, several thousand employees, and turned over $1.15 billion this year. The times of uncertainty and controversy of the 1960s 1970s and 1980s are largely forgotten. This chapter outlines the history and development of the alumina feedthrough technology that underpinned the world's first bionic ear capable of speech recognition.

This achievement by Graeme Clark in 1985 came 5 years before competitor MED-EL was founded, and 8 years before competitor Advanced Bionics was founded, giving Graeme's Company (Cochlear Ltd.) a huge global advantage, which they retain today with 70% of the world bionic ear market. Of the four competitor companies in the bionic ear market today, the two main rivals for Cochlear Ltd. are Sonova/Advanced Bionics (Switzerland) and MED-EL (Austria). Two other bionic ear companies operate on a smaller scale: Oticon/Neurelec (Denmark) and Nurotron (China).

9.1.1 Bionic ear: The telectronics legacy

Graeme Clark and his chief engineer Jim Patrick led the world in the late 1970s in pioneering multiple-channel bionic ear technology and computational speech processing technology. However, these two important factors alone were not sufficient to win the race. A chain is as strong as its weakest link, and in the late 1970s, after a successful implantation in 1978, Graeme Clark's team did not have a commercially viable hermetic encapsulation system (alumina feedthrough technology), for a bionic ear implant, a component which was essential for commercialization.

Developing a titanium/alumina hermetic encapsulation system in an 18-month period in 1981/1982 was a significant factor in Graeme Clark's team winning the global race to be the first to commercialize a bionic ear. This achievement was actually a legacy of Telectronics, the company which commercialized the world's first hermetic implantable bionic pacemaker more than a decade earlier in 1971 (Chapter 8). The era of implantable bionics began in the year 1970 with the invention of the titanium/alumina hermetic feedthrough system for implantable bionics by David Cowdery of Telectronics [146,147]. Specifically the alumina feedthrough/titanium casing implantable pulse generator (IPG) concept was invented for the 1971 Telectronics P10 pacemaker, the world's first hermetic bionic implant.

In the early 1980s, Telectronics/Nucleus became the parent company to Graeme Clark's team during the bionic ear commercialization program. Telectronics was, by then, a large and thriving pacemaker company possessing advanced in-house hermetic encapsulation technology for bionic implants (Telectronics P10 pacemaker) and all the manufacturing infrastructure and engineering expertise required for bionic ear hermetic encapsulation system design and development. Thus, Telectronics gave the Cochlear Ltd. team a huge headstart over the global competition in bionic ear hermetic encapsulation development. All that was required was to adapt the Cowdery Telectronics pacemaker feedthrough technology to the bionic ear requirements, rather than the much more challenging task faced by their global bionic ear competitors, of developing this from scratch.

As discussed in Chapter 8, alumina is an essential component of the hermetic encapsulation and electrical feedthrough system used in all implantable bionics, that is., the IPG concept: titanium case containing battery and microelectronics, brazed to an alumina feedthrough, with at least one channel (usually platinum wire) passing through and hermetically bonded to the alumina feedthrough. This enables simultaneously:

- Electrical connectivity from inside the casing to the external electrodes.
- Hermetic sealing of the casing to protect the microelectronics (and battery if one is required, which it was not for a bionic ear) from fluid, gas, and ion ingress from the body cavity.
- Hermetic sealing of the casing to protect the body cavity from the toxic effects of battery and microelectronics.

These principles are discussed in detail in Chapter 8.

The titanium/alumina hermetic feedthrough system remains the industry standard for the 14 bionic implant technologies that are now on the market today, half a century later. There has been much hermetic feedthrough technology evolution since 1970. The implantable bionics industry has evolved into applications in a wide range of bionic implants comprising a global industry in the multiple $10 billion order of magnitude, and growing rapidly, thereby making implantable bionics one of the most important commercial applications for alumina today. This will be discussed in Chapter 10.

9.1.2 Bionic ear: Quantum leap in hermetic feedthrough technology

The bionic ear was the next great leap forward in bionic technology after the pacemaker. A quantum leap in alumina feedthrough technology was required, which came in the year 1982 with the invention by Janusz Kuzma (Telectronics/Nucleus/Cochlear team), of the 22-channel cofired titanium/alumina-platinum feedthrough.

What stayed the same with the bionic ear feedthrough? Like the IPG concept of the original hermetic pacemaker, the bionic ear utilized the same IPG concept with a *titanium casing*, brazed with *TiCuNi* brazing alloy to the *alumina* feedthrough.

What changed with the bionic ear? Unlike the original hermetic pacemaker feedthrough, which had a single titanium terminal incorporated into a peripheral brazing flange, brazed peripherally to the alumina feedthrough cylinder, the Kuzma bionic ear feedthrough concept had 22 channels (not 1), made of *platinum* (not titanium), and *cofired* directly into the alumina feedthrough ceramic (not peripherally brazed). Moreover, the IPG *had no internal battery*. It was driven by an external transmitter using inductive coupling.

The ceramic engineering aspects of the quantum leap in alumina feedthrough technology for the Cochlear Ltd. bionic ear are outlined in detail in Section 9.3.

9.2 A brief history of the bionic ear

9.2.1 Bionic ear: The early history

The world's first experimental bionic ear clinical trial took place in 1957, the same year as the first experimental portable pacemaker (Earl Bakken). This was a single-channel experimental bionic ear trial. Pioneered by Djourno by request from otologist Eyries, it was trialed in a human patient in 1957 by Eyries [399]. This device was functional in that it stimulated a sound response for the patient, but it was ineffective in that all sounds sounded the same, since speech recognition is impossible with a single-channel device.

The next significant pioneer after Djourno was William House, an American otologist who, in 1961, conducted bionic ear clinical trials, and in 1972 implanted his first single-channel Cochlear implant. He was a significant presence in the US otology/bionic ear fraternity. With backing from 3M, House was the first to obtain FDA approval for a bionic ear, in 1984, a year before Cochlear Ltd. The bionic ear of William House was incapable of speech recognition and therefore rapidly faded from the market, however, not before about 1000 patients had been implanted [400].

Multiple-channel trials were pursued by various researchers. The first was by the Stanford University team of otolaryngologist Blair Simmons and engineer Robert White. Beginning with bionic ear clinical trials in 1962, they then went on to implant a six-channel bionic ear device in 1964 [401–403]. In 1970, otolaryngologists Robin Michelson and Robert Schindler, together with neurophysiologist Michael Merzenich, at UC San Francisco began bionic ear implantation clinical trials [404]. This early research evolved later into the bionic ear company Advanced Bionics, which was founded in 1993, and acquired by Swiss company Sonova in 2010. In 1973–76, France was again a major participant in bionic ear research with French otolaryngologist Claude Henri Chouard, who implanted six patients with a seven-channel bionic ear [405–407]. He developed a precommercial bionic ear, sold the rights to Neurelec in 1987, who commercialized it (but only in France) in 1991. Neurelec was acquired by Oticon Medical in 2015.

However, single-channel bionic ear research continued to see significant focus throughout this period. In Vienna in 1977, Erwin and Ingeborg Hochmair at the Vienna University of Technology designed and manufactured a passive single-channel broadband analog implant, which was first implanted in 1977 [408]. This began a long process to multiple-channel systems that ultimately led to the founding of MED-EL in Innsbruck, Austria, in 1990, with Ingeborg Hochmair as CEO. MED-EL is now the world's second largest bionic ear company after Cochlear Ltd., which still retains a 70% market share.

9.2.2 Brief history of Cochlear Ltd.

Invention and commercialization of the world's first hermetic bionic implant, the Telectronics P10 pacemaker, was a lightning fast process, invented by David Cowdery and commercialized by Telectronics in an 18-month period in 1970/1971.

In contrast, the bionic ear by Cochlear Ltd. involved dozens of engineers and other professionals, working cooperatively over many years. The invention and commercialization process took place over a 15-year period, from 1970 to 1985, and there has been an ongoing continuous improvement program ever since. However, at the heart of this invention, there were two central people:

> *Graeme Clark.* The Inventor. Graeme Clark is an otologist whose entire life has been devoted to the invention of the bionic ear. This was born of a childhood ambition to cure deafness, inspired by his hearing impaired father.
>
> *Jim Patrick.* The Engineer who made it work. Jim Patrick was the first Engineer (appointed 1975) employed by the bionic ear project. His entire professional career has been devoted to the bionic ear as the engineering innovator and project manager behind the whole development and commercialization process. He remains Chief Scientist and Vice President of Cochlear Ltd. today.

In addition to these two central people, the 15-year invention and commercialization process involved a large number of scientists, engineers, business people, audiologists, surgeons, and managers. The 250,000 bionic ear recipients globally are indebted to them all.

This book is about alumina ceramics. From an alumina feedthrough point of view, there have been two very important inventors:

> *David Cowdery.* Inventor (1970) of the hermetic titanium/alumina feedthrough (Chapter 10) and a consultant to the bionic ear project, which adapted his titanium/alumina feedthrough invention to the bionic ear.
>
> *Janusz Kuzma.* Engineer who adapted (1982) the hermetic titanium/alumina feedthrough concept to the bionic ear.

9.2.3 Cochlear timeline

The history of the invention and commercialization of the Cochlear Ltd. bionic ear spanned a 15-year period from 1970 to 1985, but it has its early roots in the early life of inventor Graeme Clark. The essential timeline is as follows [146,409]:

1957: Graeme Clark completes his Medical Degree at University of Sydney.

1958: Graeme Clark commences as Medical Resident at Royal Prince Alfred Hospital Sydney.

1961: Graeme Clark is appointed Medical Registrar in Ear Nose and Throat Surgery, Royal Prince Alfred Hospital Sydney.

1967–69: Graeme Clark (Now an Otolaryngologist) does a PhD at Sydney University (Australia) on electrical stimulation of the cochlea.

1970: Graeme Clark is appointed Chair Professor in Otolaryngology at the University of Melbourne.

1970–74: Graeme Clark builds up a team of researchers and conducts an extensive program of basic research into the question of one channel or many channels and how to generate speech recognition by cochlea stimulation.

1974 (October): First "crowd funding" event. TV "Telethon" $87 k ($1 million 2017).

1975 (January): Jim Patrick, PhD Electrical Engineering (Communications Systems) is the first salaried engineer appointed to the project.

1975 (December): Second Telethon raises more money than the first.

1976 (May): Third Telethon raises $125 k the most of all ($1.5 million 2018).

1977: Jim Patrick becomes bionic ear project manager.

1977: First prototype bionic ear is completed. Silicone-coated gold-plated box encapsulation system; 10 electrodes.

1978: First implantation (Graeme Clark) patient Rod Saunders. Speech recognition is proven.

1979. Australian Government announces tender for Grant to Commercialize Cochlear implant which Telectronics/Nucleus wins.

1979–85 Australian Government invests $4.7 m ($50 m 2018) in bionic ear commercialization via Telectronics.

1981 (28 October): Paul Trainor (Owner of Telectronics/Nucleus) appoints David Money Project Leader of Telectronics/Nucleus development of commercial Cochlear Implant device. Jim Patrick and Peter Crosby his two deputies. David Cowdery and Carl Doring expert consultants. These five are the management team of proto-Cochlear.

1981–82: Commercial bionic implant developed in an intensive 18-month effort including titanium/alumina hermetic encapsulation system, 22-electrode feedthrough, and inductively coupled external speech processor.

1983: Cochlear Ltd. is founded. David Money is the founding CEO.

1983–84: Extensive clinical trials for US FDA PMA commercialization approval.

1985: US FDA PMA approved. Global commercialization of the world's first bionic ear capable of speech recognition commences.

2018: Cochlear Ltd. retains 70% of the world bionic ear market, $1.15 billion turnover, and ∼4000 employees. Over 250,000 bionic ears implanted since 1980s.

9.2.4 Summary of the key innovations underpinning the bionic ear

- Multiple electrode (rather than single electrode) to enable speech recognition (Graeme Clark).
- Soft silicone-coated microcoiled electrode lead, with tapered and coiled end, capable of curving as it enters the cochlea so as to curl into the cochlea helix without damaging the cochlea neural tissue (Graeme Clark).
- The 22 electrodes engineered as platinum foil strips bonded to the soft electrode lead that is implanted into the cochlea (Quentin Bayley/Jim Patrick).
- External "smart device" (miniaturized microprocessor speech processor incorporating battery and microphone) providing inductive transfer of power and information to subcutaneous "receptor" implant via an inductive external transmitter coil connected to the "smart device" interacting with a receiver coil on the implant. This innovation involved a large team led by David Money, Jim Patrick, and Peter Crosby.
- The titanium/alumina hermetic encapsulation system. Adapted from the Cowdery pacemaker innovation to the bionic ear requirements by Janusz Kuzma.

The Cochlear Ltd. bionic ear concept has become the template for all competitor bionic ear technologies. In essence, 22 platinum electrodes are located sequentially on a microcoiled electrode lead with its 22 electrodes (Fig. 9.1) that coils into the cochlea helix, stimulating the cochlea neural tissues in 22 different zones in the cochlea helix, which correspond to pitch (highest frequencies are nearest the entrance

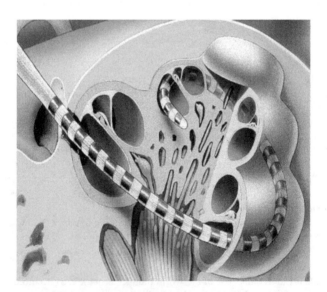

Fig. 9.1 The intra-cochlea 22-electrode-array implanted inside the cochlea. The highest frequency sounds correspond to the neural tissue at the proximal end of the electrode support, while the lowest frequency sounds correspond to the neural tissue at the distal end of the electrode support, deepest inside the cochlea helix.
Image courtesy of Cochlear Ltd.

to the cochlea helix, lowest frequencies are at the distal end of the cochlea helix). Thus, the 22 zones enable pitch control. Amplitude control is easy to achieve through stimulation intensity. Speech requires precision pitch and amplitude control. Therefore, with multiple intra-cochlea electrodes, and advanced computational power, sophisticated speech processing is possible. This helical electrode support is driven by very advanced technology, summarized in three main innovations:

(1) Advanced miniaturized alumina feedthrough technology adapted from the pacemaker.
(2) Advanced computational power for the speech processor adapted from the (then) newly evolved microprocessor technology.
(3) The use of a very slim implant (no unsightly bulge beneath the scalp) due to the absence of a battery or complex speech processor in the implant, made possible through the innovation of inductive transcutaneous power and information transfer from an external battery/microphone/speech-processor unit worn on the ear.

9.3 Development of the alumina feedthrough technology for the bionic ear

9.3.1 Titanium/alumina versus all-alumina casing

The Telectronics pacemaker discussed in Chapter 10, the template for all subsequent bionics, involved a titanium case with alumina feedthrough of a single titanium terminal. The initial plan during the development of the Cochlear Ltd. bionic ear was to use an all-alumina (totally ceramic) casing and eliminate titanium entirely. The reason for this was the need to have an induction pick-up coil, and the preference was for this coil to be inside the implant casing, rather than external. Obviously a titanium casing is a faraday shield and so with the use of a titanium casing, it is necessary to have the inductive receiver coil external to the implant, as shown in Fig. 9.4.

Fig. 9.2 shows that the commercialized Cochlear Ltd. bionic ear implant involved a titanium casing with external coil, and the same is seen in Figs. 9.3 and 9.4, the later and current forms of the Cochlear Ltd. Bionic ear. Obviously, the decision was ultimately taken in 1981/1982 to use the original 1970 Cowdery concept of titanium casing/alumina feedthrough, and this has remained the template ever since (Figs. 9.4 and 9.5).

However, in the 1990s, Cochlear Ltd. did experiment with developing an all alumina casing and two of their competitors (manufacturer Medel released the Pulsar CL 100; also manufacturer Neurolec experimented with alumina casings). It was found by all three companies that the reliability just is not there with all-alumina casings. In the titanium/alumina system, the alumina feedthrough is retained in residual compression by the titanium wrapped around it, due the slightly higher thermal expansion coefficient of titanium ($8.5\,\mu/mK$) compared to alumina ($8.1\,\mu/mK$) [391]. This was part of the genius behind the 1970 titanium/alumina feedthrough innovation discussed in Chapter 8. Without this residual compression from the titanium, and with a large exposed surface, an all-alumina implant casing is vulnerable to failure under impact,

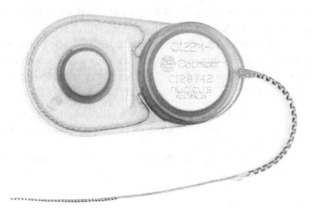

Fig. 9.2 The Cochlear C122M, also known as the "mini implant," the second bionic ear implant that Cochlear put on the market, and the first one with the magnetic transmitter-implant magnetic alignment system (note the magnet in the center of the coil). Also, the first one small enough to implant in children. Note the silicone electrode lead that is connected to the alumina feedthrough (implant exit point on the right-hand end of the implant) and its 22 sequential electrodes on the end of the lead, this is the section that is coiled into the cochlea. Note the inductive receiver coil, silicone-coated, attached to the left-hand end of the implant. This CI22M was the primary Cochlear commercial implant until the 1990s with 17,000 sold, before it was replaced by the next-generation Cochlear implant.
Image courtesy of Cochlear Ltd.

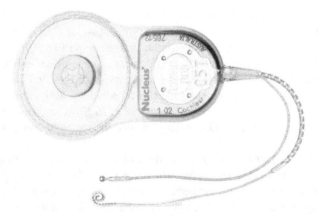

Fig. 9.3 The Cochlear C124RE third-generation implant. Still available in several market regions today. This utilized a similar alumina feedthrough to the first-generation C122M shown in Fig. 9.2. Note the microcoiled electrode lead with its 22 electrodes, coiled for better matching to the cochlea internal helix. Note also a second lead containing the "hard ball electrode" on its end.
Image courtesy of Cochlear Ltd.

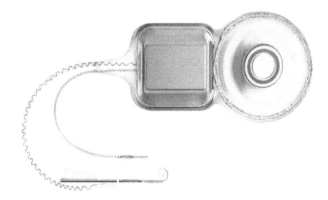

Fig. 9.4 The latest generation Cochlear C1532M, which is an implant that incorporates the "platinum comb" of Fig. 9.11 and matching titanium casing of Fig. 9.12. Note the microcoiled electrode lead with its 22 electrodes is coiled for better matching to the cochlea internal helix. Note also a second lead containing the "hard ball electrode" on its end, but it has a different shape, now long and cylindrical rather than a ball as was the case for the CI24RE. Image courtesy of Cochlear Ltd.

bearing in mind that a blow to the head of the implant recipient can impact on the implant itself.

The concept of the all-alumina casing involves:

(1) Alumina case, brazed to a titanium ring of inside diameter $ID = X$
(2) Alumina feedthrough brazed to a titanium ring of outside diameter $OD = X$
(3) The ring for the feedthrough is brazed into the closely fitting ring of the case, with the OD (feedthrough ring) a close fit to the ID (case ring).

The differential thermal expansion of titanium and alumina means that the ring around the feedthrough provides stable compression, the same as occurs with the standard titanium casing/alumina feedthroughs of IPG bionics. In contrast, the ring inside the alumina casing shrinks away from the alumina case during cooling, which can potentially compromise hermeticity, and gives no compressive mechanical support for the alumina casing. Moreover, given that 96% alumina is the standard composition, with its significant silicon content, the casing is susceptible to moisture-enhanced static fatigue in vivo, with no compression fit to prevent slow crack growth, unlike the feedthrough itself which is constrained in compression by the titanium perimeter ring.

Thus, while the feedthrough is well protected, the case is physically very exposed to mechanical impact, and unprotected against mechanical impact and static fatigue.

With about half of all Cochlear recipients being children, who are physically very active, and many of the adult bionic ear recipients also very active people, impact to the head is a significant consideration. Moreover, given that there is a requirement for the alumina to be only ∼96% pure, with a significant silicon percentage, in order to bond with the platinum (see Section 9.3.6), it is an unavoidable fact that an all-alumina implant of ∼96% purity will be susceptible to moisture-enhanced static fatigue.

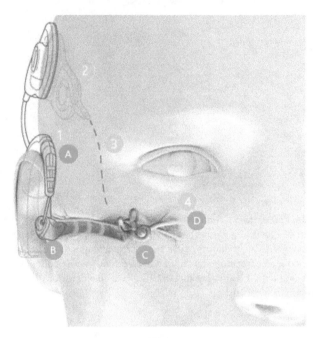

Fig. 9.5 The Cochlear implant system. (1) The external sound processor incorporating microphone and battery which hooks over the ear and captures sound and converts it into digital signals; (2) the sound processor sends digital high-frequency signals to the implant via the inductive coupling coil of the implant shown in Fig. 9.3; (3) the implant converts high-frequency signals into electrical energy, sending it to an electrode array inside the cochlea (shown in Fig. 9.1); (4) the nerve hearing response caused by electrical stimulation inside the cochlea (the small bony coil seen just above the letter C) is sent to the brain which combines them into a perceived sound.
Image courtesy of Cochlear Ltd.

This is all the more reason to have an alumina feedthrough in residual compression from a titanium case, rather than an all-alumina case.

In short, the all-alumina casing solves a minor problem (the inconvenience of having an external induction coil), and in doing so creates a major problem (mechanical unreliability and static fatigue susceptibility of the casing).

9.3.2 22-Electrodes versus 1-electrode

The original titanium/alumina casing-feedthrough system invented as the P10 Telectronics pacemaker had a single titanium terminal passing through an alumina feedthrough, engineered such that the casing and the titanium terminal were TiCuNi brazed onto the periphery of the alumina feedthrough cylinder thereby placing the alumina in residual compression, and naturally opposing any tendency for leaks at the brazed seal as shown in Fig. 8.3. In 1981/1982 when the Cochlear Ltd. bionic ear feedthrough and casing was being developed, the Telectronics pacemaker was the same conceptually as it had been in 1971 when first invented, that is, single titanium terminal, peripherally brazed to alumina feedthrough.

The bionic ear required 22 channels in its feedthrough for its 22 electrodes, and there was a much greater imperative for miniaturization than for the pacemaker. A pacemaker IPG implant simply lies beneath the skin in the abdominal cavity. Any bulge beneath the skin is hidden by with clothing, and the implant is interfacing with soft tissues. Miniaturization, though preferred, is not essential. A bionic ear IPG implant is an ultraslim casing placed beneath the scalp in a bony depression in the mastoid region close to the ear. It is important for obvious cosmetic and esthetic reasons to avoid an unsightly bulge under the scalp.

Thus, not only did the bionic ear IPG feedthrough requires many and much smaller channel wires than the Telectronics pacemaker, it also needed to be as slim as possible.

9.3.3 The telectronics environment: Ideal platform for bionic feedthrough innovation

Commercialization of the bionic ear began after parent company (Nucleus holdings which owned Telectronics and several other biotech companies) won the tender to set up manufacturing of the cochlear implant. Subsequently, Nucleus was given access to the first tranche of the large Government Grant for the bionic ear development.

In 1981, Paul Trainor (Owner of Telectronics/Nucleus) appointed David Money to the role of Project Leader of the Telectronics/Nucleus development of commercial bionic ear implant device (in that era David Money was Head of R&D for Telectronics). In effect Money was appointed acting CEO of proto-Cochlear, and he became actual CEO upon incorporation of Cochlear Ltd. in 1983. Jim Patrick and Peter Crosby were appointed as the 2 deputies to David Money, with David Cowdery (in that era, a Director of Telectronics) and Carl Doring expert consultants. These five were, in effect, the management team of proto-Cochlear.

It was the role of David Cowdery to interview and hire most of the key staff. Janusz Kuzma was one of those recruited by Cowdery. Kuzma was recruited in 1981 for the role of feedthrough design and development. Kuzma was ideal for the position as he had a strong background in hermetically sealed packages from working in the semiconductor industry in Europe.

It was fortunate for proto-Cochlear that David Cowdery, the inventor of the hermetic encapsulation/feedthrough system and chief hermetic/feedthrough expert for Telectronics, was on hand, as well as the team of Telectronics engineers, during the challenging development of the bionic ear IPG feedthrough in 1981/1982. This meant that once Kuzma began work, he had at his disposal complete access to the expertise of Cowdery, the Telectronics engineers and the telectronics technicians, all of whom had much experience in hermetic feedthrough technology, it being a decade since the first hermetic pacemaker was commercialized by Telectronics.

Telectronics was by then one of the largest pacemaker companies in the world. Moreover, Kuzma had access to the extensive facilities of Telectronics including:

(1) Complete in-house alumina ceramic manufacture (from raw powders to finished ceramic) of a suitable alumina composition compatible with the brazing alloy, and a manufacturing process that could be adapted to different shapes of alumina ceramic.
(2) The brazing ovens and techniques including molybdenum heat radiators used to house the items being brazed.
(3) The brazing alloy required and the use of outside compression design in the braze.
(4) The knowledge of a suitable metal case and ferrule material.
(5) The welding techniques and equipment, including the techniques to create welds while limiting heat conductivity into the electronics.

Kuzma made good use of the many advantages at his disposal and designed a practical hermetically sealed implant using the three materials in current use for Telectronics pacemakers: the high alumina ceramic, the TiCuNi brazing alloy, and the pure titanium casing. The reader is referred to Figs. 9.6–9.10 to supplement the following outline of the Kuzma feedthrough development in 1981/1982, which is based on interviews conducted with David Cowdery on the specifics of the feedthrough invention process [392]. In addition, the author has a technical familiarity with the system through his long association with Cochlear, and Telectronics going back to the 1990s, and by informal conversations with his extensive network of colleagues who are past and present employees of Cochlear Ltd.

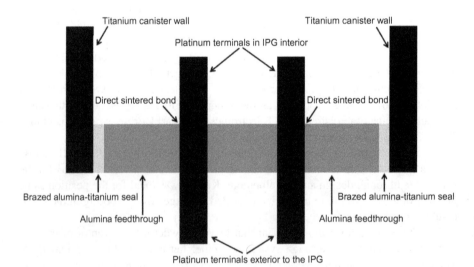

Fig. 9.6 The next evolution in alumina feedthrough technology. The platinum-channel wires were, in the Kuzma concept, small platinum tubes to which electrodes could be later attached. They were placed in the green alumina ceramic and cofired into the alumina. This alumina-platinum feedthrough was then brazed into the titanium casing. This is the essence of the innovation of Janusz Kuzma in 1981/1982. The IPG is the implantable pulse generator, the titanium/alumina Cochlear implant.

The bionic ear 267

Fig. 9.7 The prototype of the first Cochlear feedthrough shown prior to welding. It comprises a 14.5 mm diameter, 1.5-mm thick alumina disk brazed to a titanium flange using TiCuNi brazing, the same brazing alloy as used in the Telectronics pacemakers. Note the 22 platinum tubes are cofired into the alumina disk in two concentric rings, 14 in the outer ring and 8 in the inner ring. Image courtesy of David Cowdery.

Fig. 9.8 First Cochlear feedthrough with titanium flange after welding to the titanium case that housed the electronic circuitry.
Image courtesy of David Cowdery.

- Because the cochlear implant required 22 channels instead of one, Kuzma used small platinum tubes placed in the green alumina in closely fitting holes. Tubes were used rather than solid wires to prevent tensile cracking of the alumina during firing.
- Holes were punched in the unfired green ceramic and the platinum tubes were placed into these 22 holes.

Fig. 9.9 The electrode side of the first Cochlear feedthrough with polyimide insulator and parylene-coated platinum electrode wires soldered into the platinum tubes.
Image courtesy of David Cowdery.

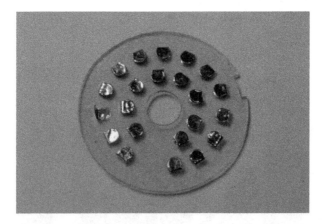

Fig. 9.10 Silicone-mounted platinum strips for contacting the 22 platinum tubes on the first Cochlear feed through.
Image courtesy of David Cowdery.

- The tubes were polished so they became flush with the alumina surface after firing.
- When the alumina was fired, the large diametral firing shrinkage of the alumina caused the soft platinum tubes to form a tight seal without the need for any brazing material. The finished diameter of the platinum tubes was about 1.2 mm OD × 0.6 mm ID.
- This process is known as alumina-platinum cofiring, that is, placing the platinum into the green alumina and sintering it in, using the heat, the glass formation, and the diametral

shrinkage to create a hermetic seal by two physical mechanisms: (1) platinum-alumina solid-state diffusion bonding and (2) platinum-glass-alumina bonding.
- The alumina disk containing the platinum tubes was very different in dimensions to the Telectronics feedthrough. It was 14.5 mm in diameter and 1.5 mm thick. This was necessary because the Cochlear implant needed to be as slim as possible.
- The slim ceramic disk with the 22 platinum tubes cofired into it was then brazed around its circumference to a round titanium casing. This is shown in Figs. 9.7 and 9.8.
- Final hermetic closure was achieved using similar welding procedures as used on the Telectronics pacemakers.
- The platinum tubes were finally sealed by soldering wires into the hollow tubes at one end. This is shown in Fig. 9.9.
- The 25 µm platinum microelectrodes were attached to the platinum tube terminals at the "electrode side" of the feedthrough (the external implantable-electrode connection side) and the internal electronics attached at the "chip side" of the feedthrough (the internal microelectronics side).
- The other end was connected electrically by an array of spring loaded platinum contacts set in a sheet of silicone elastomer. This is shown in Fig. 9.10.

So, in summary, in 1982, Janusz Kuzma adapted the Telectronics hermetic feedthrough technology into the bionic ear feedthrough:

- 22 platinum tubes (final size: 1.2 mm OD; 0.6 mm ID)
- 14.5 mm diameter, 1.5-mm thick alumina disk, shown in Fig. 9.7.

This was achieved by fitting platinum tubes into the green alumina disk by inserting them into holes in the green alumina disk, and sintering the piece, a process known as cofiring. Platinum tubes were used rather than solid wires so as to enable shrinkage of the alumina feedthrough disk during firing, and thereby avoid cracking of the alumina disk. This was an innovative and successful solution to a difficult ceramic engineering problem.

In one decade, feedthrough technology had made the jump from a single 1.5-mm titanium electrode passing through a 9.6-mm alumina cylinder to 22 platinum-channel wires (tubes 1.2 mm OD; 0.6 mm ID) cofired into a 14.5-mm diameter alumina disk. This highly evolved alumina feedthrough for the bionic ear was made possible by the fact that parent company Nucleus/Telectronics had paved the way a decade earlier, and the expertise of David Cowdery and other Telectronics engineers was on hand, as well as all the technical resources of Telectronics.

There are anecdotal accounts of the Telectronics/Nucleus engineers expressing concerns that cofiring platinum tubes into an alumina feedthrough was a crazy idea [146]. While this may have been the perception of nonexperts in feedthrough technology, cofiring is a common technique in use since the 1950s in the electronics industry, for making electronic components such as capacitors, resistors, and even semiconductors, and Kuzma had a strong background in hermetically sealed packages from working in the semiconductor industry in Europe, and industry in which cofiring is a standard technology. So cofiring was a technology Kuzma would have been familiar with.

Moreover, the viability of alumina-platinum high-temperature solid-state bonding has been known since the 1970s as pressure-assisted presintered-alumina/platinum couples [410]. This is significantly different to platinum-alumina cofiring (pressureless sintering green alumina/platinum couples), but at least the scientific literature of 1982 demonstrated the chemical compatibility of alumina/platinum couples at high temperature, and the capability of alumina/platinum solid-state bonding at high temperature. As an expert in cofiring and hermetic semiconductor technology, Kuzma would have been aware of the 1970s alumina/platinum research, and he would have had significant cofiring expertise from his semiconductor background. So his was not a crazy idea, this was a leading expert applying his knowledge to a new and unique problem. Alumina-platinum diffusion bonding is discussed in detail in Section 9.3.6.

The first Cochlear feedthrough adopted the following innovations that David Cowdery developed for the Telectronics P10 titanium/alumina hermetic feedthrough system:

(1) The use of alumina as the feedthrough ceramic.
(2) The use of a titanium casing (manufactured by deep-drawing) for the microelectronics.
(3) The complex and sophisticated "ship in a bottle" technology for brazing an alumina feedthrough to a titanium casing that contained sensitive microelectronic circuitry.
(4) The use of TiCuNi brazing alloy for brazing the alumina to the titanium.
(5) The use of the titanium casing as the reference electrode.
(6) The use of silicone coating of the casing.

Janusz Kuzma and the bionic ear team also contributed three innovative breakthroughs that represented an incremental advancement of the alumina feedthrough technology over the Telectronics P10 pacemaker. So, in addition to the six innovative aspects of the Telectronics feedthrough concept listed above, the first Cochlear feedthrough development of Kuzma and the Cochlear engineers involved the following three new innovations, two of which were specific to the alumina feedthrough:

Innovations for the alumina feedthrough
(1) The 22 channels (instead of just 1) were platinum (rather than titanium).
(2) The platinum channels were tubes cofired into the alumina ceramic (rather than a single titanium macroterminal channel brazed to the alumina periphery).
Innovation for managing ultrathin (25 μm) platinum electrode wires
(3) The use of parylene insulation on the 25 μm ultrathin electrode lead wires.

25 μm is half the diameter of a human hair. This was a significant brazing/soldering/welding challenge at that time.

Commercial production of Kuzma's innovation was developed by the Ceramic Corporation Pty Ltd., the company of Miroslav Kratochvil [411]. This made Kratochvil the world's first manufacturer of cofired platinum-embedded alumina miniaturized ceramics for bionics, now a routine process for the millions of microarray bionic implants manufactured worldwide.

9.3.4 Commentary on the feedthrough innovation and related issues

The first important question to explore is why use platinum-channel wires in the feedthrough rather than titanium? Neither platinum (9.4×10^6 S/m) nor titanium (2.5×10^6 S/m) has particularly high electrical conductivity values compared to copper (59×10^6 S/m) or silver (62×10^6 S/m) [393]. Platinum is 3.75 times more conductive than titanium, but 6.6 times less conductive than silver. So while platinum is an improvement over titanium in conductivity, it is by no means one of the best electrical conductors, but rather the best combination of biocompatibility, corrosion resistance, and electrical conductivity of the proven biomaterials. The electrical conductivity of platinum and titanium could be described as moderately high, but their biocompatibility and bioinertness are outstanding. However, when it comes to the ultrafine 25 μm diameter electrodes used in a Cochlear implant, the higher electrical conductivity of platinum did become important.

Undoubtedly, the most significant contribution of this Cochlear feedthrough innovation to the world of bionics was the Kuzma concept: cofiring platinum channel wires (as tubes) directly into the alumina (innovations 1 and 2). Fortuitously this proved to be technically feasible from a ceramic engineering viewpoint. Moreover, this has been the enabling technology of subsequent high-electrode density bionic devices, for example, the extraordinary electrode feedthrough achievements in bionic eye technology in the current decade (see Chapter 10), with the world record currently standing at 1145 platinum electrodes cofired into a 7 mm diameter alumina disk [375], set by former Cochlear engineer Gregg Suaning, an innovation in which the author was involved [375,412–414]. The ceramic engineering aspects of alumina-platinum cofiring will be discussed in detail in Section 9.3.6.

The second important question to explore is the thermal expansion issue. A thermal expansion appraisal shows that Kuzma was very fortunate that his platinum-tube/alumina cofired feedthrough was hermetic. The thermal expansion coefficients of the respective materials are [391]:

Alumina: 8.1 μ/mK
Titanium: 8.5 μ/mK
Platinum 9.0 μ/mK

Of course dilatometry curves show that thermal expansion coefficients are not necessarily linear with temperature, and therefore the thermal expansion coefficients quoted here are indicative of probable mismatch only, for the sake of illustration, not a quantitative indication over the entire cofired temperature range of 20–1500°C. Nonetheless, the following analysis based on these thermal expansion coefficients demonstrates the potential problem.

In the case of linear thermal expansion coefficients over the entire sintering range, an internal platinum tube would theoretically shrink away from alumina on cooling, as calculated below:

- 14.5-mm alumina disk
- 1200 μm OD, 600-μm ID platinum tube
- 25-μm platinum electrode soldered in

Cooling from 1500°C, across a 1200 μm span, *alumina* will contract:

$$(8.1\,\mu/\text{mK}) \times (1500\,\text{K}) \times (1.2 \times 10^{-3}\,\text{m}) = 14.6\,\mu$$

Cooling from 1500°C, across a 1200 μm span, *platinum* will contract:

$$(9.0\,\mu/\text{mK}) \times (1500\,\text{K}) \times (1.2 \times 10^{-3}\,\text{m}) = 16.2\,\mu$$

Differential diametral contraction is therefore 16.2–14.58 = 1.6 μ, corresponding to an 800 nm (0.8 μ) radial clearance between the 1.2-mm platinum tube outer wall and the inner wall of the hole in the alumina.

Given that the grain size of the alumina is in the order of a μm (1000 nm), 800 nm is a significant amount in the context of the alumina-platinum interface and could potentially be sufficient to break the hermetic seal sufficient for helium permeation in the helium leak test, and for water/gas/ion diffusion in vivo.

However, history shows that it did work, and Kuzma is to be commended for his innovation, without which there would have been no bionic ear ready for clinical trials at the end of 1982. Kuzma delivered a functional feedthrough, and the rest is history. Sometimes good fortune is an essential component of technology innovation. This problem arose again to a more serious extent with the development of alumina retinal implants containing hundreds of cofired embedded platinum channel wires, and an innovative thick-film hot-pressing solution was developed by bionic eye feedthrough innovator Gregg Suaning, in collaboration with the author and others, as discussed in Chapter 10.

In retrospect the 1981/1982 development of the bionic ear commercial implant was an extraordinary achievement, producing, in just 18 months, a titanium-encapsulated hermetic 22-channel bionic ear with alumina feedthrough, and complex computing power for a portable external speech processor for 22 channels of auditory output in real time, an incredible challenge for the early 1980s. Moreover, this device needed to be miniaturized more than had been achieved for the contemporary pacemaker, since the bionic ear IPG unit needed to be located beneath the scalp near the ear, and there was a strong imperative, for obvious cosmetic reasons, for it to be as slim as possible so as not to have an unsightly bulge under the scalp.

In short, developing the bionic ear, and in such a short time, was an extraordinary achievement. It was made possible by a number of reasons.

First, the bionic ear initiative was managed by Telectronics/Nucleus. Thus, the now prosperous global pacemaker company Telectronics, as it had become a decade after the launch of the world's first hermetic pacemaker, became the big brother to the fledgling bionic ear company. This provided both financial resources by lending its credibility to the $4.7 m ($50m 2017) in Government grants, and technological/intellectual resources, and Telectronics "know-how" in bionics manufacture and design.

Second, the alumina feedthrough headstart. One of the most challenging aspects of developing the Cochlear implant was developing an alumina feedthrough. In total, 22 channels were required, and the alumina feedthrough disk needed to be as slim as possible. The importance of the headstart the bionic ear team received in having the viable, proven, and optimized Telectronics feedthrough technology as a starting point was a factor that proved very important in solving the Cochlear feedthrough challenge in such a short time. Mpreover, Janusz Kuzma was an engineer with a strong background in hermeticallysealed packaging from working in the semiconductor industry in Europe. He was the right man for the job.

Third, the recent invention of the microprocessor made complex speech processing possible, for the first time, in a miniaturized speech processor device that was worn externally. Nonetheless, it was a major challenge to develop and perfect this in the time, and it is a testament to the strength of the electrical and computing expertise in the Cochlear team, led by Jim Patrick since 1975, that they succeeded in the time available, and with the microprocessor technology then available.

Finally, the problem of batteries was a major potential roadblock. With such device miniaturization required, and limited battery miniaturization technology available in that era, a novel solution was called for. This novel solution was inductive powering. The microprocessor/microphone/battery external device mounted on the ear was inductively coupled to the Cochlear implant, which resided beneath the scalp near the ear. This solved the battery problem and the implant miniaturization problem in one solution.

9.3.5 Subsequent evolution of bionic ear feedthrough technology

Ultimately the Kuzma feedthrough concept needed to be further optimized. In 2002, two decades after the Kuzma innovation, a Cochlear patent was filed on a new feedthrough, which was granted in 2008 [415]. Cochlear engineers redesigned the titanium/alumina feedthrough from the ground up, evolving it into the powder-injection-molded "ceramic comb" concept shown in Fig. 9.11 comprising an alumina square-cross-section bar with 18 platinum channel wires cofired into the alumina. The latest generation Cochlear implants contain two of these "platinum comb" feedthrough units, one on either side of the titanium casing shown in Fig. 9.12, giving a total of 36 terminals, 22 of which are used for the Cochlear implant electrode, and the remainder used for various proprietary support functions. These two feedthroughs are brazed into the titanium casing as shown in Fig. 9.12, creating residual compression of the titanium onto the alumina, just as in the original Cowdery innovation of 1970. The schematics of this design from the patent are shown in Fig. 9.13.

The key inventive step in the "platinum comb" patent appears to be that the platinum channel wires that pass through the alumina feedthrough do not pass directly through (linearly in the shortest path) as was the case for the Kuzma concept: 1.2-mm platinum tubes cofired into the 1.5-mm thick alumina disk. Instead, this "platinum comb" patent involved wires with a convoluted pathway through the alumina, for example, in a zigzag, or as a screw-thread, to maximize the leakage path length along the platinum/alumina interface passing along the channel-wire surface through the alumina [415]. The schematics for this concept, from the patent, are shown in Fig. 9.14.

The wires were also as thin as technically feasible. This was because leakage through the feedthrough was believed to arise predominantly from the alumina-platinum interface, due presumably to the thermal expansion mismatch of alumina and platinum described in Section 9.3.4, and incomplete alumina-platinum bonding, rather than due to flaws in the alumina-titanium brazing, or microstructural flaws or porosity in the alumina itself. The rationale behind the patent was the relationship

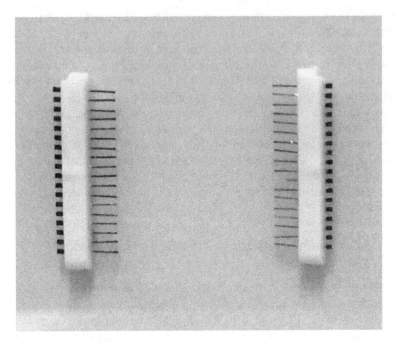

Fig. 9.11 The "platinum comb" alumina-platinum feedthrough system of the current generation of Cochlear bionic ear implants. The 18 platinum channel wires are cofired into the alumina just as they were in the original Cochlear feedthrough, giving a total of 36 terminals, a big increase over the old 22-terminal feedthrough. The cochlea insert still has just 22 electrodes. The remaining 14 terminals have various proprietary uses.
Image courtesy of Cochlear Ltd.

below, which is essentially a measure of the total capacity of the platinum/alumina interface for leakage by fluid diffusion [415]:

$$H = f(L, 1/A, 1/t)$$

where H = hermeticity; L = the length of the channel wire as it travels between the front and rear faces of the insulator (alumina); A = the cross-sectional area of the channel-wire (platinum); t = the time that the implant is exposed to the fluid, such as body fluids.

Thus, the objective of this patent was to claim the IP rights over a feedthrough characterized by practical strategies for maximizing "L" and minimizing "A," where "t" is a given, and is potentially many decades, equal to the life of the implant patient. This was achieved by the following strategy:

- Preparing a construct of multiple platinum wires (nonlinear channel wires) characterized by zigzag, or coiling, or similar systems for maximizing the mean path length of the wire from front to rear of the feedthrough.

Fig. 9.12 The titanium casing into which the two "platinum comb" alumina-platinum feedthrough components, shown in Fig. 9.11, are brazed, into the housings seen on the left- and right-hand sides of this casing.
Image courtesy of Cochlear Ltd.

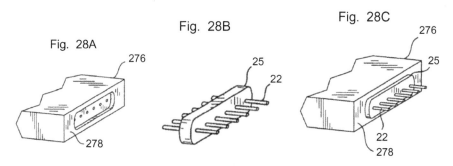

Fig. 9.13 "Platinum comb" device schematics from the originating patent: "FIG. 28a is a partial perspective view of an implantable medical device to which a feedthrough according to the present invention is electrically coupled; FIG. 28b is a perspective view of an embodiment of a feedthrough according to the present invention; FIG. 28c is a partial perspective view of an implantable medical device after a feedthrough according to the present invention has been electrically coupled" [415].

- The diameter of the wires was as thin as feasible, around 50–175 μm. This was both to assist the hermeticity formula described above and also to prevent cracking during shrinkage of the alumina during debinding and sintering. This was the main reason the Kuzma concept used tubes, with a view to allowing shrinkage of the alumina without cracking. However, these were very thick tubes. An order of magnitude thicker than the wires in this patent.

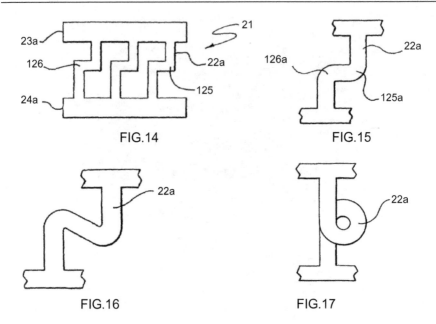

Fig. 9.14 "Platinum comb" schematics of the convoluted pathways for the platinum channel wires through the alumina ceramic feedthrough, from the originating patent. "FIG. 14 is a plan view of another embodiment of an electrically conductive structure for use in the method according to the present invention; FIG. 15 is an enlarged plan view of a single conductive member of another embodiment of a conductive structure according to the present invention; FIG. 16 is an enlarged plan view of a single conductive member of another embodiment of a conductive structure according to the present invention; FIG. 17 is an enlarged view of a single conductive member of another embodiment of a conductive structure according to the present invention" [415].

- Placing the construct of multiple platinum channels wires into a powder-injection-molding mold.
- Filling the mold with alumina by powder injection molding.
- With such thin wires, in a delicate coil or zigzag, it was necessary to use a sacrificial support for the wires, and also to fill the mold gently, by gradually opening it during the injection process so as not to damage disturb or displace the delicate thin zigzag/coiled platinum wires. A great deal of specialized know-how is embodied here.
- Removal of the workpiece from the mold and debinding at 150–200°C.
- Pressureless sintering at 1700°C.

From an alumina ceramic engineering point of view, the diameter of the wires should be as thin as feasible, not just to assist in the hermeticity formula described above, but also to prevent cracking during shrinkage of the alumina during debinding and sintering. This was the main reason the Kuzma concept used tubes, with a view to allowing shrinkage of the alumina without cracking. However, Kuzma used very thick tubes in comparison to this patent as he was constrained by the technology available in 1982.

This 2008 Cochlear "platinum comb" patent is a very innovative solution, obviously born out of long experience in feedthrough design and manufacture. Clearly, a great deal of skill and experience is required in supporting the fine platinum coils/zigzags during filling the mold, so as not to damage disturb or displace the delicate thin zigzag/coiled platinum wires. For example, if incorrectly filled or supported, the coils/zigzags could be stretched or deformed and the wires could come into contact with one another and short circuit. The patent talks about sacrificial component support in this regard. The patent claims and descriptions encompass a broad range of manufacturing engineering possibilities with regard to forming the coil/zigzag channel wires, the sacrificial support system, and the alumina injection molding system. However, the inventive step is very clear.

Another major evolution to the Cochlear implant technology in the decades after 1982 was the reference "ball electrode." Thus, in addition to the titanium casing being a reference electrode, there is also a reference ball-electrode,[1] which is on the end of a long (many centimeters in length) silicone-coated lead and enabled up to a sixfold improvement in battery life due to more energy efficient current flow in the cranial cavity. This can be seen in the later generation Cochlear implants shown in Fig. 9.2 (Cochlear C124RE) and Fig. 9.3 (Cochlear C1532 which also utilized the platinum combs of Fig. 9.11).

9.3.6 Research into cofired platinum-alumina interfaces

9.3.6.1 The state of the art in alumina-platinum bonding in 1981

Alumina-platinum feedthroughs are now very important commercially, underpinning a ~$25 billion industry in bionic implants (see Chapter 10), not just the pacemaker and bionic ear, but 14 bionic implants all based on the titanium/alumina-platinum hermetic IPG, but with different electrode and programming configurations, and implanted for different disease states. This is discussed in Chapter 10. Hermetic bonding in alumina-platinum interfaces is essential to prevent fluid diffusion into the implant and exposure of the body cavity to the toxins in the implant electronics. It is therefore not surprising that there is a sizable body of literature on the topic. However, this was not the case in the early 1980s when the alumina-platinum feedthrough was being developed for the bionic ear.

The first report of alumina-platinum bonding was in 1977, which documented pressure-assisted alumina-platinum bonding experiments using 25 μm platinum foil pressed at 100 kPa between 12.5-mm thick presintered (1500°C) alumina disks up to a temperature of 1055°C, and suggested that theoretically this could be viable up to 100°C below the metal melting point (platinum melts at 1769°C) [410]. This is significantly different from what Kuzma did. He used green alumina, pressureless sintering, and a temperature >500°C higher than the reported 1055°C, and his alumina was in the form of thick-walled tubes, not foil. Nonetheless, the 1977 pressure-assisted platinum-alumina bonding experiments suggested that a solid-state alumina-platinum

---

[1] The author has a long association with Cochlear in a range of R&D collaborative projects. See "About the Author".

bond could form, and the fact that there were probably no unexpected Al-Pt-O binary or ternary phases or Al-Pt-O volatilization products produced that could have made such an approach impossible [410]. Therefore, though the de Bruin work [410] was of little relevance to Kuzma's cofiring methodology, its chief value was in the phase equilibria deductions one could make from it, deductions that would have given Kuzma reassurance that, from a phase equilibrium point of view, he was on safe ground with his cofiring strategy.

Kuzma was otherwise flying blind in that there were no published binary or ternary phase diagrams in the Al-Pt-O system in 1982. The literature contains no published Al-Pt-O ternary phase diagram to date. The Al-Pt binary phase diagram was published in 1985 [416], and the Al-O binary phase diagram was also published in 1985, by different authors [417]. These two binary phase diagrams came too late for Kuzma. Moreover, there is nothing further in the literature on platinum-alumina bonding after 1977 until 1983, when more pressure-assisted presintered-alumina/platinum-foil experiments were published [418], which went into much more detail than the 1977 work [410], exploring bonding temperature, bonding pressure, and their effects on bonding strength, and reporting that an optimal bond was achieved at 1700°C, for 10 h at 2 MPa pressure. Again, this was pressure-assisted bonding of presintered alumina, not cofiring, and therefore of only peripheral relevance to Kuzma's cofiring methodology other than the phase equilibria deductions one could make from it.

Finally, the 1977 study involved high-purity alumina [410]. Kuzma was using 96% alumina with a significant glassy component. Glass-platinum interactions could have proven problematic indeed. Fortuitously, as it turned out, the glass helped rather than hindered.

9.3.6.2 The state of the art in alumina-platinum bonding today

There is now a large body of literature on alumina-platinum bonding, including a paper the author coauthored on cofired alumina-platinum feedthroughs for retinal implants in 2014 [375]. While metal-ceramic bonding is a well-established field, alumina-platinum is something of a special case. Platinum is such an extremely oxidation-resistant metal that even in extreme oxidation conditions, such as air at 1500°C, the Pt-Pt bond cannot be broken and simple adsorption of oxygen at the platinum surface occurs. However, a monolayer of oxidized platinum is found at the platinum surface, which allows a strong Pt-Pt-O-Si-O bond to form with glass [375].

The key finding of relevance, from the body of literature on this subject, is that glass seems to perform a vital role in hermetic alumina-platinum bonding, both when cofired with green alumina as done by Kuzma and by Cochlear today, and during pressure-diffusion bonding with presintered alumina.

A 1992 study [419] investigated the diffusion bonding interface between 99.5% presintered alumina and platinum exploring diffusion bonding at 1450°C and 5 MPa pressure for 12 h. Even though this was 99.5% pure alumina, much lower in glass than feedthrough alumina (typically 96% pure), transmission electron microscopy (TEM) analysis identified glass regions about 150 nm thick and 600 nm long that were occasionally found at the alumina-platinum interface. This was a very pure alumina, and it was not cofired, and so this suggests that in the case of cofired 96% alumina with platinum, interfacial glass could potentially be significant.

Atomic-resolution TEM of the diffusion-bonded alumina-platinum interface was reported in a 1993 study [420]. Again, like the 1992 study [419], this was not a cofiring study but a pressure-bonding study of high purity sintered alumina (single crystal in this case). Platinum and alumina single crystals were diffusion bonded at 1200°C and 1.8 MPa for 12, 24, for 100h in air or 24h in argon. Atomic-resolution TEM identified a 10–100 nm glassy interlayer at the alumina-platinum interface, confirming the finding of the 1992 study [419].

Interestingly, the 1993 study [420] also looked at a diffusion-bonded alumina-titanium interfaces. This is of significance to bionics in that cofired titanium-alumina feedthroughs may have been an option considered by Kuzma in 1981 considering that the titanium feedthrough of Telectronics was an industry standard at that time, albeit the Telectronics titanium terminal was 1.5 mm thick and passed through a 2.3 mm hole in the alumina (noncontact). Given that the sintering temperature of alumina is 1500°C or more, and the melting point of titanium is 1668°C, alumina-titanium cofiring is not ruled out on temperature criteria. The 1993 study [420] conducted alumina-titanium diffusion bonding at up to 800°C for 3 h in argon, of a titanium film sputtered onto alumina. Electron diffraction showed an interlayer, which contained more than one type of intermetallic compound. This proves what one would expect that titanium-alumina cofired feedthroughs are not feasible unless the titanium is of substantial cross section, as a lot of titanium parent metal will be lost through intermetallic compound formation. Therefore, it is not suitable for the miniaturization required for a bionic ear feedthrough.

The first study of a cofired alumina-platinum interface using 96% alumina (feedthrough grade) was relatively recent in 2012 [421]. It is surprising that the first such study was so recent given that cofired alumina-platinum feedthroughs had been in use for 30 years by then. It is obviously of much greater relevance to commercial feedthrough technology to explore the possibility of significant glass at the alumina-platinum interface for a standard alumina feedthrough of the typical $\sim$96% purity used in industry with its significant glassy content, much like the alumina used in the original Kuzma concept.

The 2012 study [420] investigated a platinum pin (0.1 mm diameter; 5 mm length) cofired in an alumina feedthrough disk of typical $\sim$96% alumina feedthrough composition (1–2 wt% SiO_2, 1–1.15 wt% CaO, <1 wt% MgO, <1 wt% Na_2O, <1 wt% Fe_2O_3). It was cofired at 1550–1650°C. Two TEM specimens from the alumina-platinum interface were prepared by focused ion beam milling (FIB) as 0.1 × 5 × 20 μm wafers, and SEM was done of the FIB-exposed interface. The TEM image demonstrated that the glass wets both the platinum and the alumina, and an intimate bond, on the nanometer scale, was formed at the platinum-glass-alumina sandwich interface. This finding is obviously very significant in terms of hermetic alumina-platinum seals.

An important finding by this 2012 study [421] was that residual stress may be present in the glassy phase boundaries due to the absence of annealing in the feedthrough, or perhaps due to differential thermal contraction between the glass matrix and secondary phases. This is a finding well worth exploring in further depth as the idea of adding an annealing stage in cofired alumina-platinum feedthrough sintering cycles for stress relief, and the issue of using very slow cooling rates below the glass softening point for differential thermal contraction stress management, could be commercially important.

The 2012 study [421] also investigated a glass-free platinum-alumina interface. Glass was not ubiquitous to all alumina-platinum interfaces in 96% alumina. Direct metal-ceramic bonding was found at an apparently glass-free alumina-platinum interface, thereby demonstrating that hermetic bonding does not necessarily require a glass. There was a suggestion that a possible interdiffusion zone existed in the glass-free alumina-platinum interface. However, it was unclear from the TEM analysis whether or not this was the case. This is certainly an area worthy of further future study.

Further light was shed on these issues by a collaboration of the author with the bionic eye team led by Gregg Suaning (bionic eye issues are discussed in detail in Chapter 10) and a resulting paper [375]. Thomas Guenther was lead researcher, and Martin Svehla and Hong Lu of Cochlear were also involved in this collaboration. In this study, rather than cofiring monolithic platinum with alumina, platinum paste was used, which was printed into ultrafine channel wires using the laser-patterning/screen-printing approach discussed in Chapter 10. A typical 96% feedthrough alumina was studied (2 wt% SiO_2, 0.5 wt% MgO, 0.25 wt% CaO, and other trace additives). FIB-prepared $0.1 \times 10 \times 18$ μm wafers from the alumina-platinum interface were studied using TEM-EDS, and TEM with selected area electron diffraction (SAED) in order to study both the composition and the crystallographic orientation of the grains at the alumina-platinum and alumina-glass-platinum interfaces. This was the first atomic-resolution study of an interface formed with platinum particles rather than monolithic platinum and revealed new information regarding the role of lattice matching in direct alumina-platinum interfaces.

In the alumina-glass-platinum interface, it was found that the formation of glass pockets filled the gaps created at the grain interfaces where multiple grains join each other. This is obviously an important contributor to hermeticity. At glass-free alumina-platinum interfaces, TEM revealed a multitude of dislocations within the grains. Dislocations are an indicator for residual stresses within the system, due to mismatches of the varying coefficients of thermal expansion of the materials. Residual stresses were identified at the interface, as reported previously by the 2012 study [421]. However, a new finding was also revealed by the Suaning team, which was a lattice matching of the alumina and platinum at the interface, related to a direct (non-glassy) bonding of alumina to platinum [375]. This study therefore established that both vitreous bonding and direct alumina-platinum bonding coexist. However, the question as to whether the relative hermeticity of an alumina-glass-platinum double interface is better or worse than a direct alumina-platinum bonded interface remains an open question.

9.3.6.3 Alumina-platinum: Conclusions

Several decades of research have now established that cofired alumina-platinum feedthroughs, using a standard feedthrough 96% alumina, generate hermetic interfacial bonds: both alumina-platinum and alumina-glass-platinum hermetic interfacial bonds. A question mark remains over the issue of residual stress in the glassy phase, and the risk that may pose to hermeticity, and the possible need for annealing and slow cooling to manage that. A question mark also remains as to which type of bond is better for hermeticity: alumina-platinum or alumina-glass-platinum. Industry experts the author has spoken to universally believe that some glass is essential for a hermetic bond.

The final question, which has been a concern since the original Kuzma innovation, is the issue discussed in Section 9.3.4 regarding the differential thermal contraction of platinum to alumina, which can result in a propensity for the platinum channel wires to shrink away from the alumina to an amount that is potentially in the 1 μm order of magnitude in a 1 mm diameter platinum channel wire. This problem has now been resolved by a novel solution developed by Gregg Suaning, in which the author was a participant: hot-pressed stepped feedthroughs prepared by the laser-patterning/screen-printing approach. This approach has the added benefit of enabling channel densities of up to 2500 per square centimeter, which is essential for bionic eye retinal implants. This innovation will be discussed in detail in Chapter 10.

Finally, it should be noted that platinum, platinum-iridium, and the noble metals in general ultimately completely supplanted titanium as a channel-wire biometal in the bionics industry. Titanium was the Telectronics pacemaker feedthrough conductor wire from the early 1970s. Niobium was the conductor wire of choice in pacemaker feedthroughs from (United States) from the mid-1970s onward, in the era when Telectronics was ranked number 3 in the world. Telectronics continued to use titanium channel wires right up to 1996, 26 years after the invention of the titanium/alumina hermetic implantable bionic system, and 15 years after the first use of platinum by sister company Cochlear Ltd. However, in 1996, Telectronics was shut down and its assets sold to St Jude Medical. In the 21st century, niobium and titanium were completely supplanted by platinum and the noble metals as bionic feedthrough channel-wire material.

In the 21st century, channel wires of platinum and other noble metal alloys (particularly platinum/iridium) became an industry norm for the 14 different bionic implants in the world today, following in the footsteps of industry leader Cochlear, the company that pioneered the platinum channel-wire concept. Titanium has remained in use as the casing, and alumina has remained in use as the feedthrough material, for all hermetic bionic implants over the half-century that has elapsed since the invention of the titanium/alumina hermetic feedthrough system.

Alumina in bionic feedthroughs: The bionic eye and the future

10

This chapter begins with a brief overview of the 14 implantable bionics on the market today, all of which have a hermetic titanium/alumina feedthrough. We then will take a look at what the future holds for the hermetic titanium/alumina feedthrough, with a particular focus on the bionic eye. A bionic eye with reading and object recognition capabilities is close to a commercial reality today, and this is in no small part due to a quantum leap in alumina feedthrough technology pioneered by bionics innovator Gregg Suaning in the last decade which will be overviewed in detail. This chapter will then conclude with a glimpse at the brain computer interface (BCI), also most likely involving alumina feedthroughs, and how it may influence the shape of things to come.

The implantable bionic industry of today is significantly advanced in scope and miniaturization from where it was at the end of Chapter 9, that is, where it was over three decades ago in 1985, when the world's second bionic device was launched on the market—the bionic ear. In that era, the pacemaker was the well-established bionic implant, and the bionic ear the new entrant. Pacemakers in that era most commonly had either a titanium conductor wire or a niobium conductor wire in their alumina feedthrough. The bionic ear was the first bionic implant to have multiple platinum channel wires in its alumina feedthrough. However, in recent decades, all bionics have generally become standardized to the bionic ear feedthrough concept: alumina feedthrough brazed to a titanium casing, with multiple platinum wires in the alumina feedthrough, The platinum/iridium alloy is commonly used for this purpose.

Since 1985, 12 more implantable bionic devices have come onto the market. All of them use the titanium/alumina-platinum hermetic capsule/feedthrough system, that is, the implantable pulse generator (IPG) concept, originally invented by David Cowdery in 1970, although substantially miniaturized today. The bionic implants all look remarkably similar, they comprise a slim titanium can (IPG) with alumina feedthrough and a terminal for connecting the leads of electrodes, usually 10 or more feedthrough channels are involved, often tens of channels.

The field of implantable bionics began with the development of the pacemaker (see Chapter 8). Development outside of the pacemaker field was relatively slow throughout the 1970s, but the widespread availability of microprocessors from the 1980s onward, led to an explosion of sophisticated bionic innovation in the late 20th century, and early 21st century, which continues to this day, from which the numerous bionic implantable systems listed below arose. In essence the three drivers for bionic innovation were as follows:

(1) The invention of the transistor in 1947 kick-started the field of implantable bionics in the 1950s and enabled the development of the world's first wearable pacemaker in 1957 (Earl Bakken).

(2) The invention of the titanium/alumina hermetic feedthrough concept in 1970 (David Cowdery) made implantable bionic medical devices a viable technology thereafter.

(3) The invention of the microprocessor in 1968, and its subsequent wide availability and usage that arose in the 1970s, and spawned the microcomputer industry, had the effect of turbocharging the field of implantable bionics from the 1980s onward.

Thus a clear lesson emerges from the history of bionics. Each leap forward was made possible by the invention of a key platform technology: the transistor, the titanium/alumina hermetic feedthrough, and the microprocessor. As a consequence, the bionics explosion since the late 20th century has brought us now 14 different types of commercial hermetic bionic implants, commonly known as IPG devices in current use, and a number under development for the future.

10.1 Alumina in bionics today

It is interesting to reflect, half a century after the invention of the hermetic titanium/alumina-feedthrough bionic implant system by David Cowdery in 1970 [146,147], that all 14 implantable bionic technologies on the market today that require external electrodes (and therefore a feedthrough), still use the same Cowdery titanium/alumina hermetic feedthrough concept. These are further summarized, categorized by the physiological systems they apply to: heart, sensory organs, brain and central nervous system, peripheral nerve system, functional implant, and the medical conditions treated are itemized also:

Heart
   (1) Cardiac pacemaker (heart rhythm)
   (2) Cardioverter defibrillator (heart defibrillation)
   (3) Cardiac pacemaker/defibrillator (heart rhythm, heart defibrillation)

Sensory organs
   (4) Cochlear implant (hearing)
   (5) Bionic eye (vision)

Brain and central nervous system
   (6) Deep-brain stimulator (Parkinson's disease, essential tremor, dystonia, OCD)
   (7) Spinal cord stimulator [(SCS) back pain, angina pain, peripheral artery disease pain]

Peripheral nerve stimulators
   (8) Vagal nerve stimulator (epilepsy, depression)
   (9) Occipital nerve stimulator (migraine, brain trauma)
   (10) Sacral nerve stimulator (SNS) (incontinence, pelvic pain, ED)
   (11) Gastric stimulator (obesity, gastroparesis, IBS)
   (12) Pulmonary stimulator (respiratory support)
   (13) Functional electrical stimulation (FES)/peripheral nerve stimulation (PNS) (foot drop, mobility for spinal cord injured patients)

Functional implants
   (14) Implantable drug delivery pumps

It should also be noted that there are also some temporary bioelectronic devices used in the body cavity that do not require external electrodes (therefore no feedthrough needed), such as the glass-encapsulated pillcam which is swallowed and films the gastrointestinal track during its slow passage through. Such devices do not involve alumina and are therefore beyond the scope of this book.

However, all 14 hermetic implantable bionics listed above share the same characteristics. The bionic implant shown in Fig. 10.1 is for the application of the pacemaker based on the Cowdery concept: the hermetic titanium/alumina capsule, which is commonly known today as the IPG, implanted in a convenient location, and connected to electrode leads which travel to the remote location for neural tissue stimulation. In the case of Fig. 10.1 this is deep within the heart. Whether pacemaker, deep brain stimulator (DBS), or any of the 14 implants listed above, every IPG looks much the same, it is just a question of its programming, total channel number, and which neural tissue its electrodes stimulate. The platform technology: IPG, leads, and remote electrodes,

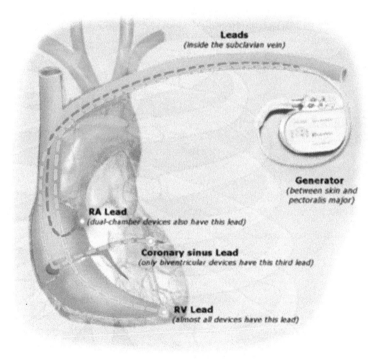

Fig. 10.1 The hermetic bionic implant concept, involving the titanium/alumina-platinum implant itself, commonly known as the IPG, and labeled above as the "Generator," connected to electrode leads which travel to a remote tissue location. Here the hermetic bionic implant concept is shown for the case of the pacemaker, with the IPG (generator) located beneath the skin in a convenient location for ease of access and removal, and the leads connected to remote electrodes, in this case deep within the heart.
Image Author Nicholas Patchett [381]. Reproduced from Wikipedia (Public Domain Image Library).

forms the basis of this wide range of bionic implant technologies on the market today, characterized by the following general features:

- titanium casing for the hermetic bionic implant (IPG)
- hermetic alumina feedthrough brazed to the titanium casing
- multiple platinum-conductor-based channels passing hermetically through the alumina feedthrough
- hermetic-bionic-implant device (IPG) implanted in a convenient location, usually just beneath the skin, to enable ease of removal, wireless remote recharging and programming/control
- electrode leads travel some distance from the hermetic bionic implant (IPG) to the stimulation site
- platinum-based stimulation electrodes in the neural tissue

All of these principles, other than wireless recharging and programming, were available in the earliest pacemakers. Induction control came in with the bionic ear. The wireless charging and control, programming capability, miniaturization, and the broadening range of medical conditions treatable with implantable bionics have all increased exponentially in the 21st century. The total market size for hermetic bionic implants is now around $25 billion globally.

Bionics global market size 2020: $Billion/PA:

- Pacemakers/defibrillators: $12
- Spinal cord stimulators (back pain): $5
- Cochlear implants (bionic ear): $2
- SNS (bowel, bladder): $1.8
- DBS (Parkinson's disease, dystonia, OCD): $1.6
- Vagus nerve stimulator (VNS) (epilepsy, depression): $1
- Gastric electric stimulator (various): $0.4
- All other bionic implants (as per the list below): $1
- Total implantable bionics market: $25 billion

All utilize the titanium/alumina-platinum hermetic system (Pt or Pt/Ir) (Fig. 10.2).

This book is about alumina. Space does not permit going into exhaustive detail on all these spin-off bionic technologies. They all use the titanium canister with alumina feedthrough. This platform technology was invented in 1970 and is now highly refined and in a very standardized format of the titanium/alumina-platinum IPG connected to electrode leads and remotely placed electrodes. However, I will highlight some of them as they demonstrate the enormous impact that alumina, in its central role in hermetic bionics, has had on some very serious disease states that cannot be treated effectively by any other means:

- Heart arrhythmia (pacemaker, defibrillator)
- Sensorineural deafness (bionic ear)
- Drug-resistant Parkinson's disease (DBS)
- Drug-resistant epilepsy (vagus nerve stimulation)
- Chronic drug-resistant pain (SCS)
- Urinary and bowel incontinence (SNS)
- Mobility after spinal cord injury (FES)

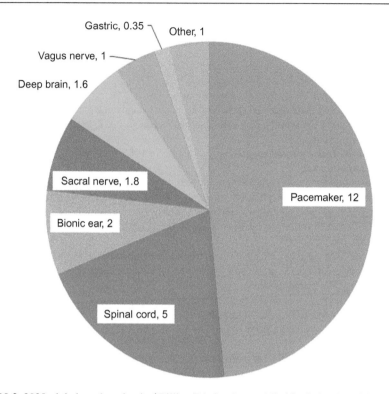

Fig. 10.2 2020 global market size in $Billion/PA for the world's bionic implantable devices. Note that pacemaker includes pacemaker, defibrillator, and pacemaker/defibrillator.

10.1.1 The bionic pacemaker, defibrillator, and pacemaker/defibrillator

The history of the pacemaker has been reviewed in Chapter 8 which documents its invention. Pacemakers today are available in magnetic resonance imaging (MRI)-compatible versions as well as standard pacemakers (not MRI compatible). The main advance in pacemaker technology other than miniaturization has been the introduction of the implantable cardioverter defibrillator (ICD), and the combination unit pacemaker/defibrillator. Thus three bionic devices are now on the market for the heart: pacemaker, ICD defibrillator, pacemaker/ICD combination unit. Around 1 million pacemakers are implanted every year, at a cost of $10,000–$30,000 each. The global market in 2020 for all three cardiac bionic implants is around $12 billion The ICD was first clinically trialed in 1980, but research into its development dates back to the pioneering work of Michel Mirowski, Morton Mower, and William Staewen, beginning a decade earlier in 1969 [422].

The ICD bionic implant can intelligently detect a cardiac arrest or a pathological heart rhythm state, and intelligently deliver an electric shock to restore normal heart rhythm: a small shock in the case of arrhythmia, a large shock in the case of the detection of fibrillation. It uses a small but powerful capacitor to charge up and deliver this shock, with charge-up times in the order of 2–6 s [423]. Pacemaker companies include:

Medtronic
USA. Founded 1949. ~90,000 employees.
Core business is medical devices. World's largest standalone medical device company. Originator of the world's first wearable pacemaker in 1957, invented by Earl Bakken (Medtronic cofounder).
Acquired Dutch pacemaker company, Vitatron in 1986

Abbott Laboratories
USA. Founded 1888. ~100,000 employees.
Core business: Health products spanning a broad spectrum from pharmaceuticals to nutraceuticals to diagnostics to medical devices.
Acquired ($25 billion in 2017) St Jude Medical, USA pacemaker company (founded 1976).
St Jude Medical acquired (1996) Telectronics, Australian pacemaker company, founded 1963, originator of the world's first hermetic implantable pacemaker, invented by David Cowdery (Telectronics Director)

Boston Scientific
USA. Founded 1979. ~30,000 employees.
Core business is medical devices

LivaNova
UK. Founded in 2015 from merger of:
Sorin Group. Founded 1956. Italian pacemaker company.
and
Cyberonics. Founded 1987. USA vagus nerve stimulation bionic company.
Core business is pacemakers and medical devices

Biotronik
Germany. Founded 1963. ~6000 employees.
Core business is pacemakers

Medico
Italy. Founded 1973. Private pacemaker company.
Core business is pacemakers

Shree Pacetronix
India. Founded 1988.
Core business is pacemakers.

10.1.2 Spinal cord stimulator

The development of SCS dates back to first use in 1967 [424], and it was FDA approved in 1989. It uses paresthesia to mask the feeling of pain in body areas innervated by the stimulated fibers, essentially overstimulating the pain-delivering nerves and thereby impeding their task of transmitting pain. Though the mechanism is not fully understood, the "gate theory" is the accepted mechanism, that is, that large nerve fibers act to turn the neurological pain gate off upon stimulation [425]. SCS technology is also useful for treating other forms of chronic pain, for example, peripheral ischemic limb pain.

IPG Location: Beneath the skin, lower back, to the side.
Electrode Location: Platinum alloy electrodes adjacent to spinal column.
First Studied: 1967.
First Clinical Trial: 1967.
First Commercialized: FDA 1989.
Other Uses: Chronic pain from peripheral artery disease.
Corporations: Medtronic, Abbott/St Jude Medical, Boston Scientific, Nevro Corp, Saluda Medical.
Second largest bionic implant in market size (after pacemaker family): $5 billion/PA.

10.1.3 The bionic ear (Cochlear implant)

The bionic ear was the world's second implantable bionic technology to be commercialized after the world's first implantable bionic device—the pacemaker (1971). The bionic ear capable of speech recognition was commercialized in 1985 by Cochlear Ltd. The inventor was Graeme Clark and the feedthrough innovator was Janusz Kuzma, who invented (for the bionic ear) the world's first co-fired alumina-platinum feedthrough in 1982. The early history of the bionic ear has been documented in Chapter 9.

Cochlear Ltd. has maintained a focus on R&D to maintain its market edge over its now four competitors and retains 70% market share today. At $2 billion a year in sales, the bionic ear is now the world's third largest hermetic bionic implant (IPG) device market in dollar value, behind the pacemaker/defibrillator family and the SCS. The five bionic ear manufacturers in the world are as follows:

Cochlear Ltd.
    Australia. Founded 1983. ~4000 employees.
    Core business is bionic ear implants.
    Currently has 70% of global market
MED-EL
    Austria. Founded 1990.
    Core business is bionic ear implants.
    Currently ranked number 2 in the global market
Sonova
    Switzerland. Founded 1947. 14,000 employees.
    Formerly called AG für Elektroakustik (1947–1965), and then Phonak (1965–2007).
    Core business is hearing aids.
    Acquired (2010) Advanced Bionics, USA bionic ear company (founded 1993)
Oticon Medical
    Denmark. Founded 1904. ~3000 employees.
    Founded by William Demant in 1904. Parent company is William Demant Holding Group.
    Core business is hearing aids.
    Acquired (2015) Neurelec, French bionic ear company (bought Chouard's bionic ear technology in 1987, and commercialized it in 1991)
Nurotron
    Headquartered in the United States (manufacturing in China). Founded 2006.
    Core business is bionic ear implants and neuromodulation devices.

10.1.4 Vagus nerve stimulator

The vagus nerve is the largest nerve in the human body. It represents the parasympathetic nervous system for autonomic regulation of the larynx, esophagus, heart, and majority of the gastrointestinal structures The VNS is the only bionic device in the world approved to treat *epilepsy* or *depression*. Mechanism of action is not entirely understood, but it has been known since the 1930s. Given that the vagus nerve is connected to the brain, it is obviously this connection with the brain that gives VNS functionality. A summary of key VNS facts is as follows [426–428]:

IPG Location: Beneath the skin just below the left collar bone
Electrode Location: Two platinum alloy electrodes wrapped around the left vagus nerve beneath the skin of the neck (left side at the front)
First Studied: 1938 (in cats)
First clinical trial: Epilepsy (1988); Depression (2000)
First Commercialized: Epilepsy (1994 EU, 1997 FDA); Depression (2001 EU; 2005 FDA)
Other Uses: Possibly Alzheimer's disease, eating disorders, and heart arrhythmias
Corporations: LivaNova (Cyberonics developed VNS, and merged with Sorin in 2015 to form LivaNova)
Over 100,000 VNS implanted worldwide to date, mainly for epilepsy.

10.1.5 Sacral-nerve stimulator

The sacral nerves are a collection of five pairs of nerves that exit the spinal column at its lowest point, the sacral vertebral area. The SNS is the only approved implantable bionic solution in the world for *urinary* or *bowel incontinence*. SNS may also have efficacy for pelvic pain and erectile dysfunction (ED). The S2, S3, and S4 nerves are the target locations for SNS electrodes as they have demonstrated optimal response to stimulation most frequently in clinical trials, S3 is particularly relevant to bladder. The electrode positioning and IPG-implant programming need to be optimized to the patient requirements, which can vary significantly as there are a wide range of urinary and bowel voiding dysfunction states suffered by patients. The Medtronic Interstim has the global monopoly on SCS [429–431]. A summary of key SNS facts is as follows:

IPG Location: Beneath the skin, upper buttock area
Electrode Location: Sacral (S3 or S4) nerve, close to the sacrum, to stimulate the neural pathway that controls bladder and bowel function
First Studied: 1979 in animals
First Clinical Trial: 1989
First Approved: Urinary incontinence (1994 CE-Mark EU; 1997 FDA); bowel incontinence (2011 CE-Mark EU, not yet FDA)
Other Uses: Appears to have some efficacy for ED and pelvic pain. Unapproved at this stage.
Corporation: Medtronic (Interstim system) has world monopoly
Over 100,000 SNS implanted worldwide to date for incontinence.

10.1.6 Functional electrical stimulation

FES, also sometimes known as PNS was actually one of the earliest bionic technologies to be developed. FES is a *mobility-stimulating bionic technology*. The very first bionic FES system was the subject of a patent filed in 1951 and granted in 1956 by Charles Giaimo [432], a wearable battery-powered walking activator. This was followed by a famous study of 1960 by Liberson [433] for stimulating ankle dorsiflexion so as to treat "*foot drop*," a common problem with multiple-sclerosis patients.

FES is most widely known for its use in exercise bike training for health and muscle/bone strengthening for paraplegics.[1] The appropriate nerves in the leg are stimulated by electrodes. For temporary use this is done transcutaneously, but it is much more effective with implants which give much more precise neural targeting and stimulation. In the future, fully implantable hermetic FES mobility systems could be the means by which spinal cord injured patients walk again unaided. Paraplegics routinely walk using FES bionic external or implantable systems, and a walking frame. Gross motor control for walking is possible with FES, fine motor control for balance is not yet. This would require a very large channel number/electrode number and very precise positioning of the plethora of electrodes, requiring a large capital investment in device development, and high ultimate purchase price. FES is still largely an experimental discipline and has thus far failed to make the leap into widespread mass-market commercialization due to practical and commercial reasons beyond the scope of this brief summary.

10.1.7 Deep brain stimulator

For drug-resistant *Parkinson's disease*, DBS has been a very important therapy for thousands of sufferers worldwide, and this is what it is best known for. DBS has been very successful in treating neurological diseases that produce involuntary movement, like Parkinson's disease, *dystonia* (subthalamic nucleus; globus pallidus) and *essential tremor* (central lateral nucleus of the thalamus) [434]. It is also effective in treating *OCD*—obsessive compulsive disorder (nucleus accumbens). DBS has just been approved in Europe for *epilepsy* (anterior thalamic nucleus), but not yet FDA approved [435]. DBS appears to have some capability for treating *depression*, and *schizophrenia*, clinical trials are under way but it is not yet approved for these.

Spanish neuroscientist Jose Delgado is the original pioneer of DBS with his seminal paper in 1952 [436]. Many researchers worked in this area in the decades to follow, but the introduction of L-DOPA treatment for Parkinson's disease in the late 1960s lead to a decline in surgical solutions. It was not until the landmark paper of Benabid in 1987 [437], in an era when drug-resistant Parkinson's disease began to

---

[1] The author's own involvement in FES began in 2001 when Glen Davies, Che Fornusek and the author led a team that produced one of the world's first FES pedal-powered recumbent bicycle systems, which evolved over the subsequent decade into training systems for spinal-cord-injured patients. This collaboration is ongoing. See "About the Author."

be recognized as a significant problem, that the concept of DBS for drug-resistant Parkinson's disease really took off, and it was commercialized within a decade.

In spite of the fact that DBS involves implanting electrodes deep in the brain, the surgical procedure is highly refined and relatively risk free. A very thin electrode is inserted, using sophisticated brain navigation systems, via a stainless steel cannula, deep into the brain tissue where the neurological disorder originates. The lead exits the top of the brain where it is connected to the cables from the IPG. The IPG bionic implant is implanted beneath the skin below the collarbone, and the electrode cables are tunneled through the neck to the top of the brain where they are connected to the electrode lead. The electrodes stimulate the localized brain region for an exhibitory or inhibitory response [438]. The final location is determined by either microelectrode recordings from the surface of the brain, or via magnetoencephalogram (MEG) [434]. Medtronic has the world monopoly on DBS, and it is a high-tech therapy that Medtronic is renowned for, especially for Parkinson's disease. DBS is one of the most high-tech commercial bionic therapies in the world today.

IPG Location: Beneath the skin, below the collarbone
Electrode Location: Depends on the condition treated, see above
First Studied: 1952
First Clinical Trial: 1950
First Approved: Essential Tremor FDA 1997; Parkinson's disease FDA 2002; dystonia FDA 2003; OCD FDA 2008; Epilepsy only CE-Mark (EU) 2017. All others also CE-Mark EU
Other Uses: Appears to have some efficacy for depression
Corporation: Medtronic has world monopoly

10.1.8 Evolution of the alumina feedthrough

It is instructive to pose the question, of the original hermetic-bionic-implant feedthrough (IPG/Leads/Electrodes) innovation of 1970, what has changed and what has stayed the same?

What has stayed the same?

- Titanium casing remains the industry standard
- Alumina feedthrough seal remains the industry standard
- Titanium casing holding the alumina seal in residual thermal contraction compression remains the industry standard
- It is still standard practice to braze the alumina and titanium together
- The titanium-based brazing alloy from 1970, TiCuNi, is still commonly used

What has changed?

- In all cases, the single titanium feedthrough of 1970 has evolved into multiple-channel feedthroughs of noble metals, platinum being the main one, either pure or as a platinum-iridium alloy.
- In some cases, instead of TiCuNi brazing of alumina to titanium, gold brazing is used, on pre-gold-coated alumina.
- In some cases, the 1970 "titanium single-channel terminal TiCuNi brazed to the alumina" concept has evolved into noble metal channel wires (commonly Pt or Pt/Ir) gold-brazed to

gold-coated alumina. This is commonly used in the case of macro-terminals when the total channel number is small, that is, up to 100 or so. This is shown schematically in Fig. 10.3.
- In some cases the 1970 "titanium single-channel terminal TiCuNi brazed to the alumina" concept has evolved into platinum channel wires co-fired into the alumina. This is commonly used in the case of large numbers of micro-channels. Up to 2500 channels/cm^2 (platinum channels in alumina) has been achieved by this strategy, using printed platinum tracks rather than wires for the channels [412]. This is shown schematically in Fig. 10.4 (it is shown for the actual bionic ear feedthrough in Figs. 9.7 and 9.8).

A common hermetic implantable feedthrough scenario today, and indeed for several decades now, is 95%+purity alumina ceramic, perforated to provide holes for platinum or platinum/iridium channels, which emerge from their braze-points in the alumina surface as terminal pins, suitable for fixing biocompatible in vivo terminals. The alumina is gold sputter coated and then gold brazed with 99.99% pure gold to both the titanium casing and the platinum/iridium channel pins or wires [439]. This is commonly used for as few as two channels right up to 100 or so channels. It is rare to encounter a system these days with <16 channels [439].

An alumina seal hermetic bionic implant (IPG) with more than about 100 channels is regarded as a high-density feedthrough system. In high-density feedthrough systems, co-firing of the channel wires with the alumina is a more viable strategy rather than brazing. In some of these feedthrough systems, particularly the high-density systems (around 100 or so channels), the platinum/iridium pins (pins brazed or co-fired into the alumina) function directly as pin electrodes for the neural tissue, for example

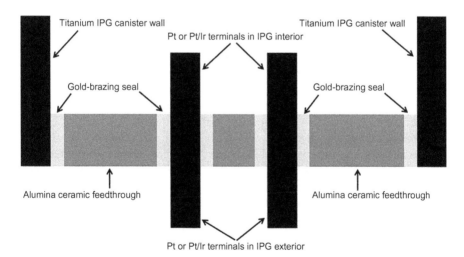

Fig. 10.3 Titanium/alumina gold-brazed feedthrough. This has been commonly used since the 1980s for hermetic-bionic-implant (IPG) devices with a small number of channels (<100), such as pacemakers. The alumina is gold coated, the channel conductor wires (two channels in this diagram) pass through holes in the alumina ceramic and are brazed to the gold-coated alumina with gold brazing. Surface gold on the alumina can then be removed by sandblasting to eliminate short circuits from surface conductivity on the alumina.

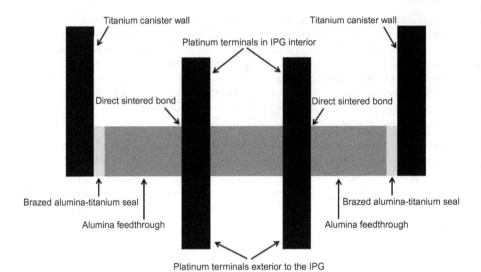

Fig. 10.4 Co-fired platinum channel wires in an alumina feedthrough (two channels in this diagram). Invented by Janusz Kuzma in 1982, this is a method used since its invention by Kuzma in 1982, and increasingly commonly used today when a very large number of channels are required, such as a bionic ear or bionic eye. The platinum channels are inserted in the green alumina and sintered in place by co-firing. The co-fired feedthrough is then brazed to the titanium casing.

this could be used in a BCI implant implanted into the surface of the brain (cerebral cortex see Section 10.5).

Essentially the titanium/alumina-platinum IPG feedthrough systems have become a standard platform technology, provided by OEM manufacturers in some cases, that can simply be adapted to new bionic implantable medical device innovations, both of the IPG type with leads to remotely located neural tissue, and as the IPG/electrode package in one, as is the case for bionic eye retinal implants.

The regulatory regime globally is very much stricter today than it was half a century ago in 1970. The materials used in tissue-contact in a bionic implant must be of "substantial equivalence" to other regulatory-approved implantable devices. This leaves manufacturers with a rather short list of biomaterials to choose from, which has shrunk in common practice for IPG implants for in vivo use, where the biomaterial interfaces with tissues or body fluids: platinum and platinum/iridium, titanium and alumina, with insulation of leads generally by parylene or silicone.

As discussed in Chapters 8 and 9, alumina is an essential component of the hermetic encapsulation and electrical feedthrough system used in all implantable bionics. The hermetic bionic implant (IPG) concept invented for the pacemaker in 1970 has now evolved into applications in a wide range of bionic implants comprising a global industry in the $25 billion order of magnitude, and growing rapidly, thereby making implantable bionics one of the most important commercial applications for alumina today.

The most significant evolution in the alumina feedthrough technology since 1970 has been miniaturization. However, very little else has changed since the 1980s when we already had the switch to using platinum and other noble metals rather than titanium for the channel wires fed through the alumina, widespread almost universal use of silicone coating, widespread use of gold brazing, and the introduction of the concept of co-firing platinum channels directly into the alumina ceramic. All of these platform technologies were developed in Australia during the development of the world's first commercial implantable pacemaker and a decade later the world's first bionic ear capable of speech recognition (see Chapters 8 and 9).

Chapter 8 outlined the origin of bionics, and specifically the crucial role of alumina in the bionic feedthrough, invented in 1970 for the world's first hermetic implantable bionic device: the pacemaker, commercialized in 1971. Chapter 8 explored in detail the invention of the world's first titanium/alumina hermetic feedthrough technology, all within the context of the development of the world's first commercial implantable hermetic bionic device: the cardiac pacemaker. It was the alumina feedthrough that was the key enabling technology for the hermetic implantable pacemaker invented in 1970.

Chapter 9 explored the quantum leap from the single-channel pacemaker feedthrough to the world's first high-channel-number feedthrough: the 22-channel bionic ear feedthrough, with the introduction of the world's second bionic technology: the bionic ear capable of speech recognition, commercialized in 1985. Feedthrough technology made the quantum jump from 1 to 22 channels, and the conductors were platinum not titanium, but nothing else changed. The hermetic bionic implant (IPG) remained as a titanium casing TiCuNi-brazed to an alumina feedthrough.

A quarter of a century went by from the mid-1980s to the end of the 2000s, during which time the main changes were iterative miniaturization, and the commercial introduction of more and more bionic devices, using the same titanium/alumina hermetic-feedthrough platform technology, for treating more and more pathologies that could be treated by bionic neuromodulation implantable devices.

And then just in the last decade, a third quantum leap forward took place, associated with the bionic eye. Feedthrough technology made the jump from tens of channels (up to a limit of about 100) to 1000+ channels, but continued to be titanium casing, alumina feedthrough, platinum-based channels, as it had been for four decades.

There have been essentially three quantum leaps made in alumina feedthrough technology in the history of implantable bionics:

Quantum Leap 1: The invention by David Cowdery (1970) of the hermetic feedthrough. This was a single-channel system developed for the pacemaker, based on a titanium/alumina feedthrough platform technology that has remained the global standard ever since.

Quantum Leap 2: Adaptation by Janusz Kuzma (1982) of the titanium/alumina feedthrough into the multiple-channel feedthrough (22 channels), developed for the bionic ear, miniaturized and utilizing platinum channels.

Quantum Leap 3: Adaptation by Gregg Suaning (2012) of the titanium/alumina multiple-platinum-channel feedthrough into the 1000+ channel stepped layered alumina-platinum-array feedthrough, as a titanium/alumina-platinum miniaturized hermetic-implantable-bionic system for bionic eye retinal implants.[2]

Quantum leap 1 formed the basis of Chapter 8. Quantum leap 2 formed the basis of Chapter 9. Quantum leap 3 forms the basis of the following bionic eye case study in this chapter. This book is about alumina. Therefore, the primary focus in this chapter is a case study on quantum leap 3: the adaptation of the alumina feedthrough to the 2500 channels/cm^2 titanium/alumina feedthrough system for the retinal implant for the bionic eye. This will constitute Sections 10.2 (bionic eye overview) and 10.3 (alumina feedthrough case study for the bionic eye).

10.2 The bionic eye

10.2.1 Bionic eye: The bionic technology concept

The bionic eye is not yet one of the well-established commercial bionic technologies that were discussed in Chapters 8 (pacemaker) and 9 (bionic ear), nor is it one of the dozen or so well-established commercial bionic technologies such as the SCS, DBS, and vagus nerve simulator that were highlighted in Section 10.1. The bionic eye today is about where the bionic ear was in the early 1980s when a rudimentary bionic ear had been FDA approved, the 3 M-William House device which did not have speech recognition capability. The 3 M-House bionic ear had very little market uptake: 1000 units, compared to the 250,000 Cochlear Ltd. Implants sold since then. This was because the 3 M-house implant did not deliver to patients what they required from a bionic ear—speech recognition. In the early 1980s, the world's first bionic ear capable of speech recognition (Cochlear Ltd.) was already a laboratory reality, and just a few years from FDA approval which came in 1985. The Cochlear bionic ear, after a long hard process of commercialization in the late 1980s, was changing lives globally on a large scale within a decade. We are in much the same place with the bionic eye today.

Currently, there is an intense global race to see who will produce the first commercial (FDA approved or at least CE Marked) bionic eye that can give reading capability (large print text), and object recognition capability, to the implant recipient. Nobody has yet won this race although the technology is now capable of it and commercial realization may come in the next decade. The first FDA approved bionic eye implant was launched in 2012 (Argus II), but like the 3 M-William House bionic ear, it is a very rudimentary device which falls far short of object recognition or reading capability. So strictly speaking, the bionic eye in a rudimentary form does exist today (Argus II), but a bionic eye with reading and object recognition capability would have to be classed as "bionics for the imminent future."

---

[2]The author was a collaborator on this project and a coauthor on the key paper. See "About The Author."

10.2.2 The Suaning retinal implant concept

Retinal implants capable of object recognition and reading capability were perfected just a couple of years ago and are the primary focus of this chapter. From here it is just a matter of investment capital, perfection of a prototype, applying for FDA, CE mark, and other regulatory approvals, and then the long hard grind of commercialization. Mainly it's about the investment capital. The science is largely done. In theory, from a technical and engineering perspective, a bionic eye with object recognition or reading capability is almost a laboratory reality today. All the elements are there, it's just a question of money and time. With my decades of experience as a participant and spectator in the long grinding process of commercializing new biomedical technology in the current era, I would forecast at least a decade before this becomes a commercial reality.

The concept of the bionic eye is that vision is planar (a two-dimensional planar pixel array). Pixel count correlates with resolution, a concept we are all familiar with in the digital camera era. Stimulation of the optic nerve will create flashes of light but it cannot create the spatial two-dimensional image that a healthy eye would see. To create an image in the patient's brain that corresponds to what a healthy eye would see, a planar retinal implant is required that stimulates the retinal tissue in a planar way, in just the same way as the incoming light stimulates healthy retinal tissue in a planar way (bearing in mind that the retinal surface is concave).

It is possible that planar visual stimulation can also be achieved with a visual cortex implant, and feasibility studies on this are ongoing (see Section 10.4). This may or may not be technically feasible and could be decades in the future if it is. In contrast, the retinal implant approach has already been commercialized in a very rudimentary form and is therefore well beyond the feasibility trial stage.

There are essentially four components to the Suaning bionic eye retinal implant innovation, and to a large extent this system mimics the long-established Cochlear Implant bionic ear system in commercial use since the 1980s[3]:

External device
    Bionic eye: Camera/battery/image processor
    (Bionic ear: Microphone/battery/speech processor)
Inductive transmitter
    Bionic eye: Coil on the scalp
    (Bionic ear: Coil on the scalp)
Inductive receiver
    Bionic Eye: Subcutaneous receiver coil
    (Bionic Ear: Subcutaneous implant incorporating microelectronics, titanium/alumina-platinum feedthrough and receiver coil)
Neural stimulation electrodes

---

[3] Prior to becoming a professor of biomedical engineering, Gregg Suaning was an engineer at Cochlear Ltd., and thus had a strong background in the highly evolved Cochlear technology, which has undergone constant enhancement and fine tuning since the 1980s. His retinal implant technology represents a quantum leap from existing Cochlear technology.

Bionic eye: Retinal implant combines titanium/alumina-platinum feedthrough with microchip inside. The platinum terminals on the outside surface of the alumina feedthrough form the planar retinal stimulating electrode array
(Bionic ear: Silicone-platinum coiled rod that forms intra-cochlea electrode array)

Apart from the obvious differences in function (microphone versus camera; retinal stimulation versus cochlea stimulation), the primary difference between these two systems is that, due to major improvements in bionic miniaturization since the invention of the bionic ear four decades ago, the retinal implant can combine feedthrough, microprocessor, and electrodes in one tiny hermetic package 10 mm in size, and a couple of millimeters thick. That is, a titanium case brazed to an alumina feedthrough, with platinum channels (simultaneously both feedthrough conductors and electrodes) co-fired into the alumina, the same basic hermetic concept as for all bionic implants that have come before. However, the "electrode side" of the feedthrough (the outside face of the alumina feedthrough) instead of being the terminal connection point is in fact truly the "electrode side" because the exposed platinum feedthrough conductors (which are therefore simultaneously channels, terminals and electrodes) are micron sized electrode "pads" flush with the alumina surface and function as electrodes, thereby stimulating the retinal tissue directly. The inside face of the alumina feedthrough is the "chip side" where the exposed feedthrough terminals couple electrically with the terminals of the microprocessor chip.

10.2.3 Retinal implant location

The various bionic teams around the world have been focusing on four locations within the eye for the implantation site of the retinal implant:

- epiretinal, tacked to the fragile retinal surface
- subretinal, embedded in the neural tissue at the back of the retina
- suprachoroidal, between the choroid (the vascular layer of the eye) and the sclera (the white of the eye)
- intrascleral, embedded within the sclera

Each of the various global bionic eye research/commercial teams listed in Table 10.2 has decided on a particular implant location (itemized in Table 10.2). These four sites are shown in Fig. 10.5, and a risk analysis is summarized in Table 10.1. There is a trade-off between risk and benefit.

Table 10.1 **Risk-benefit analysis of retinal implantation sites**

Implantation site	Resolution capability	Risk level	Surgical challenge
Epiretinal	High	High	High
Subretinal	High	High	High
Suprachoroidal	Moderate	Low	Low
Intrascleral	Low	Very low	Low

Table 10.2 A summary of global bionic eye research/commercial teams

Team/product	Implant site	Electrodes (pixels)	Progress status	Ref.
Argus II	Epiretinal	60	Developed 2009, *FDA Approved 2012*. Over 100 patients to date	[440]
IRIS II	Epiretinal	150	*CE Mark 2016*. 10 patients to date	[441]
IMI: Intelligent Medical Implant	Epiretinal	49	Human Clinical trials	[442–444]
EPI-RET	Epiretinal	25	Human Clinical trials	[445]
3D Stacked Retinal Prosthesis	Epiretinal	16	Rabbit testing	[446]
Wayne State University	Epiretinal	60	Clinical trials	[447]
Boston Retinal Implant Project	Subretinal	15	Animal testing	[448]
Retina AG	Subretinal	16	Clinical trials	[449,450]
ASR Device (Optobionics)	Subretinal	a	Clinical trials. Controversial—re its 3500 microphotodiodes rationale	[451]
Retina Implant AG	Subretinal	a	Clinical trials with 1600 microphotodiode system	[452]
Biohybrid Implant	Subretinal	N/A	Neural cells cultured directly onto a chip. Early development	[453,454]
3D Implant	Subretinal	N/A	Developing a novel 3D electrode concept	[455,456]
Nidek	Suprachoroidal	64	Animal trials	[457]
Seoul National University	Suprachoroidal	108	Animal trials	[458]
Gregg Suaning Group	Suprachoroidal	1145	Prototyped, quality manufacturing system established	[375]

[a]Microphotodiodes.

Fig. 10.5 Retinal implantation sites. (1) Epiretinal, tacked to the fragile retinal surface; (2) subretinal, embedded in the neural tissue at the back of the retina; (3) suprachoroidal, between the choroid (the vascular layer of the eye) and the sclera (the white of the eye); and (4) intrascleral, embedded within the sclera.

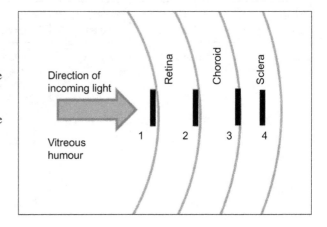

The risky BUT high-resolution options are epiretinal and subretinal. Epiretinal, tacking an implant to the fragile retinal surface, gives excellent resolution as the electrodes stimulate the retinal tissue direct, through intimate electrical contact, however the retina is fragile and easily damaged. This is also a delicate surgical procedure with problematic access. Subretinal is a more stable location as the implant is embedded within the neural tissue of the retina, enabling better fixation, and also excellent resolution as the electrodes stimulate the retinal tissue direct, through intimate electrical contact. However, there is a risk of damage of the fragile neural tissue during implantation, and long-term damage through disruption of the blood flow in the retinal tissue.

The safer, lower-resolution options are suprachoroidal and intrascleral. Access is easier, and there is no risk of damage to the retina. However, the electrodes are only stimulating the retina indirectly, current has to pass through the choroid in the case of suprachoroidal, and both choroid and sclera in the case of intrascleral. This requires a higher charge input, and results in lower resolution. Suprachoroidal certainly has a significant advantage over intrascleral for resolution, without being greatly riskier.

10.2.4 Bionic eye: A summary of the global technology status quo

Table 10.2 summarizes all the bionic eye teams around the world today and their progress. For brevity, I have confined this summary only to those that involve a retinal implant, as this is the approach that is currently achieving success.

As Table 10.2 shows, we are at the point now where the first FDA-approved bionic eye system is on the market, the 60 electrode Argus II, approved in 2012. This gives a roughly 8 × 8 pixel array. As shown in Fig. 10.6, at 60 electrodes (pixels), a bionic eye is capable of little more than detecting large objects, distinguishing light from dark, and detecting movement of large objects. For example, recognizing when a doorway is open or closed, or when a large object is moving in their field of view, such as a car or another person. But large print reading or object recognition is well beyond what 60 electrodes (pixels) can achieve. As shown in Fig. 10.6, at least 1000 pixels are required to read large print text, and object recognition (a car in this example).

Fig. 10.6 Electrode number (pixel number) and resolution. Top row is a car, lower row is large print text. It is apparent that the world's first commercial bionic eye (Argus II FDA approved 2012) with 60 pixels is only useful for discerning light from dark and movement of large objects. The 1145 pixels current world best achieved for a retinal implant by the Suaning group in 2012 [375], prototyped, but far from commercialized, is sufficient to recognize a car and read large print text. The hope for the future of 10,000 pixels, which comes close to normal vision, is probably decades into the future.

So, in summary, the world has its first commercial bionic eye, but it is a rudimentary system. It falls far short of the capability of reading or object recognition.

The retina has in the order of 100,000 receptors per square millimeter. A 7-mm diameter retinal implant such as the 1145 alumina-platinum feedthrough shown in Figs. 10.7 and 10.8 can never achieve the resolution of the natural retina, but it can enable sufficient resolution for reading and object recognition. On the retinal surface, 7 mm diameter corresponds to a pixel count in the order of a million.

However, using electrical stimulation of retinal cells, a million pixel level of resolution is probably not possible, regardless of the electrode number, due to the effect of the electric field overlapping multiple neural cells. A similar effect occurs with the stimulation of auditory neurons in the cochlea by the 22 electrodes of the Cochlear implant. Increasing the number of electrodes does not necessarily enhance tonal resolution due to electric field overlap. Indeed, it is very probable that for the foreseeable future, resolutions in the order of 1000 pixels may be a practical limit for retinal implants, and higher resolutions may require visual cortex technology (see Section 10.5), if it is even possible with that. However, as Fig. 10.5 shows, ~1000 pixels are sufficient for reading and object recognition.

It should be clear from the above discussion that retinal implants using electrode stimulation are constrained on the one extreme by the incredibly high resolution of the eye (megapixels), and at the other extreme by electrical stimulation resolution limits

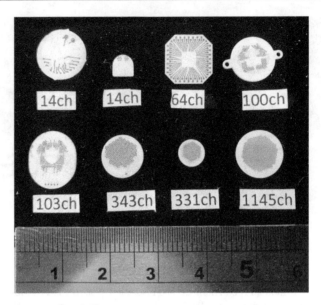

Fig. 10.7 Some of the alumina-platinum feedthrough prototypes that arose from quantum leap 3 of inventor Suaning, sponsored by the $42 million Australian Government initiative.

Fig. 10.8 The world-record breaking 1145-channel feedthrough alumina-platinum co-fired feedthrough 7-mm diameter region containing the 1145 channels. Shown ready for brazing into a titanium casing for a bionic eye retinal implant.
Invented by Gregg Suaning, with principal researcher Thomas Guenther, and other team members including Rylie Green, Charlie Kong, Hong Lu, Martin Svehla, Nigel Lovell, Christoph Jeschke, Amandine Jaillon, Jin Yu, Wolfram Dueck, William Lim, William C. Henderson, Anne Vanhoestenberghe, and Andrew Ruys [2].

for the retina. We are also constrained by the risk/resolution trade-off in the choice of implantation site. Finally, we are constrained by what is feasible in terms of electrode number that can be crammed onto the surface of a 7–10 mm diameter alumina feedthrough.

With the current world record of 1145 electrodes on an alumina-platinum retinal implant [375], alumina feedthrough technology has advanced a long way since the first hermetic pacemaker (Telectronics 1971) with its single terminal, but we are still using the same proven and trusted platform technology: invented by David Cowdery in 1970 [146,147]: titanium casing, brazed to an alumina feedthrough that is hermetically sealed to the electrical channels that pass through the alumina. What has changed is miniaturization, the co-firing technology for embedding platinum electrodes in the alumina, and some local preferences in brazing alloy.

The computational and electronic aspects of bionic eye technology are well advanced these days and fully capable of the computational processing challenges. There is little to distinguish the various competitors when it comes to visual processing software/hardware, technology, in the global bionic eye race to be the first to produce a commercial bionic eye that can enable the recipient to read.

This book is about alumina, and it is also about focusing on the technology that is leading the world. While it is true that the Argus II is a commercial product, FDA approved some time ago in 2012, this device falls far short of the 1000+ electrode array required to achieve the goal of reading large print text. There is only one team in the world that has a proven commercial prototype with the required electrode number: the team of Gregg Suaning in Australia.

It is interesting to note that Suaning's team set the world record in electrode number of 1145 in 2012 [375], within 2 years of receiving funding in 2010, and in the same year the 60 electrode Argus was FDA approved, rendering the Argus technologically obsolete in its FDA approval year. Such is the nature of the very frustratingly slow process of medical device commercialization in this era. Invention is fast. Commercialization is very slow. As Fig. 10.5 demonstrates, 1145 electrodes are sufficient for large print reading and recognition of objects. Thus the point of differentiation between the competitors is prototyped and proven retinal implant with 1000+ electrode count, and on this criterion the Suaning team is the leader. However, the Argus is a commercially available device, while the Suaning technology is years from commercialization, due to financial capital constraints.

Alumina technology is at the heart of this 1145 electrode achievement, as it was done via a retinal implant that is a titanium/alumina-platinum hermetic implant, utilizing a co-fired alumina-platinum feedthrough, a quantum leap forward from the original co-fired alumina-platinum Cochlear feedthrough with its 22 feedthroughs (see Chapter 9), or even the so-called high density co-fired feedthroughs on the market today with up to 100 feedthroughs [439]. The huge head start that the Suaning team has over its competitors in the global race to commercialize a 1000+ electrode bionic eye is the next chapter in the Telectronics/Cochlear legacy. Australian bionics research has world-leading technology capabilities, flow-on effects, and bionic research critical mass as a result of Australian company Telectronics, and its successor

Cochlear Ltd., which remains the global leader in bionic ear technology, and this provides a strong impetus for cutting edge bionics research in Australia.

It turns out that alumina ceramic engineering technology is the least developed technology in the global bionic eye race, the rate-limiting step as it were, for this race. Unlike in the 1980s when the bionic ear was commercialized, microprocessor technology then was also a significant challenge, today microprocessor technology is well beyond what is needed. In short, where the greatest technology improvements are needed is the alumina-platinum feedthrough for the retinal implant. Key imperatives are increasing the channel number in the feedthrough, and optimizing the electrode array layout for optimal interaction with retinal cells. There are also manufacturing challenges in mass production of retinal implants with hundreds, even thousands, of micron-sized platinum electrodes emanating from their high-channel-number feedthroughs. This is an alumina-focused research initiative and is one of the most cutting-edge applications for alumina in the world today, both in terms of the sophisticated ceramic engineering technology, and the huge dollar stakes involved.

10.3 Quantum leap 3 in alumina feedthrough technology: Alumina retinal implant for the bionic eye

The final episode in the three chapter story so far, on titanium/alumina feedthrough evolution since 1970, is the development of an advanced titanium/alumina-platinum microelectrode feedthrough system for use as a retinal implant for the bionic eye. The research that led to this breakthrough began as one of the three research teams funded by a 2010–16 Australian Government $42 million dollar grant dedicated to bionic eye research (ARC Special Research Initiative in Bionic Vision Science and Technology, and formerly known as Bionic Vision Australia). Funding commenced in 2010 and ceased in 2016. However, the research of Suaning continues.

Gregg Suaning was the inventor and project leader of the retinal-implant alumina feedthrough project under this grant. Suaning was at that time a professor of biomedical engineering at University of NSW and was formerly a Cochlear engineer in the 1990s. He therefore had the perfect background for this task. His co-fired alumina-platinum bionic eye retinal implant feedthrough innovation had many manifestations, shown in Fig. 10.7.

Ultimately, the Suaning retinal implant innovation set the world record in 2012 for the highest number of electrodes in an alumina feedthrough (retinal implant) of 1145, reported in two significant papers, one with Rylie Green as lead author which looked at physical and biological characterization [2], and one with Thomas Guenther as lead author,[4] which looked at the alumina-platinum co-firing ceramic engineering aspects [375].

In four decades, alumina bionic feedthrough technology evolved from a single titanium terminal in a 9.6-mm diameter alumina feedthrough (Telectronics pacemaker 1971), to 22 platinum feedthroughs in a 14.5-mm diameter alumina feedthrough

---

[4]The author was a coauthor on the Suaning paper. See "About the Author."

(Cochlear implant 1982), to 1145 platinum feedthroughs in a 7-mm diameter alumina feedthrough (Suaning Team 2014).

How this quantum leap was achieved makes a very instructive alumina ceramic engineering case study. Essentially Suaning's innovation has taken the Cochlear "platinum comb" feedthrough concept (Chapter 9) to the next level. There are three aspects to this quantum leap of Suaning, which work in combination:

(1) *The concept of the stepped feedthrough pathway* taken by the printed platinum channels, through the alumina feedthrough from the front face (electrode side) to the rear face (chip side). This overcomes the problem of platinum/alumina thermal expansion mismatch, and its potentially negative impact on hermeticity, discussed in Section 10.3.2.

(2) *The use of hot pressing* to ensure 100% hermeticity of the horizontal segments (tracks) of the stepped feedthrough pathways, discussed in Section 10.3.2.

(3) *Novel manufacturing process* involving printing the platinum channels into the alumina via tape casting, laser micromachining, screen printing, and hot pressing (thick-film technology), discussed in Section 10.3.3.

In addition to enabling huge numbers of channels in a small alumina feedthrough, this Suaning process has other advantages. Chief among them is enhanced hermeticity through hot pressing. Also, it is possible to align the channels to the microprocessor chip (on the chip side of the feedthrough) and it is also possible to link channels via "electrode clustering," for example, the concept of the seven-electrode linked hexagonal cluster, to increase the retinal stimulation efficiency of each electrode, so as to manage the "edge effect" in retinal stimulation [459]. This reduces the pixel count by a factor of 7, but enhances the retinal stimulation efficiency.

However, the primary purpose of this process is maximizing channel number for resolution, and obviously in this scenario, the channels are all independent and not linked.

10.3.1 The alumina feedthrough state-of-the-art today

To summarize, in four decades, the titanium/alumina feedthrough had evolved as follows:

1970. Original invention of the titanium/alumina feedthrough by David Cowdery Telectronics/Nucleus
 1 channel in a 9.6-mm alumina disk.
 $\sim$1.4 feedthroughs/cm^2

1982. Janusz Kuzma. Feedthrough for the world's first commercial bionic ear Telectronics/Nucleus/Cochlear
 22 channels in a 14.5-mm alumina disk.
 $\sim$13 channels/cm^2 (10-fold increase)

2013. Suaning Team
 1145 channels in a 7 mm diameter on the alumina disk.
 $\sim$2500 channels/cm^2 (1800-fold increase).

The expression "standing on the shoulders of giants" applies here as there is a direct lineage from Telectronics to Cochlear to the Suaning Team. While Australia may not be a global leader in as many areas as the United States, UK, continental Europe, or Japan, in bionic feedthroughs it has been a global leader since the invention of the world's first hermetic implantable bionic feedthrough in Australia in 1970 by David Cowdery.

During these four decades, the core technology has remained the same, and this applies to all implantable hermetic bionics with an electrical feedthrough in the world today: alumina feedthrough, titanium casing brazed to the alumina using TiCuNi brazing. The composition of the feedthroughs has changed (from titanium to platinum) and the means of passing the feedthroughs through the alumina has changed.

It is unlikely that this 1145 record will be exceeded for the foreseeable future as the $42 million research funding that financed it is exhausted, and no other competitors have come closer than 360 electrodes [460]. However, the Suaning technology is ripe for commercialization with a quality manufacturing system having been established. This is now being advanced to the next level at the University of Sydney.[5]

10.3.2 The stepped platinum micro-conductor pathway: The via/track concept

Essentially Suaning's innovation has taken the "platinum comb" concept (Chapter 9) patented by Cochlear Ltd. in 2008 [415] to the next level. The platinum comb involved a co-fired alumina-platinum feedthrough using thin platinum conductor wires in which the conductor pathway was a step, or a zigzag, or a coil, rather than a straight-through path as per the Kuzma concept. The purpose of this was to maximize hermeticity by maximizing the length of the platinum-alumina interface, and therefore maximizing the mean travel pathway for leakage.

In developing his bionic eye retinal implant feedthrough, Gregg Suaning applied some lateral thinking to this issue (literally). He reasoned that if the conductor pathway was designed as a simple step geometry, as shown in Figures 10.9. and 10.10., and if the co-sintered feedthrough was hot pressed, rather than simply pressureless sintered, thermal contraction in the lateral region of the conductor (internal track) would be completely suppressed. The alumina-platinum interface in the lateral zones (tracks) would be free of residual tensile stresses, and therefore of the highest attainable hermetic potential. Thus while thermal contraction and residual interfacial stress may exist in the longitudinal regions of the feedthrough pathway, as with the Kuzma concept and the "platinum comb" feedthroughs, the lateral zone of the Suaning concept could guarantee hermeticity absolutely.

The obvious question to ask here is does this invalidate all feedthroughs that came before? History tells us that it does not. The Kuzma concept and platinum comb feedthroughs were high-channel-density feedthroughs for their era at around 10–15 channels/cm^2. Thirty-five years of successful use of these bionic ear feedthrough

---

[5]Professor Gregg Suaning relocated to the University of Sydney in 2017.

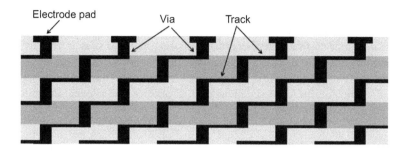

Fig. 10.9 A five-layer ($n=5$) feedthrough showing the stepped pathway of each platinum channel through the alumina feedthrough, and identifying the "electrode pad", the vertical "via," and the horizontal "track."

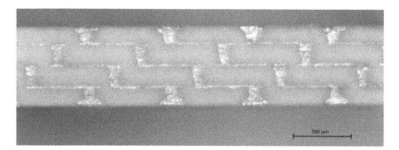

Fig. 10.10 Cross-sectional image of the platinum-alumina Suaning feedthrough. This is a four-layer system, that is, $n=4$. Each channel (conductor) has four vias, and three internal tracks [2]. This is the physical manifestation of the concept shown in the schematic of Fig. 10.9.

systems has proven that at 10–15 channels/cm^2, the Kuzma concept, and its iterative improvement the platinum comb system, absolutely are hermetic in vivo over very long time periods. However, at 1000+ channels, a higher level of sophistication is required.

The Suaning bionic eye retinal implant feedthrough involves 2500 channels/cm^2, 200 times higher conductor feedthrough density than the bionic ear systems. This is 2500 $\{H=f(L, 1/A, 1/t)\}$ leakage pathways[6] per square centimeter instead of only 13. Suaning wisely surmised that increasing feedthrough density 200-fold from the platinum comb, as required by bionic eye retinal implants, required a fundamental rethink of the co-fired alumina-platinum feedthrough concept. He then applied his long experience in biomanufacturing to devise an ingenious solution to manufacturing alumina-platinum feedthroughs of 2500/cm^2, with the stepped conductor pathway, a major quantum leap forward in feedthrough technology.

Suaning first proposed his innovative hot-pressed stepped-platinum-track feedthrough concept in 2006 [461]. As a one-time Cochlear engineer, Suaning was

---

[6] See Chapter 9 under platinum comb for explanation of this terminology.

mindful of the problem of the higher thermal expansion coefficient of platinum compared to alumina, which results in the platinum shrinking away from the alumina-platinum co-fired interface during cooling, and the subsequent risk of loss of hermeticity, This issue is discussed in detail in Chapter 9. Moreover, the literature demonstrates clearly that pressure enhances a platinum-alumina diffusion bond [410,418], which is also discussed in detail in Chapter 9. The relevant thermal expansion coefficients are as follows [391]:

Alumina: 8.1 μ/mK
Platinum: 9.0 μ/mK

Of course dilatometry curves generally show that thermal expansion coefficients are not linear with temperature and therefore the thermal expansion coefficients are indicative of probable mismatch only, for the sake of illustration, not a quantitative indication over the entire co-fired temperature range of 20–1500°C.

The Suaning concept involves a stepped pathway, shown in Fig. 10.9, manufactured from powdered alumina and powdered platinum by tape casting and laser machining, and sintered by hot pressing, such that the horizontal sections of the platinum conductor pathway (tracks) are hot pressed into the alumina thereby eliminating any possibility of loss of hermeticity in the horizontal segment of the pathway. The perpendicular segments of the stepped pathway (vias) do not have the benefit of being parallel to the hot-pressing pressure and therefore have the same interface as a pressureless sintered feedthrough, such as the Kuzma concept or the platinum comb.

It is critical to avoid any drying shrinkage cracks in the tape-cast alumina sheets as these cracks would become platinum filled during the screen printing stage resulting in many short-circuits. Tape casting of alumina is discussed in detail in Chapter 4.

So that the following discussion does not become confusing, the following terms need to be clarified in terms of their meaning in the context of advanced alumina feedthroughs.

Feedthrough: A hermetic alumina ceramic seal, through which channels comprised of platinum wires (thick wires, thin wires, tubes, or printed tracks of platinum) pass. The alumina seal is brazed to a titanium casing. The platinum channels pass through the alumina and are hermetically bonded to the alumina in a continuous ubiquitous alumina-platinum solid-state diffusion bond. These channels pass from the front face to the rear face of the alumina ceramic. At the rear face (the chip side) they connect electrically with the internal electronics of the device (microchip). This is also the side that is brazed to the titanium casing. At the front face "electrode side" (the retinal tissue side) the exposed platinum channels function as electrodes and interface directly with retinal tissue. Note: This is a direct evolution from the early hermetic pacemakers, in which titanium was used for the feedthrough, and it was peripherally brazed to the alumina, not diffusion bonded (see Fig. 8.2).

Co-fired: Platinum channels (thick wires, thin wires, tubes, or printed tracks of platinum) are embedded in green alumina, and the entire system is sintered, bonding the platinum to the green alumina by a solid-state diffusion bond developed during the co-firing process (see Chapter 9).

Via: Refer to Fig. 10.9. In the stepped feedthroughs of the Suaning concept, a "via" is a segment of the channel pathway that is perpendicular to the front and rear faces of the feedthrough. A "via" does not have the benefit of being parallel to the hot-pressing pressure and therefore has the same alumina-platinum co-fired interface as a pressureless sintered feedthrough, such as the Kuzma concept or the platinum comb.

Track: Refer to Fig. 10.9. The track is a key inventive step in the Suaning innovation. Tracks are the horizontal segments in the channel pathway. Being parallel to the hot-pressing pressure, tracks are hot-pressed intimately into the alumina thereby eliminating any possibility of loss of hermeticity in the horizontal "track" segment of the platinum feedthrough pathway. So, the lateral segment of the stepped channel pathway is the "track." There are external and internal tracks. There is always only 1 external track (or electrode pad), on the front face, for connecting to the semiconductor chip. There is at least 1 and frequently >1 internal tracks, which are the critical aspect of this concept as the hot pressing process ensures that internal tracks have intimate and residual-stress-free diffusion bonds, while the vias, like the feedthroughs in a bionic ear feedthrough, depend on pressureless sintering and have residual tensile interfacial stresses which are a potential problem for hermeticity. In its simplest two layer manifestation ($n=2$) where layer number is n ($n=2$ layers), a Suaning feedthrough has 2 vias and one internal track. In a multilayered system (where $n>2$), there will be n vias, n tracks, and $n-1$ internal tracks. For all values of n, there is only 1 external track.

Electrode "pad": The widened end of the "via" at the feedthrough surface where it acts as an electrode, interfacing electrically with the neural tissue. This is shown in Figure 10.11. It is well known that increasing the surface area of an electrode reduces the impedance and therefore the required voltage for a given current injection into the tissue, thus increasing the power efficiency of the feedthrough [462,463].

10.3.3 The retinal implant titanium/alumina-platinum feedthrough manufacturing process

The following process is summarized from the publications of lead researchers Thomas Guenther, and Rylie Green [2,375].

(1) *Alumina composition*: A powdered 96% pure alumina was used containing 2 wt% SiO_2, 0.5 wt% MgO, 0.25 wt% CaO, and other trace additives.

(2) *Tape-casting the powdered alumina*: A tape casting slurry was prepared by mixing the alumina powder with a proprietary wax binder using a terpineol solvent. Tape thickness was 200 μm.

(3) *Designing the vias*: AutoCAD was used to design the via and track positions for the feedthrough, saved as a dxf file which could be read by the Nd:YAG (DPL Genesis Marker, ACI, Nohra, Germany) laser patterning machine.

(4) *Laser machining the vias*: The platinum tapes were cut into 20×20 mm pieces, attached to a microscope slide using adhesive tape, so as to enable precise alignment in the laser machining unit. The vias comprised vertical holes laser ablated in the 200-μm thick alumina tape, and subsequently filled with platinum paste.

(5) *Preparing the platinum screen-printing paste*: A platinum paste was prepared using the same proprietary wax binder that was used to prepare tape casting slurries of the powdered alumina.

(6) *Preparing the platinum paste-printing stencil*: A screen printing stencil was prepared from 50-μm thick stainless steel foil with a hole cut into it of a diameter matching the desired feedthrough and thickness matching the desired track thickness.

(7) *Paste printing the powdered platinum*: The stainless steel stencil was placed on top of the laser-machined alumina tape and the platinum paste was screen printed onto it using a

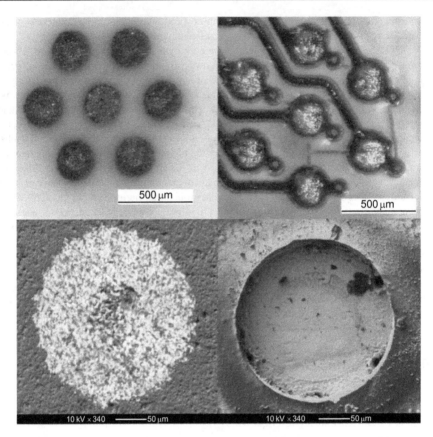

Fig. 10.11 Stereoscopic (top) and SEM (bottom) images of ceramic-Pt electrode pads (left) compared to laser micromachined PDMS-Pt electrode pads (right) [2].

rubber squeegee. The result was platinum-filled holes in the alumina tape (vias), and a thin surface layer of platinum paste on the surface of the alumina tape. The platinum paste was then dried for 15 min at 130°C.

(8) *Laser machining the platinum tracks*: Using the same AutoCAD feedthrough design file and the laser patterning machine, the platinum paste layer was patterned by selectively laser ablating the platinum paste layer leaving behind only the laser tracks connected to the vias for the feedthrough design.

The result was a 200-μm thick alumina tape, with vias, that aligned with tracks on its top surface, patterned in accordance with the AutoCAD feedthrough design file. A feedthrough needed a minimum of two layers, and in most cases more than two layers, typically around five layers.

Steps 4–8 were therefore repeated to create duplicate layers.

(9) *Aligning the layers*: The duplicate layers were stacked on top of each other using a proprietary micron-precision alignment system.

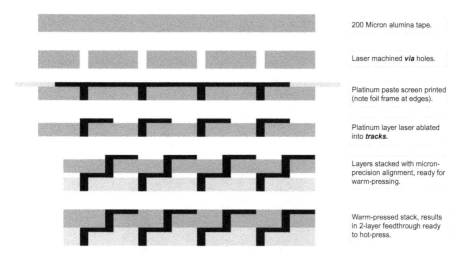

Fig. 10.12 Flow chart of the co-fired alumina-platinum feedthrough manufacturing process. Note, for simplicity, this example did not note the fact that the bottom layers have tracks and the top layers have an electrode pad (widening of the via at the point where it is flush with the alumina surface) as can be seen in Fig. 10.9. This is easily achieved by laser-ablating a T-shaped hole on the correct side of a layer that is to be on the electrode side of the final feedthrough.

(10) *Lamination by warm pressing*: The stack of aligned layers was warm pressed at 80°C and 50 MPa, in a hot press which melted the proprietary wax binder and fused the assembly into a coherent laminate.

(11) *Co-firing by hot pressing*: The laminate was sintered at 1500°C.

This feedthrough manufacturing process is shown schematically in Fig. 10.12. Note that the platinum features could reproducibly be made to a minimum line width of 20 µm, and a minimum center-to-center distance (pitch) of 40 µm [2], and a maximum terminal density of 2500 channels/cm^2 [375]. This not just a quantum leap forward from Kuzma of 1982 (13 channels/cm^2), but also three times higher than the nearest competitor, Schuettler et al. [460] who in 2010 reported a record hermetic feedthrough with 360 channels.

The assembly involving the alumina feedthrough, microchip, and titanium casing was assembled with a brazed bond using a proprietary process that protected the microchip from heat damage. This process, and its deliverable outcomes, are shown in Figures 10.13. to 10.17. Electrode pads on the "electrode side" of the feedthrough were produced by laser-ablating a T-shaped hole (see Fig. 10.9) on the correct side of a layer that was to be on the "electrode side" of the final feedthrough, and tracks on the "chip side." Fig. 10.12 shows a simple two-layer feedthrough manufacturing process for illustration purposes, it does not go into the detail of the electrode pads, and the full process of making a five-layer feedthrough that was commonly done in practice (Fig. 10.14).

Cell culture testing using neural cells has validated that the feedthrough itself, and all the materials used to produce the feedthrough (as listed above) did not impact on

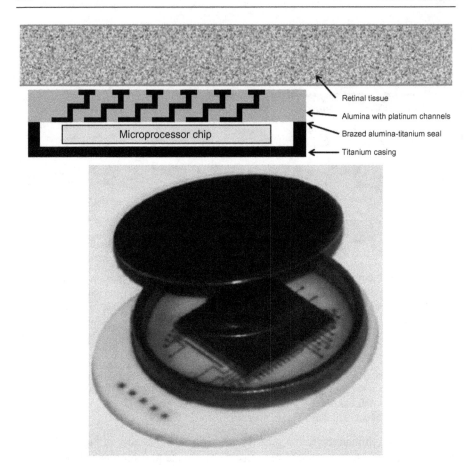

Fig. 10.13 Above: Schematic of the retinal implant hermetic feedthrough concept based on the Suaning concept of stepped printed platinum micro-channels in the feedthrough. The hundreds to 1000+ channels are platinum (six channels are shown in this schematic for illustration purposes), and terminate on the electrode side (top side) in widened electrode pads for direct electrical interaction with neural tissue. The platinum stepped channels pass through the alumina in a stepped configuration for maximum hermeticity and terminate on the chip side (bottom side) in engineered tracks to interface with the microchip. Below: The actual components ready for brazing. Alumina feedthrough with chip on top. Black ring and disk are the two pieces of the titanium casing for forming the hermetic seal with the alumina feedthrough. The titanium casing/alumina feedthrough brazed bond is achieved with a proprietary technology not disclosed here that enables protection of the chip from heat damage [2].

the cell growth and survival properties [2]. It is important to note here that while alumina implants have been used in dentistry, orthopedics, and bionics for half a century, this retinal implant in which alumina/platinum feedthrough surface is in direct contact with the neural tissue (rather than silicone platinum as for the bionic ear and other bionic implants) is a relatively unique case of an alumina platinum/neural tissue

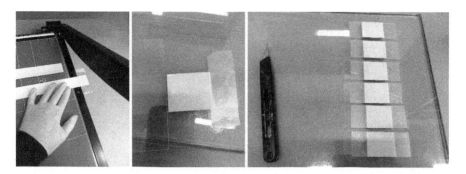

Fig. 10.14 The alumina tapes during the process. Left: The alumina tape being cut into 20 × 20 mm pieces; Middle: The alumina tape attached to a microscope slide using adhesive tape, so as to enable precise alignment in the laser machining unit; Right: Alumina tapes for the entire six-layer feedthrough lined up ready for laser processing.
(Image courtesy of Hayat Chamtie)

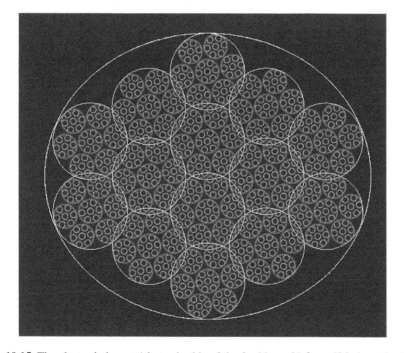

Fig. 10.15 The electrode layout (electrode side of the feedthrough) for a 686-channel retinal implant manufactured by researcher Hayat Chamtie. Each channel (electrode pad) is an orange circle. Channels are grouped in sets of 7 (called flowers) and flowers are grouped in sets of 7 (called a hex). 14 "hexs" make up the total electrode surface: 7 × 7 × 14 = 686 channels.
(Image courtesy of Hayat Chamtie)

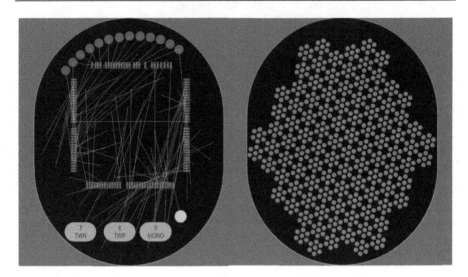

Fig. 10.16 Retinal implant feedthrough. Left: Chip side; Right electrode side. Image courtesy of Hayat Chamtie.

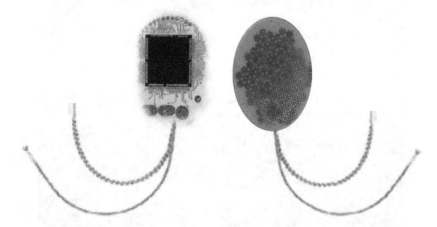

Fig. 10.17 The 686-channel feedthrough after manufacturing ready for hermetic mounting into the titanium casing to produce the final hermetic retinal implant. Left: the "chip side" of the 686-channel feedthrough with the microprocessor chip attached; Right: the "electrode side" with the 7-electrode "flowers" electrodes clearly visible in the 10-mm wide 686-channel feedthrough. Image courtesy of Hayat Chamtie.

direct-neurostimulation system. However, given alumina's outstanding reputation as the most bioinert and biocompatible material in established clinical usage, and the long-proven biocompatibility of platinum with 35 years clinical use in direct platinum/neural-tissue contact for the bionic ear, this biocompatibility is unsurprising (Figs. 10.15–10.17).

10.4 The future for alumina in bionics

It seems probable that the 14 bionic implant types on the market today, with their hermetic titanium/alumina-platinum packaging systems, and their high-channel-number capabilities, all have a long and successful future for many decades ahead, and that many more disease states may become treatable by the existing bionic implants. Moreover, it is likely that more bionic implants for new commercial neural tissue applications that may currently be a research topic, or not yet imagined, will come onto the market in due course using the same trusted, safe and proven titanium/alumina-platinum IPG hermetic packaging systems.

We now turn to the closing topic, the BCI.

10.4.1 The implantable BCI

The most futuristic bionic implant technologies today are the BCI multi-pin implants, which currently are purely for experimental use by direct implantation into the cerebral cortex (top surface of the brain), or motor cortex, by an open craniotomy. The small sharp pins penetrate brain tissue, and there is some associated risk of inflammatory response, a risk which could potentially outweigh the benefits in some cases. They do not have a titanium/alumina hermetic long-term stable encapsulation system at this stage as this is highly experimental technology. For long-term implantation, a hermetic durable and biocompatible encapsulation system would be needed, and the tried and proven gold standard of the last half century, titanium/alumina-platinum hermetic-implantable bionic system, perhaps with the gentler electrode system of the retinal implant (platinum electrode pads flush with the surface of an alumina feedthrough) would be a wise lower-risk approach to take for the future. Whether this is how the future develops is beyond our current event horizon, but it seems a sensible strategy to explore.

Connecting the brain to a computer is not a new idea, the earliest BCI (BCI) study was in 1929 when Hans Berger discovered electroencephalography (EEG) a recording of the electrical activity along the scalp [464]. This was done using cutaneous scalp electrodes (adhesive to skin), still a common procedure, which suffers from a low signal to noise ratio due to the electrical impedance of the skull and scalp. However, with the development of the advanced computer and its wide availability since the late 20th century, the number of BCI studies began to increase in the early 1990s, and increased dramatically from about 2003 as demonstrated by a PubMed analysis of the field shown in Fig. 10.18.

This 2003 date coincides surprisingly closely with the filing by Brown University in 2002 of a paradigm-shifting patent on a BCI cerebral cortex implant, which was granted in 2007 [465]. In 2012, a ground-breaking paper was published in the journal *Nature* [466] by the same Brown University team on success with BCI-driven thought-controlled robotics for paralyzed patients using the patented BCI implant device (called BrainGate™) [465]. Two paralyzed patients fitted with a BCI implant

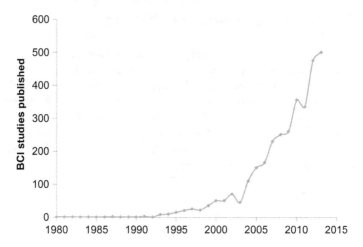

Fig. 10.18 A PubMed analysis of brain computer interface (BCI) studies published since 1980.

successfully used thought control to move a robotic arm in a meaningful way. One of them famously, and on camera, picked up a bottle with the robotic arm to drink from it [467].

Fig. 10.19 shows the schematic of the BCI implant, much like the one implanted, which was a 4 mm × 4 mm 96-channel microelectrode array, with the pin electrode system as shown in Fig. 10.19 together with a schematic of one of the actual implant pins. Fig. 10.20 shows the implantation site of the cerebral cortex. Fig. 10.21 shows a dummy with the BCI driver coupled to the BCI, which can then be connected to the robotic arm.

There are of course many other potential applications for the BCI, Theoretically, such technology could be used to drive an exoskeleton, for example, by a tetraplegic

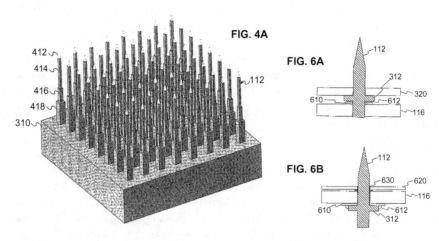

Fig. 10.19 The BrainGate™ cortical implant, from the original patent by the BrainGate™ pioneers at Brown University USA, filed in 2002 and granted in 2007 [465].

Alumina in bionic feedthroughs: The bionic eye and the future

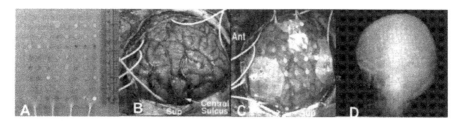

Fig. 10.20 Electrocorticography, recording the electrical activity of the cerebral cortex by means of electrodes placed directly on it, beneath the skull: (A) is the electrode grid to be implanted; (B) shows the exposed cortical surface of the brain of a human patient with epilepsy, before placement of the electrode grid; (C) shows the placement of the electrode grid on the exposed cortical surface of the brain of the patient. For orientation purposes, the reference "Ant" refers to the anterior (front) of the subject's brain; (D) is an X-ray of the patient's skull from one side, showing the electrode grid in place after surgery.
From the Washington University BCI patent of Leuthardt [468].

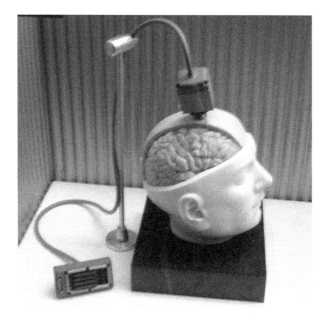

Fig. 10.21 Dummy unit illustrating the design of the BrainGate™ interface. The BrainGate™ implant in the cerebral cortex (top surface of the brain) interfaces with the robotic arm via the external device shown here.
Author Paul Wicks [469].

patient. Sufferers of advanced motor neurone disease, such as the late Stephen Hawking, and sufferers of other such degenerative conditions, would benefit greatly from such technology. Such a multi-tasking activity might require multiple BCI implants for the different motor regions of the cerebral cortex, and extensive brain/machine training. There are also other nonmotor applications potentially possible with a

BCI, such as synthetic telepathy (transmitting thoughts via a BCI implant), using a BCI for implantation of memories directly into the brain of a recipient, and BCI implants for the visual cortex as a futuristic solution to visual impairment. The future possibilities for the BCI are potentially revolutionary.

BCI visual cortex research is very new and highly experimental technology and it is difficult to forecast at this stage whether such bionic implants would restore sight at all, and if so whether they would be an improvement over the current and proven technology of retinal implants. Again, how the future develops is beyond our current event horizon, but it appears that interesting times lie ahead.

While experimental short-term BCI prototypes have not necessarily involved alumina feedthrough systems (as indeed experimental prototypes of the bionic ear and bionic eye did not), *long-term BCI implants will depend on alumina feedthroughs*, and probably also inductive coupling systems for transcutaneous power and information transfer, such as those used in the bionic ear and bionic eye. Moreover, rather than the invasive pin-electrode technology (penetrating neural tissue) of the experimental BCI implants, long-term BCI implants are likely to use the noninvasive high-channel density "electrode pad" technology (surface contact with neural tissue) outlined in Section 10.3.2.

In the coming decades, it is likely that we will see long-term BCI implants becoming as commonplace as the bionic ear is today, almost certainly utilizing alumina feedthroughs and inductive transcutaneous power and information transfer, much the same as the bionic ear and bionic eye. However there is a significant competitor technology to the implantable BCI, invented two decades later in 2016, that is, very recently, which is much less invasive, and may prove to be the future for BCI technology: the Stentrode.

10.4.2 The Stentrode

A much less invasive option than the BCE is the Stentrode, that is, the *Stent*-Elect*rode*. Unveiled in Nature Biotechnology in 2016 [470], the Stentrode was invented at the University of Melbourne almost two decades after the Brown University implantable multi-channel-pin BCI. The Stentrode is a stent that embodies electrodes, connected to a lead, thence a hermetic connector, and thence a transcutaneous wire link. This was just a temporary solution for the sheep trials. Ultimately, a hermetic bionic feedthrough with inductive coupling could be the ideal system to use with a Stentrode.

The Stentrode, being a stent,[7] can be percutaneously implanted in the brain vasculature via a safe day-surgery procedure. It is a very new development, one of the most

---

[7] A stent is a small spring-like wire tube, in the order of a few millimeter in diameter and tens of millimeter long. It is a long-established technology, originally commercialized in the early 1990s for propping open clogged coronary arteries. Stents are implanted percutaneously (percutaneously—access to inner organs or other tissue via needle-puncture of the skin) via a thin steerable catheter, usually fed in through the femoral artery in the upper leg, and guided by a fluoroscope. It is a simple outpatient procedure that was a huge improvement on the previous approach of open-heart bypass surgery. Fitting Stentrodes would also be a simple low-risk procedure.

advanced developments in bionics at the time this book was written, and it had only got to the animal trials at this stage (sheep) [470]. Being a stent, the Stentrode is a much safer and more convenient way to implant electrodes directly in the brain, no open brain surgery, no pins penetrating brain tissue. Moreover, it has less limitation in terms of the brain locations that are accessible by this means. Surgically implanted multi-pin brain implants, such as BrainGate™ can only be safely implanted on the outer surface of the brain, whereas the Stentrode can be implanted in many locations both deep within the brain or near the surface, anywhere where the brain vasculature is large enough in diameter to accommodate it, which, given the scale of the vascularization in the brain, gives substantial scope. The intracranial stent is already a well-established treatment for blockages in brain vasculature. The coronary stent has been in use for three decades now and is a very safe and proven technology.

The electrodes on the Stentrode clinically tested in sheep were 750-µm diamond-cut platinum electrodes. An electrode lead was run from the Stentrode at the implant site, through the brain vasculature to the Jugular vein, where it was passed through a perforation in the vein, tunneled subcutaneously to a custom-made hermetic connector secured to the muscle, and exited the skin via a flexible transcutaneous lead, which terminated in a plug. The specifics of the hermetic connector are not yet publically disclosed, but the possibility exists that alumina may be involved, if not at the animal testing stage, possibly in the commercial stage with an inductive coupling rather than a transcutaneous lead. This would be an obvious fit for the standard titanium/alumina-platinum mini-implant package used in a bionic ear. One could envisage Stentrode systems as being much like the bionic ear, with remote electrodes deep in the brain vasculature (Stentrode) connected to a lead running through the vasculature, connected at the other end to the feedthrough of a hermetic titanium/alumina-platinum implant beneath the skin that communicates transcutaneously, by induction, with an external processor.

At this point we have gone as far into the future as it is reasonable to go, without stepping into the realms of science fiction and speculation. Clearly, the rest of the 21st century is going to be an exciting time for implantable bionics, and alumina remains at the forefront of bionic hermetic packaging technology at the end of the second decade of the 21st century.

Alumina in lightweight body armor

Lightweight body armor is one of the most important areas, in both commercial and humanitarian terms, in which ceramic engineering and biomedical engineering intersect. Body armor of animal skins and leather dates back to prehistory, and by 1500 BCE metal body armor was well evolved functionally. About 3000 years later, around 1500CE, with the advent of effective battlefield handguns, traditional metal body armor, began to become obsolete very rapidly, and by 1700 vanished almost entirely. From 1700 until the late 20th century society came to accept the fact that military and paramilitary personnel would be exposed in the line of duty, unprotected, to high-velocity munitions, an unsatisfactory state of affairs. Ceramic body armor, revolutionized that. The ceramic composite armor concept was invented by Lt Commander Andrew Webster in 1945 (USA), and developed into a functional wearable armor system by the US military over the period 1945–1963. First deployed to aircrews (1965), it was not until 1996 that it was mass-deployed to ground troops, and in the early 21st century the world has seen a massive up-armoring program across all major armies globally, deploying lightweight ceramic body armor to ground troops. Alumina was the original ceramic in lightweight body armor, and remains one of the two main ceramics used in this role today, with SiC (silicon carbide) the other main ceramic. B_4C (boron carbide) is also occasionally used in spite of its problems of high cost and premature failure due to the B_4C shock-loading problem. This chapter overviews the invention, evolution, deployment, and underlying technology of lightweight body armor, for which alumina is the original and still a major player (SiC and B_4C are also discussed briefly), as well as lightweight ceramic vehicle and aircraft armor, in which alumina is dominant. Soft body armor (Silk, Nylon, Kevlar, and Dyneema), which is suitable only against soft low-velocity projectiles and shrapnel, is also briefly discussed.

This chapter concerns the very important role of alumina in the life-saving technology of body armor, and the related applications in vehicle and aircraft armor. Traditional industrial applications and biomaterials applications for alumina are easily categorized to biomedical or to industrial. Lightweight body armor uses of alumina, and its spin-offs in lightweight vehicle armor, are neither a traditional industrial application, nor a biomaterials application. They do involve protecting the human body, combining materials science and the study of traumatology, and have delivered an enormous humanitarian benefit to society with millions of lives protected, and many saved. Therefore body armor and its spin-offs have been categorized in this book as belonging to the biomedical engineering category, although obviously they are not biomaterials, unlike the orthopedic bionic, dental, and scaffold applications.

It must be said that in the context of the two life-saving biomedical applications for alumina (pacemakers and body armor), while alumina is the ONLY option for pacemaker bionic feedthroughs, it is not the only option for ceramic body armor. However, alumina was the original body armor ceramic (1965), from which all ceramic body armor derives, and today just two ceramics (alumina and silicon carbide) dominate the body armor industry, with boron carbide also significant. Alumina dominates lightweight vehicle and aircraft armor applications.

Alumina remains the affordable body armor ceramic that dominates the civilian market (law enforcement personnel, private security personnel, private individuals), and has an important role to play for military body armor in economically disadvantaged nations. Alumina as an armor ceramic is therefore the great equalizer, protecting millions of ordinary, often disadvantaged people, whose work exposes them to the hazard of gunfire.

11.1 Ballistic armor—The sociocultural context

One of the most important areas (in both commercial and humanitarian terms) in which ceramic engineering and biomedical engineering intersect, is the field of body armor. Body armor is a field in which the author has been active as a consultant since 1988.[1] Alumina body armor is important both from a commercial perspective, being one of the major markets for alumina ceramics, and from a humanitarian perspective in that many lives have been protected and saved by alumina body armor.

Alumina is actually used both in body armor and vehicle armor (land, air, and sea). Alumina is the primary armor ceramic used in lightweight vehicle and aircraft armor as it is the only armor ceramic commercially viable for the multihit mosaic panels. However, the primary focus of this chapter is on body armor as that is the biomedical engineering perspective, and also the largest use of armor ceramics in the world today. Nonetheless, this chapter will also take a brief look at alumina vehicle armor applications as this also involves alumina saving lives, albeit by an indirect means.

Armor saves lives. While all would agree that an ideal world is one in which military personnel, police officers, and security guards are not shot at in the line of duty, inevitably they may be shot at. Today, they have the same right to wear protective apparel as does the fireman (hazard suit), the welder (goggles), the concrete cutter (ear and eye protection), the fiberglass worker (dust mask), and other workers in hazardous environments requiring protective apparel. In our 21st century workplace health and safety (WHS) litigious era, the prevailing view is that if the nature of the job puts a person in harm's way, then their access to protective apparel is a fundamental human right. Accordingly, it is the responsibility of the biomedical engineer, the materials engineer, and other relevant experts to develop the best possible protection system that current technology and know-how allows.

Moreover, it is the responsibility of an employer to provide that safety equipment, and enforce its usage. In the body armor realm, this right is now seen to encompass all military personnel at risk, regardless of rank and role. In the civilian sector, this includes law enforcement officers, private security personnel, other paramilitary

---

[1] See "About the Author."

personnel, war correspondents, VIPs, bodyguards, and other workers exposed to the risk of gunfire. This may seem like a statement of the obvious, however, while lightweight alumina body armor was perfected in 1963, in the United States, and first deployed to elite US military forces in 1965, it was not for another three decades, a mere two decades ago, that the US regular infantry were first given ceramic body armor: in 1996. It was 1999 in Australia; the timing was similar in other western nations. As soon as the US regular infantry were protected with life-saving lightweight ceramic body armor, all the other armies of the world were compelled to follow suit, or else find themselves at an enormous disadvantage. Thus, the US military set the global agenda.

Therefore, widespread use of ceramic body armor has been the case globally for little more than a decade. Prior to that, flak jackets were relatively common for US infantry, even Kevlar battle jackets after 1983, but unlike ceramic body armor, soft body armor fabrics such as Kevlar cannot stop rifle bullets. Only soft handgun and shotgun projectiles, and shrapnel. As mentioned above, elite US military personnel have been privileged to wear ceramic body armor for much longer, with the first usage being half a century ago in 1965.

It has been estimated that 75% (over 8 million) battlefield deaths would have been prevented if soldiers in World War I had worn body armor based on the silk body armor technology available at that time. In World War II, many lives were indeed saved among *aircrews* who wore body armor (flak jackets), likewise in the Vietnam war. However, for virtually the entire 20th century, regular infantry did not have body armor capable of stopping rifle bullets, that is, ceramic body armor. Ceramic body armor was first deployed for elite US forces in 1965, and regular US forces 31 years later in 1996. Therefore, ceramic body armor for regular infantry is essentially a 21st century phenomenon.

11.2 A brief history of body armor

Note: Some of the citations in this section come from military reports. All classified military reports cited here have been declassified. Nothing is cited here that remains under security classification.

11.2.1 Body armor up to 1500CE

Body armor of animal skins and leather dates back to prehistory. From the advent of the bronze age around 3000 BCE, metal body armor began to evolve, and benefitted from the superior technology of iron with the advent of the iron age around 1200–600 BCE. Archeological records show that by 1500 BCE metal body armor was highly evolved functionally, and not very much different in concept or functionality to the well-known metal body armor of Europe 3000 years later, that is, around 1500 CE at the end of the middle ages. Throughout that 3000-year period, only the wealthy and social elite could afford metal body armor, and so body armor was not just a form

Fig. 11.1 Traditional metal body armor on display in the Metropolitan Museum of Art, New York. Left: suits of armor for horse and rider. Right: light cavalry armor.
Left: Image Author Mattes [471]. Right: Image Author Unknown [472]. Reproduced from Wikipedia (Public Domain Image Library).

of protection, but also a symbol of wealth and social position. Historic metal body armor is shown in Fig. 11.1.

The turning point came around the year 1500 CE, with the advent of effective battlefield handguns. Traditional metal body armor, after >3000 years of usage, became obsolete very rapidly, and by 1700 with the widespread deployment of battlefield handguns and the advent of the bayonet, vanished almost entirely, with only brief reappearances in low-cost modern manifestations in the industrial era, up until the paradigm shift in body armor in the mid to late 20th century.

11.2.2 The failure of metal body armor: 1500 CE onward

In the era post-1500 CE, and certainly by 1700CE, the balance was tipped so heavily in favor of the gun that metal body armor was just not practical in terms of the weight/protection equation. Nonetheless, there were a number of attempts to adopt metal body armor solutions in this era.

In the US Civil war of the 1860s, metal body armor made a reappearance. The military did not provide it, but metal body armor vests were widely available to soldiers for private purchase from the various body-armor peddlers who sold body armor vests to freshly recruited soldiers at the recruiting posts. Made of cast iron, the manufacturers claimed they were capable of stopping a 45 caliber bullet. The lightest version, the "soldiers bullet proof vest" made by Cook & Co weighed just 1.6 kg, comprising two overlapping left and right metal plates sewn into a button-up military vest. It cost only about $6 ($130 in today's money) and could possibly have saved many lives. However, an ethos prevailed at the time of ridicule toward those who wore the armor, and so most soldiers threw their vests away on the march to the war front due to peer pressure. It is significant to note that a mere few hundred years earlier, metal body

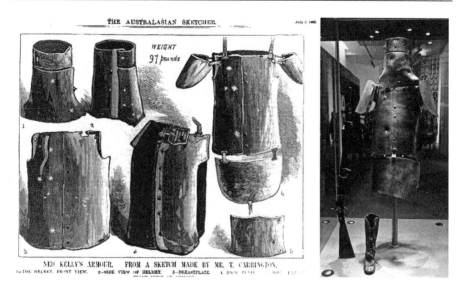

Fig. 11.2 Homemade metal armor of the mid-19th century made and worn by famous Australian outlaw Ned Kelly. Left: sketch of the armor from The Australian Sketcher 1880. Right: armor on display in the State Library of Victoria, Australia. The original bullet impact indentations are quite visible.
Left: Image Author Unknown [473]. Right: Image Author Chensiyuan [474]. Reproduced from Wikipedia (Public Domain Image Library).

armor had been a status symbol, but in the 1860s US Civil War it was a stigma. Today, body armor is simply regarded as part of a soldier's kit.

Home-made metal armor was also used in this era, as shown in Fig. 11.2. Ned Kelly (and his gang), a famous Australian outlaw had a shootout against overwhelming police numbers in the 1860s, while wearing a helmet and body armor made in a "bush forge" using 4.8-mm mild steel plate roughly forged from plowshares. The world's first feature-length movie[2] was about the Ned Kelly legend, that Ned walked into a hail of police bullets and did not die. It is true that the life of Ned was saved by his body armor, which stopped every impacting bullet. At 40 kg, it was a very heavy suit of armor, and while 4.8 mm of mild steel could protect against the weapons of the 1860s, it would be of little use against the modern assault rifle which would effortlessly penetrate twice this thickness of mild steel. An equivalent suit weighing over 100 kg would be required for Ned Kelly today. There are other less dramatic episodes of outlaws using home-made metal armor, but universally the ballistic protection came at such a weight disadvantage, that it was impractical.

In World War I, metal body armor was available in principle, from the armies of most nations, but saw very limited use due to its extreme weight. For example, the Brewster Body Shield of the US army (see Fig. 11.3), which resembled medieval

---

[2]Movie: "The Story of the Kelly Gang" 1906.

Fig. 192. Brewster body armor, 1917-1918

Fig. 11.3 World War I era metal body armor—Brewster body armor 1917–18 [475].

armor, comprising a plate covering the abdomen and a slit helmet, both made of chrome nickel steel. Weighing about 18 kg, it could protect against high-velocity 30 caliber Lewis machinegun bullets. Its weight and design made it impractical for general service use.

Having said that, metal armor continues to dominate vehicle armor systems for land and sea, although metal is as obsolete for aircraft armor as it is for body armor.

11.2.3 The rise of soft body armor

11.2.3.1 Silk

The use of silk body armor in Asia dates back over 1000 years. It was introduced into Europe from Japan a couple of centuries ago, and rose to some prominence in Europe in the late 19th century/early 20th century. Comparable in ballistic protection to ballistic nylon [476], silk was very expensive. Archduke Franz Ferdinand of Austria owned a silk body armor vest when he was shot in Sarajevo in 1914, the assassination which initiated the World War I. The silk vest the Archduke may have been wearing at the time of the assassination was a popular item with European dignitaries in this era.

The Royal Armories has had replicas of the Archduke's vest made to the original patent specification of manufacturer Casimir Zeglen, and tested them with a gun and projectile comparable with the 1910 Browning semiautomatic pistol used in the assassination. These tests suggest that the vest the Archdduke was maybe wearing that day may have been able to stop the bullet [477]. However, it was a bullet to his unprotected neck that took his life.

While effective against shrapnel, there are conflicting reports as to the reliability of the silk vests of the late 19th/early 20th centuries in protecting against handguns. Silk is somewhat biodegradable, and its ballistic performance is reduced when wet. However, in general, with appropriate design and testing, silk, ballistic nylon, and the modern high-tech successors Kevlar and Dyneema, can all be rendered effective against shrapnel and handguns, although not assault rifles. The issues of comparison are principally price, weight efficiency, and volume efficiency (minimal bulkiness).

It is now well known that soft body armor of any sort is ineffective against armor-piercing projectiles (whether fired from a handgun or rifle) or against high-velocity rifle projectiles of all types. However, against soft handgun bullets, shotgun pellets, and shrapnel, soft body armor is effective. The only contemporary use for silk in body armor today is for ballistic underwear, due to the fact that the "canvas" texture of Kevlar and Dyneema is not comfortable in direct skin contact for prolonged periods. While silk is ballistically inferior to Kevlar or Dyneema, it is more comfortable in direct body contact than Kevlar or Dyneema.

The great majority of casualties in World War I were from shrapnel or low-velocity bullets. Retrospective analysis has suggested that about 75% of combat casualties in World War I would have been prevented if soft body armor of the type then available (silk) had been worn. Something akin to the flak jacket of World War II. Given that 11 million soldiers died in battle, and another 6 million died of disease in World War I, this represents over 8 million lives lost needlessly through not having a flak jacket.

The above discussion relates to silk from the mulberry silkworm (*Bombyx mori*). However, spider silk is possibly the toughest of all ballistic fibers, three times tougher than Kevlar, and five times stronger than steel on a weight basis (specific strength) [478], a truly high-toughness fiber, possibly superior even to Dyneema (see Section 11.2.3.4). The problem with spider silk is developing a commercial-scale manufacturing process [479]. Spiders are not amenable to farming and milking. Genetic engineering approaches are under investigation. However, silk is currently of little relevance to alumina armor, other than its role in the historical pathway to modern armor, and the future possibility of spider silk ceramic claddings.

11.2.3.2 Ballistic nylon (Nylon-66)

Ballistic nylon is the common name for Nylon-66 fabric. In World War II, ballistic nylon flak jackets were introduced by the US military and became ubiquitous for aircrews of the allied forces. Ballistic nylon, as the Nylon-66 fabric produced by Du Pont, remained the material of choice for US body armor vests until 1983. Ballistic nylon flak jackets did indeed save many lives. Very effective against shrapnel, not very effective against bullets, they were ideal for aircrews since most injuries for flight crews were due to shrapnel from antiaircraft fire. The early flak jackets comprised manganese steel plates sewn into a ballistic nylon vest covering the abdomen. They had a quick release tab for emergency removal in the event of bailing out.

In 1945, ballistic nylon flak jackets containing fiberglass plates (Doron ballistic fiberglass: see Section 11.3.1) were used as body armor for ground troops at the battle of Okinawa, making Okinawa possibly the first military assault by a force of ground troops wearing body armor since the Middle Ages, and certainly the first for some centuries. Ballistic nylon vests with Doron plates went on to be used to a significant extent in the Korean war. As of early 1953, 90,000 had been dispatched to Korea [480]. Anectodal evidence from combat troops attests that the ballistic nylon/Doron vests stopped shrapnel, handgun projectiles, and one solidier attested that his Doron vest protected him when he smothered a grenade with his body.

The US military continued to use ballistic nylon vests for its regular infantry forces right up until the 1980s, in various upgrades of which a few representative examples are cited below. However, ballistic nylon is of little relevance to alumina armor other than its role in the historical pathway to modern armor.

- M-1943 Field Jacket (1943 to early Korean War).
- M-1951 Field Jacket (late Korean War to early Vietnam War). Ballistic nylon incorporating fiberglass or aluminum plates.
- M-1965 Field Jacket (M-65): Vietnam War to 1983. Ballistic nylon infantry combat jacket.

As Table 11.1 shows, the ballistic properties of ballistic nylon are much the same as silk, however ballistic nylon is in the order of a 100 times cheaper. So, essentially ballistic nylon was a low-cost mass-market silk body armor equivalent.

11.2.3.3 Kevlar

In 1964, Stephanie Kwolek of Du Pont invented poly(p-phenylene terephthalamide) PPTA fibers (para-aramid), commonly known as aramid fiber or by its Du Pont tradename of "Kevlar." Some years of development followed and it was first marketed by Du Pont in 1971 as Kevlar [481]. Aramid fiber is also sold under the tradenames Twaron (Tejin) and Heracron (Kolon). At a density of 1.44 g/cm^3, it is 50% heavier than its 21st-century successor Dyneema, but it was a huge leap forward from ballistic nylon and silk.

The US military began to introduce Kevlar battle jackets to its regular infantry in 1983 as the 1983–98 Personnel Armor System for Ground Troops (PASGT). This was the world's first Kevlar infantry flak jacket: 13 ply 14 oz Kevlar 29 fabric, total weight ~4 kg. The PASGT was designed in 1975, when Kevlar had just come onto the market, and introduced from 1983. The PASGT system was replaced by the Interceptor Kevlar battle jacket program in 1998, which continues to this day to be the standard battle jacket for the US regular infantry.

Compared with silk, Kevlar is very cheap. The 1914 silk body armor vest of Archduke Franz Ferdinand cost $20,000 in today's money, a Kevlar vest of equivalent protection is just a few hundred dollars. Moreover, compared with both silk and ballistic nylon Kevlar is greatly ballistically superior as Table 11.1 shows: Kevlar has three times the tenacity, more than three times the specific strength, and nearly 10 times the specific modulus of silk and ballistic nylon.

Alumina in lightweight body armor 329

Table 11.1 **Properties of ballistic fibers** [486,487]

Fiber	Tenacity index	Tensile modulus (GPa)	Tensile strength (MPa)	Failure strain (%)	Density (g/cm³)	Specific strength	Specific modulus
Silk	0.6	10	740	10–30	1.25	590	8
[a]S-Glass	N/A	90	4750	1.5	2.49	1910	36
Nylon-66	0.8	10	910	15–20	1.14	800	9
Kevlar-29	2.1	74	3000	3.5	1.4	2140	53
Kevlar-49	2.0	105	2900	2.5	1.44	2010	73
[b]Dyneema	3.1	171	3000	3.1	0.97	3100	176
[c]Zylon	3.3	169	5200	3.3	1.56	3330	108
[d]M5-2001	2.3	271	3960	2.3	1.74	2280	156

[a]S-Glass is not used as a fabric for soft body armor but as the fiber component of FRP backings in ceramic composite armor, for which it is the most cost-effective option.
[b]Spectra-1000 grade. Spectra is the trade name for Dyneema produced under license by Honeywell.
[c]Zylon has been withdrawn from the market due to its rapid biodegradability.
[d]M5 is the trade name of PIPD—poly(diimidazo pyridinylene (dihydroxy) phenylene), a new polymer fiber under development in the USA, not yet commercialized.

Essentially, Kevlar was the first commercial fabric in the world capable of reliably stopping soft handgun and shotgun projectiles. Kevlar was the game changer that made body armor a mass-market product in both the military and civilian sectors. In the late 1970s, Kevlar put personal body armor into the hands of the civilian populace for the first time in history (see Section 11.3.7). However, Kevlar has two major flaws:

(1) Kevlar has a short service life as it is subject to degradation overtime.
(2) Kevlar is quite heavy, 50% heavier than Dyneema for the same protection level.

But on the plus side, Kevlar is very cheap. More importantly, Kevlar fiber does not burn or melt, but decomposes above 450°C. Thus, like wool, Kevlar is a nonflammable fabric. It is likely Kevlar will be with us for a very long-time servicing the low-cost and fire-resistant segments of the market. It is a cheap and effective fabric for soft body armor, albeit with a short shelf life. Kevlar has significant relevance to alumina armor as will be discussed later.

From our 2020 perspective, the true game changer in soft body armor is proving to be Dyneema.

11.2.3.4 Dyneema (UHMWPE-polyethylene)

Dyneema (DSM) also marketed under the tradename Spectra (Honeywell) is a fiber made of ultrahigh-molecular-weight-polyethylene (UHMWPE), the most advanced of the engineering plastics. UHMWPE has been used since the 1960s as the bearing surface in prosthetic joints (hip, knee, elbow, shoulder, ankle, etc.) due to its outstanding toughness, strength, wear resistance, and biocompatibility (Chapter 6). Like the parent UHMWPE material, the fibrous form of UHMWPE (Dyneema) is also characterized by its outstanding properties.

The data in Table 11.1 show that Dyneema is the most advanced of the fibrous ballistic materials. It has the highest tenacity, highest specific strength, highest specific modulus, and lowest density of all the commercial alternatives. In fact, although to date Dyneema has not been produced with a modulus above 170 GPa, its theoretical modulus (crystal modulus) is 220 GPa, which would give it a theoretical specific modulus of 220 GPa, more than three times higher than Kevlar [482]. As it is, its specific modulus of 176 is 2.5 times higher than Kevlar. The tenacity and specific strength of Dyneema are 50% higher than Kevlar.

Dyneema is one of the 20th century's most world-changing materials science inventions. Apart from its position as premier fibrous armor material, it is also extremely effective for bomb disposal suits, bomb containment blankets, high-performance rope, and sailcloth. It is also very biocompatible and has applications as a futuristic biomaterial. The only theoretical rivals for Dyneema as a ballistic fabric are two currently unviable alternatives: spider silk (difficult to mass produce) and Zylon, which was used as a ballistic fabric in the late 20th century but was discontinued after deaths resulted from ballistic failure, owing to its biodegradability.

Dyneema was invented by Albert Pennings of DSM (the Netherlands) in 1968 [483]. However, DSM management did not grasp the potential of this invention and the project languished for some years. Ultimately Paul Smith and Piet Lemstra of DSM perfected the manufacturing process and patented Dyneema in 1979. It is an unfortunate fact that Albert Pennings, the original inventor, was left off the patent. The Dyneema project continued to languish for some time after it was patented until a few years later Allied Signal (now Honeywell) independently invented and patented the same material, calling it Spectra. This spurred DSM into committing serious resources to the Dyneema project and licensing Dyneema to Allied Signal (Honeywell) for the US market. In the United States, Honeywell has the manufacturing license [484]. Honeywell sell it under the brand name Spectra.

Dyneema was commercialized in 1990, but it took until the early 21st century for Dyneema to begin to make a dent in the market of the much ballistically inferior Kevlar. This process continues today with Dyneema moving very slowly toward a position of total market dominance of the top end of the ballistic fabric market. The main obstacle has been the low cost of Kevlar, and the production capacity of Honeywell to meet demand.

Dyneema was as big a game changer in the 1990s as was Kevlar in 1971. At a density of only $0.9\,g/cm^3$, and comparable performance on a thickness basis to Kevlar, this equates to Kevlar being 50% heavier than Dyneema for the same protection level.

Apart from the ballistic efficiency advantage, Dyneema has a huge practical advantage in that Dyneema has excellent water and chemical resistance, and is much less photodegradable than Kevlar. For this reason, unlike Dyneema, Kevlar has a relatively short service life and is photodegraded quite rapidly by UV light, and this causes a major logistical challenge for military, and law enforcement, body armor management systems, with strict obsolescence management programs required for Kevlar body armor vests.

However, on the negative side, Dyneema is flammable and melts at just 150°C, whereas Kevlar, like wool, is nonflammable. Flammability is the one Achilles heel for Dyneema.

Dyneema has significant relevance to alumina armor, as will be discussed later.

11.2.3.5 Other fibers

Dyneema remains the highest performance ballistic fabric in the world today. There is no superior technology imminent for commercialization. A more stable version of Zylon, a commercialized spider silk, some futuristic use of the much talked about carbon nanotubes, or some new miracle fabric discovery in the foreseeable future could surpass Dyneema sometime in the future, but for the foreseeable future Dyneema seems set to dominate the high-performance body armor market, and Kevlar the low-cost (and fire-resistant) body armor market.

Theoretically carbon nanotubes have extraordinary ballistic properties. Numerical modeling suggests that a 0.6-mm thick body armor fabric made of carbon nanotubes could defeat a bullet with muzzle energy of 320 J [485], which is approximately the energy of a 9-mm handgun projectile. However, this is just a theoretical prediction.

Whether it translates into reality, and if so how many years that may take to develop, is the subject of pure speculation at this stage. Nonetheless, it is a possibility for the future.

In our current era, the two closest rivals for Dyneema are Zylon and M5.

Zylon has the highest specific strength and tenacity of any known ballistic fiber, impressive mechanical properties, as can be seen in Table 11.1, however, it is unstable. It is degraded by UV and even by visible light, and by hot/wet environments. It was used for soft body armor in the 20th century, but its use in body armor ceased after incidents in which police officers died or were seriously injured when their Zylon vests failed.

M5 is the trade name of PIPD—poly(diimidazo pyridinylene (dihydroxy) phenylene), a new polymer fiber under development in the United States. It has not yet been commercialized although commercialization is imminent. M5 has some promising properties as shown in Table 11.1. It has reportedly the highest specific strength of any known fiber (3330), ahead even of Dyneema (3100) although its specific modulus (108) is substantially behind fiber (176). Overall, M5 could be comparable with Dyneema as a ballistic fiber, and if it was lower in cost could become a serious competitor in the future. Time will tell.

11.2.3.6 Relevance of ballistic fabric to alumina

The ideal fiber for body armor has a high tensile strength and tensile modulus, a low density, and a failure strain below about 5%. Table 11.1 lists eight ballistic fiber materials in current or past usage. Specific strength (strength/density) and specific modulus (modulus/density) give an indication of ballistic weight efficiency, while tenacity index gives an indication of ballistic effectiveness. On these criteria, Zylon is a clear winner, but it deteriorates rapidly and is therefore unreliable. Dyneema therefore reigns supreme as the highest performance commercially available ballistic fiber.

This book is about alumina. The question that needs answering at this point is: what do Dyneema and Kevlar have to do with alumina? The answer is that they are critical to ceramic body armor, for two main reasons:

(1) Alumina and indeed all ceramic body armor breastplates are never used in isolation, but always as inserts in a soft body armor vest, to upgrade the vest from merely shrapnel and handgun protection to assault rifle protection. The vest is usually Kevlar, sometimes Dyneema.
(2) Alumina and indeed all ceramic body armor breastplates are never used as naked ceramic tiles, but must be wrapped in fiber-reinforced polymer (FRP) cladding. The fiber can be fiberglass (S-glass), Kevlar, or Dyneema. Dyneema is by far the best backing, and also by far the most expensive.

The 2020 present, and the foreseeable future, for body armor is armor ceramics (primarily alumina and SiC, also B_4C to a lesser extent) with Dyneema the deluxe ballistic fiber, Kevlar the cheaper heavier alternative, and S-Glass the cheapest heaviest alternative. The niche for alumina is that it is the most cost-effective armor ceramic by far. However, it is extraordinary to note that while ceramic composite armor was invented by the US military in 1945, and first deployed in 1965, three more decades were to elapse before this game-changing armor technology was deployed to regular US infantry forces in 1996. Alumina ceramic armor was deployed among aircrews from 1965 onward. Finally, in 1996, the US military began deploying ceramic armor to regular infantry. The Soviet Union deployed ceramic armor with regular infantry much earlier than the United States (1980s). The pathway from 1945 to 1996 is a most instructive history.

11.3 The origin of alumina ceramic composite body armor

11.3.1 Doron: The precursor to ceramic body armor

Efforts were made during World War II to develop body armor for ground forces. Initially, the technology for the time made most options impractical due to weight. This all changed with the Doron innovation, a 1942 initiative of Brigadier General Georges Doriot, after whom Doron was named [480]. Doriot had the breakthrough idea of combining the recently invented fiberglass fabric, with recently invented thermosetting polymer resins, in order to produce a lightweight armor material. This "Doron" early fiberglass technology was the world's first FRP armor. Fiberglass became commercially available from Owens Corning in 1935, phenolic resin had been available since 1910, polyester resin became commercially available from DuPont in 1941, and epoxy resin from Ciba in 1946, so fiberglass (glass FRP composites) was a very recent innovation from a 1942 perspective.

Doriot coordinated a research project that ultimately led to Dow Chemical Company producing a prototype pressure-cured dense fiberglass laminate plate in 1943. These "Doron" plates, 3 mm thick and 125 mm × 125 mm in size were inserted in pockets in a ballistic nylon flak vest. They were a substantial weight improvement over manganese steel plates.

Doron proved to be effective against shrapnel, and even against a handgun projectile, as attested by Lt. Commander Andrew Webster, who made a public demonstration at Quantico, USA, at which he shot Lt. Commander Lyman Corey with a 45 caliber pistol. Corey was protected by a Doron vest [480]. This was a landmark event in body armor development.

Doron was the breakthrough that led to the invention of ceramic composite armor.

11.3.2 Invention of ceramic composite armor (1945)

The first recorded use of ceramic armor was vehicle armor, specifically for tanks. This was an innovation introduced by the German army a century ago during World War I due to the discovery that hard-faced enamel coatings on tank armor-enhanced ballistic performance. This gave a small but significant improvement in performance, but the technology did not progress beyond that for a long time, indeed it was more than half a century later that advanced ceramic tank armor was to be developed. Ceramic vehicle armor is discussed in Section 11.6.

Lt. Commander Andrew Webster of the US Navy is the inventor of the ceramic composite armor concept [488]. While Webster's invention was in 1945, it took

20 more years of intensive research to perfect the concept into the first military deployment of ceramic composite armor in 1965 [489]. Webster was one of the key engineers of the 1942 Doron ballistic fiberglass innovation (see Section 11.2). In 1945, Webster experimented with plate glass backed by Doron and found that this ceramic composite armor configuration was capable of stopping 30 caliber rifle projectiles, at a lower weight than steel armor [488]. This was a paradigm-shifting discovery.

The use of ceramic composite armor from the mid-1960s onwards is well known. However, to the best of the author's knowledge, Webster has never been publicly acknowledged for his original 1945 world-changing invention of the ceramic composite armor concept. His original invention was not patented but remained hidden for many years in the original 1945 military report [488]. This report was security classified at the time, and by the time it was declassified many years later, the proliferation in the world's scientific literature of ceramic armor papers generally saw the 1965 (first military implementation of ceramic composite armor) as day zero in this technology. The 1965 deployment was in fact preceded by a 1945 invention with the interim time period comprising a 20-year intensive research and development program.

Now, seven decades later, it is appropriate to correct the historical record for inventor Lt. Commander Andrew Webster (US Navy) in this regard. The importance of Webster's invention cannot be overemphasized. Ceramic composite armor was the paradigm shift that brought life-saving advanced armor technology back to the battlefield since its departure five centuries earlier in the 15th century.

What Webster invented in 1945 was the world's first ceramic composite armor, albeit the plate glass was significantly less hard than alumina ceramic, and Doron FRP an early form of ballistic fiberglass (see Section 11.3.1) was nowhere near as tough as todays SpectraShield Dyneema-based advanced FRP backings (see Section 11.5.1), and there was no spall plate. However, in spite of that, Webster's plate-glass ceramic composite armor was capable of stopping 30 caliber rifle projectiles, at a lower weight than steel armor, which had never been achieved before, and has been the enabling technology for virtually all lightweight hard armor innovation in the ensuing seven decades.

Inspired by Webster's 1945 invention, an extensive range of testing was done later that year by the US military on plate glass backed by modeling clay and it was found that for a given glass thickness, residual penetration into the modeling clay was the same, around 50mm, regardless of projectile velocity, within the range 365–580m/s [490]. This was a surprising finding for the time, and over the next 10 years a range of ballistic tests were done leading to the conclusion of Alesi in 1955 "that composite armor through proper selection and combination of components would provide significantly greater protection than any single armor material" [491]. This is a principle that underlies the weight efficiency of ceramic armor. Alesi studied composite armor experimenting with several different hard materials as the front plate: glass, titanium, aluminum, and PMMA. He was not able to improve upon the 1945 Webster concept of glass/Doron in weight efficiency [491].

11.3.3 Ceramic composite armor—The great leap forward: Alumina (1962)

The Webster invention was never patented by Webster or his associates. The first ceramic armor patent was filed 5 years later in 1950 by Robert Eichelberger (US Army), and granted in 1967 [492]. It concerned glass elements (plates, spheres, and sheets) sandwiched between metal plates. It was somewhat, although not exactly, following the ceramic composite armor concept of Webster. It did, however, adopt Webster's innovation of glass as a ceramic armor component.

Finally, 17 years after Webster's 1945 innovation, the next great leap forward came from the US Army Natick Laboratories in 1962, when an alumina ceramic, a very new and highly experimental advanced ceramic at that time, was tested in the Webster composite armor concept of Fig. 11.4, with 6 mm of Doron as the backing, and a polyurethane coating as the spall shield. The weight efficiency of the resultant armor was instantly doubled [489]. This Natick alumina ceramic innovation was the world's first high-performance ceramic composite armor, and the first version of the Webster ceramic composite armor concept that was capable of stopping 30 caliber armor-piercing projectiles at close range.

The first ceramic armor patent based on the Webster ceramic composite armor concept was filed 1 year later, in 1963, by Richard Cook (Goodyear Aerospace Corporation), and granted in 1970 [493]. The classic ceramic composite armor concept can clearly be seen in the schematics from the patent, shown in Fig. 11.5. It was the classic ceramic composite armor concept: alumina ceramic, fiberglass backing (polyester or phenolic resin), and spall cover. The patent notes that the alumina should be at least 85% pure, preferably >90% pure, and that "excellent results have been obtained with 94% pure alumina." It notes that ¼ to 3/8 in. thickness of alumina (6.4–9.5 mm) was sufficient for 30 caliber armor piercing protection, with a ballistic weight efficiency

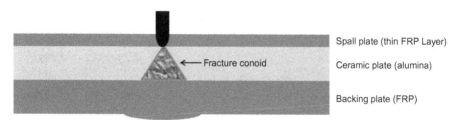

Fig. 11.4 The ceramic composite armor concept based on a combination of ceramic plate and FRP plate. Shown in the classic scenario of the impact of a hard sharp-nosed bullet into ceramic composite armor. The fracture conoid in the ceramic plate, contained by the FRP spall plate and FRP backing plate, transmits the concentrated impact load of the projectile tip into a much more diffuse load over the large area at the rear of the conoid, the energy of which is absorbed by the tough backing plate, with only minor deformation. Webster's 1945 innovation [488] was the first manifestation of this concept, originally with no spall plate, plate glass as the ceramic, and "Doron" (fiberglass) as the backing.

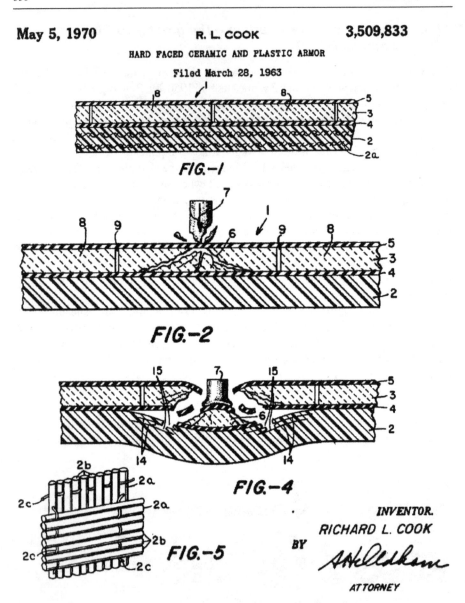

Fig. 11.5 Schematics from the original ceramic composite armor patent, with alumina ceramic, fiberglass backing, and "flexible adhesive" as both spall cover and alumina/fiberglass bonding layer. Filed in 1963 by Richard Cook (Goodyear Aerospace Corporation), and granted in 1970 [493].

four times greater than steel armor. It notes that "armor protection of this type is desired for aircraft and helicopter protection against ground troop fire" [493].

The Cook patent refers to all the classic principles of ceramic armor: multiple alumina tiles as a mosaic for multihit capability, the fracture conoid, the importance of the ceramic-backing bond, armor piercing bullet resistance, and shaped charge resistance. The Cook patent also claims the ceramic composite armor concept as the original bulletproof glass manifestation, with the ceramic plate as transparent glass, the backing as transparent polymer (PMMA is proposed), bonded with a transparent adhesive (MMA), and the concept of a glass tile mosaic to maximize multiple-hit capability [493]. Andrew Webster is not cited or mentioned anywhere in this patent, presumably because the Webster US Navy report was still classified at the time. The Cook patent has therefore been seen ever since as year zero in ceramic armor. Thus Lt. Commander Andrew Webster was lost to history.

Strictly speaking, the first ceramic composite armor patent filed was that of Harry King (Aerojet-General Corporation, USA), originally filed December 21, 1962, abandoned and then refiled (continuation-in-part) on April 26, 1965, and granted in 1969 [494]. Schematics of the King armor concept can be seen in Fig. 11.6. Thus, Cook has the earlier priority date, but King was the first to file. King claims alumina, nickel-plated alumina, and B_4C as his ceramic component. King claims tile mosaics or spheres or elongated cylinders (embedded in plastic) as his ceramic component, and a metal backing. The King ceramic armor manifestation can be seen as the precursor invention to the classic ceramic-metal composite armor systems used in ceramic armor for land vehicles, whereas the Cook patent is the classic lightweight ceramic composite body armor system, also used in body armor and aircraft armor. King cites fiberglass only as a means of containing the tiles, spheres, or rods, not as a lightweight backing. However, King [494] addresses the front spall layer in even more detail than Cook who simply notes an elastomer cover layer [493]. Spall covers have proven to be essential in ceramic armor in general. The spall cover serves two important functions: (1) protection of nearby personnel from high-velocity ejected spall debris and (2) containment of the pulverized ceramic inside the fracture conoid which aids projectile defeat (see Sections 11.4.2 and 11.6.3) (Fig. 11.6).

Several more significant ceramic armor patents were filed in the prolific period of the mid to late 1960s. Cook followed up with another patent filed in 1967 and granted in 1979 on the concept of alumina spheres, with alumina backing [495], however,

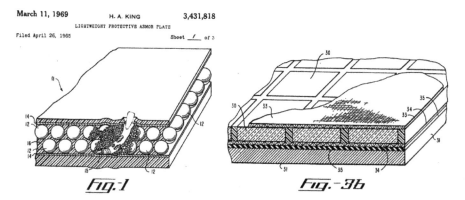

Fig. 11.6 Schematics from the ceramic composite armor patent filed in 1962 then again in 1965 by King on metal-backed alumina tile mosaics, or alumina spheres [494].

King had the earlier priority date on spheres [494]. Ferguson filed a patent in 1965, granted in 1978, on a three-layer ceramic armor system: alumina/beryllium/fiberglass [496]. Beryllium is a very lightweight metal, but unfortunately given its toxicity, this patent had no commercial future. Alliegro [497] filed a patent in 1967, granted in 1972, on the use of metal lateral containment strips to minimize edge effects in ceramic/fiberglass composite armor involving alumina and other ceramics.

With the 1962 alumina breakthrough, a number of innovations quickly followed at Natick including the use of ballistic nylon as the spall shield, using a mosaic of curved 150×150 mm alumina ceramic tiles, changing the backing from Doron to a more advanced fiberglass composite developed at US Army Picatinny Arsenal, and forming alumina ceramic into a single large contoured breastplate [489]. By 1965, a working prototype body armor insert breastplate was developed, the T65-2 involving a carrier vest that held two 85% alumina ceramic plates, front and back, backed by the Picatinny fiberglass, with ballistic nylon spall-cover, each shaped as a contoured breastplate [498].

French multinational company Saint-Gobain was the supplier of the alumina ceramics that were used in the original Natick alumina ceramic composite armor of 1965 [22]. Saint Gobain no longer manufacture alumina armor ceramics having switched to SiC since that time.

11.3.4 First operational use of ceramic body armor: Alumina (1965)

The world's first operational ceramic composite body armor insert panel was the T65-2 alumina/fiberglass breastplate developed by Natick, over the long 20-year development process beginning with the original invention of the ceramic composite armor concept (glass/Doron) by Lt. Commander Andrew Webster in 1945 [488], and culminating in the 1963 patent of Richard Cook [493]. The Natick alumina/fiberglass breastplates were immediately trialed with US helicopter crews in the Vietnam War, beginning in 1965. Subsequently composite armor was also developed for the pilots legs, and a diaper-like garment for the groin and buttocks.

The early development work for the Natick ceramic breastplates was all done with alumina ceramics. Ballistic testing showed that the alumina ceramic needed to be at least 6 mm thick in order to stop the 30 caliber (7.62 mm) projectile, and that there was little benefit in making it thicker.

The US helicopter crews were the first military personnel in history, and the only military personnel in the Vietnam war to use ceramic composite armor in action: alumina-ceramic/fiberglass contoured body armor breastplates. The breastplates were "insert plates" designed to be inserted into the large pockets in the matching flak jacket to provide protection against high-velocity bullets (see Fig. 11.7). The conventional ballistic nylon flak jacket could only protect against shrapnel. Helicopter crews would routinely land and takeoff in the combat zone while delivering or rescuing military personnel. They spent a lot of time exposed to hails of bullets, a very hazardous occupation.

Fig. 11.7 Souvenir "chicken plate" alumina-fiberglass body armor insert plate from the Vietnam War [3].

The US helicopter crews of the Vietnam War gave the Natick T65-2 alumina/fiberglass insert plates the nickname "chicken plate." The phrase "chicken plate" is a welders joke. A "chicken plate" is a metal plate you weld on top of your weld because you do not trust your welding. Thus, the ceramic armor insert plate was the "chicken plate," the extra plate they had to use because they could not trust their flak jackets alone. The front and rear ceramic insert plates were held in a ballistic nylon carrier vest with Velcro fasteners and a snap fastener for emergency removal. Vintage "chicken plate" has now become a collector's item.

There prevailed at the time some aircrew compliance issues with wearing the "chicken plate" with some complaining that the flak jackets and "chicken plate" were too hot to wear in the hot tropical climate, and anecdotal stories of aircrew using their "chicken plate" as a seat. Pilots had an armored seat so they did not need the back plate and preferred to remove it. The other crew members, who did not have an armored seat, often sat on their back plate, while wearing their front plate, as bullets were most likely to come from in front or below. An evaluation of the combat performance of these Natick alumina/fiberglass plates was conducted for 72 aircrew casualties from mid-1966 to mid-1967 which showed that 76% of aircrewmen wore the armor, and that only 3.6% of their injuries were in the area of the torso protected by the armor [499].

Sometime after 1964, SiC and B_4C ceramic tiles were studied by the US military. The carbide ceramics were more expensive and slightly lighter: in the 1960s manifestations, 10% weight reduction and a doubling of the cost with SiC, and a 22% weight reduction and a quadrupling of the cost for B_4C [489]. In 1967, B_4C breastplates manufactured by Norton were introduced for the air crews. These were supplied only to navy, air force, and marines, while army crews only ever received the Saint Gobain alumina plates [500]. In 1969, Norton introduced the first hot-pressed B_4C ceramic-composite armor helicopter seats, backed by Kevlar-29 which had just been developed by DuPont. The B_4C/Kevlar helicopter seat was the pinnacle of the 1960s ceramic armor technology, years ahead of its time, not surpassed for some decades.

By the end of the 1960s, the first dedicated US armor ceramics company Ceradyne had been founded. Ceradyne ultimately bought out Norton armor ceramics and became a leader in the US armor ceramic fraternity. Richard Palicka was one of the key armor ceramic engineers in this early era, serving as Vice President and General Manager of Ceradyne from 1970 to 1985. When he left Ceradyne, Palicka founded Cercom which manufactured B_4C, SiC, and titanium diboride armor ceramics, along with other advanced ceramics for nonarmor applications. The focus of both Ceradyne and Cercom was nonoxide ceramics, not alumina. In the alumina armor ceramics realm, Saint Gobain and CoorsTek were the early pioneers.

11.3.5 Military sector—Ceramic body armor: 1965–96

The Natick ceramic composite body armor T65-2 "chicken plate" panel, reflecting as it did the Cook patent [493] went on to become the standard for all ceramic body armor since then. The basic ceramic body armor breastplate "insert plate" concept since 1965 has comprised a contoured ceramic breastplate around 6–9 mm thick with thin FRP fiberglass spall shield and thick fiberglass or other FRP backing, generally wrapped in a layer of ballistic nylon. This "insert plate" or breastplate is inserted into the pocket/pouch of a flak jacket, originally ballistic nylon, but the US ballistic nylon flak jacket of 1965 evolved into the Kevlar flak jacket beginning in 1983 (see Section 11.2.3.3).

In 1968, Carborundum was contracted to supply B_4C plates to Natick for deployment to US ground forces in Vietnam. This would have been the first deployment of ceramic body armor to any army anywhere in the world. This was a significant escalation of the US ceramic armor program implemented for helicopter aircrews just 3 years earlier. However, after such a flying start, it all came to a sudden halt in 1970 when the 30k vest order was canceled by the US leadership [501]. This coincided with the beginning of the withdrawal of the US from the Vietnam war which occurred gradually from July 1969 until November 1972, with the armistice coming in early 1973.

It was to be 29 years before the US military gave ceramic body armor to their ground forces. This began in 1996 with the ISAPO plate. The reasons for this delay were not technological nor could they be due to ceramic armor reliability. In 1967, 2 years after the first US military deployment of ceramic body armor, and 29 years before ISAPO, it had already proven very reliable and effective [499]. Nor does it appear that weight, cost, or personnel compliance with wearing the ceramic insert plates were significant factors. By all accounts the T65-2 "chicken plate" 1960s ceramic body armor program was very successful.

It seems most probable that the 26-year delay was related to the changing priorities of US leadership. From 1970 to 1979 US military developmental activity in the armor ceramic area was dormant. From 1979 to 1982, there was a period of renewed US government-sponsored ceramic armor research under Defense Advanced Research Projects Agency (DARPA). This was primarily focused on land vehicle armor, with a small focus on aircraft armor. This was followed by another period of downsizing in the early 1990s after the fall of the Soviet Union and the end of the Gulf war [501].

As a result, it was the Soviet Union, not the United States, who first deployed regular infantry with ceramic body armor. During the 1980s Afghanistan war, the Red Army was the first army in the world to use ceramic body armor on a large scale. The Soviet Union issued Red Army troops in Afghanistan with battle jackets that contained pockets into which were placed SiC 100×100 mm plates, individually clad in fiberglass backing. The withdrawal from Afghanistan, and the demise of the Soviet Union shortly after in 1989, brought this Soviet ceramic armor initiative to an end.

11.3.6 Ceramic body armor becomes a mass-market technology: 1996 to present

Aside from losing the ceramic body armor initiative between 1970 and 1996, the US military has largely been the trendsetter in the body armor realm since World War II, when it pioneered ballistic nylon flak jackets, Doron, and invented the ceramic composite armor concept all in the 1942–45 period. US troops at Okinawa 1945 were probably the first ground troops since the middle ages to be clad in body armor, a US trend which continued in the Korean War of the early 1950s, followed by several evolutions of US infantry soft body armor programs, through to the current era, as discussed in Sections 11.2.3.2 and 11.2.3.3.

In 1996, the US military again seized the initiative and deployed ceramic body armor to ground forces: originally as ISAPO then from 1998 as the small arms protective insert (SAPI) body armor program. Wide-scale deployment of ceramic body armor to US ground forces began in 1996 in the last 2 years of the PASGT system. The ceramic insert plates used for PASGT were known as ISAPO plates, and with the launch of the Interceptor infantry battle jacket in 1998, the SAPI ceramic plates were introduced and ramped up significantly thereafter, an initiative that continues to this day.

The SAPI plate was an NIJ4 ceramic plate (30 caliber armor piercing—see Section 11.3.7), most commonly SiC, clad today in the latest generation backings (e.g., Dyneema-based SpectraShield), although there has been significant evolution in ceramic plate and backing material since 1996 when it was first deployed as the ISAPO ceramic insert plate in the PASGT system. SAPI does not use fiberglass backing which had begun to fall out of favor in military use in the 1990s, with Kevlar backings, and ultimately SpectraShield replacing it. However, in low-cost markets, fiberglass continues to be used widely today (see Sections 11.3.7 and 11.5.1).

The ISAPO/SAPI program was a paradigm shift: the mass market-scale formal state-sponsored armoring of regular infantry with the body armor commensurate with that of elite military forces. In 1996, the WHS principle that everyone should have an equal right to the best protective apparel available (in this case available to elite forces since 1965) reached the US regular infantry.

This trend toward up-armoring of regular infantry with the best body armor available, initiated by the US military in 1996, was rapidly emulated worldwide. In 1999, the Australian Defense Force (ADF) ground forces first received ceramic body armor, which served them well in the 1999 East Timor conflict. The alumina ceramics used for this late 1990s Australian ADF body armor program were manufactured by Taylor Ceramic Engineering (TCE), following a ceramic armor development program led by David Taylor (Managing Director: TCE). TCE is the leading alumina manufacturer in the Southern Hemisphere, founded in 1967, and profiled in Chapter 1 (Section 1.3.2.5) and in Chapter 12 (Section 12.2). Other western nations also began to introduce ceramic body armor to regular ground forces around the same time. In the last decade, many developing countries such as China and India have also adopted this trend.

This was not really a WHS revolution, it was just common sense. Once US regular ground forces were all protected with ceramic body armor, and the leading western nations followed suit, the need to up-armor all troops around the world by all nations that fielded a viable army, became a matter of necessity. The alternative would be to send unprotected troops up against ceramic-armored "super soldiers" and the inevitable disadvantage this would produce.

As a consequence, the ceramic body armor industry made a dramatic shift in the early 21st century from elite boutique product to mass-market product. In parallel, in the 21st century, alumina ceramics have made the switch from a boutique product in the armor realm to becoming one of the leading uses for alumina ceramics in the world. The same applies to SiC armor ceramics.

The egalitarian and humanitarian significance of ceramic body armor, and the US late 1990s ISAPO/SAPI program, cannot be over-emphasized. For the first time since the early bronze age, ordinary soldiers were sent onto the battlefield protected against the munitions of the era. For over 4000 years, from the bronze age until the obsolescent of metal armor in the 1500s–1700s, only the elite went into battle wearing metal armor. From the 1500s until the late 20th century, armor technology was inferior to the munitions of the day. From the 1700s onward, and greatly escalating in severity from the late 19th century after the invention of the machine gun, millions of soldiers were sent unprotected into gunfire, and died in the millions.

In November 2001, a pivotal international conference on ceramic armor was held at Maui (Hawaii). It was organized by Bill Gooch, the doyen of ceramic armor in the United States, who has been involved in ceramic armor since the era of first implementation of ceramic armor during the Vietnam War, and remains active to this day. The timing of this conference was very significant: 2 months after 911, that is, the beginning of the Afghanistan war; 5 years after the first US deployment of ceramic body armor to regular infantry (ISAPO 1996); 2 years after the first Australian army deployment of ceramic body armor. In general, this was the era in which many nations were actively developing ceramic body armor programs for their armies. Hundreds of delegates from nations around the world (including the author[3]) attended this event where the future of ceramic armor for the 21st century was laid out. The conference proceedings make interesting reading, albeit somewhat historical, it being nearly two decades ago [502].

11.3.7 Ceramic body armor in the civilian market

Ceramic body armor found its way into the civilian realm gradually, beginning as the privately purchased alumina-fiberglass insert plates for use in privately purchased Kevlar body armor vests in the 1980s. There is virtually no historical record on the penetration of alumina ceramic armor into the civilian sector. However, the author was an active participant in this process,[4] and can therefore offer some insights.

The driver for ceramic body armor finding its way into the civilian sector was the commercialization of Kevlar fabric by Du Pont in the 1970s, and the rapid rise in public awareness of the Kevlar body armor vest in the late 1970s. Soft body armor fabricators entered the market in the 1970s. The commercialization of Kevlar in the 1970s, made affordable low-bulk bullet-proof soft body armor commercially viable for the mass market for the first time in history. Kevlar was fundamentally a very much cheaper, and significantly less bulky version of the early 20th century silk body armor. While silk body armor a century ago cost about $20,000 in today's money, a Kevlar vest is just a few hundred dollars, and much less bulky for the same protection level.

American Body Armor was founded in 1969 and produced the first civilian body armor product, essentially mimicking the ballistic nylon military battle jacket, incorporating steel plates. This was marketed under the trade name "Barrier Vest." It was not really much different in concept to the "soldiers bullet proof vest" (steel plates in a jacket) made by Cook & Co during the US Civil war of the 1860s, a century earlier. Although the nylon was a noticeable advance as was the quality of the steel.

In 1975, American Body Armor was the first company to introduce a civilian Kevlar vest (the K-15 with 15 layers of Kevlar fabric), which contained a single 125×200 mm steel plate to protect the heart [503]. In 1976, Richard Davis founded his low-cost body armor company, Second Chance Armor, capitalizing on the newly

---

[3] The Author was a member of the delegation from Australia. See "About the Author."

[4] The author has been involved in the ceramic body armor field since 1988. See "About the Author."

available low-cost Kevlar fabric. The dramatic publicity stunts of Davis, who famously donned his Kevlar vest and shot himself with a handgun on camera, brought the idea of mass-market body armor technology to public awareness. However, it was the low cost that really drove Kevlar soft body armor vests to experience rapid market penetration from the late 1970s onward.

So the question is, how did the commercialization of Kevlar lead to ceramic armor entering the civilian domain in the 1980s? Kevlar vests provide the same limited protection as silk: that is, *only shrapnel and soft low-velocity bullets* (handgun slugs and shotgun pellets), albeit Kevlar is 50 times cheaper than silk and less bulky for the same protection. Kevlar led to a boom in body armor usage, particularly in the United States initially. With the widespread prevalence of military assault rifles (NIJ3) in the well-armored US populace, especially the AK47 Kalashnikov and the M16/Styr, Kevlar vests did not provide sufficient protection. This was also the case in Australia up until the prohibition of guns in 1995.

Therefore, in the United States in the 1980s, and Australia before 1995, alumina body armor insert plates began to appear as an optional extra from pioneer body armor companies.[5] This ceramic body armor was essentially reverse-engineered "chicken plate": contoured alumina ceramic breastplate, with fiberglass backing and spall plate. There was nothing complex about reverse engineering "chicken plate." Contoured alumina plates are easily fabricated on a small scale by slipcasting (Chapter 4), or mass produced on a large scale by dry pressing with an automatic or semiautomatic press (Chapter 4). Any fiberglass fabricator is capable of fiberglassing the alumina with backing and spall cover to 1960s fiberglass backing technology standards. Souvenired "chicken plate" was and remains a collectible item in the post-Vietnam War era, that is, after 1973, so the coincidence of the commercialization of Kevlar and the end of the US involvement in the Vietnam War, both occurring in the 1970s, led to a marriage of alumina/fiberglass ceramic body armor insert plates, with Kevlar vests in the civilian sector in the 1980s. Alumina breastplates retailed for as little as $25–$50 each, and fiberglassed could cost less than $100 each. This is significantly cheaper than a Kevlar vest, which was several hundred dollars.

The civilian market is primarily law enforcement and licensed private security personnel. In the early era of body armor in the civilian sector, long before body armor was part of the issued kit, many police officers and security guards bought their own body armor on the private market. Many security guards continue to do so today. In Australia, since guns were outlawed in 1995, guns have been very rare, and assault rifles virtually nonexistent. Therefore, there was little call for assault rifle protection in Australia post-1995, that is, ceramic body armor. While soft armor vests are widely used by Australian police and security guards today, ceramic armor insert plates are primarily used only by SWOT teams. In the United States where military assault rifles

---

[5] In the era in which ceramic body armor first entered the public domain the author was active in the field. In 1988, the author was a consultant to ESK Germany (ceramic armor company) and in 1992 a consultant to Signal 1 International, Australian pioneer in civilian ceramic body armor. In 1996, the author was appointed director of Modern Ceramic Company, one of the two leading global companies in reaction-sintered silicon carbide armor ceramics in that era. See "About the Author."

are prolific, ceramic body armor remains a significant product in the civilian sector. In other countries, the situation depends on assault rifle prevalence, private body armor possession laws, and the average spending power of private individuals.

Alumina continues to this day to be the armor ceramic of choice where cost is concerned, that is, virtually all civilian sector usage, and military usage in some developing countries. SiC and B_4C are used by most western military forces, while the slightly heavier but much cheaper alumina is the standard for military body armor where cost is a factor.

In 1970, the US National Institute of Justice (NIJ) initiated a body armor development program for law enforcement personnel. From this was to arise the standard NIJ ratings for body armor: NIJ1 (lower) and NIJ2 (higher) with NIJ3A the highest level for soft body armor, capable of protecting from soft low-velocity handgun and shotgun projectiles. Above NIJ3A ceramic armor is required. NIJ3 is for jacketed 30 caliber rifle projectiles, and NIJ4 for armor piercing 30 caliber (see Section 11.5.3). NIJ4 is the highest protection provided by body armor.

11.4 Ceramic armor design principles

Beginning in the late 1960s, pioneer Mark Wilkins of the Lawrence Livermore Laboratory (the United States) developed much of the theory for understanding, designing, and modeling ceramic armor [504]. Alumina-ceramic/aluminum-metal backing was the primary experimental system used in this extensive work by Wilkins. Questions investigated included why a fracture conoid forms, the forces and mechanisms involved, and how to design a ceramic armor system to optimize the ballistic response. Much numerical modeling and finite-element modeling has since followed. Many of the principles that we take for granted today were developed by Wilkins. In summary, perforation of ceramic composite armor systems involves the following sequence of three events [505]:

(1) Shattering
(2) Erosion
(3) Catching

Shattering: The softer (softer than the ceramic) sharp-pointed projectile traveling at high velocity (high kinetic energy) is shattered against the harder ceramic. The compressive stress at the point of impact is typically tens of GPa, sufficient to cause the ceramic (compressive strength well below 10 GPa) to suffer compressive failure and fracture into the "fracture conoid." Tensile wave reflections through the ceramic immediately follow, resulting in some flexing of the ceramic tile and backing plate. The crack propagation velocity in alumina ($\sim$3 km/s) is significantly slower than the shock wave propagation velocity ($\sim$10 km/s) which means that brittle fracture is unable to keep pace with the shock wave, and therefore the resistance of the ceramic at the shock-wave wavefront corresponds to its theoretical strength.

Erosion: The shattered remains of the projectile, still retaining high kinetic energy, pass through the debris of the locally shattered ceramic (fracture conoid) and are eroded by the hard abrasive ceramic debris.

Catching: Momentum transfer between the high-energy debris (ceramic fracture conoid/shattered projectile) and the backing plate. The backing is a soft but extremely tough material. If the projectile was not first shattered by the ceramic, the projectile would effortlessly penetrate the backing as the sharp hard high-velocity object that it arrived as. However, the large cross-sectional area (rear fracture cone diameter) blunt mass of debris at the back of the fracture cone, with its high kinetic energy, is easily stopped by the soft tough backing.

The best way to visualize ceramic armor is a ceramic-lined "catchers mitt" that catches a hard bullet, after first shattering it.

For small-arms projectiles which have a short aspect ratio (length/diameter ratio) of 3–5, the ceramic is thin, and shattering and catching are the primary mechanisms. 99% of ceramic armor systems are of this type. This is the situation for all ceramic body armor and most ceramic vehicle armor. The exception to this rule is long-rod penetrators [the armor-piercing discarding sabot (APDS) see Section 11.6.3] with aspect ratios up to 30. Here, the ceramic needs to be very thick and erosion is the primary mechanism. The APDS is a heavy caliber munition primarily targeted against battlefield tanks and therefore irrelevant to lightweight body armor.

With this threefold mechanism in mind, there are a number of principles by which ceramic armor can be designed and optimized.

11.4.1 Mechanisms of projectile defeat

Quantifying and optimizing ceramic armor systems require attention to the following factors where RHA armor steel is the reference material in all cases:

- *AD*: Areal density measured in kg/m^2. The lower this is, the more weight efficient the armor.
- *V50*: Projectile velocity (for a given projectile type, armor system, and AD value) at which 50% of impact events correspond to penetration, and 50% to projectile defeat.
- *EM*: Equivalent mass performance, AD(steel armor)/AD(ceramic armor). For example, EM = 2.0 means that steel armor would be double the weight of the ceramic armor for the same protection. The higher the EM the more weight efficient the ceramic armor.

11.4.2 Defeat mechanisms utilized in ceramic armor design

Fracture of an armor ceramic is primarily dependent on [505]:

- compressive loading of the ceramic directly beneath the impacting projectile and
- maximum flexure of the ceramic plate and tensile stress/strain at the rear of the ceramic plate.

The above two factors are heavily influenced by the side and rear containment material and stiffness. Associated with this are a number of important principles that underlie the design of a ceramic composite armor system.

- *Ceramic composite armor concept*: Ceramic layer as the projectile "disruptor" that blunts/shatters/erodes the projectile; backing layer as "catchers mitt" that both mechanically supports the ceramic (prevents it from flexing under ballistic impact) and absorbs the projectile energy. Spall layer contains the ceramic and projectile debris (see Fig. 11.4).
- *Fracture conoid*: Transfers concentrated penetrative force at the tip of a hard projectile into a diffuse load on the backing at the rear of the fracture conoid (see Fig. 11.4).
- *Mosaics*: Multiple-hit protection with ceramic armor is achieved through the use of mosaics of ceramic tiles (Fig. 11.8, 11.10), generally hexagonal tiles, creating a honeycomb array, or square tiles in a square array. The destruction of one mosaic tile cannot cause crack propagation to the rest of the ceramic armor panel. Mosaics are expensive to manufacture and mainly used for light vehicle and aircraft armor. In monolithic ceramics (e.g., body armor breastplates), crack propagation travels far from the impact point, raising the impact separation distance to a significant extent. Mosaic tile size is calculated to be small enough to maximize multihit, but not so small that the tile forms an inadequate fracture conoid and can be punched through the backing as a quasi-projectile. In practice, it has been found that there is very little loss in ballistic performance for an impact at a mosaic tile-tile border
- *Convex surfaces*: In mosaics, it is common to have a convex tile surface which causes the projectile to "yaw" (tilt from the perpendicular) on impact, except in the rare case of it hitting the tile dead center. A yawing armor-piercing projectile massively loses its penetrating power (Fig. 11.8).
- *Stepped surfaces*: Destabilization/yawing can also be induced by other surface effects, such as a step-function surface discontinuity effect (Fig. 11.5).
- *Hardness*: It is generally only necessary for the ceramic to be harder than the projectile it is stopping, but in practice it should be significantly harder. At NIJ3, for which the hardest penetrator is the AK47-7.62/39 soft-steel-core at around 2 GPa hardness, 8 GPa ceramic hardness [lithium-alumininosilicate (LAS) glass ceramic] is sufficient but even better is 9.5 GPa (AD85 alumina). For NIJ4 (8 GPa hardened steel armor piercing core) ceramic hardness must be significantly over 8 GPa, which means that high-purity alumina (14 GPa), SiC (24 GPa), or B_4C (25 GPa) are all more than adequate. A recent alarming development is the M993 7.62 tungsten carbide projectile (STANAG-3: see Section 11.6.2) which is around ~12 GPa hardness. For the M993, the highest performance alumina (17 GPa) is adequate, B_4C is a more risky choice due to the uncertainty associated with the B_4C shock-loading problem (see section 11.5.2.4), leaving SiC as the lowest risk option for M993, and ideally SiC-N not reaction-sintered SiC (RSSC). This M993 projectile is very rare, and only found on the battlefield, but it is an alarming case of the body armor/munitions arms race that continues inexorably.
- *Brittleness*: It is essential that the ceramic be sufficiently brittle to be pulverized by the projectile (rather than undergoing ductile deformation as would a metal) and thereby form a

Fig. 11.8 Surface effects in mosaic-tile systems for destabilizing (yawing) an incoming projectile. Left: convex surface on mosaic tiles and right: step function produced by mosaic tiles with a stepped raised perimeter. Refer also to Fig. 11.10.

fracture conoid. This is why ceramic armor even works with soft ceramics (such as glasses) against soft projectiles (such as NIJ3), as per the original Webster ceramic armor invention of 1945.
- *Dilatancy*: Dilatancy enhances erosion. Dilatancy is a rheological principle whereby a moving fluid occupies a greater volume than the static fluid. In the case of slurries, this can be due to hard angular particles, for example, wet sand. In ballistics, a well-contained zero-porosity ceramic expands as it is pulverized into hard angular particles by a projectile, thereby creating a volume increase and enhancing resistance to projectile penetration and enhancing projectile erosion. This mechanism is reduced or lost when an edge impact occurs, that is, when there is no adjacent parent ceramic tile (monolithic breastplate ceramic) or mosaic tile (mosaic ceramic) to provide lateral constraint (Fig. 11.9).
- *Edge effect*: If the projectile impact point is close to the tile edge, by an amount approaching the fracture conoid radius, the V50 will be significantly less, resulting in probable penetration. In body armor, there is no solution to this problem, and it is well understood that the perimeter region provides inadequate protection. In advanced vehicle armor, metal containment (constraint cells) is sometimes used to overcome this problem (see Fig. 11.4 for the fracture conoid concept).
- *Density*: Ideally the ceramic density is as low as possible for minimum AD and maximum EM.
- *Cost*: Ideally as low as possible. Alumina is by far the lowest cost armor ceramic (Fig. 11.10).

11.4.3 Designing and optimizing ceramic armor systems

End users always want weight, bulk, and price to be minimized: lighter, less bulky, and cheaper. Rather like the NASA adage of "faster better cheaper," an unrealistic demand is always going to require some compromise. In the case of ceramic armor, lighter is the main win. Often ceramics are bulkier than metal armor. Ceramic armor is always more expensive than conventional RHA steel armor, but alumina/S-glass systems are certainly the cheapest option.

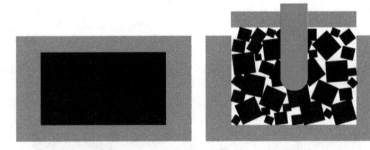

Fig. 11.9 Dilatancy. Left: armor ceramic (black) in metal containment cell (gray) before ballistic impact. Right: after ballistic impact by red projectile. The hard angular ceramic fracture particles, resulting from pulverization of the ceramic, occupy a larger volume than the intact ceramic. This causes an expansion that resists further penetration by the projectile and causes abrasive erosion of the projectile. *Note*: the spall plate (lid of the metal containment cell) has dislodged, for illustration purposes. An advanced vehicle armor system would be designed so that expansion of the metal containment cell would require great force, which enhances the power of the dilatancy mechanism.

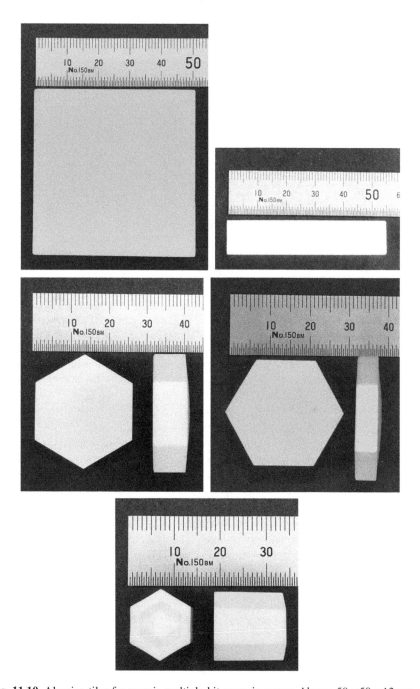

Fig. 11.10 Alumina tiles for mosaic multiple-hit ceramic armor. Above: 50 × 50 × 12 mm alumina square flat tile for STANAG-3/4 (14.5-mm armor piercing) light-vehicle/helicopter mosaic armor, top view and side view. Below left: 25 × 9 mm alumina hexagon with convex tile top face for STANAG-2 (equivalent to NIJ4) light-vehicle/helicopter mosaic armor. Below right: 25 × 6 mm alumina hexagon with convex tile top face for STANAG-1 (equivalent to NIJ3) light-vehicle/helicopter mosaic armor. Bottom: 13 × 15 mm hexagonal mosaic tile with a raised perimeter for step-function surface discontinuity effect. Note: the convex surfaces for the hexagonal tiles which are around 1 mm thicker in tile center with an approximately 100-mm sphere radius on tile surface profile. Refer also to Fig. 11.8 for the surface profiles these tiles generate.

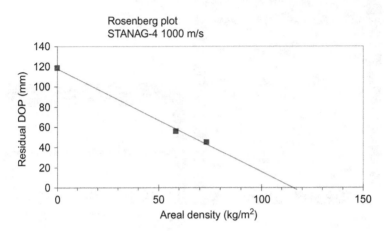

Fig. 11.11 Rosenberg plot for an experimental armor ceramic (AusShield: developed by Modern Ceramics Company MC^2), using aluminum as the reference RDOP material, at 1000 m/s at STANAG-4, showing that V50 at 1000 m/s is achieved at $AD = 115 \text{ kg/m}^2$. Given that RHA has an AD of 320 kg/m^2 at STANAG-4, this corresponds to an EM of $320/115 = 2.8$ as an applique armor, a very high EM for an armor ceramic. If it was standalone armor, then an extra AD of about 10 would be required for SpectraShield cladding making an armor system AD of 125 kg/m^2 and EM of about 2.55 [507].

In the 1980s, Rosenberg developed the residual depth of penetration (RDOP) method now widely used for calculating V50 for a given ceramic armor system [506]. Essentially numerous duplicate test samples of an armor system are mounted in front of a thick slab of reference material (such as aluminum) and shot by a given projectile at a range of velocities. The RDOP into the aluminum is plotted against AD (for a given projectile velocity) in a Rosenberg plot, from which the V50 for zero RDOP can be extrapolated. This is a standard laboratory tool for developing a ceramic armor system from scratch, or evaluating a new armor ceramic. This is shown in Fig. 11.11.

11.4.4 Traumatology benefits of ceramic body armor

The weight advantage of ceramic body armor over steel armor, in the order of two to three times lighter for the same protection level, is its most well-known benefit. However, less well known is the greatly reduced blunt trauma of the body when protected by ceramic armor, versus metal armor such as armor steel. This is because metal armor defeats a projectile by plastic deformation, while ceramic composite armor defeats a projectile by a complex energy absorption and load-spreading mechanism, as discussed in Section 11.4. The result is that the energy transfer to the body (blunt trauma on the human body) from ceramic composite armor is an order of magnitude less severe (tens of times less) than for metal armor. This translates into greatly reduced incidence of broken ribs, internal injuries, and other tissue damage for the wearer of ceramic composite armor compared with the wearer of metal armor. This has substantial humanitarian, and practical injury management implications, for military, law enforcement, and security personnel.

Ceramic performance for body armor and vehicle armor systems is initially tested and optimized using an RDOP test. However, final commercial-ready body armor systems (generally the ceramic insert plate/Kevlar vest combination) are ballistically tested while placed against a block of preconditioned and appropriately contoured modeling clay (precisely conditioned). This is a standardized way of measuring the blunt trauma experienced by the human body during the ballistic impact, according to the NIJ standard testing criteria. Traumatology validation has developed the optimal temperature and conditioning preconditions for modeling clay such that it matches human soft tissue. The NIJ standard specifies a maximum trauma depth permitted, determined by measuring the indentation depth in the modeling clay by the rear of the body armor system.

11.5 The state of the art in alumina and other ceramic body armor systems

Note: This section is largely devoid of citations as it primarily constitutes industry knowhow, not information from published sources. All knowledge disclosed here was obtainable by a civilian consultant involved with the industry. No security classified secrets are disclosed here.

11.5.1 Backing materials: Body armor/lightweight vehicle armor

Doron was simply a name given to an early form of ballistic fiberglass. Fiberglass belongs to a class of composite materials known as fiber-reinforced polymer, or FRP. Fiberglass is the cheapest and by far the heaviest of the FRP's. The FRP essentially involves two components:

Fiber reinforcement
- Glass fiber (S-glass)—very heavy at about $2.5 \, g/cm^3$.
- Carbon fiber—too brittle for armor backings.
- Kevlar fiber (polyaromatic amide)—light at $1.44 \, g/cm^3$
- Spectra/Dyneema fiber (UHMWPE)—ultralight at 0.9–$1.0 \, g/cm^3$.

Kevlar and dyneema were discussed in detail in Section 11.2.3. S-Glass fiberglass "S-Glass" (64% SiO_2, 24% Al_2O_3, and 10% MgO) is the strongest grade of fiberglass fibers, and the fiberglass grade exclusively used in ballistics. Fiberglass is NOT used for soft body armor, it is used as the fiber component of FRP backing for ceramic composite armor. S-Glass was used exclusively for FRP backings for ceramic composite armor up until the 1990s, and was gradually supplanted first by Kevlar in the 1990s/2000s, and ultimately by Dyneema/Spectra since the 2000s. S-Glass is an order of magnitude cheaper than Kevlar or Dyneema. S-Glass still sees widespread use as the low-cost backing of choice, and is widely used in the civilian body armor sector, as is alumina ceramic: alumina/ S-Glass, the original "chicken plate" ceramic composite body armor panels, is the civilian concept for ceramic body armor still in widespread use today.

Matrix "glue" that bonds the fibers together
- Thermosetting (curable liquids): epoxy, polyester, polyurethane, phenolic.
- Thermoplastic (meltable polymers): Polyethylene film (HDPE and LDPE).

The "Doron" and other fiberglass backing systems continued to be used as the FRP backing material up to the end of the 20th century. Today, in civilian body armor applications, fiberglass is still the default system. In its most highly evolved form, ballistic fiberglass comprises S-Glass fiber reinforced by a ductile tough resin, either polyurethane resin or rubber-toughened epoxy resin.

However, for military applications, since the late 20th century, fiberglass backings were gradually replaced by more advanced backings such as Kevlar with the ductile thermosetting resins such as polyurethane resin and rubber-toughened epoxy resin, and then evolving to Kevlar reinforced by the ductile thermoplastic resin LDPE and HDPE. Kevlar saw very limited use in this role. Rather like the fax machine, it was a short lived transitional technology, in this case the transition from S-Glass domination of the late 20th century to SpectraShield domination of high-end applications in the 21st century.

Ultimately, in the last decade, the technology evolved to the premier backing of them all which was SpectraShield, a Honeywell trade name for UHMWPE (Spectra and Dyneema are the two branded products) with thermoplastic polyethylene as the matrix material. It is hot-pressed at around 115°C.

SpectraShield ballistically performs very well as a standalone armor against NIJ1 and NIJ2, and as a backing is the state of the art in the world today. It combines the toughest fiber known, with one of the toughest matrix materials known, and is the lowest density of any ballistic FRP at $0.9–1.0 \, g/cm^3$. SpectraShield hot pressing is a very delicate temperature control task as the temperature needs to be high enough to melt the polyethylene matrix material (as high above 110°C as possible), but not high enough to recrystallize the Spectra/Dyneema fibers (never exceed 138°C). Given the fact that polyethylene is a thermal insulator, temperature gradients are inevitable in large preforms, and the 18°C window is a technical challenge that requires many thermocouples and a lot of skill. FEA modeling also helps.

Honeywell provides SpectraShield raw materials and as a relevant guide to the precision required, it is instructive to show the Honeywell general processing guidelines for optimal fabrication by armor manufacturers, with the Honeywell disclaimer that *these are general processing guidelines for Honeywell SpectraShield ballistic materials. Specific process parameters should be optimized for the equipment and materials used by our customers* [508].

Autoclave process
- Insert thermocouple into center of the stack of material to be processed.
- Place assembled material onto caul sheet and vacuum bag.
- Connect vacuum line, pull vacuum, and seal leaks in vacuum bag if required.

- Insert materials in autoclave and close door.
- Ramp autoclave temperature to processing temperature of 240°F (115°C) to 280°F (138°C).
- Monitor autoclave temperature and material centerline temperature.
- Once the centerline of material reaches processing temperature, apply pressure (100–200 psi) and hold at temperature under pressure for 30–60 min.
- After 30–60 min at pressure and temperature, begin cool down of autoclave under pressure.
- Once the centerline of the material reaches 100°F (38°C), release autoclave pressure.
- Release vacuum.
- Open autoclave.

Compression molding process
- Preheat the platens (or mold) to processing temperature 240°F (115°C) to 280°F (138°C).
- Layup the material to the desired areal density.
- Insert a thermocouple into the center of the stack of material to be processed.
- Open the press.
- Place the stack of material (with the thermocouple at the centerline) between the platens.
- Close the platens on the material, applying minimal pressure (this allows the material to preheat).
- Once the centerline of the material stack reaches the processing temperature 240°F (115°C) to 280°F (138°C), apply tonnage and hold under pressure and temperature for 15–30 min.
- Cool, under pressure, until the centerline temperature is <100°F (38°C).
- Open press.
- Remove molded sample from press.

Honeywell Disclaimer: These are general processing guidelines for Honeywell SpectraShield ballistic materials. Specific process parameters should be optimized for the equipment and materials used by our customers.

A Kevlar version of the SpectraShield concept is also sometimes used as a backing for ceramic armor (hot-pressed Kevlar/LDPE), but since Kevlar (1.44 g/cm^3) is much heavier than Spectra/Dyneema (0.9–1.0 g/cm^3), this makes the backing much heavier: ~1.2 g/cm^3 or 30% heavier than SpectraShield. Therefore, Kevlar never really took much market share from S-Glass or Dyneema. The Kevlar version does have the advantage that it is easier to fabricate. While SpectraShield hot pressing is a very delicate temperature control task because of the thermal sensitivity of Spectra/Dyneema, Kevlar is stable to temperatures above 350°C, and so there is no sensitivity required in the hot-pressing process with Kevlar/HDPE.

SpectraShield has become such a mass-market product in the last decade that even the civilian market is now sometimes using it as standalone armor NIJ and NIJ2 armor panels and as a backing for alumina ceramics in specialty vehicle armor applications.

Today, SpectraShield dominates the body armor systems used in the military realm, and is also making inroads into the top end of the civilian body armor and vehicle armor realm. Kevlar backings are rare, as Kevlar is not much different in price to Spectra/Dyneema, but significantly inferior on a weight basis. Low-tech S-Glass backed body armor plates are the default civilian body armor technology, and in

Fig. 11.12 Contemporary ceramic body armor breastplates, which are likely to embody the manufacturing technology described above.
Image Author James Craig of Craig International Ballistics [509]. Reproduced from Wikipedia (Public Domain Image Library).

the military sector in nations and applications where cost is an issue. Contemporary high-performance body armor panels, for which SpectraShield would be the appropriate cladding, are shown in Fig. 11.12.

11.5.2 Ceramics used

The most important material properties for armor ceramics are hardness, density, and elastic modulus: ideally high modulus, low density, and a hardness significantly greater than the projectile. Toughness and flexural strength are also very important. However, toughness is much the same for alumina, SiC, and B_4C (LAS glass ceramic is somewhat lower). Moreover, measured strength for ceramics is a very subjective property which depends on flaw population, surface finish, specimen dimensions as tested, test configuration, strain rate, and other specimen- and test-dependent factors (Chapter 4). For alumina, SiC and B_4C strength is much the same (LAS glass ceramic is somewhat lower). Table 11.2 summarizes the key properties of the five armor ceramics used in body armor.

A typical 250×300 mm ceramic contoured breastplate is approximately \$25–\$50 in alumina, depending on size purity and quality, \$50–\$100 in RSSC depending on density, over \$100 in direct-sintered SiC (DSSC), and perhaps \$200 in B_4C. The general rule of thumb is: RSSC = alumina $\times$ 1.5; DSSC/HPSC = alumina $\times$ (2–3); B_4C = alumina $\times$ 4.

Table 11.2 **The key properties of the five armor ceramics used in body armor [28,507,510]**

Ceramic	Density (g/cm³)	Hardness (GPa)	Elastic modulus (GPa)
LAS (glass ceramic)	2.4	8	86
AD85 alumina	3.4	9.4	221
Pure alumina	3.9	14.5	386
RSSC	3.1	24	400
SiC-N	3.2	24	460
B_4C	2.5	25	460

In density, in comparison with high-purity alumina, for the same thickness (which is usually not the case), AD85 alumina is 12.5% lighter, SiC is 20% lighter, and B_4C and LAS are 35% lighter.

11.5.2.1 Alumina

Alumina is the budget ceramic armor choice. It dominates the civilian sector of the ceramic armor industry, and the affordable segment of the military sector, both body armor and vehicle armor, but is no longer used in elite military body armor applications. At a density of 3.9 g/cm³ for pure alumina (hardness ~14 GPa) which is required for armor-piercing projectiles (NIJ4, see Section 11.5.3), down to 3.4 g/cm³ (hardness ~9.5 GPa) for 85% pure alumina (Coors AD85) which is softer but still much harder than LAS and performs comparably with high-purity alumina against nonarmor piercing 7.62 projectiles (NIJ3, see Section 11.5.3). AD85 and its equivalents are widely used for NIJ3 due to the unbeatable cost-weight dual advantage. For NIJ4, high-purity alumina is used, and most body armor insert plates used in military applications are NIJ4, whereas for civilian use NIJ3 is the common requirement, as armor-piercing bullets are relatively uncommon, a small but significant risk on the battlefield, and a negligible risk in civilian situations.

Alumina's dominance is simply because its price is so much lower than the competition: SiC and B_4C. Of course it is heavier than the carbides. The density of pure alumina is 3.9 g/cm³, while RSSC is 3.1 g/cm³ and SiC-N is 3.2 g/cm³ (20% lighter) and B_4C is 2.5 g/cm³ (35% lighter). This is why the CoorsTek Alumina AD85 armor ceramic, and its equivalents, are an attractive option. While less hard than pure alumina, its density of 3.4 g/cm³, is only 10% heavier than SiC, and very much cheaper, suitable only for NIJ3 (Section 11.5.3.1), it is popular in that role.

In general, for civilian applications NIJ3 alumina AD85 backed by fiberglass is the budget choice, or AD85 alumina backed by SpectraShield for the more expensive option. It is uncommon to use the heavier pure alumina for NIJ3, as the weight penalty is not justified.

11.5.2.2 LAS glass ceramic

A throwback to the original Webster invention, which used glass as the ceramic component of ceramic composite armor, LAS glass ceramic, and even some other glass ceramics, are occasionally used for body armor plates for nonarmor piercing projectiles (NIJ3, see Section 11.5.3.1). LAS is not very hard, at around 8 GPA, but at a density of 2.4 g/cm^3 it is even lighter than B_4C. It is not widely used because of the perception of a risk of failure with LAS that is often considered too high to justify the weight saving compared with AD85 alumina and cost saving compared with equivalently lightweight but much harder (25 GPa) B_4C. However, LAS has been used successfully in viable and very lightweight NIJ3 commercial body armor insert plates.

11.5.2.3 SiC

SiC (SiC) is the next tier up from alumina, at a density of 3.2 g/cm^3, and a hardness of about 22 GPa. It comes in three forms:

- Reaction-sintered SiC (RSSC).
- Direct-sintered SiC (DSSC)
- Hot-pressed SiC (HPSC)

HPSC is quite popular, especially in the CoorsTek SiC-N grade. The hot-pressing technology is very expensive and so this is an extremely expensive ceramic at over $60/kg (over $100 per breastplate), but with the B_4C shock-loading problem, SiC-N is increasingly seen as the premier low-risk high-performance armor ceramic in the world today. There are two cheaper alternatives: DSSC which is midway between SiC-N and RSSC in both cost and ballistic performance, and RSSC, the much cheaper (and only slightly ballistically inferior) alternative to DSSC. DSSC and SiC-N are very expensive ceramics to produce as they require submicron SiC which is a very expensive raw material as it is so expensive to fine-grind high-purity SiC. Being a very hard material, SiC tends to erode the grinding media, and so energy costs in grinding, purification costs for the ground powder, and attrition of the grinding media all contribute to a very expensive raw-material powder.

The other factor which contributes to the high cost of DSSC is sintering temperature, which is in the 2000–2200°C range. This is ~500°C higher than alumina. Furthermore, SiC requires a vacuum or inert atmosphere (alumina is sintered in ambient air), requiring specialized furnace technology.

RSSC, on the other hand, is the only serious competitor for alumina in the cost-performance equation. RSSC uses very low-cost raw materials (green SiC abrasive grit in the 200–1000 mesh range), and the only costly aspect is the high sintering temperatures involved, again in the 2000–2200°C range. DSSC raw materials cost is even more significant than sintering cost in the price equation. RSSC is made by mixing SiC abrasive grit with a carbon precursor such as phenolic resin, followed by pyrolysis to carbonize the resin, then infusion with molten silicon metal above the silicon melting point (1414°C) and ideally above 2000°C. The molten silicon reacts with the carbon to form alpha silicon carbide, and any residual porosity is filled with hard (11 GPa)

Alumina in lightweight body armor 357

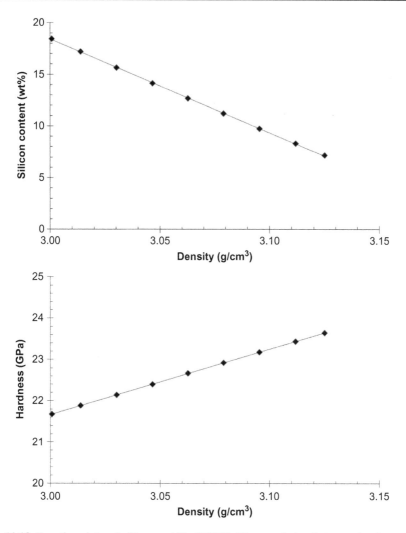

Fig. 11.13 Reaction-sintered silicon carbide (RSSC). The correlation between density and properties. Top: silicon content (wt%). Bottom: theoretical hardness (GPa) based on SiC:Si volumetric ratio.

silicon metal. The microstructure is therefore SiC-matrix cermet, with small isolated islands of hard silicon metal.

The objective with RSSC manufacture is to minimize silicon content, and therefore maximize density. RSSC of density below $3.05\,g/cm^3$ is marginal in ballistic performance. From 3.05 to $3.10\,g/cm^3$, it is ballistic grade NIJ4. Above $3.10\,g/cm^3$ RSSC is comparable with DSSC in ballistic performance, but yet half the price. The density effect is shown in Fig. 11.13. Making RSSC is a highly complex manufacturing process, and a great deal of industrial knowhow is required to get the density above

$3.10\,\text{g/cm}^3$, but when optimized (i.e., density above $3.10\,\text{g/cm}^3$), RSSC is an unbeatable combination of low cost and low weight.

The supposed ballistic edge of DSSC over RSSC is mostly a market perception issue. There are many grades of RSSC on the market. The highest grades with a density over $3.10\,\text{g/cm}^3$, and certainly those over $3.12\,\text{g/cm}^3$ are actually ballistically equivalent to DSSC, although much cheaper. Because there have been RSSC products on the market with densities as low as $3.05\,\text{g/cm}^3$ and even below $3.05\,\text{g/cm}^3$, these have created the perception that RSSC is ballistically inferior to DSSC. RSSC with density below $3.05\,\text{g/cm}^3$ is suitable only for NIJ3 body armor. High-grade RSSC with a density over $3.10\,\text{g/cm}^3$ has grown in popularity now that the market has been educated in this regard.

In the 2000s, the two world leaders in RSSC technology were M-Cubed (United States) and MC^2—Modern Ceramics Company (Australia) [505]. The author was appointed Technical Director of MC^2 in 1996,[6] and has a deep knowledge of RSSC as a result, which is a discussion beyond the scope of the present book. MC^2 were also active in alumina ceramics, but it was RSSC that they were renowned for.

11.5.2.4 Boron carbide—B_4C

Benefits of B_4C

At a density of $2.5\,\text{g/cm}^3$ and a similar hardness to SiC, B_4C used to be seen as the ultimate armor ceramic. Only LAS is lighter, but LAS is only NIJ-3. For NIJ-4, B_4C was once seen as the premier choice. Most B_4C is hot pressed, although direct sintered B_4C is becoming increasingly common. Like DSSC and HPSC, B_4C must be sintered at temperatures of 2000–2200°C under vacuum or inert gas. Moreover, when hot pressed, there is the added cost (in terms of throughput, molds, and furnace capital outlay) of the hot-pressing technology. However, the main reason for the high price of B_4C remains the extreme cost of fine-ground B_4C raw materials. B_4C is inherently a much more expensive raw material than SiC, to which must be added the extreme cost of fine grinding. Being a very hard material, B_4C tends to erode the grinding media, and so energy costs in grinding, purification costs for the ground powder, and attrition of the grinding media all contribute to B_4C feedstock being the most expensive of all the armor ceramics by a large margin.

If it was simply a question of cost, given the "money no object" mindset of NATO and other developed nations toward advanced military hardware, B_4C would be a huge seller. However, it is somewhat under a cloud due to the B_4C shock-loading problem. This is the reason why SiC and alumina essentially dominate the ceramic armor industry today. Although even more expensive than RSSC and only marginally ballistically superior, and much heavier than B_4C, SiC-N is the highest performance low-risk armor ceramic for advanced ceramic armor. The cost of SiC and B_4C is also a significant reason why alumina still has a major role in the vehicle armor market.

B_4C remains popular for NIJ-4 body armor, which is the standard military body armor used by NATO and most advanced nations today. The B_4C shock-loading

---

[6] See "About the Author."

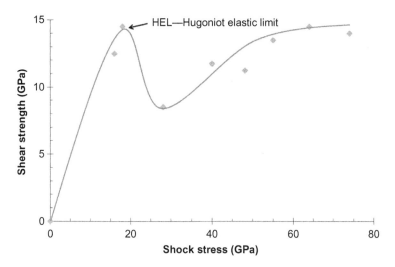

Fig. 11.14 Shear strength correlation with shock stress for B_4C [511].

problem is not fully understood and so the actual risk factor with B_4C shock-loading failure in NIJ4 body armor and in vehicle armor remains an open question. Risk factors aside, B_4C represents the most weight efficient NIJ-4 body armor option, regardless of its extremely high cost, which is much higher than SiC-N.

The B_4C shock-loading problem

It is been well known since the early 2000s that B_4C suffers from a premature failure mechanism and behaves in a more glass-like manner than alumina or SiC under high-velocity impact. This underperformance problem is generally considered an acceptable risk when weight is the highest priority. Essentially, the shear strength of B_4C in the shocked state decreases rapidly above the Hugoniot elastic limit (HEL), as shown in Fig. 11.14. The mechanism is believed to be shear localization under shock loading [511]. However, it is still not well understood. Other theories include phase transition and microcracking mechanisms [512,513]. The discovery of the B_4C shock-loading problem has caused extremely expensive (although lightweight) B_4C to fall from favor. This is primarily because of the fact that the mechanisms and management of this B_4C problem are not well understood, resulting in uncertainty as to risk factor, and risk management. This has significantly enhanced the profile of alumina in the armor market given that it now has only one strong competitor, which is SiC.

11.5.3 Contemporary body armor systems

In ceramic breastplate cost terms, alumina is the base price ($25–$50 for a breastplate depending on purity), RSSC is about $1.5 \times$ the price of premier alumina ($50–$100), SiC-N is about double to triple the premier alumina price ($100–$150), and B_4C more than four times the premier alumina price (over $200). S-Glass cladding is very cheap (similar to an alumina plate), while SpectraShield cladding is very expensive (similar to a B_4C plate cost) but outstandingly lightweight. Kevlar cladding is not much lighter than S-Glass and not much cheaper than SpectraShield and thus rarely used.

These quoted costs are manufacturing costs to the armor fabricator, that is, the wholesale cost of buying ceramic plates from a ceramic fabricator, for the purpose of FRP cladding them and selling them as armor insert plates. The end-user sale price is another matter entirely.

11.5.3.1 NIJ3

NIJ3 is 7.62 assault rifle (non-armor piercing). This is the highest civilian risk level. At NIJ3, ~ 8 mm RHA armor steel is required, which sets the EM reference point at an AD of approximately 65kg/m^2.

A 6 mm ceramic and 6–12 mm FRP backing is typical for NIJ3 ceramic body armor systems. In the author's experience, high-density ($>3.10 \text{g/cm}^3$) RSSC ceramic has been proven to work for NIJ3 at thicknesses down to 4.5 mm with a thicker than normal backing of SpectraShield. One can surmise that probably the same is possible with alumina and SiC. It would be more risky with LAS or B_4C. LAS because of its marginal hardness, and B_4C because of its shock-loading problem (section "The B_4C shock-loading problem").

In budget NIJ3 applications, alumina is mated with fiberglass. In higher performance applications, alumina or RSSC are used in combination with SpectraShield, with more ceramic and less SpectraShield. For "money is no object" applications, such as elite military uses, the SpectraShield is the thickest, the ceramic the thinnest, and SiC-N or B_4C is used. Table 11.3 lists various NIJ3 systems, and associated AD and EM values, as well as an evaluation of the cost and risk factor. Risk mainly relates to the ceramic type, thickness, and suitability for NIJ3. SpectraShield is slightly safer than S-Glass for the same thickness, and significantly lighter.

Table 11.3 NIJ3 ceramic body armor systems

Ceramic type/ thickness (mm)	Backing type/ thickness (mm)	AD (kg/m^2)	EM	Cost	Risk
AD85 Alumina 6	S-Glass 10	37	1.75	V low	Med
AD85 Alumina 6	SpectraShield 10	31	2.1	V low	Low
LAS 6	S-Glass 10	32	2.0	Low	High
SiC-N 6	SpectraShield 10	29	2.25	High	v low
RSSC 6	SpectraShield 10	28.5	2.3	Med	V low
LAS 6	SpectraShield 10	24.5	2.65	Med	Med
SiC-N 4.5	SpectraShield 12	24.5	2.65	High	High
RSSC 4.5	SpectraShield 12	24	2.7	Med	High
B4C 6	SpectraShield 10	24	2.7	V high	Med

Ranked by EM.

NIJ3 has a smorgasboard of options. Alumina is the cheapest option, and it does seem wasteful to combine very cheap AD85 alumina plates with expensive SpectraShield cladding although this is a very low risk and relatively low-cost option. One would not need high alumina at NIJ3, only AD85, as the extra hardness of pure alumina is not necessary and comes with a major weight penalty.

LAS competes very well at NIJ3, but it is a somewhat risky ceramic. The best combination of cost, risk, and EM is RSSC. SiC-N is not a good investment at NIJ3, nor is B_4C even though it does have the best EM/risk performance, and at NIJ3 its shock-loading problem may be less of an issue.

However, on EM/risk criteria, with greatly reduced cost, LAS and RSSC both compete extremely well with B_4C such that the small EM advantage B_4C has is very difficult to justify on cost and risk grounds.

11.5.3.2 NIJ4

NIJ4 is specifically the APM2 7.62/51 NATO hardened steel core armor-piercing projectile. This is the highest body armor threat level in the NIJ standard, and essentially confined to the military realm. Armor-piercing projectiles are extremely rare, almost mythical, in the civilian realm, and when encountered, they are usually armor-piercing handgun projectiles (low velocity) which can probably be stopped by NIJ3 armor, but certainly not by NIJ1/NIJ2/NIJ3A soft body armor. At NIJ4, 12-mm RHA steel armor is required, which sets the EM reference point at an AD of approximately $95 \, kg/m^2$.

Today, most military body armor insert plates around the world are NIJ4, and those that are not are soon to be upgraded from NIJ3. In the 21st century, many armies which had no body armor at all are making the jump to the 21st century in a single leap from nothing to NIJ4.

For NIJ4 performance, hardness needs to be well above 10 GPa, which rules out LAS and AD85. Ceramic thickness needs to be at least 8 mm, sometimes as much as 9 mm ceramic, and SiC-N would be considered lower risk than both RSSC (SiC-N is slightly harder than RSSC) and B_4C (shock-loading problem). High-purity alumina and SiC-N are the best choice at NIJ4 and both would function well at 8 mm, while RSSC and B_4C should ideally be thicker, perhaps 9 mm. For this reason, it is economically viable to combine NIJ4 alumina with SpectraShield (Table 11.4).

Table 11.4 NIJ4 ceramic body armor systems

Ceramic type/ thickness (mm)	Backing type/ thickness (mm)	AD (kg/m²)	EM	Cost	Risk
High Alumina 8	S-Glass 10	49.5	1.9	V low	Low
High Alumina 8	SpectraShield 10	42	2.25	Low	Low
RSSC 9	SpectraShield 10	38	2.5	Med	Med
SiC-N 8	SpectraShield 10	35.5	2.7	High	Low
B_4C 9	SpectraShield 10	32.5	2.9	V high	High

Ranked by EM.

Overall SiC-N/SpectraShield is the best system for NIJ4, even though B_4C is slightly lighter, this weight edge comes with the shock-loading uncertainty for B_4C and thus a raised risk level and a significantly higher cost. Alumina with SpectraShield is a strong competitor on EM/cost criteria, not far behind RSSC, and lower in risk. Thus, alumina/SpectraShield is a very good choice at NIJ4, and SIC-N is the best high-cost choice.

11.6 A brief summary of vehicle armor

Vehicle armor is a large and highly secretive field of research and development, and successful applications. Only a brief summary will be included here. This is primarily because of the high level of military secrecy associated with vehicle armor and the complexity of the systems developed. Moreover, while body armor is a biomedical engineering technology (a primary focus of this book), vehicle armor is more in the mechanical engineering realm, albeit it does save lives which is a worthy endeavor. Therefore, for technical and historical context, its humanitarian relevance, and the fact that alumina is the leading armor ceramic for lightweight multiple-hit vehicle and helicopter armor, an overview of alumina vehicle armor is appropriate.

Ceramic body armor is fundamentally very simple. Only three ceramics are involved (alumina, SiC, and B_4C), and with the B_4C shock-loading problem, this leaves two low-risk ceramics: alumina (low cost) and SiC (high cost, somewhat lighter). The ceramic thickness is 6–9 mm, the backing is around 10 mm of FRP (S-glass, Kevlar, or Dyneema), and a thin spall shield of FRP is on the front. This has been public domain knowledge for some decades and is easily fabricated (using alumina and S-Glass) by anyone with training in fiberglassing and alumina manufacture. SiC manufacture is more expensive and complex as is SpectraShield cladding.

Vehicle armor, on the other hand is highly complex, and a broad range of materials are used: metals, ceramics (many more than the three standard ceramics used in body armor), and FRPs. Many vehicle ceramic armor systems have been developed, trialed, and used with success, most of them veiled by either military secrecy or proprietary knowhow, or both. However, in ceramic quantity terms, the great majority of ceramic vehicle armor is used in lightweight well-known, often public domain, ceramic armor systems for light vehicles and helicopters. Ceramic armor in heavy vehicles (tanks) is a smaller part of the ceramic armor market in volume terms, and is highly technologically advanced and largely secret.

The first recorded use of ceramic armor was introduced by the German army a century ago during World War I due to their discovery that hard-faced enamel coatings on tank armor-enhanced ballistic performance. This gave a small but significant improvement in performance, but the concept was not revisited for many decades. The next appearance of ceramic vehicle armor was an innovation of the US military in the 1950s Korean war involving steel-armor cast around a silicate core [514]. In the 1960s, after the invention of ceramic body armor, a major and secret ceramic vehicle armor program began independently in many leading nations, that bore fruit from the 1980s onward.

The US military did very little with ceramic body armor between 1970 and 1996, which was conceptually very little changed from the 1970 T65-2 plate when it returned as the 1996 ISAPO plate, although SAPI has somewhat evolved since 1996. In contrast, the US military were very hard at work since the 1960s developing advanced ceramic armor systems for aircraft, particularly the floors of helicopters, upgrading the hull of light armored vehicles, and developing extraordinary projectile defeat capabilities for heavy armored vehicles (tanks), and to a limited extent, naval vessel armoring.

Likewise in Europe with Germany (ESK) initially leading the way with B_4C/Kevlar vehicle armor, and the United Kingdom leading the way with the famous Chobham tank armor, developed under conditions of great secrecy in the United Kingdom, and used with great success in the Gulf war.

Vehicle armor systems can be classified according to whether they apply to light vehicles [light-armored vehicle (LAV); armored personnel carrier (APC); helicopter] or heavy vehicles (tanks).

11.6.1 Premier role of alumina in lightweight vehicle armor

Alumina is the dominant ceramic in lightweight vehicle armor. Multiple-hit armor systems are commonly made using mosaics of small tiles, generally hexagonal (Figs. 11.8, 11.10, and 11.15). Alumina is the only armor ceramic that can be cost-effectively mass-produced into small mosaic tiles. While this is theoretically possible with silicon carbide and boron carbide, the costs are prohibitive being orders of magnitude higher. Alumina mosaics can be as cheap as $200 per square meter. For the carbides, it can be in the multiple $1000s order of magnitude per square meter. This is why the majority of aircraft and light vehicle armor systems involve alumina mosaics, and if they are SiC, large tiles are used and multiple-hit performance is poor.

11.6.2 Civilian lightweight vehicle armor

The main application here is in armored money vans, for which alumina ceramics dominate the market. This is generally achieved with alumina/fiberglass panels, to about STANAG-1 level (equivalent to NIJ-3, see Section 11.5.3.1) although alumina with SpectraShield backings are increasingly popular in wealthier markets, such as VIP vehicles, diplomatic vehicles, and private vehicles of the wealthy, or those living in hazardous locations. RSSC has seen a small amount of experimental use in this role also, with SpectraShield backings.

11.6.3 Military lightweight vehicle armor (LAV, APC, helicopter)

Moderate threat levels, as defined by the STANAG (NATO standardization agreement) vehicle armor standard:

- STANAG-1—same as NIJ3 (jacketed assault rifle)
- STANAG-2—same as NIJ4 (7.62/51 APM2 armor piercing: hardened steel core)
- STANAG-3—M993 7.62 extreme armor piercing (tungsten carbide core)
- STANAG-4—14.5 mm (armor piercing)

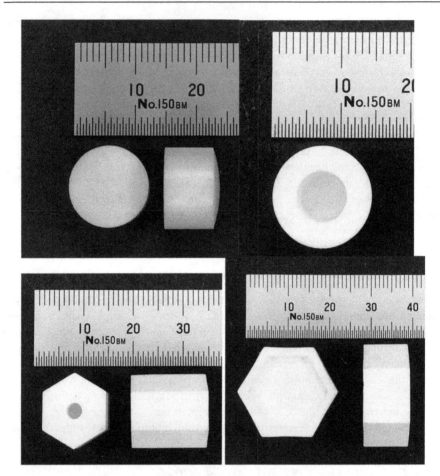

Fig. 11.15 Precision-engineered alumina "tiles" for highly engineered mosaic vehicle armor systems. Note the convex surface on the top left round tile, and the surface step-function profile generated by the raised perimeter of the hexagonal tile. See also Fig. 11.10.

Weight efficiency and multiple-hit capability are critical. Alumina mosaics of hexagonal tiles with convex or stepped frontal profile are a very common strategy here (see Figs. 11.8, 11.10, and 11.15). Indeed mosaics are almost exclusively alumina based. SiC is not easily or affordably produced in small mosaic tiles of high-dimensional tolerances. So, when SiC is used, it is generally used as large tiles, and is therefore much inferior in multiple-hit than alumina (mosaics). Poor multiple-hit does not really compensate for the small weight advantage of SiC. In contrast, alumina mosaic tiles of complex shapes and high-dimensional tolerances can be mass produced very cheaply. B_4C is somewhat under a cloud due to the shock-loading problem (see section "The B_4C shock-loading problem"). This is why alumina dominates multiple-hit lightweight vehicle armor. Dyneema and Kevlar backings are generally used for weight efficiency, rather than metal backings, especially for helicopters.

In short, in practice with lightweight vehicle and helicopter armor: minimal weight means SiC. Maximum multiple-hit means alumina. B_4C is somewhat under a cloud.

It is quite common with light vehicle armor to simply attach the ceramic panels (wrapped in an FRP spall layer) to the metal hull of an APC or LAV using Velcro. This is known as applique armor.

Examples:

Vehicle hull STANAG-2: Conventional armor requires $\sim$12 mm of RHA steel, or $\sim$35 mm of aluminum armor, an areal density of about 95 kg/m². Armor aluminum and armor steel have the same areal density against armor piercing rounds but aluminum is $\sim$3 times thicker for the same protection level. With 99%+alumina (8.5 mm) backed by Dyneema/LDPE (10 mm) the areal density is about 44 kg/m² (EM=2.2). With SiC this reduces slightly to 36 kg/m² (EM=2.6), the cost goes up by a factor of perhaps 3, and the multiple hit is greatly reduced.

Vehicle hull STANAG-4: Conventional armor requires $\sim$40 mm of RHA steel, or $\sim$120 mm of aluminum armor, an areal density of about 320 kg/m². For reasons of military secrecy, the author cannot disclose the alumina (or SiC) tile and backing thicknesses, but suffice to say that the EM values for STANAG-4 ceramic armor are similar to STANAG-2.

Helicopter floor armor: In helicopters, floor armor is much the same as light-vehicle hull armor at STANAG-1 or STANAG-2. For example, the STANAG-2 alumina/SpectraShield helicopter-floor ceramic armor systems that cover a 5 square meter floor area at an areal density of around 40 kg/m², which adds 200 kg to the helicopter weight. 200 kg is the weight of two passengers with kit, which is a small price to pay for protecting all occupants against small-arms ground fire.

Transparent alumina armor (vehicle windows): Single crystal synthetic sapphire is widely used for high-end bullet-proof glass for armored vehicles. It is used as a laminate with a transparent impact-resistant polymer. Submicron transparent polycrystalline alumina is also used in this role as is aluminum oxynitride and magnesium aluminate spinel [515,516].

11.6.4 Heavy armored vehicle (tank)

The main focus has been on using ceramics in defeating two extreme threats, each capable of penetrating in the order of a meter of steel armor under optimal conditions:

APDS: Dense long-rod penetrator, usually depleted uranium $\sim$25 mm diameter $\sim$0.5 m long, velocity up to 1600 m/s, fired from a tank gun of up to 120 mm caliber, with discarding sabot. Also common in small caliber APDS version, such as the 30 mm sabot for the 30 mm gun.

Shaped charge: Explosively ejects a copper metal jet at $\sim$7000 m/s that penetrates metal armor by hydrodynamic flow. Ceramics are very much more difficult to penetrate hydrodynamically. Commonly comes in the form of the high-explosive anti-tank (HEAT projectile). Note: explosive reactive armor (ERA), external metal mesh premature-detonation systems, and other first line of defense systems are also used against shaped charges.

Given that the threat levels are extreme, and weight is not so critical for tanks as it is for body armor, helicopters, and light vehicles, weight efficiency for tank armor is not as important as maximal ballistic performance. Even heavy ceramics like tungsten carbide are sometimes used. Alumina is a significant component in these systems as is SiC. B_4C is a higher risk option due to the B_4C shock-loading problem (see section "The B_4C shock-loading problem"). For this reason, ultra-high-purity alumina and high-performance SiC are very important ceramics usually used in extreme heavy vehicle armor. There are two well-known mechanisms by which ceramics defeat these extreme threats:

Ceramic in metal constraint cells: The important mechanism here is dilatancy (see Section 11.4.2 and Fig. 11.9) and maximum "dwell" time. Dilatancy means that the volume of the ceramic increases when it is pulverized, that is, when a zero-porosity brittle armor ceramic is pulverized, the hard angular particles of the pulverized powder occupy a larger volume than the dense parent ceramic. This effect is optimized with a robust metal spall cover, and when the ceramic is contained in residual compression in a metal constraint cell, and the ceramic has zero porosity (maximal dilatant expansion) as shown in Fig. 11.9. The dilatant expansion opposes penetration, and enhances grinding and erosion. A thick ceramic, and a dilatant response, maximizes the "dwell time," that is, the time during which a projectile penetrator can be eroded inside the pulverized ceramic. This is very important for APDS long-rod penetrators so as to maximize the erosion and grinding down of the fast moving projectile as its shattered remains travel through the abrasive grit of the pulverized ceramic in the fracture conoid, that is, the ceramic grit expands into the shattered remains of the fast moving projectile and thereby impedes its motion and enhances its erosion. This is a much more important mechanism in heavy vehicle armor (rather than light vehicle or body armor) because for the ADPS, the projectile can be very long, the ceramic must be very thick. Robust metal spall layers and containment cells must be used. Dilatancy is still important, although much less so, in body armor where the ceramic is very thin and the projectile very short, and the thin FRP spall layer is not robust. In the early days, metal constraint cells involved casting the ceramic into the metal "box." Hot isostatically pressed metal/ceramic constraint cells are a state-of-the-art in this regard [505].

Ceramic-metal laminates: Ceramic/metal laminates disrupt the copper jet of the shaped charge. Moreover, while penetration by hydrodynamic flow of the armor by the $\sim$7000 m/s copper jet is possible through RHA steel armor, it is problematic through ceramic armor. Glass was used in the early ceramic/metal armor systems for shaped charges. Today, advanced ceramics such as alumina are more common.

Other secret and proprietary constructs involving ceramics are also used, which cannot be disclosed here.

The UN coalition tanks in the 1990/1991 gulf war had ceramic armor. Reportedly not a single ceramic-armored tank of the UN forces in the gulf war was destroyed by enemy fire. Against the munitions of Iraq, they were "invincible." Ceramic-based heavy vehicle armor was seen as nothing short of miraculous when first deployed in the Gulf war. Unfortunately, this scenario simply precipitated a further arms race between increasingly potent munitions and increasingly advanced armor.

11.7 Notable armor ceramic companies

11.7.1 Alumina armor ceramics

- CoorsTek (United States). The world's leading pioneer and current leader in alumina armor ceramics. Also specializes in many other armor ceramics.
- CUMI Murugappa. One of the world's leading manufacturers of high-quality low-cost alumina armor ceramics. The factory in India operates by German specifications, and

manufactures a wide range of alumina ceramics for both industrial applications and armor including a wide range of small alumina tiles for multihit alumina mosaic armor.
- Kyocera (Japan). Alumina armor ceramics are among their huge ceramic product portfolio.
- Morgan Advanced Materials. Alumina armor ceramics and nonoxide armor ceramics are among their huge ceramic product portfolio.
- CeramTec (Germany/United Kingdom). Alumina armor ceramics are among their huge ceramic product portfolio.

Apart from the above mega-companies, a great deal of alumina armor ceramics production now takes place in a plethora of SME ceramic companies around the world, many of which are in China. These supply the civilian sector in both western nations and developing countries, and the military sector of developing countries.

11.7.2 Other armor ceramics

- Saint Gobain (France). Formerly a pioneer in alumina ceramics. Now specializes in SiC, and to a lesser extend B_4C.
- 3M (ESK) (Germany/United States). World leader in nonoxide armor ceramics.
- ADA (Australia). Developed proprietary B_4C technology in the 2000s.
- BAE Advanced Materials. Have been involved in many armor ceramics over the years.
- Ingenieurbüro Deisenroth (IBD) (Germany). Large ceramic vehicle armor company with a global footprint and trademarked product MEXAS™.

Many SME companies also operate in this space globally.

11.7.3 Pioneer armor ceramic companies: Acquired/onsold

There has been a major trend toward acquisitions and consolidation of almost all pioneer armor ceramic companies, as the following nonexhaustive list demonstrates.

- Norton (United States). Produced the original nonoxide armor ceramic of the 1960s. Acquired by Ceradyne.
- Cercom (United States). Incorporated 1985. Hot pressed B_4C, TiB2, SiC, WC. Acquired by BAE.
- Ceradyne (United States). Founded late 1960s. Acquired Norton. In 1971 began producing B_4C ceramic armor using ESK's B_4C powders. RSSC B_4C TiB2 in the 1980s. Acquired by 3M.
- Lanxide (United States). Founded 1986. Developed novel DIMOX technology (directed metal oxidation) and spawned Ceramic Protection Corporation (CPC) and M-Cubed before fading from the scene in the 1990s.
- CPC (Canada). From the 1990s until the end of the 2000s, this was a leading global manufacturer of alumina armor ceramics, and also an armor fabricator fabricating the actual FRP-backed end-user body armor breastplates. Shutdown in the mid-2000s.
- Simula (United States). Body armor, helicopter seats, applique armor, and aircraft armor. Acquired by Armor Holdings in 2003, which was acquired by BAE Systems in 2007.
- United Defense Limited Partnership (United States). One of the world's largest armored vehicle manufacturers, including tanks with ceramic armor. Acquired by BAE Systems Land Armaments.

- General Motors Defense Systems (United States). Numerous ceramic-armored vehicles. Acquired by General Dynamics in 2003.
- Modern Ceramics Company (MC^2)[7] (Australia). Founded in 1996, in the 2000s it was one of the world's two leading RSSC armor ceramic manufacturers [505]. It was restructured and reconstituted as Military Ceramics Corporation MCC in 2009.
- Military Ceramics Corporation (MCC) (Australia). Continued the activities of MC^2 after 2009. It went offshore in 2012.
- M Cubed Technologies, Inc. (United States). Founded in 1993 as a subsidiary of Lanxide Corporation, and was acquired by II–VI Incorporated in 2012. Best known as the leading manufacturer of RSSC SiC in the United States, and with MC2 one of the world's two RSSC leaders [505].

---

[7] The author was a Director of MC2 from 1996 onward, see "About the Author."

Alumina as a wear-resistant industrial ceramic 12

Alumina is the most widely used wear-resistant ceramic, due to its unbeatable combination of low cost, high wear resistance, and high corrosion resistance. For example, in the wear-resistant ceramics realm, silicon carbide is about three times the cost of alumina and tungsten carbide about four times the cost. Alumina is tougher than silicon carbide but less tough than tungsten carbide. For most applications discussed in this chapter, the toughness of alumina is more than sufficient. Tungsten carbide is used in expensive niche applications where toughness is critical. Moreover, as discussed in Chapter 4, zirconia-toughened alumina (ZTA) rivals the toughness of tungsten carbide, at a much lower cost than tungsten carbide, but a higher cost than pure alumina. The primary commercial application of ZTA is in orthopedics although it is beginning to see some use in dentistry and industrial wear-resistant applications. In time ZTA may significantly supplant tungsten carbide in the wear-resistant industry, but it will never supplant pure alumina, which dominates the wear-resistant industry because of its unique combination of low cost and high performance. This chapter begins with a detailed treatise on the tribology of alumina. This is followed by a case study based on the industrial alumina technology of Taylor Ceramic Engineering (alumina specialty manufacturer for over 40 years). Alumina industrial wear and corrosion resistant ceramics service a diversity of industries, particularly mining, but also including textiles, wire drawing, food and beverage, papermaking, heavy clayware industry, pumps, seals and general engineering.

Alumina has outstanding resistance to wet-chemical corrosion: acidic, caustic, and saline. This has been known since the early 20th century when it was demonstrated that concentrated sulfuric acid, hydrochloric acid, nitric acid, phosphoric acid, and 20% sodium hydroxide all dissolved less than 0.02% of a 30×35 mm alumina crucible within 6h [517]. However, even alumina has its limits. As Chapter 3 shows, aluminum hydroxide dissolves above 150°C in concentrated sodium hydroxide during Bayer refining. However, Bayer refining is an extreme corrosion situation. In practice, alumina has a very important function in containment in corrosive industrial environments, acidic, caustic, and saline, but its role in wear resistance is its central role. Its corrosion resistance is merely an added bonus. The highest corrosion resistance alumina needs to be above 99% purity, because silicate impurities (the most common impurities) dramatically reduce alumina corrosion resistance [518].

The use of wear-resistant ceramics is a highly evolved discipline today. In the mining industry, the largest user of wear-resistant ceramics, the essential driver for wear-resistant ceramics usage is the enormous cost of production downtime, which can run into millions of dollars a day. Therefore, the engineering operating principles are:

(1) Use the cheapest ceramic that will do the job.
(2) Design the system so that the wear-resistant lining throughout has the same service life. This means in high wear areas use the highest wear-resistant ceramics, in moderate wear and low wear areas use moderate and low wear ceramics respectively. This means a single scheduled maintenance event can service the entire system.
(3) If corrosiveness is a factor, use corrosion-resistant wear-resistant ceramics.

Alumina is capable of servicing this whole spectrum, as it is a cheap ceramic that is highly corrosion resistant and can be made in low wear grades (85%) up to ultra-high-wear grades (99.9% fine grained). The one exception is areas of high impact, where tungsten carbide or ceramic-coated metals are required. Silicon carbide is quite commonly used at the top end of the wear resistance range, but its high cost limits this use to niche applications.

Alumina wear-resistant ceramics are used in numerous *industrial wear-resistant applications* of which the following list gives some indication of the diversity of uses today:

Wear and corrosion resistance uses for alumina

- Hydrocyclone components and linings (mineral processing)
- Spigots (wear and corrosion)
- Straight and curved piping (wear and corrosion)
- Reducers (wear and corrosion)
- Orifice and baffle plates (wear, corrosion, papermaking)
- Dart valve plugs and seats (wear and corrosion)
- Tiles for ore chutes (standard, engineered and weldable)
- Knife-edge blades (conveyors, wine industry)
- Mill linings and milling media (wear and corrosion)
- Thread guides (textiles)
- Wire drawing step cones and dies (wire)
- Brick and heavy claywear (brick cores, sleeves, die boxes & shaper caps)
- Augers and trough liners (numerous industries)
- Nozzles (wear and corrosion)
- Water-faucet valves
- Rotary seals
- Bearings and sleeves

This chapter concerns the wide range of uses for alumina for wear-resistant linings and components, and also corrosion resistant linings and components (chemical processing, acids, etc.), and in some cases the application requires both wear resistance and corrosion resistance.

Electrical applications, of alumina are discussed in Chapters 13 and 14. Refractory and niche industrial applications are discussed in Chapter 15. Biomedical applications were discussed in Chapters 5–11.

12.1 Tribology

Tribology is the study of friction, wear, and lubrication. A discussion of the tribology of materials, and specifically the tribology of alumina ceramics, is essential for an understanding of the reasons underlying alumina's role as the world's leading general-purpose industrial wear-resistant ceramic. The literature does not contain many published papers with actual test data from wear testing of alumina. In the mid-to-late 20th century, the era of alumina science, a relatively small number of papers were published on ceramic wear resistance, most of them on micro-indentation, numerical modeling, simulation, and other theoretical studies, a small number on actual laboratory wear testing, and a smaller number still on alumina laboratory wear testing.

Wear testing in the laboratory is conducted in a number of ways, including:

- Sliding abrasion simulation: pin-on-disk, pin-on wheel
- Slurry erosion: rotating pin in a slurry
- Impingement: grit blasting

Wear tests attempt to simulate the complex reality of the industrial wear environment, to varying degrees of success. These tests are more meaningful in giving relative wear performance than quantitative absolute wear performance. Therefore, the general principle followed is that a control material of standard formulation should be tested alongside the test material so that the relative wear performance of a test material can be meaningfully evaluated. Ninety-six percent sintered alumina is commonly used as the control ceramic.

Among the most important papers published on laboratory wear testing of alumina are the dry pin-on-disk study of Moore and King [519], the pin-on-disk alumina study by Gee and Almond [520], and the dry-impingement alumina study of Wiederhorn and Hockey [521]. However, wear-resistant research is extensive in the industrial domain. Company archives of the ceramic manufacturers hold a wealth of data on wear testing. Most ceramic manufacturers who supply wear-resistant ceramics have extensive in-house databases on wear testing that underpins their R&D, manufacturing, and sales strengths.

The author has been involved in industrial wear-resistant applications of ceramics, including laboratory wear testing in various forms, since 1996.[1] Therefore, this chapter will also provide insights from the author's own experience in this field. Moreover, this chapter will also involve a case study from the leading alumina manufacturer in the southern hemisphere [Taylor Ceramic Engineering (TCE)], a company that has specialized in wear-resistant alumina ceramics for four decades (profiled in Section 1.4.2.5).

Table 12.1 compiles a nonexhaustive list of the tribological parameters of relevance to the commercial application of wear-resistant materials.

---

[1] See "About the Author."

Table 12.1 Tribological parameters for wear-resistant materials applications

Wear-resistant material—Property	Requirement for low wear rate
Hardness	High
Toughness	High
Strength	High
Porosity	Low
Coefficient of friction	Low
Grainsize	Small
Corrosion resistance	High
Environment property	**Requirement for low wear rate**
Applied Pressure	Low
Chemical environment	Inert
Velocity of Movement	Low
Lubrication	Present
Abrasive material—Property	**Requirement for low wear rate**
Abrasive hardness	Low
Abrasive size	Small
Abrasive shape	Rounded
Abrasive particle movement	Rolling

12.1.1 Modes of wear

There are many different modes of wear that can be encountered in real-world scenarios. The following nonexhaustive list gives some indication of the range of wear scenarios that can be encountered:

Sliding wear:
- Sliding abrasion between surfaces
- Fixed abrasive particles, for example, grinding using a grinding wheel
- Rolling abrasive particles, for example, rounded grit particles between bearing surfaces

Impact wear:
- Micro-impact (impingement)—impact by small hard objects, for example, ore particles in a moving slurry
- Macro-impact, that is, impact by large objects such as large chunks of ore

Other types of wear:
- Adhesive wear, that is, when the "abraded" material is removed by adhering to the "abrading" surface. This occurs in the case where the adhesive force between two articulating surfaces exceeds the inherent material properties of either material.
- Chemical wear, either as pure corrosion, or as one or more of the above mechanisms corrosion enhanced.

- Third-body wear. This is when a hard particle becomes embedded between two articulating surfaces. This is a common scenario when a ceramic is articulating against another surface, and a fragment of the ceramic spalls off its surface and becomes lodged between the articulating metal and ceramic surfaces and plows the surfaces of the ceramic and its articulation partner.

The above mechanisms apply to wear-resistant applications of alumina (mining industry, textile industry, paper industry, and other industrial wear applications) involving sliding abrasion of alumina against various surfaces or impingement onto alumina of mineral particles of significantly lower hardness than alumina. It also applies to the use of alumina as cutting tools for machining purposes, as a high-load sliding-abrasion scenario. It also applies to the machining of alumina, which is generally done with diamond-embedded grinding wheels (small sharp diamond abrasive particles), and the polishing of alumina, which is generally done with diamond polishing pads. It also applies to the sandblasting of alumina (air-entrained impingement) and water-jet cutting (water-entrained impingement), both of which commonly involve impingement of SiC particles ejected from boron carbide nozzles at high velocity: sandblasting involves air-entrained impingement; water-jet cutting involves water-entrained impingement.

Not all of these mechanisms are equally important to all classes of materials. Adhesive wear is most common with soft polymer surfaces. Corrosive wear is most common with metals, and only encountered with ceramics under extreme conditions such as highly acidic, or highly caustic environments. Ceramics cannot be used in situations of macro-impact due to their low toughness. Of the above list, the most important types of wear of industrial relevance to wear-resistant ceramics such as alumina are:

- sliding wear
- impingement wear

These are shown schematically in Fig. 12.1.

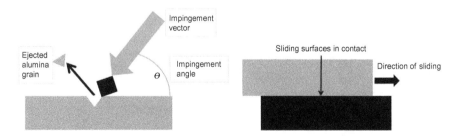

Fig. 12.1 Left: Impingement wear, with incoming high-velocity particle (black), entrained in a high velocity gas or liquid, which impacts on the surface of the alumina ceramic (gray) with sufficient energy to eject a grain (grain pop out) or spall particle of the alumina into the entraining liquid or gas. Impingement can also cause material ejection by localized fracture, or simply cause localized microcracking. Note the impingement angle (θ), which is the angle of the velocity vector, relative to the alumina ceramic surface, from 0° to 90° (θ=45° in this example). Right: Sliding wear between the surface of an alumina ceramic (gray) and the opposing surface (black) which may or may not be alumina.

12.1.1.1 Impingement wear

Impingement wear is a common mining and mineral-processing mode of wear. Examples of dry impingement include sand-blasting and fan-blades extracting dust. Wet impingement generally involves mineral slurry processing, when mineral slurry flows over the surface of alumina, and water-jet cutting. For example slurry flows over a wear-resistant tile or through a wear-resistant pipe, or as the slurry vortex in a hydrocyclone. The impingement angle, as shown in Fig. 12.2, can vary between from 0° and 90°. Low-angle impingement occurs where the slurry flow is parallel or nearly parallel to the tile surface, as for example in a pipe transporting the mineral slurry. Moderate-angle impingement occurs at pipe bends, in a hydrocyclone inlet/vortex finder or spigot, and other situations where a significant change in the flow direction of the slurry occurs. High-angle impingement occurs at a pipe elbow, for example a 90° elbow, or a pipe T-Junction, or with a nozzle or pipe inlet directing the slurry flow perpendicularly, or nearly perpendicularly into the alumina surface. In general, wear is more severe as impingement angle increases from 0° to 90°.

Impingement wear is a complex scenario, depending on many material properties, not just hardness, but also toughness, grain size, and other microstructural parameters. For this reason, different wear-resistant materials have different wear/impingement-angle correlations. Fig. 12.3 shows the relative impingement wear rate for alumina of varying purity levels. The sigmoidal correlation holds true for the entire purity range of industrial alumina, from 85% to 100%, the variation is in the absolute value of wear for a given impingement angle. Impingement also depends on the mineral slurry itself, for example, particle size, particle hardness, slurry flow rate.

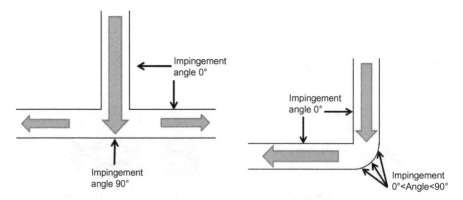

Fig. 12.2 Left: Pipe T-Junction showing the 90° impingement angle region. There is significant turbulence around this region where the impingement is somewhat chaotic. Further from the T-junction, where lamellar flow occurs in the pipe, impingement angle is 0°. Right: Pipe Elbow Junction showing the region where 0° < impingement angle < 90°. There is significant turbulence around this region where the impingement is somewhat chaotic. Further from the Elbow junction, where lamellar flow occurs in the pipe, impingement angle is very low approaching 0°.

Alumina as a wear-resistant industrial ceramic

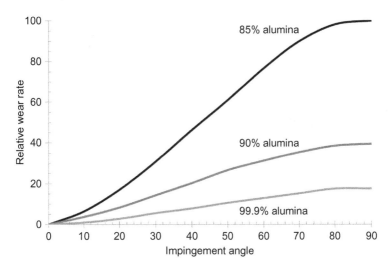

Fig. 12.3 Impingement wear of alumina: relative wear rate as a function of impingement angle. Shown for high-purity, medium-purity, and low-purity alumina ceramics.

12.1.1.2 Sliding wear

Cost-effective performance in sliding wear resistance is another reason why alumina is the most widely used industrial wear-resistant ceramic. This can be clearly seen in a comprehensive study of sliding wear of alumina, represented in Figs. 12.4–12.6. These figures involve analysis of the data from Moore and King [519], one of the few comprehensive studies in the literature on alumina ceramic sliding-wear resistance. The study involves comparison of a number of grades of alumina and a wide range of competitor wear-resistant materials. This study involved dry pin-on-disk tests of ceramic pins against disks of either 384 µm flint abrasive, or 84 µm SiC abrasive. The pin on disk test is one of the standard tests for measuring abrasive wear. A cylinder (pin) of test material is pressed against a spinning disk. The "pin" has a flat base, which is abraded while parallel to the surface of the spinning disk. The following test parameters are pertinent:

- composition of the test ceramic "pin"
- composition of the abrasive surface of the "disk"
- surface roughness or grit size of the abrasive surface of the "disk"
- speed of movement of the disk at the point of contact with the ceramic
- test conditions: dry or wet, that is lubricant supplied to the disk surface
- load (in Newtons) of the test "pin" as it presses against the spinning disk
- the loss rate of material from the "pin," measured as quantity of material per unit time

From these parameters, a measure of the wear rate can be derived. This test, and others like it, such as the pin on wheel test, attempt to mimic real-world sliding abrasion scenarios. However, real-world conditions are very difficult to quantify, let alone replicate, in laboratory tests. Therefore, such tests are most meaningful when used to

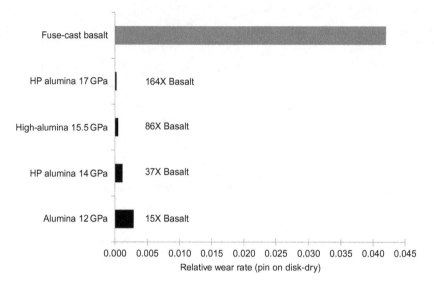

Fig. 12.4 Relative wear rate for sliding wear (pin-on-disk dry) of four different grades of alumina with differing hardness, as well as fuse-cast basalt for comparison. The "pin" in the pin-on-disk test is the test ceramic and the spinning "disk" is the abrasive material (84 μm SiC). Microhardness of each alumina type is listed in GPa.
Data from [519].

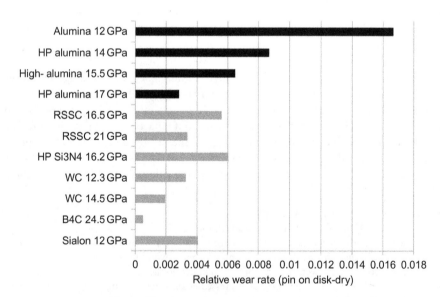

Fig. 12.5 Sliding abrasion wear resistance against a SiC (84 μm) surface for alumina ceramics of four different hardness values, and a range of high-cost alternative wear-resistant ceramics. The "pin" in the pin-on-disk test is the test ceramic and the "disk" is the abrasive material (84 μm SiC). Microhardness of each ceramic type is listed in GPa.
Data from M.A. Moore, F.S. King, Abrasive wear of brittle solids, in: Proceedings of the International Conference of the Wear of Materials, American Society of Mechanical Engineers, Dearborn Michigan, ASME, USA, 1979, pp. 275–285.

Alumina as a wear-resistant industrial ceramic 377

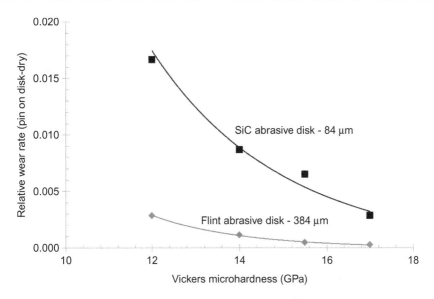

Fig. 12.6 Sliding abrasion wear rate (pin-on-disk dry) as a function of Vickers microhardness, for alumina ceramics of varying hardness. This is shown for a hard abrasive surface (SiC) and for an abrasive surface of comparable hardness to most mineral processing situations (flint). The "pin" in the pin-on-disk test is the test ceramic and the rotating "disk" is the abrasive material (84 μm SiC).
Data from M.A. Moore, F.S. King, Abrasive wear of brittle solids, in: Proceedings of the International Conference of the Wear of Materials, American Society of Mechanical Engineers, Dearborn Michigan, ASME, USA, 1979, pp. 275–285.

compare different materials under identical test conditions. This is precisely what the Moore and King [519] study does, comparing four grades of alumina with a number of other wear-resistant materials, testing pins of these materials against disks of both flint and SiC. The flint is the most representative of real-world mineral-processing wear situations since flint is comparable in abrasiveness to most industrial ores. The significance of the SiC-disk data is that it demonstrates the difference in wear rate against flint and the much harder SiC abrasive.

In Fig. 12.4, alumina is compared to fuse-cast basalt, which is a very commonly used low-cost wear-resistant ceramic for wear-resistant linings in the mineral processing industry. The data show that the wear resistance of alumina is one or two orders of magnitude higher than basalt, ranging from 15 to 164 times more wear resistant depending on the hardness of the alumina.

In Fig. 12.5, the sliding-abrasion wear resistance against a SiC surface is shown for a range of alumina ceramics of different hardness values, as well as a range of much higher-cost alternative wear-resistant materials: RSSC (reaction-sintered silicon carbide), tungsten carbide, sialon, and boron carbide. The highest hardness alumina (17 GPa Vickers) has much the same wear resistance of RSSC, tungsten carbide, and sialon. Given that silicon carbide is about three times the cost of alumina, and tungsten carbide four times the cost, Fig. 12.6 demonstrates why alumina is considered the best value for money in wear-resistant applications. Only boron carbide is significantly higher in wear resistance: approximately five times more wear resistant than the hardest alumina grade. Boron carbide is also by far the most expensive wear-resistant ceramic.

Tungsten carbide (WC) is a very expensive cermet, approximately four times the cost of alumina. Being a cermet, it has a higher toughness (6–13 MPa m$^{1/2}$ depending on metal content) than alumina (4.1 MPa m$^{1/2}$) [522], making it the essential choice in high-impact high-wear applications, such as excavator and jaw crusher liners. Hardness decreases and toughness increases with increasing metal content in tungsten carbide (tungsten carbide is a cermet). So the toughest tungsten carbide grades have the lowest hardness and vice versa. The wear resistance of the highest grades of alumina is comparable to tungsten carbide, as shown in Fig. 12.5.

Recently invented ZTA is just beginning to make an appearance in the industrial wear resistance realm, which is a realm that is very slow to adopt new technology. The toughness of (ZTA) 7 MPa m$^{1/2}$ is comparable in toughness to high hardness/low-metal tungsten carbide. ZTA is slightly more expensive than pure alumina, but much cheaper than tungsten carbide and so it may begin to cut into the tungsten carbide market sector in the future. There are no useful comparative wear resistance studies for ZTA of the type shown in Fig. 12.5 and so this expectation is not yet backed up by a body of published wear-resistance data.

The other ceramics have their niche applications, which are beyond the scope of this chapter. However, on the whole, the mining and mineral processing industries are dominated by alumina, for general wear applications, and tungsten carbide or ceramic-coated metals for high-impact/high-wear situations. Fuse-cast basalt is common in low-wear low-cost applications, as is white cast iron.

12.1.2 Material properties and wear resistance

All material properties have relevance to wear, but the most important of these for industrial wear-resistant ceramics, are those itemized in Table 12.1. That is:

- Hardness
- Toughness
- Strength
- Porosity
- Grain size
- Chemical inertness
- Corrosion resistance
- Thermal conductivity
- Chemisorption

12.1.2.1 Hardness

As discussed in Chapter 4, hardness is essentially a measure of the resistance of a material to penetration by another material. Thus it is a measure of resistance to surface pressure and has the SI unit GPa. Hardness in ceramics is commonly measured as microhardness, by indentation with a diamond indenter, either spherical (Rockwell), square pyramid (Vickers) or rhombohedral pyramid (Knoop). Indentation is done at a precise metered load in Newtons, producing a plastic indenter impression in the test specimen surface, with micron dimensions. Hardness is then determined by measuring, with a microscope and scale bar, the dimensions of the plastic impression formed in the test material surface at the measured load (typically 5–100 g), where hardness (pressure) is equal to (test calibration constant) × (load/area). For example, Vickers hardness is calculated in accordance with ASTM E384 [136]:

$$Hv = 1854.4 \left(\frac{P}{d^2}\right)$$

where Vickers hardness is Hv in GPa, P is load P in Newtons, and d is the diagonal dimension of the indentation in microns. See Fig. 12.7 for the geometry of this test.

Diamond is used as the indenter since it is the hardest known material, and therefore there is no risk that the indenter will be deformed or distorted by the material it is penetrating, although there is always the risk of chipping of the indenter tip. Therefore, the indentation thus produced is a true reflection of the plastic deformation of the test material surface at the measured load. At the low loads involved, the indentations are measured in microns, and commonly microhardness indentations test the properties of a single grain of material.

The association between hardness and wear resistance is so obvious that it is common for technologists to assume that hardness is the sole determinant of wear resistance. Hardness is certainly a very important wear-resistant-determining parameter in general, and for alumina specifically, as the data of Moore and King [519]

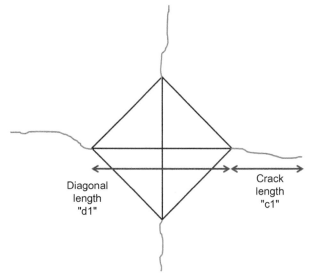

Fig. 12.7 Vickers test geometry. The two diagonals d1 and d2 are measured where "d" is the average of d1 and d2 (for simplicity only d1 is labeled here). The length of each of the four corner cracks c1, c2, c3, and c4 are measured where "c" is the average of c1, c2, c3, and c4 (for simplicity only c1 is labeled here). The higher the indentation load, the larger is the value of both "d" and "c."

demonstrate in Figs. 12.4–12.6. Indeed, hardness is a key parameter for predicting wear resistance for alumina, correlating exponentially (exponential decay), with wear-resistance, approaching an asymptote for high hardness values. This holds true for both hard SiC abrasives and for softer flint abrasives, as shown in Fig. 12.5. The absolute value of the wear is significantly higher with the hard SiC abrasive but the exponential-decay correlation holds true in both scenarios.

Hardness is by no means the sole determinant. Nonetheless, for dense ceramics, such as alumina wear tiles, hardness is probably the most important determinant, with fracture toughness number 2, and grain size number 3 in importance. This correlation, seen in Fig. 12.6 as exponential, was found to be linear in a different study by Rice and Speronello [523] of surface machining of alumina, which found that machining difficulty, which is the reciprocal of wear rate, correlated linearly with hardness.

It is important to note that the hardness of alumina can be enhanced by Cr_2O_3 additions [137,138]. Alumina with 3% Cr_2O_3 is about 10% harder than pure alumina, but hardness decreases at higher Cr_2O_3 content. A formula was proposed to correlate hardness with Cr_2O_3: Hardness $= 1945 + 15.23$ (mol% Cr_2O_3) for a 500 g microindenter load [138].

12.1.2.2 Toughness

It is well known that the high modulus and lack of plastic flow of ceramics means that stresses are concentrated at the point of application, with the result that a localized high surface load, such as an impingement, can cause a localized surface crack to form in the ceramic. However, this problem is mitigated by the fact that most abrading materials in industrial applications are softer than wear-resistant ceramics, which means that quite large loads are required to induce surface cracking. It also goes some way toward explaining why harder particles such as SiC create greater wear than softer particles such as flint, and that sometimes this occurs to an extent that is out of proportion to the difference in particle hardness.

Therefore, in spite of the clear correlation between hardness and wear resistance for alumina shown in Fig. 12.6, a phenomenon that is common to wear-resistant ceramics, hardness alone is not a sufficient indicator of wear resistance. This is because microindenters measure hardness by plastic deformation of a very small surface area. In reality, wear of alumina and other hard ceramics generally occurs by limited local plastic flow and plastic surface plowing at low-energy abrasion, trending toward surface microcracking and surface chipping for high-energy abrasion.

This is exactly the same phenomenon as is seen in ceramic microindentation hardness tests. Microindentations in ceramics commonly include visible cracks emanating from the plastic indentation. The higher the indentation load, the more and longer the cracks. In Vickers and Knoop indentations, these cracks emanate from the corners of the plastic indentation square or rhomboid indentation. This is because, in accordance with the Griffith Crack Theory, the corners represent stress concentrators. Since the crack length correlates with indentation load, it is common to use microindentation as a means of measuring fracture toughness, by correlating average crack length (measuring the four corner crack lengths and taking the average) with load. For example, K_{Ic} fracture toughness is calculated by Vickers indentation as follows [139]:

$$K_{\text{ic}} = A \left(\frac{E}{H}\right)^n \left(\frac{P}{c}\right)^{1.5}$$

E is the young's modulus, H is the Vickers hardness, P is the indentation load, c is the average corner crack length, and A and n are the relevant material constants.

Thus toughness is extremely important in wear resistance, particularly in the higher energy impingement and abrasion situations, which are commonly encountered in mineral processing. A good example is transformation toughened zirconia, which has a lower hardness (12 GPa) and a higher toughness (13 MPa m$^{1/2}$) than alumina (14.5 GPa and 4.5 MPa m$^{1/2}$) [28], and in some applications outperforms alumina in wear resistance. ZTA (Chapter 6) has a similar hardness to alumina (14.5 GPa) but 50% higher fracture toughness (7 MPa m$^{1/2}$) of pure alumina [28], and significantly outperforms pure alumina in wear resistance in higher energy sliding abrasion and impingement environments. The correlation between wear rate and toughness is shown in Fig. 12.8.

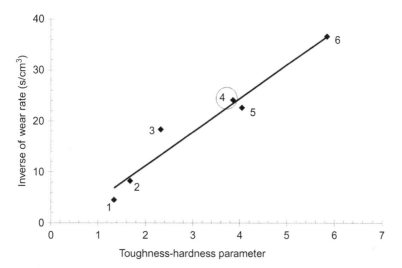

Fig. 12.8 Correlation of wear rate with a toughness-hardness parameter that combines toughness and hardness: $K_c^{2/3} \cdot H^{1/2}$ (units $10^9 \text{N}^{7/6} \text{m}^{-2}$). The higher the inverse wear rate, the lower the wear. The data demonstrate that wear rate decreases with increasing toughness and hardness, for a range of ceramics. This data are from a range of ceramics: $1 = \text{MgO}$; $2 = \text{Spinel}$; $3 = \text{ZrO}_2$; $4 = \text{Al}_2\text{O}_3$; $5 = \text{Si}_3\text{N}_4$; $6 = \text{B}_4\text{C}$. Note that alumina (circled) is below only B$_4$C and sits right on the line of best fit [524].

12.1.2.3 Strength

Higher strength has been shown to enhance wear resistance [523]. However, it must be said that strength of brittle ceramics is a very difficult parameter to precisely quantify, since the measured strength depends on many factors including:

- Surface finish (flaw population and severity).
- The proportion of material surface subject to tensile loads during testing. Measured strength is highest with greatest scatter for three-point bend test (smallest surface area subject to tensile forces), medium with moderate scatter for four-point bend test (moderate surface area subject to tensile forces), and lowest with smallest scatter for hydrostatic loaded hollow cylinders (entire surface subject to tensile forces)
- Porosity
- Grain size

These factors are discussed in detail in Chapter 4. Therefore, strength effects need to be seen within the context of grain size, porosity, and test method used.

12.1.2.4 Porosity

The first and most obvious problem with porosity in wear-resistant ceramics is that high porosity makes a ceramic friable. Any amount of porosity is still a problem, even if it is low. This is because, according to fracture mechanics theory, pores act as flaws, that is, stress concentrators, and therefore toughness is reduced when porosity is present. Moreover, porosity makes a surface more open to destruction at low impingement angles, and indeed all angles, by creating a high angle of attack for an impinging particle, regardless of impingement angle. In other words, low-angle impingement onto a smooth nonporous surface is likely to result in a glancing rebound, whereas a surface pore is at risk of catching a low-angle impinging particle in its lip representing a high impingement angle impact and high probability of crack initiation, fracture, or chipping. The same principle applies with sliding abrasion of a porous ceramic against a high roughness surface, or in third-body wear, that is, when an abrasive particle is trapped between the two sliding surfaces.

12.1.2.5 Grain size

Alpha alumina (corundum), the standard crystalline form in polycrystalline alumina ceramics, is a rhombohedral crystal structure, which is characterized by anisotropic properties in individual grains, for example, linear thermal expansion differs by 10% in the c-axis compared to the basal plane [525]. This anisotropy means that polycrystalline alumina tends to fracture along the grain boundaries. This is believed to be the reason why the wear mechanism for polycrystalline alumina ceramics is generally by grain pull-out or pop out of an intact grain, rather than by losing microparticles from the surface of a grain.

The significance of this grain pop out mechanism is that the larger the grain size of the alumina, the more material is lost for every grain pop out event. A typical alumina has a grain size from a few microns to a few tens of microns in size, so the loss of a

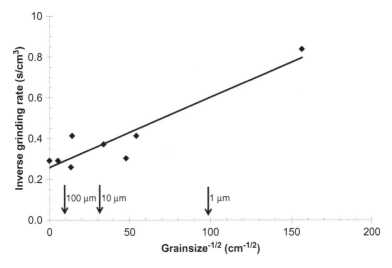

Fig. 12.9 The effects of grain size on grinding rate. The higher the inverse grinding rate, the better the wear resistance. The higher the G-1/2 parameter, the finer the grain size (as annotated) [523].

large grain of around 50 μm leaves a significant surface pore which makes the alumina surface highly susceptible to further wear as per the porosity mechanism outlined in Section 12.1.2.3. Moreover, such a large grain can cause significant third-body wear in a sliding wear situation, such as rotary seals.

Therefore, the finer the grain size of the alumina, the higher the wear resistance as shown in Fig. 12.9.

12.1.2.6 Chemical inertness and corrosion resistance

Corrosive conditions are often found in real-world wear situations, whether acidic, caustic, or saline. A wear-resistant material that is susceptible to corrosion is going to be more affected by surface corrosive attack, which can render the surface more easily worn away. Alumina has outstanding corrosion resistance, which is another of its strengths in wear applications.

There is also evidence that wet environments (water or alcohol) have an effect on the wear behavior of alumina not related to lubrication, but related to an environment effect such as that seen in moisture-enhanced static fatigue [526–528]. Moisture-enhanced static fatigue is discussed in detail in Chapter 4.

12.1.2.7 Thermal conductivity

Ceramics generally have much lower thermal conductivities than metals. Therefore, frictional heat buildup on ceramic surfaces can cause wear. For example in pump rotary seals when there is a risk that the pump could periodically run dry. This can cause localized thermomechanical stress, with the associated risk of thermal-shock-induced microcracking. Surface microcracking is therefore a possible mechanism for wear in rotary seals and other high-friction applications. The solution to this problem is a high level of surface polish, which reduces friction for a given rotation speed. Thus alumina rotary seals are generally produced with a high level of surface polish.

Fereira and Briggs [529] formulated an expression for the modulus of dimensional stability "S" to quantify the resistance of a material against strain induced by thermomechanical stress:

$$S = \frac{Ek}{\alpha}$$

A second parameter of relevance for quantifying thermomechanical stress tolerance is the thermal stress factor "B" formulated by Mayer [530]:

$$B = \frac{\sigma k(1-\upsilon)}{E\alpha}$$

where E is Young's modulus, k is thermal conductivity, α is the thermal expansion coefficient, σ is the fracture strength, and υ is the Poisson's ratio.

Based on the above, an ideal wear-resistant material in situations where frictional heating is a risk, will have a low thermal expansion, a high thermal conductivity, a high strength, and moderate Young's modulus given that a high E enhances dimensional stability but diminishes the thermal stress factor. Compared to metals and cermets, alumina is an unbeatable performer in its combination of high corrosion resistance, high wear resistance, high dimensional stability, with its only weakness being a modest thermal stress factor, though not dramatically inferior to metals and cermets. The thermal conductivity of alumina is high for a ceramic (Chapter 4). These attributes translate to unbeatable performance in rotary seals and plain bearings, as discussed in Sections 12.2.8 and 12.2.9.

12.1.2.8 Chemisorption

The crystal structure of alumina is rhombohedral (Chapter 4), with HCP oxygen layers and aluminum ions on the octahedral sites, which means that the surface of the alumina crystalline structure comprises exposed unsaturated oxygen ions [531]. This means that in a wet environment, alumina is capable of chemisorption of water molecules, or carboxylic acid molecules, to the exposed unsaturated oxygen ions with unsaturated bonds on the surface of the alumina crystal structure, through van der Waals bonding. These bonds are stable up to 200°C and create a protective layer that can theoretically reduce the wear rate to zero for smooth well-aligned surfaces [531]. This represents a significant advantage for alumina wear components in a sliding situation in a wet environment, for example, in bearings or pump components.

12.1.3 Mechanisms of the wear of alumina

As discussed in Section 12.1.2, the rate of wear of alumina decreases in relation to the important material properties as follows:

- Hardness: increasing hardness decreases the rate of wear
- Toughness: increasing toughness decreases the rate of wear
- Strength: increasing strength decreases the rate of wear
- Porosity: decreasing porosity decreases the rate of wear
- Grain size: decreasing grain size decreases the rate of wear

Moreover, the high chemical inertness of alumina assists in wear resistance, and the moderate thermal conductivity of alumina (high for a ceramic) can assist wear resistance in situations where surface heating is involved, such as rotary seals.

The wear mechanism depends very much on the nature of the abrasive process. If it is impingement, then wear is dependent on the size, velocity, and hardness of the impinging particles. Coarse particles and higher velocities tend to cause surface fracture and chipping. In contrast, fine particles at low loads tend to polish the surface by plastic processes. In the case of sliding wear the difference between polishing and chipping depends on the surface roughness, and the hardness of the opposing material. In the case of impingement, the difference between polishing and chipping depends on the size and hardness of the impinging particles. In many ways, impingement mechanisms mimic ballistic impact, as a special case of ballistic impact onto a very thick ceramic by a very small projectile. Ballistic impact mechanisms are described in detail in Section 11.4. In the case of impingement, the wear mechanism from the impact of an impinging particle is identical to the effect of a microindentation. That is, a single impact point, with associated subsurface cracking.

The mechanism for alumina abrasive wear is more complex, but it is essentially the same as that observed in microindentation. Localized micron-scale plastic deformation of a brittle material, with associated cracking, is the standard response of alumina, and ceramics in general, to hardness testing by microindentation. It is the measurement of this residual plastic indentation that enables hardness to be quantified [136]. It is the measurement of the length of the surface cracks emanating from the indentation that allows fracture toughness to be quantified [139]. The specific mechanism for the plastic deformation in alumina was studied in the early era of alumina science by a study of sapphire and determined to be basal slip and prismatic slip of the alumina crystal structure [532]. A number of subsequent alumina science studies have investigated the slip mechanisms and dislocations in the plastic zone of alumina [533–541]. Some of the key findings were as follows:

- Plastically deformed surface layers are produced in alumina surfaces subject to polishing, abrasion, machining, or grinding [540].
- The average slip depth of the plastic surface deformation has an inverse correlation with hardness [540].
- The plastically deformed surface layer for alumina has an average basal slip depth in the order of 10–30 µm [537,540,541].
- Failure occurs when a slip band intersects a grain boundary [538].

Sliding wear occurs in many situations including grinding, machining, polishing, lapping, ore chutes, and rotary seals. Regardless of which real-world application, the sliding wear principle is the same. A sharp particle penetrates the surface, and while penetrated, moves across the surface creating a scratch. Therefore, In the case of sliding wear, the abrasive opposing surface creates a "scratch" in the ceramic surface. A wear scratch involves a surface plastic deformation zone, with associated subsurface cracks, which can result in spallation of surface chips. However, in the case of standard microindentation, it is a point loading (one dimensional), that is, an isolated micro-dent. With a wear scratch, it is a linear plastic surface defect (two dimensional), that is, a linear scratch. A number of papers have been published in the literature exploring this phenomenon. This principle is shown schematically in Fig. 12.7.

This concept of the scratch with subsurface cracks has been experimentally simulated with a microindenter by a number of researchers [524,527,538,542]. This involves a two stage process:

- Descending the microindenter onto the alumina surface as per the customary microindentation process, so as to create a microindentation.
- While leaving the indenter in the descended position, moving the micrometer controls of the microindenter in a controlled way so as to create a linear scratch in an alumina surface at a given load, that is, turning the static "indentation" into a scratch.

This simulates, in a highly controlled laboratory situation, the scratching action of an abrasive surface. The following key findings have arisen from this alumina science in controlled surface scratching of alumina, as shown in Figs. 12.10 and 12.11.

- A linear relationship exists between the depth of damage and the indenter load [527].
- The depth of damage is at least an order of magnitude greater than the depth of the surface scratch itself, for example, if the scratch depth is tens of microns, crack depth is hundreds of microns [527].

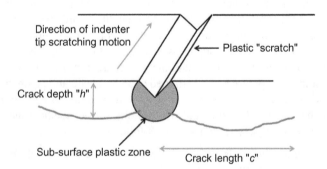

Fig. 12.10 Schematic of the controlled abrasion "scratch" in the surface of a ceramic produced under controlled conditions by the diamond tip of a microindenter under a given load. Note the key features: subsurface plastic zone, subsurface cracks emanating from the plastic zone and curving up toward the surface, characterized by length "c" and depth "h." Spallation of ceramic material corresponds to the volume between the upper surface of the crack and the surface of the ceramic [524].

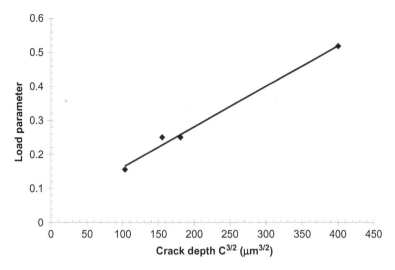

Fig. 12.11 Correlation between crack depth and load parameter from cracks induced in alumina by a sliding indenter. The indenter "load parameter" is $P/(\tan\Psi \cdot \pi^{3/2})$ where P is the indenter load and Ψ is the half angle of the indenter [538].

- The damage is manifested as lateral cracks which emanate from the plastically deformed zone beneath the scratch as shown in Fig. 12.10 [538].
- The depth of these emanating cracks as (crack depth)$^{3/2}$ correlates linearly with indenter load, as shown in Fig. 12.11. [538].
- The actual wear process involves spallation of alumina from the surface as a result of these subsurface cracks curving up to the surface, allowing alumina surface material to spall off as chips, flakes, or as a single grain "pop out."

In summary, loss of material from the surface of abraded (scratched) ceramic is believed to occur by subsurface lateral cracking, with the cracks emanating from the plastic zone beneath the scratch, and ultimately curving up to the surface, resulting in micro-spallation. The volume of ceramic material lost is related to the average depth and length of these cracks. Given that with alumina, cracks tend to follow the grain boundaries, this micro-spallation commonly results in grain pop out, though it can also involve microparticles spalled from a grain.

- The crack depth is believed to relate to the hardness of the alumina and to the indenter load. In real-world situations, this equates to the load exerted by the scratching abrasive particle in the opposing sliding surface.
- For a given load, increasing hardness results in a decrease in the depth of the plastic zone beneath the scratch.
- The crack length is due to the toughness of the alumina.
- For a given load, increasing fracture toughness results in shorter cracks.

Thus, for a given load (whether this is the indenter or an opposing abrasive sliding surface) hardness and toughness control the severity of cracking, and therefore the extent of the damage, and therefore the rate of wear. The highest hardness and highest toughness ceramic has the greatest wear resistance. This principle is not confined merely to alumina, but to wear-resistant ceramics in general.

In real-world sliding wear applications (such as grinding, machining, lapping, polishing, ore chutes, rotary seals), the situation is highly complex as multiple two-dimensional linear scratches are introduced, often overlapping, and with differing associated loads and depths. Nonetheless, the above discussion demonstrates the mechanisms involved, and the importance and role of hardness, toughness, grain size, and porosity in the wear resistance of alumina.

A similar principle applies with impingement wear, but in this case it is multiple one-dimensional surface impacts, not two-dimensional scratches. Nonetheless, the same surface deformation and subsurface cracking mechanisms are involved, with ballistic mechanisms applicable at high impact energies, and the same importance is applicable to hardness, toughness, grain size, and porosity.

12.1.4 Machining of alumina

The machining of alumina is generally done with diamond-embedded grinding wheels. These contain small sharp diamond abrasive particles. The polishing of alumina is generally done with diamond-impregnated soft polishing pads. The above discussion should make it clear that the key issue with the machining of alumina is:

- Load exerted by the machining tool. The higher the load, the more severe the induced subsurface cracking. Therefore, a lower load and a longer machining time is wise.
- The hardness of the alumina. The harder the alumina the more wear resistant it is, and therefore the more resistant it is to machining.
- The toughness of the alumina. The tougher the alumina the more wear resistant it is, and therefore the more resistant it is to machining. More importantly, the tougher the alumina, the more resistant it is to cracking.
- The grain size of the alumina. Finer grain size alumina is more wear resistant and therefore more difficult to machine.
- The porosity of the alumina. The more porosity, the more friable the alumina, and therefore the easier it is to machine, but the more susceptible it is to cracking, chipping, and spalling.

It is wise to polish alumina after machining to remove deep surface scratches which can act as stress concentrators in accordance with the Griffith Crack Theory.

In practice, alumina machining is a very commonly used postproduction step and much in-house expertise exists with alumina manufacturing firms. One of the most important requirements is tight machining tolerances, which means careful attention to the wear of the diamond grinding wheel itself.

Commonly a range of grit-size wheels are used in alumina machining, from the coarse cutting grades, to the finer lapping/finishing grades, leading ultimately to a polishing stage. Cutting grit sizes typically range from about 200 µm down to about 50 µm. These coarse grits produce a blend of chipped surfaces with chips and plowed grooves much like that shown in Fig. 12.10. Lapping/finishing grit sizes then range from 50 µm down to a few microns, producing a smooth flat surface, but not a polished surface as even lapping can result in plowing, subsurface cracking, and grain pop out.

The specific grades and times need to be finely managed, as does the flow-rate of lubricant/coolant. It is common for this process to be automated, which can make it very cost effective. One important difference between machining alumina and machining metals is the cost of the diamond machining components which are more expensive than the WC and other cermet cutters used in conventional metal CNC machining.

The final stage is polishing the machined alumina. This involves micron to submicron diamond grits in soft polishing pads. For the best results, lower loads and longer times are best, making polishing a slow process, but it can be highly automated. The low loads are essential to avoid subsurface cracking, and the consequence of low loads is long polishing times. As discussed in Section 12.1.3, submicron diamond particles at low loads tend to polish the alumina surface by plastic processes. This is because the low energy of the low-load small diamond particles is insufficient to produce cracking, but the local stress induced in the alumina surface is sufficient to cause plastic flow of the alumina surface. This requires submicron diamond grits as coarser grits may be big enough to produce surface loads that will cause cracking.

12.2 Wear-resistant alumina in industry

In keeping with the dual scholarly and commercial-relevance focus of this book, and having outlined the tribology of alumina, it is instructive to look at the actual industrial applications of alumina in the wear-resistance/corrosion resistance realm today. There are many industries that use alumina as a wear resistant and/or corrosion resistant material. The most important of these, in terms of commercial use, are as follows:

(1) Mineral processing and mining
(2) Textile industry
(3) Wire drawing industry
(4) Paper industry
(5) Food industry
(6) Heavy clayware industry
      In addition, there are many industrial processes where ordinary engineering components such as bearings, washers, pipes, rods, seals, augers, and so on, need to have extreme wear resistance. This creates the seventh area of importance.
(7) Alumina as a platform technology for making high-wear bearings, washers, pipes, rods, tubes, seals, and so on.

Undoubtedly the mining and mineral processing industries are the largest consumer of wear-resistant materials worldwide. The world's 40 largest mining companies had a combined 2016 revenue of $500 billion. The world's top three minerals in global mine productions are:

(1) Coal
(2) Iron ore
(3) Bauxite

Notably China, Australia, and India are in the world top four for each of these leading minerals, making them the world's top three mining nations, with China (1) and Australia (2) clearly in the top two positions. Brazil, United States, and India are also significant in the top listing. However, mineral processing is not necessarily done in the country of mining. Usually some is done onsite, but in many cases raw ore is exported and processed in the destination country.

World's top four producers of the top three globally mined minerals

Mineral	Rank 1	Rank 2	Rank 3	Rank 4
Coal	China	USA	India	Australia
Iron ore	China	Australia	Brazil	India
Bauxite	Australia	China	Brazil	India

The $500 billion global mining industry, handling hundreds of millions of tons of abrasive ore a year, therefore has an enormous consumption of wear-resistant materials.

In wear situations involving impact, such as excavation, jaw crushers, and handling of large heavy ore fragments, tungsten carbide is often the only option. However, it is at least four times more expensive than alumina, and only used when absolutely necessary. Ceramic-coated metals are the cheaper alternative to tungsten carbide in impact applications, but they often have a very short service life, limited by coating thickness. White cast iron is also used in high-wear high-impact situations but its wear resistance is modest in comparison.

Much of the high-wear aspects of the mineral processing industry involve processing fast-moving streams of abrasive slurries (impingement). Sliding wear from abrasive ore is also significant. In these two scenarios, alumina is unrivaled in its cost-benefit equation.

In some cases, wear-resistant coatings on metal surfaces suffice, and there is a large industry in flame-spray coating, and to a lesser extent plasma coating, of metal components with wear-resistant coatings for the mining and mineral processing industries. Metal coating processes are expensive and alumina is very price competitive with coated metals. Moreover, in the mining industry, it is very common for the wear rate to be so high that a hard coating has an unviable short service life. In such scenarios, a bulk material of high wear resistance and significant thickness is required. Alumina is the material of choice for such situations as it combines extreme wear resistance with a much lower cost than the very expensive alternatives—silicon carbide, boron carbide, and tungsten carbide. Moreover, alumina is not much more expensive than the lower wear-resistant alternatives such as fuse-cast basalt and white cast iron, but alumina is a great deal more wear resistant, as shown in Fig. 12.4.

In short, the widespread use of alumina in the mining and mineral processing industry is not a coincidence. It is because alumina represents the best value for money industrially, in terms of cost and service life for a wear-resistant material. This is why alumina dominates the industry.

The following case study overviews industrial wear resistance/corrosion resistance, applications for alumina ceramics supplied by TCE.

TCE was founded by ceramic innovators and pioneers David Taylor and Julie Taylor half a century ago in 1967 and is now under the leadership of General Manager Alyssa Taylor. TCE has been the leading manufacturer of advanced alumina ceramics in the southern hemisphere for four decades, specializing in wear-resistant applications since 1982, servicing the huge Australian mining industry, and many other industries. TCE is profiled in Chapter 1, Section 1.4.2.5.

TCE have also been involved in the design and development of alumina ceramics for biomedical applications, both bionics (Chapter 8) and orthopedics (Chapter 6). TCE was the first company to supply armor ceramics (alumina) to the Australian military when Australia made the move to ceramic body armor in the late 1990s (Chapter 11). TCE have also pioneered architectural applications for alumina, specifically high-alumina tiles for cladding the sails of the Sydney Opera House (Chapter 15). Finally, TCE are particularly renowned for not simply providing standard inventory alumina ceramics, but also designing and developing a wide range of custom-manufactured alumina components, driven by user requirements.

A unique aspect of TCE is that they manufacture their own alumina powders, thereby maintaining extremely high and reproducible quality, and purity. They also specialize in very large monolithic alumina ceramics. Because the product range in advanced alumina ceramics of TCE is very extensive and comprehensive, and the range of industries serviced by TCE industrial alumina is very broad, the following case study on the TCE wear-resistant alumina ceramics product range is helpful in giving an instructive A-Z of wear-resistant alumina ceramics in the world today.

The following is a list of industries that TCE currently supply with alumina ceramic components. This can be considered a general representation of the industries worldwide that utilize industrial alumina ceramics.

- Cement
- Chemical manufacturing
- Food and beverage processing
- Microwave
- Mining metals and minerals
- Heavy clay
- Laboratory and scientific
- Oil and gas
- Paper and pulp
- Power generation
- Rail
- Textile
- Waste, wastewater, and recycling
- Welding
- Wine

12.2.1   Alumina for the mining and mineral processing industries

Alumina and other ceramics have been used in wear-resistant linings in the mineral processing industry for over half a century, since the early 1960s. Alumina has been the dominant wear-resistant ceramic used in this field, with fuse-cast basalt, AZS (fuse cast alumina zirconium silicate) silicon carbide (nitride-bonded, RSSC siliconized, and direct sintered) and zirconia also seeing significant use in this area, as well as the carbide cermets, particularly tungsten carbide. In general, basalt and AZS have inferior wear resistance to alumina, with basalt the lowest cost and lowest wear-resistant ceramic. Zirconia and silicon carbide are comparable to alumina but more expensive. The high end of alumina ceramic overlaps with the moderate end of the silicon carbide wear resistance spectrum. The cermets are used where significant impact is involved.

Alumina used in wear-resistant applications spans the whole 85%–99.9% purity range, with anything over 96% considered high performance.

The principal four uses of wear-resistant ceramics in the mineral processing industry, in terms of volume of alumina ceramics used, are:

- Chute linings
- Hydrocyclones
- Pipelines
- Pumps and valves

These applications all involve only low-to-moderate impact and stress loadings, which is ideal for alumina. The severity of the wear environment, and corrosion environment, varies depending on the ore, but is generally also low to moderate.

The largest consumption of ceramic linings in the mineral processing industry is in chute linings. Wear severity is lower for wet sliding wear than dry sliding wear, and the larger the ore particles, the more severe the wear due to impact effects. While fuse cast basalt is commonly used because of its low cost, the overall cost-benefit equation, including production down time for relining, favors alumina. Silicon carbide is used in severe wear situations, but at three times the cost of alumina, the cost-benefit equation again favors alumina. Much the same situation prevails with pipelines. Basalt is commonly used but alumina is dominant because the cost-benefit equation favors alumina.

Basalt is unsuited to hydrocyclones which experience severe impingement wear, and pumps which can experience severe multimode wear. Alumina is the primary liner for hydrocyclones and pumps, and in some cases silicon carbide is used where extreme wear justifies the cost.

12.2.1.1   Alumina-lined hydrocyclones

A hydrocyclone is a high-throughput gravity separation device used for separating slurry particles based on particle weight. For example, particles of similar size but different specific gravity, or particles of different size but identical specific gravity. Cyclones are also commonly used for dewatering of slurries given that they are much faster throughput than the alternatives, such as the traditional filter press. The operating principle of the hydrocyclone is that of a conical vortex-generating chamber. Liquid slurry enters at the top of the conical wall of the hydrocyclone through a vortex

Alumina as a wear-resistant industrial ceramic 393

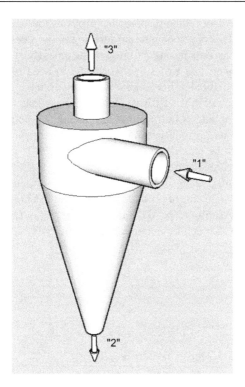

Fig. 12.12 Hydrocyclone schematic. Slurry enters, at a high flow rate, at the "feed" point (1) as shown, into the vortex finder, and forms a powerful vortex inside the conical chamber. The coarse particles settle out and exit at the "underflow" (2) and the fines remain in suspension and exit out the "overflow" (3).
Image Author VanBuren [543]. Reproduced from Wikipedia (Public Domain Image Library).

finder which creates tangential flow and therefore a strong vortex in the hydrocyclone. The slurry in the cyclone spins in a high velocity vortex. The fines capable of remaining in suspension exit out the top central pipe (the overflow), the coarse particles spiral down to fall through the underflow at the narrow vortex base, as shown in Fig. 12.12. Each particular cyclone is designed for a specific separation threshold: the cutoff between oversize and undersize (overweight and underweight) particles. In some cases total solids removal (slurry dewatering) is the objective.

Hydrocyclones are very widely used for high-throughput coarse/fine particle separation in the mineral processing industry using a vortex effect, or simply slurry dewatering. They don't suffer from the rate-limiting steps of the alternatives:

- Sedimentation settling: rate-limiting step is the sedimentation rate.
- Separation by sieving: rate-limiting step is screen clogging and cleaning.
- Filter pressing: a very slow process.

A hydrocyclone can handle very high slurry throughput rates, and often an entire battery of hydrocyclones can be found at a mineral processing facility. The advantage of fast processing by a hydrocyclone is somewhat offset by the associated high impingement wear rate of the hydrocyclone internal surface lining by the fast-moving abrasive slurries. For this reason, wear-resistant linings of the pipes (inlet and overflow), vortex finder, cyclone chamber, and exit spigot (underflow) need to be lined with wear-resistant materials, usually alumina ceramics. Moreover, the impingement angle varies for different areas within the hydrocyclone. For practical servicing reasons, it is desirable that the wear rate is uniform, which means that ceramics of higher wear resistance might be used for high-impingement-angle areas such as the top wall near the vortex finder, and the underflow spigot.

Refurbishing of hydrocyclones represents one of the highest volume uses of alumina in the mineral processing industry. Ideally, it should be done as infrequently as possible due to the huge cost of production downtime in a mineral processing plant, which can run into the $millions per day range in large plants. Cyclone components of the cyclone assembly that can be manufactured in alumina ceramics include:

- Cylindrical and reducing liners
- Inlets
- Outlets
- Spigots
- Inserts
- Upper, mid and lower cone sections
- Vortex finders
- Tiles (ideally curved) for lining the interior wall

Linings of the curved interior wall are a large-volume ceramic requirement given the size of the vortex chamber of an industrial hydrocyclone. This can be either a single monolithic liner (small cyclone), or tiles. If tiles, ideally they should be curved to match the contours of the hydrocyclone interior. When flat tiles are used, this leaves a series of flats radially around the internal surface of the cyclone which interrupts material flow due to changing impingement angle and increases impingement wear on the highest impingement-angle areas of the tiled surfaces. Obviously the spigot and piping of the hydrocyclone also require ceramic linings.

Alumina is the most common material used for hydrocyclone linings. Some examples of this are shown in Figs. 12.13 and 12.14. In rare cases silicon carbide is used, however it is much more expensive than alumina, in the order of three times the cost, depending on the respective grades, for only a marginally improved service life. So silicon carbide is only justifiable in the cost-benefit equation in special cases. It is unwise to use ceramic-coated metals in a hydrocyclone. The wear-resistant coatings are soon gone and then very rapidly the fast-moving slurry erodes the unprotected metal wall of the hydrocyclone.

In general, alumina gives an unbeatable combination of low price and high wear life for hydrocyclones and this represents one of the highest volume uses of alumina ceramics in the world, in tonnage terms, though not in dollar terms. In dollar terms, industrial wear-resistant alumina ceramics are at the bottom end of the alumina price scale with armor ceramics and refractories, where biomedical alumina is at the top end, and electrical and specialty wear components are in the middle.

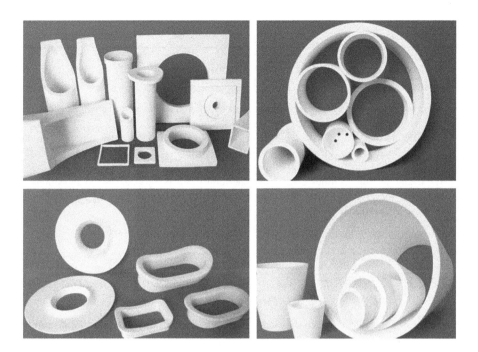

Fig. 12.13 Various alumina ceramic components used for lining hydrocyclones. Note: curved tiles are shown in section "Alumina tiles for ore-chutes".
Produced by Taylor Ceramic Engineering. Image courtesy of Taylor Ceramic Engineering [31].

Fig. 12.14 Freshly refurbished hydrocyclone interior, small, and large examples shown. Note the smaller hydrocyclone on the left has a monolithic liner at the base (the region of highest wear). Wear resistance is significantly enhanced when joint lines are absent, as is the case with a monolithic liner.
Refurbished by Taylor Ceramic Engineering. Image courtesy of Taylor Ceramic Engineering [31].

Hydrocyclone refurbishment is a large industry and commonly happens in one of three ways:

- Alumina parts are purchased from the alumina supplier and delivered to the mine site to use for onsite refurbishment of hydrocyclones.
- The alumina supplier offers a cyclone-refurbishment service. Frequently mine sites do not have the facilities or expertise for cyclone refurbishment. The cyclone is trucked to the cyclone-refurbishment workshop of the alumina supplier.
- OEM engineering facilities offer a cyclone-refurbishment service. The hydrocyclone is trucked to their facility. They order in the alumina parts from an alumina supplier, and refurbish the hydrocyclone (Fig. 12.14).

TCE manufactures alumina parts, both standard and custom, and also does hydrocyclone construction and refurbishment at their manufacturing facility.

12.2.1.2 Other alumina wear-resistant components and plumbing for mineral processing

Commonly associated with hydrocyclones and also found in wear-resistant plumbing for slurry processing in other areas of the mineral processing industry, and also the chemical processing industry, are the following items:

- Distributor spigots, spigots
- Reducers (concentric, eccentric, round square)
- Orifice and baffle plates
- Dart valve plugs and seats
- Mill linings
- Cylinders/tubing (full and partial)
- Curved pipe sections (pipes and chutes)
- Tiles for ore chutes

Alumina distributor spigots, spigots (straight, reducing, and flanged); reducers (concentric, eccentric, round square)

The word "spigot" is commonly used to refer to a flanged conical or cylindrical pipe, used as a join between pipes of similar or dissimilar diameter, and also commonly used at the underflow exit port of a hydrocyclone. Conical pipes, when the flow is toward the narrow end (e.g., underflow spigot in a hydrocyclone), experience much higher impingement erosive wear than parallel pipes, due to a higher impingement angle. This is further exacerbated if slurry, or corrosive liquid, or both, are involved. Moreover, even if flow is toward the larger end, the turbulence generated at the join is much more erosive than for parallel pipes. High-purity alumina is ideal in this role. Some examples are shown in Fig. 12.15.

Similar in concept to the spigot, a reducer is used to join piping of dissimilar diameters and geometries, for example square to round, and diameter changes. Just as for spigots, if the flow is toward the narrow end, erosion rates are much worse than for parallel pipes due to a higher impingement angle. Even if flow is toward the larger end, the turbulence generated at the join is much more erosive than for parallel pipes. Square to round joins complicate the turbulence further.

Fig. 12.15 Left: various spigots. Right: a round (large) to square (small) reducer. Produced by Taylor Ceramic Engineering. Image courtesy of Taylor Ceramic Engineering [31].

Alumina cylinders/tubing (full and partial); curved pipe sections (pipes and chutes)

These are widely used for general plumbing applications in high-wear situations in mineral processing facilities for example the plumbing for hydrocyclones. Also used in other industries in which slurries, or corrosive liquids, or corrosive slurries are running through plumbing, for example, in the food and chemical processing industries. In many cases, the processing conditions may be both high wear and corrosive, for example, acidic chemical or food slurries. Both full cylinders and split cylinders can be produced in alumina. Some examples are shown in Fig. 12.16.

Alumina valves and seats

Liquid control valves come in many forms: butterfly valve, ball and seat valve, disk-valve, piston-sleeve metering valve, and dart valve, to name but a few. Alumina has been used in many industrial valves. Water faucet valves of the standard disk-on-disk configuration are very common and are discussed in Section 12.2.8. Since they share almost all the same features of pump rotary valves.

Dart valve plugs and seats are a fluid-flow-control component. When used in the mineral processing industry, or in other industries where slurries, or corrosive liquids, or corrosive slurries are flowing, these valve/plug systems need to be highly wear resistant, especially the plug which can be particularly exposed to the flow of the erosive/corrosive fluids. An example of a dart valve and plug is shown in Fig. 12.17. Alumina valves are an increasingly common technology in general.

Alumina tiles for ore chutes

Ore chutes (gravity driven) are widely used for moving dry or damp ore at mine sites and mineral processing plants, coal at power stations, and other situations where large quantities of powdered minerals are moved. Obviously, erosive wear is substantial with ore chutes, and as long as the particle size is fine enough that impact is not problematic, alumina is ideal. Impact is inevitable with chutes, which means that thicker tiles are preferred, and if the tiles are thin, they need to be well supported to prevent flexion under impact. In high-impact situations, ceramics are not ideal, and either

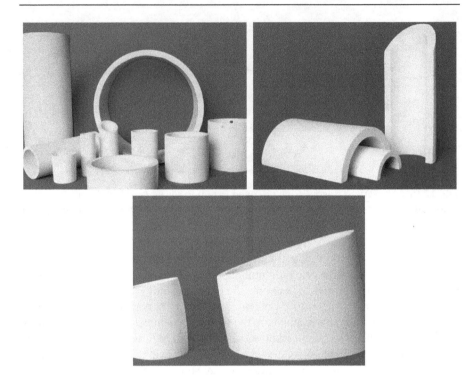

Fig. 12.16 Top left: Alumina tubes of various diameters, lengths and cutoff angles. Top Rights: Alumina split tubes of various diameters, lengths and cutoff angles. Bottom left: Alumina tri-arc linings, pipe, and chute linings.
Produced by Taylor Ceramic Engineering. Image courtesy of Taylor Ceramic Engineering [31].

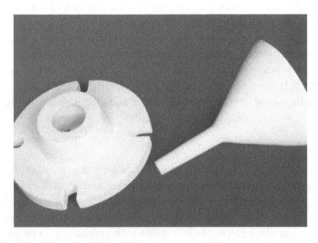

Fig. 12.17 Alumina dart valve plug and seat.
Produced by Taylor Ceramic Engineering. Image courtesy of Taylor Ceramic Engineering [31].

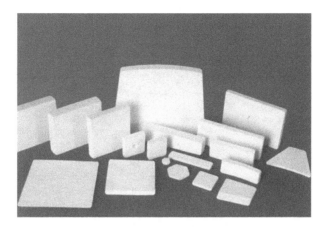

Fig. 12.18 A representation of some of the various flat, curved, plain, and perforated alumina tiles used for chute linings, hydrocyclones, and various other applications. Produced by Taylor Ceramic Engineering. Image courtesy of Taylor Ceramic Engineering [31].

tungsten carbide, a coated metal, or one of the hard metals, such as white cast iron would be required.

Alumina tiles are ideal to line ore chutes. Low cost is the primary driver for chute linings which means fuse-cast basalt is sometimes used in this role, and low-purity alumina is the more cost-effective choice for many chute applications. For long-life chute linings, high purity alumina is the sensible choice.

Alumina wear tiles can also be used in applications outside the mining industry. One important example is slide ways on machinery where a common problem is wear from machining swarf and other machining detritus.

Another common industrial product used in wear-resistant linings is the rubber-backed alumina tile mosaic, involving a thin rubber base to which is bonded a mosaic of small alumina tiles, usually square in shape, generally in the order of centimeters in size. Alumina square mosaic tiles are shown in Chapter 11 in Fig. 11.10A and B. Depending on the mosaic tile spacing, these rubber-backed tile mosaics can be attached not only to flat surfaces but also to curved surfaces. A range of different alumina tiles are shown in Fig. 12.18.

Alumina knife-edge blades

Dry or damp ore at mine sites and mineral processing plants is also commonly moved by motorized conveyor belt. In the case of conveyors, the knife-edge blade is a common system used to clean conveyor belts. Alumina knife edge blade technology is shown in Fig. 12.19. The sharp edge receives a great deal of abrasion and needs to be extremely hard and wear resistant, a role for which alumina is well suited. Alumina can have up to five times the service life of tungsten carbide in this role, and is also significantly lower in cost.

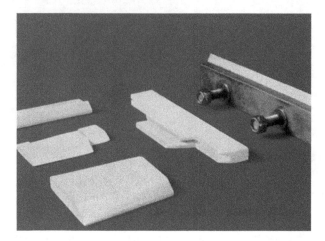

Fig. 12.19 Alumina knife-edge blades. Produced by Taylor Ceramic Engineering (Image courtesy of Taylor Ceramic Engineering) [31].

Knife-edge blades have other industrial uses including slicing and scoring products, and cleaning drum filters. Alumina knife-edge blades also have an important application in the *wine and food industry,* discussed in Section 12.2.5.

Alumina mill linings

Alumina is commonly used as both lining and media (balls, short cylinders, or rods) in ball mills, and other mills such as attrition mills and vibratory mills. Milling is a common part of mineral processing, and also ceramic production plants. Alumina combines three ideal attributes for this role: hardness, which translates into effective milling rates; low contamination (mill pickup) rates; high density, so that the media have minimal buoyancy in water to enable maximum impact on the mill charge during the milling operation.

The primary criteria for mill linings are low cost and low levels of contamination. 85% alumina linings are commonly used in ceramic and glaze ball milling, and in some cases even lower alumina linings are used such as aluminous porcelain. However, where minimal contamination and maximum mill liner life are required, high alumina is the best choice. Some examples of alumina mill linings are shown in Fig. 12.20.

12.2.2 Alumina ceramics for the textile industry

Alumina thread guides in the textile industry was the first industrial wear-resistant application of alumina. This application dates back nearly 70 years to the early 1950s, the very beginning of the alumina commercialization era. Indeed it was the second industrial application for alumina after spark plug insulators.

Fig. 12.20 Alumina mill linings.
Produced by Taylor Ceramic Engineering. Image courtesy of Taylor Ceramic Engineering [31].

Thread guides are therefore an example of alumina as an enabling technology, rather than as merely an enhancing technology. This is because there was such a strong imperative to improve upon the wear rate of existing thread guide technology of the mid-20th century, especially with the introduction in that era of synthetic fiber types.

Thread speeds in the textile industry are commonly 10 m/s and faster, and synthetic threads can be very abrasive, especially at these high velocity feed rates. Moreover, synthetic threads contain abrasive additives such as TiO_2. Vast quantities of thread guides are required in textile manufacturing facilities. Prior to the advent of alumina, thread guides were made of hardened steel, or conventional ceramics such as porcelain, and at standard thread speeds, they had a service life sometimes as little as only a few days, making stable production highly problematic. A worn surface in a thread guide can cause thread breakage. Significant costs are involved in production downtime, in a textile operation, due to a single thread-guide failure. Alumina thread guides have proved capable of providing stable continuous service for a decade or more. Therefore, the benefits of alumina to the textile industry cannot be overemphasized.

The service life of alumina thread guides can be in the order of years. Moreover, the smooth low-friction surface of alumina does not damage the thread, unlike metal thread guides. The friction between the alumina and the thread is dependent on the alumina surface finish both macroscopically and microscopically, both in the as-manufactured state, and after long wear times. Fine-grained alumina is essential which should be macroscopically smooth, but at the microscopic level, smooth well-rounded grains are required (smooth convex surface) so as to minimize the contact area with the thread. This can be achieved by controlled sintering, utilizing surface tension at high temperatures to produce spherical surface grains, with no sharp edges, just microscopic convex bumps, ideal for the low-friction high-speed passage of thread. Proprietary machining and thermal annealing techniques are also used. Corrosion resistance is also important since the threads can contain corrosive liquids.

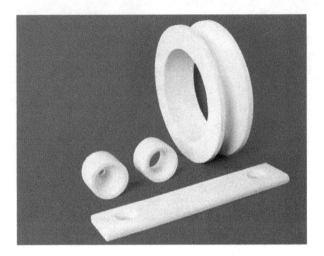

Fig. 12.21 Alumina wire drawing and thread guides. Produced by Taylor Ceramic Engineering. Image courtesy of Taylor Ceramic Engineering [31].

Thread guides are usually open cylinders with internal diameters which can range from less than 1 mm up to hundreds of millimeters. In some cases they are pulley shaped. The wire or thread guide orifice or aperture can be manufactured over a very large range of sizes, from submillimeter to hundreds of millimeters. They can also be precision machined where required. They are also commonly polished for maximum smoothness.

Alumina has proven to be the premier material for thread guides. Indeed there are some textile manufacturing processes that would not be possible without the use of alumina thread guides, where extreme thread velocities, or corrosive conditions, or both, are involved. Alternative wear-resistant ceramics and cermets, almost all of which are more expensive than alumina, have not shown any improvements over alumina in their performance as thread guides. Alumina thread guides are shown in Fig. 12.21.

12.2.3 Alumina ceramics for the wire drawing industry

Wire drawing is one of the earliest industrial applications for alumina ceramics, dating back about 60 years to the early 1960s, and the technology is now highly evolved.

The alumina technology used in the wire drawing industry shares some common elements with thread guides in the textile industry, that is, abrasive filament running through open cylinders and over pulleys. Corrosion can also be an issue from the wire-drawing coolants. As for the textiles industry, the wire drawing industry was one of the first industries to adopt alumina wear-resistant ceramics, in the early 1960s. Alumina has about 10 times the service life of hardened steel in wire drawing. Even longer if corrosive conditions are involved, for example, molten salt-bath processing.

A step cone is a pulley cluster of pulleys of incrementally decreasing pulley size, thus it appears like a stepped cone. A wire drawing die is an orifice through which wire is drawn to simultaneously lengthen it and narrow its diameter. The step cones simultaneously maintain tension and crank the wire through the wire-drawing die. This creates severe abrasion on the tracks of the step cone pulleys, and on the orifice of the drawing die. Drawing speeds can be 50 m/s or higher, causing very high wear rates on the dies and step cones. This causes extreme wear conditions, for which metal step cones are simply inadequate. Moreover, metal step cones can cause adhesive wear of the wire, which degrades it. Alumina step-cones do not cause adhesive wear of the wire, which is an added benefit. This is in addition to the obvious benefit of greatly enhanced service life of alumina step cones compared to metals used in this role.

A moderate coefficient of friction is ideal in step cones so as to maintain the optimal wire tension. Alumina is ideal in this role as it allows less wire slip, and it is particularly effective with thick wires with their high associated mechanical loading.

Alumina is widely used in wire drawing industry, with the more expensive competitor ceramics tetragonal zirconia and silicon carbide also used. Zirconia is only suitable for thin wires. Most commonly the step cones are assembled from alumina pulleys bolted together as a composite assembly with steel, aluminum, or polymer flanges, for cost minimization. Monolithic alumina systems do exist but are very expensive. The other advantage of the modular design is that a single worn pulley can be replaced without needing to replace an entire monolithic system. The smaller pulleys in the step cone tend to have the higher wear rates.

Just as for the thread guides for the textile industry, a wide range of component dimensions, orifice sizes, tapers, and curved surface cross-sections/radii can be engineered in alumina, with precision machining and polishing available if required. Examples are shown in Fig. 12.21.

12.2.4 Alumina ceramics for the paper industry

Papermaking is one of the earliest industrial applications for alumina ceramics, dating back to the mid-1960s, more than half a century ago, and the technology is now highly evolved.

The origin of the extreme wear problem in paper making is similar in principal to the problem in the textile industry. That is, a very fast-moving relatively soft material which can be both abrasive and corrosive and creates extreme wear: thread moving through thread guides at over 10 m/s in the case of textile industry, cellulose pulp flowing at around 20 m/s in the case of the paper industry. In papermaking, extreme wear results from the high cellulose pulp throughput rates, in the order of 20 m/s, and the fact that paper pulp can contain significant amounts of clay (the more clay the more glossy the paper) and hard mineral impurity particles. In addition, corrosion resistance is important as paper pulp suspensions can be acidic or basic. Thus the fast-moving pulp can cause very aggressive wear rates which is why the use of alumina wear-resistant ceramics in paper-making machinery, like in the textile industry, dates back to the early era of the alumina ceramics industry.

Alumina was first used in papermaking machinery around the early 1960s, and since the 1980s has become a standard wear-resistant material in papermaking machinery. The introduction of alumina made a significant impact on the production cost and efficiency of paper making, through enhanced production rates, reduced downtime, and enhanced paper quality. Alumina components used in papermaking include forming boards (segmented or continuous), foils and foil-cleaning devices, and covers for suction boxes. The alumina ceramic supports a high-speed ($\sim$20 m/s) continuous plastic filament screen, and needs a highly polished surface finish so as to minimize screen wear and frictional energy losses, given that the screen is up to 10 m wide.

It is of course not possible for alumina to be produced in 10 m segments, nor for it to be load bearing in this role, so the alumina segments are bonded to a metal or FRP support beam, and diamond ground once bonded, into a flat smooth continuous alumina surface for the length of the support, with appropriate leading-edge chamfering. Generally 99.5% alumina is used, but it can be as low as 96%. Tetragonal zirconia and silicon carbide ceramics are sometimes used for this role, as competitors for alumina, both of which are more expensive than alumina. Recently invented ZTA is also beginning to be used.

Alumina is also used for the slitting and sizing knives, used for cutting the paper into the desired end-product width, at the end of the papermaking process. The clay, hard impurity particles, and corrosive liquid component of pulp cause severe wear conditions for slitter knives. Metal slitters have a very short service life between sharpenings. Alumina has been a major improvement over metals in this role.

TCE supply numerous alumina components to the papermaking industry, including hydrocyclones, bearings, and liners.

Orifice and baffle plates are found in plumbing for liquid and slurry flow in many industries. For example, mineral processing, food and chemical processing, including papermaking. In cases where abrasive slurries, corrosive liquids, or corrosive slurries are flowing, these orifice and baffle plates need to be highly wear resistant. Alumina orifice and baffle plates are shown in Fig. 12.22.

12.2.5 Alumina ceramics for the food industry

The high chemical inertness of alumina, and indeed its biocompatibility, makes it the ideal lining material for food-processing equipment, as alumina does not introduce a "metallic taste," or indeed any taste to food products, unlike metal containment systems. This is a principle everyone is familiar with. Food is almost always served in ceramic bowls, plates and cups, not metal. Moreover, alumina is much more wear and corrosion resistant than alternative ceramic linings, such as glass or porcelain, making alumina an ideal long-life lining material for food-processing systems. Thus alumina is ideal for many applications in the food industry, for example, in processing equipment for working with acidic juices or food slurries, and for salt grinders.

A significant application of alumina wear-resistant ceramics in the wine industry is knife-edge blades for cleaning drum filters. Alumina blades provide not only the benefit of long wear life, but also being highly chemically inert, and wine being an acid liquid, the use of metal knife-edge blades runs the risk of acid dissolution, and hence

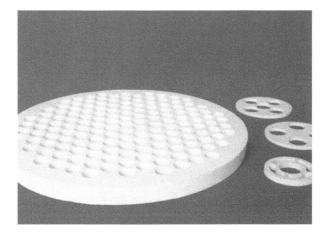

Fig. 12.22 Alumina orifice and baffle plates.
Produced by Taylor Ceramic Engineering. Image courtesy of Taylor Ceramic Engineering [31].

contamination of the wine with metallic flavors and impurities. This is why, this is an application that is ideal for alumina.

Alumina knife-edge blades are shown in Fig. 12.19.

12.2.6 Alumina ceramics for the brick and heavy clayware industry

TCE developed the first ceramic-lined extrusion die for the Australian brick industry in 1980. This lasted 14 years in service. This has now become a common technology in the industry. Alumina has also seen use in wear-resistant protection systems in metal extrusion dies. Alumina wear-resistant components for brick production equipment include:

- Auger linings
- Bridge sleeves
- Chute liners
- Core buttons
- Die boxes
- Gouging tips
- Pug knives and mixing blades
- Scraper blades
- Shaper caps

The auger structure can be manufactured predominantly from alumina with or without a metal shaft insert, or can simply be tiled on the edge as in the case of an auger shoe. When required (e.g., specialty industries like microwave), the auger can be entirely manufactured from alumina. The troughs can be lined with alumina tiles, which can be contoured to the trough geometry. Examples of alumina ceramics for the brick and heavy clayware industry are shown in Fig. 12.23.

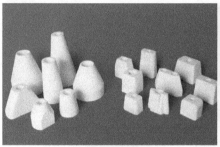
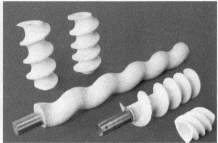

Fig. 12.23 Left: Alumina brick cores, sleeves, die boxes and shaper caps for brick extrusion. Right: Alumina augers with and without a metal shaft insert.
Produced by Taylor Ceramic Engineering. Image courtesy of Taylor Ceramic Engineering [31].

12.2.7 Alumina nozzles

Alumina nozzles can be manufactured with nozzle apertures ranging from less than 1 mm up to more than 100 mm, and almost any industrial nozzle type can be made from alumina including slurry, sampler, venturi injector, and gas. These nozzles can be used for wear-resistant applications for a wide range of gases and liquids, including slurries, at very high flow velocities. The one application to which alumina is not suited is nozzles for sand blasting. Alumina nozzles are shown in Fig. 12.24.

12.2.8 Alumina seals and pump components

Alumina seals are one of the earlier industrial applications for alumina ceramics, appearing more than half a century ago, and the technology is now highly evolved.

In its simplest form, the alumina pump-seal is a polished flat ring of alumina (like a large washer) that articulates against a matching ring-like surface, which may be alumina or may be a softer sacrificial material, such as a carbon-phenolic seal ring. Sleeve bearings, which can be simple bushes with tight tolerances, are also common. Alumina impellers, plungers, and other pump components are also widely used. High wear, corrosive conditions, and combined wear and corrosion conditions are common scenarios that call for the use of alumina in pumps and seals, for example, corrosive liquids, saline water, slurries, and water with solids contaminants. The benefits are longer service life, which can be many years, low maintenance, and reliable sealing in the long term. Alumina seals for pumps, and alumina pump components produced by TCE are shown in Fig. 12.25.

Seals for rotary pumps were one of the earliest industrial applications for alumina ceramics. The standard configuration is a stationary ring (seat) which is alumina articulating against a rotating ring (washer) which can be alumina, carbon phenolic, Teflon, cermet, or metal. In alumina-alumina seals, lubrication is required due to the thermal-shock wear mechanism discussed in Section 12.1.2.7, given that loads can be in the order of 10 MPa. However, mere water lubrication is sufficient for alumina

Fig. 12.24 Alumina ceramic nozzles of various dimensions, nozzle apertures, and characteristics. Produced by Taylor Ceramic Engineering. Image courtesy of Taylor Ceramic Engineering [31].

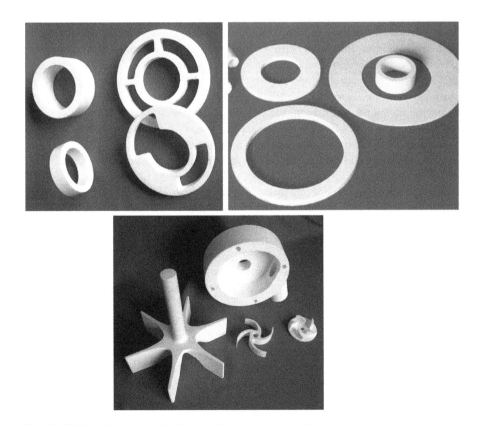

Fig. 12.25 Top: Alumina seals. Bottom: Custom-designed alumina pump components. Produced by Taylor Ceramic Engineering. Image courtesy of Taylor Ceramic Engineering [31].

seals, unlike metal seals which require oil lubrication. This gives alumina a significant advantage over traditional metal seals. Moreover, dirty water containing sediments is not a problem for alumina seals given its high wear resistance, which is another significant advantage for alumina seals.

Alumina-alumina seals require very precise machining, lapping, and polishing. Diamond machining is used to produce flatness. This results in many microscopic peaks and valleys in the alumina surface which require lapping and polishing. Flatness is essential due to the high elastic modulus of alumina. Without perfect flatness, the alumina cannot yield to accommodate poor flatness, and scuffing wear will occur in certain areas of the seal face causing high friction and noisy operation.

The degree of lapping and polishing determines: (a) the seal—a smoother surface is a better seal; (b) the friction coefficient—a smoother surface has a higher contact area and therefore a higher friction coefficient; (c) the access for the water lubricant—too smooth a surface precludes water lubricant penetration between the surfaces, which can result in noisy operation and overheating. An advantage of the alumina/non-alumina seal combination is that the alumina finishing can be less precise. This is a cost advantage. However, the service life is always going to be better with alumina-alumina seals, which makes them the high-end product.

Alumina seals have been used in many common domestic applications including automotive water pumps and air-conditioner pumps, dishwasher pumps, clothing washing machines, domestic central-heating pumps, and alumina-alumina long-life tap faucet seals capable of over a million opening/closing cycles without maintenance. Domestic central-heating pumps utilizing alumina impeller rotating in alumina plain bearings, and water lubricated, have proven very successful. In all, 85% alumina seal rings articulating against carbon-phenolic rings have been used in automotive water pumps since the 1970s and are one of the most significant market uses for alumina rotary seals. Apart from its long wear life in general, alumina is also corrosion resistant, and resistant to third-body wear by corrosion detritus in the cooling water, both of which are essential in an automotive cooling system.

Surface finish is critical. For example, in automotive water pumps, a coarse finish is prone to leakage but a supersmooth finish can raise the required rotational torque to a problematic level for efficient operation, or make the required torque too high for starting the engine when cold or the pump is dry. These principles apply to all alumina seals, not just automotive water pumps. A significant amount of industrial knowhow is embodied in the lapping of the alumina seal surface such that the surface imperfections are too small for leakage but large enough that a liquid film can form and lubricate the seal during operation. In general surface flatness tolerance and surface roughness need to be in the order of just a few microns, requiring precision lapping and polishing.

Alumina seals have also seen use in many industrial pump seal applications in the mining, power generation, food, pharmaceutical, paper, chemical, and petrochemical industries. While the domestic applications involve small alumina rings as low as 85% alumina and just a couple of centimeters in diameter, the industrial applications can involve quite large alumina rings up to 99.5% pure. However, not all industrial pumps incorporating alumina are large. Small pumps for pumping corrosive chemicals work well with alumina, using magnetically driven impellers running in alumina plain bearings.

In addition to seals, alumina has been used in a range of pump components for pumps handling abrasive or corrosive liquids, including plungers, linings, and shaft sleeves. One of the important areas for this is the oil industry. Alumina pump plungers are important when pumping water that is saline and may contain sand or sediment. Alumina pump linings are very important in pumping drilling muds which are clay suspensions used to lubricate and cool oil drilling bits. Drilling-mud pumps require very large alumina pump liners, 10 or 20 cm in diameter and more than half a meter long, requiring great care in alignment and thermal shrink fitting into the metal housing, but they bring huge improvements in pump life.

12.2.9 Alumina bearings, sleeves, and shafts

A plain bearing is the simplest type of bearing. It involves a shaft/bushing arrangement with no rolling components (balls/rods). Alumina plain bearings can run using water as a lubricant, unlike metals which require oil. They also have a very long wear life. This gives them a significant advantage over metal bearings.

Alumina bearings were one of the earliest industrial applications of alumina ceramics, dating back over 60 years to the 1950s. Both alumina plain bearings and matching shafts are manufactured by TCE. Tolerances of a few microns are commonly required for very accurate shaft location. Lubrication holes or channels can be added to the bearing as well as location flats and keyways to prevent the bearing spinning during use. Bearing types include split bearings (made in matching halves), full (cylindrical) bearings, and flanged bearings (both split and full bearings can be flanged). Alumina plain bearings components are shown in Fig. 12.26.

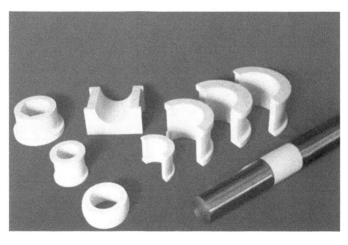

Fig. 12.26 Alumina plain bearings: full bearings, split bearings, both flanged and unflanged. Note also the alumina shaft sleeve fitted to the metal shaft.
Produced by Taylor Ceramic Engineering. Image courtesy of Taylor Ceramic Engineering [31].

12.2.10 Other alumina wear/corrosion resistant applications: Spacers, tubes, and custom components

Numerous alumina spacers, tubes, washers, and miscellaneous components are used in the myriad of plumbing needs and other engineering requirements in mineral processing facilities and chemical processing plans, and other industrial environments where corrosion or wear, or both, are an issue. Industry requirements frequently extend beyond off-the-shelf generic components and require custom-designed alumina components. A wide range of custom alumina components are manufactured by TCE. Some examples are shown in Fig. 12.27.

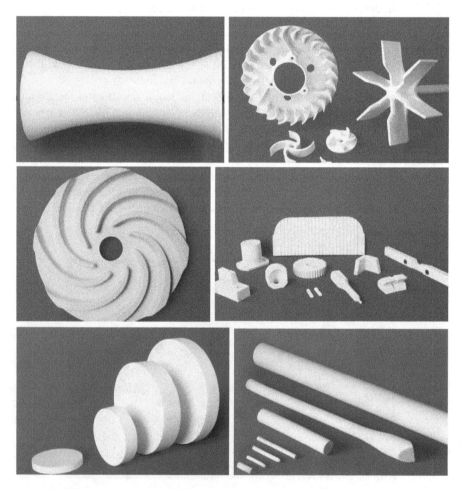

Fig. 12.27 The six images demonstrate the range of alumina ceramic components, both standard items and custom designed.
Produced by Taylor Ceramic Engineering. Image courtesy of Taylor Ceramic Engineering [31].

12.3 Concluding remarks

The diversity of contemporary industry usage of alumina ceramics demonstrates the substantial progress that has been made over the 70 years that alumina has been used in industrial applications. Alumina usage has increased dramatically over 70 years, driven by the growing sophistication of alumina fabrication technology, and the growing awareness of the economic benefits of using a relatively low cost and versatile ceramic in high-wear and/or corrosive manufacturing environments, in place of the traditional choice of metals.

The principal benefits have been greatly extended service life, cost savings through reduced downtime for refurbishment of worn parts, and the ability to operate in industrial environments that are much more highly wear inducing, or corrosive, or both, than was possible to operate in 70 years ago.

Alumina as an electrical insulator

In electrical applications for alumina, it is predominantly used as an insulator. Indeed the first commercial application for alumina was electrical: spark plug insulators, an application that alumina continues to dominate to the present day. Alumina is an outstanding electrical insulator because it uniquely combines eight key attributes: (1) It's electrical resistivity is one of the highest of all known materials; (2) the dielectric loss tangent of alumina is among the lowest of all known materials; (3) high mechanical strength; (4) thermal stability up to 1700°C; (5) outstanding corrosion resistance; (6) the moderately high thermal conductivity of alumina makes it an effective insulating heatsink; (7) thermal expansion matching with silicon which is important when used as a semiconductor substrate (SOI—silicon on insulator); (8) low cost—compared to the alternative materials of comparable outstanding electrical resistivity such as aluminium nitride, silicon nitride, and beryllium oxide. There are three main industrial electrical roles in which alumina is used: (1) Macro-insulators (spark plug insulators, power transmission systems); (2) Micro-insulators (insulating substrates in microelectronic devices); (3) Insulating electrical feedthrough seal in sealed advanced electrical devices. There are also numerous minor applications for alumina as an electrical insulator. This chapter will overview the electrical properties of alumina, and explore in detail applications (1) Macro-insulators and (2) Micro-insulators. Insulating electrical feedthrough seals and the associated area of alumina-metal bonding will be explored in Chapter 14.

The main industrial applications of alumina can be broadly classified into the following areas:

- Biomedical: orthopedics, bionics, dentistry, lightweight body armor, and niche biomedical uses (Chapters 5–11).
- Wear/corrosion-resistant applications (Chapter 12).
- Refractory and other specialist applications (Chapter 15).
- Electrical (this chapter and Chapter 14).

The focus in this chapter is on industrial applications of alumina as an electrical insulator. These are extensive and span the entire spectrum from macro applications as insulators down to micro applications in microelectronics. Indeed the first commercial application for alumina was electrical: spark plug insulators.

In the history of alumina advanced ceramics, electrical applications were the "first cab off the rank." The scientific research underpinning the development of electrical applications for alumina ceramics was the earliest research field for alumina ceramics, arising in the 1930s, with the first commercial applications emerging in the 1950s, the period from the 1950s to the 1990s was one of intensive research and development. Since the 1990s, the focus has been on commercial activity, with much of the research on alumina in electrical applications in the form of confidential corporate in-house applications and process-focused technology development, rather than published

Alumina Ceramics. https://doi.org/10.1016/B978-0-08-102442-3.00013-0
Copyright © 2019 Elsevier Ltd. All rights reserved.

scientific papers. Some areas for electrical alumina are in relative technological stasis (spark plug insulators and other macro-insulators). However, some applications are major technology growth areas such as thin-film and thick-film microelectronic substrates.

In electrical applications for alumina, it is predominantly used as an insulator. There are three main industrial electrical roles in which alumina is used:

- Macro-insulators: spark plug insulators, power transmission systems
- Micro-insulators: insulating substrates in microelectronic devices
- Insulating electrical feedthrough

Macro-insulators: Spark plug insulators are the largest global market for alumina macro-insulators. Spark plugs are a global $3 billion industry and alumina is used almost exclusively in this role, and is a major component of the spark plug embodied value. In the macro-electrical realm, alumina is not widely used as an insulator in power transmission systems. Electrical porcelain is dominant in this role, although alumina has some market share in this realm.

Micro-insulators: Alumina substrates, typically tapecast ~1-mm thick wafers, are widely used in thin-film silicon-on-insulator (SOI) substrates for microchips made by vapor deposition/nanolithography, and thick-film substrates for robust microelectronic devices made by micro screen printing, commonly multilayer devices, sometimes involving >50 interconnected layers. Alumina thin-film substrates underpin 10% of the $500 billion semiconductor market. This means alumina underpins the majority of the $50 billion SOI microchip market. The extreme heatsink roles are filled by AlN and BeO. Alumina thick-film substrates underpin the market in robust thick-film microelectronic devices, particularly in automotive, sensor, and microwave applications, a substrate market in comparable dollar value to the thin-film substrate market.

Insulating electrical feedthrough: This application is so large in scope and commercial impact that it will be the subject of a dedicated chapter: Chapter 14.

This chapter will therefore focus on macro-insulators and micro-insulators. For context, Table 13.1 summarizes the key global market applications for electrical insulators, and the materials used in each application.

Table 13.1 flags the following key points:

- Alumina dominates in thick-film microelectronics substrate applications.
- Alumina dominates in thin-film applications. These are used in about 10% of the microchips from the $500 billion semiconductor industry.
- In large-volume low-tech low-temperature applications such as large high-voltage insulators for power transmission, these would ideally be made from alumina, but cost considerations favor the lower-cost porcelains.
- Alumina dominates the spark plug insulator market, where temperatures up to 800°C are involved.
- Where extreme heatsink properties are required, much more expensive BeO or AlN are favored. BeO is not just expensive, it is also toxic.
- Where extreme thermal shock resistance is required, such as forming rods for wire heating element coils, low thermal expansion aluminosilicates are used, such as sillimanite and spodumene.

Table 13.1 Materials used in the global electrical insulator industry

Application	Material used
Microelectronics	
Substrates for microelectronic packaging	High alumina
Insulating ultra-heatsinks for microelectronic packaging	High alumina, BeO, and AlN
Power transmission and management	
Load-bearing powerline insulators	Aluminous/siliceous porcelain (nonporous, glazed)
Low-voltage insulators	Siliceous porcelain (maybe nonporous and glazed)
Fuse bodies—high power	Cordierite
Fuse bodies—small/low power	High alumina
Specialty insulators	
Spark plug insulators	High alumina
Coil formers—low power	Aluminous/siliceous porcelain or steatite (glazed)
Coil formers—high frequency	High alumina
Coil formers—high temperature (heating elements)	Sillimanite, spodumene, and other low-expansion aluminosilicates

13.1 The electrical properties of alumina

Electrical current can be conducted through a ceramic by the movement of electrons or the movement of ions. Most ceramics are ionic conductors at high temperatures, but show little or no ionic conduction at room temperature. Furthermore, most ceramics, apart from the few that are semiconductors, undergo little or no electron conduction. Most ceramics, including alumina, are insulators at room temperature, but become increasingly electrically conductive with increasing temperature. This has consequences for electrical applications as an insulator.

Ionic conduction can involve anions, as for example oxygen ions in zirconia, or cations, as for example alkali metal ion-doped alumina. In alumina, the principal mode of ionic conduction is the alkali ion contaminants. Therefore, purity level in alumina, especially alkali metal content, is critical to its insulating properties, as is operating temperature. The same applies to the widely used aluminosilicate electrical ceramics, such as porcelain, for which increasing alkali ion conduction at increasing temperature is the primary mode of conductivity.

Most of the time alkali metal ion conductivity issues are a problem to be managed or overcome. However, in the case of beta alumina, which has the approximate composition $Na_2O \cdot 11Al_2O_3$, alkali metal ion conductivity issues are commercially exploited. Beta alumina is a very effective industrial ionic conductor above about

300°C, and its primarily commercial use is in molten salt electrochemical cells, a unique application in which few other materials can operate. This is a small but significant application for beta alumina. Beta alumina is also sometimes used in glass contact refractories (Chapter 15).

The four general electrical properties of ceramics that dictate their potential value as an electrical ceramic are:

(1) Electrical resistivity (the reciprocal of conductivity)
(2) Dielectric loss (alternating current loss tangent)
(3) Relative permittivity (dielectric constant)
(4) Dielectric breakdown voltage

Alumina is the pre-eminent material with regard to the combination of *high electrical resistivity*, *low dielectric loss*, and *low cost*, and its large-scale commercial use as an electrical ceramic derive from this preeminence.

Alumina is a strong performer, but by no means pre-eminent, with regard to permittivity and dielectric breakdown voltage, and therefore these two properties are not significant in terms of commercial applications of alumina as an electrical ceramic.

Testing, reporting, and standardization of electrical properties are governed by various standards, of which the International Electrotechnical Commission (IEC) standards are the primary ones. Deutsche Institut fur Normung (DIN), British Standards (BS), and American Institute for Testing and Materials (ASTM) are also significant. The IEC standard classes electrical ceramics by their IEC class. For example, alumina >99% is IEC class C-799, alumina 95%–99% is C-795, alumina 86%–95% is C-786, alumina 80%–86% is C-780, and porcelains have many classifications in the C-110–C-130 range.

13.1.1 Electrical resistivity of alumina

Alumina has one of the highest resistivity values of any known material at room temperature and elevated temperature, and has by far the best combination of low cost and high resistivity of all known materials. Cost-effective extreme resistivity is one of its most important attributes in terms of its commercial exploitation. Other than spark plugs, the majority of commercial applications for alumina insulators involve temperatures below 250°C. Some of the key commercial applications for alumina as an electrical insulator, and the operating temperatures involved, are summarized in Table 13.2.

13.1.1.1 Electrical resistivity: Alumina and its competitors

Alumina has one of the highest resistivity values of any known material at room temperature and elevated temperature. Its primary competitors in this regard are all rare or expensive ceramics:

(1) Toxic and expensive BeO, which has the same resistivity as alumina up to 200°C, but is slightly higher in resistivity than alumina at 600°C, and higher in thermal conductivity.
(2) $BaAl_2Si_2O_8$ (celsian), a rare feldspar mineral, which has the same resistivity as alumina at room temperature, but is higher than alumina above 200°C.

Table 13.2 Operating temperature range for various commercial applications for alumina as an electrical insulator

Market	Market size	Alumina role	Temp (°C)
Bionic implants	$25 billion	Feedthrough	40 (wet)
Thin-film SOI semiconductor	$50 billion	Substrate	<125
Electrical vacuum valve	Shrinking	Feedthrough	<250
Spark plug insulator	$3 billion	Insulator sheath	500–800
Thick-film microelectronics	>$10 billion	Substrate	<900
Heater coil former	Small	Insulating core	<1300
Oxygen sensor	$Millions	Hermetic sheath	300–1600
Thermocouples	Small	Insulating sheath	<1800°C

(3) AlN which has a similar resistivity to alumina but is 5–10 times the cost of alumina and even more difficult to process. It is also higher in thermal conductivity.

(4) BN which has a similar resistivity to alumina but is 5–10 times the cost of alumina and even more difficult to process. Its thermal conductivity is somewhat higher than alumina but well behind AlN and BeO.

Thus, alumina is not the only advanced ceramic to have an extremely high resistivity. AlN, BeO, BN, and $BaAl_2Si_2O_8$ each have a comparable resistivity of $\sim 10^{12}\,\Omega m$ at room temperature. However, all of them struggle to compete with alumina on a commercial footing. AlN, BeO, and BN are all very expensive ceramics in comparison with alumina. AlN is 5–10 times the cost of alumina, BN is similar in cost to AlN, but much inferior in thermal conductivity. BeO is both toxic and expensive. Moreover, both AlN and BN are nonoxides and require hot pressing in a controlled atmosphere to attain high density. $BaAl_2Si_2O_8$ is a feldspar mineral that is not widely used or available, since it offers no real advantages over alumina, and falls far short of alumina in commercial viability.

Alumina is low in cost, and it is certainly nontoxic. Indeed, it is the most biocompatible material in clinical use. Moreover, it can be pressureless sintered to full density in air. Therefore, alumina delivers the best value for money in terms of being the lowest cost highly resistive universally applicable electrical insulator. Another way to see it is that alumina is in the middle. Too expensive for mass-market low-tech insulators in powerlines for mass electricity transportation, a role filled primarily by porcelain. Not as strong a heatsink as BeO or AlN, although an order of magnitude cheaper, and therefore not used in the extreme "money no object" microchip heatsink roles. Everything in between the mass-market low-tech (porcelain) and extreme heatsink (AlN or BeO) ends of the spectrum is the domain of alumina. This means in practice, total dominance of spark plug insulator role, the majority of microchip substrates, and dominance of the thick-film microelectronics sector.

Resistivity is the most important property driving the commercial use of alumina as an electrical ceramic, given that alumina's primary electrical application niche is that of electrical insulator. This is a role in which alumina is seen as the pre-eminent choice of all available materials, on overall performance/cost criteria. As mentioned in

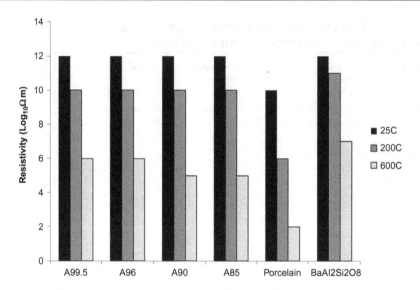

Fig. 13.1 Alumina-based electrical resistors. Approximate DC resistivity defined as $Log_{10}(\Omega m)$, as a function of temperature, for alumina of various purity levels, with aluminosilicates electrical porcelain and $BaAl_2Si_2O_8$ (a barium-based aluminosilicate mineral that is not widely used or available) included for comparison.

Section 13.1.1, alumina is an ionic conductor, and its ionic conductivity increases both with increasing temperature, and with increasing alkali metal ion content. Thus, purity and temperature are critical parameters in determining the resistivity of alumina.

Fig. 13.1 summarizes the resistivity of various alumina grades, and at various temperatures, demonstrating the practical significance of these two key parameters: purity and temperature. For comparison, aluminosilicates electrical porcelain and $BaAl_2Si_2O_8$ are also included, given that porcelain is alumina's main low-cost competitor and $BaAl_2Si_2O_8$ has similar resistivity to alumina.

The following conclusions can be drawn from Fig. 13.1.

- Alumina is substantially superior to electrical porcelain in terms of resistivity at all temperatures: by two orders of magnitude at room temperature and up to four orders of magnitude at high temperature.
- Purity of alumina only has a significant effect on resistivity at high temperature.

Given that electrical resistivity is one of the most commercially important properties of alumina, it is important to explore the factors that affect its resistivity.

13.1.1.2 Theoretical basis of alumina resistivity

Electrical conductivity in alumina can occur by either ionic transport or electron transport. Alumina is an ionic conductor, and therefore ionic conduction is the dominant mechanism for electrical conductivity in alumina. Impurity atoms in its structure primarily account for its electrical conductivity. Principally it is alkali metal impurities, commonly sodium and potassium, that cause alumina to be an ionic conductor. However, theoretically pure alumina has a limiting electrical resistivity, which is a result of two forms of conduction:

- Ions (ionic conductivity)
- Electrons (electronic conductivity)

Where *(ionic conductivity) + (electron conductivity) = total electrical conductivity.*

Essentially, alumina is an ionic conductor at low temperatures and an electron conductor at high temperatures. The electrical conductivity of high-purity alumina is low, even at elevated temperatures. So low in fact, that it is in the same order of magnitude as the surface electrical conduction and electrical conduction through the convective surface gas phase [544,545]. This makes accurate measurement of the electrical resistivity of alumina problematic, especially at high temperatures. However, in essence, electrical conductivity of alumina is dependent on four key parameters:

- Temperature
- Impurity concentration
- Grainsize
- Oxygen partial pressure

These four issues are graphically explained in Figs. 13.2–13.4.

The data in Fig. 13.2 are for ultrahigh-purity alumina at very high temperatures. In Fig. 13.2, temperature is graphed as 10,000/K. However, for clarity, actual temperatures in degree Celsius are also annotated on the graph, showing that the temperature span is from about 600°C to about 1800°C.

Log(Resistivity) generally correlates linearly with the reciprocal of temperature. This can clearly be seen in Fig. 13.2. For the purposes of illustration, the single-crystal data from Fig. 13.2 are extrapolated to 600°C, and the data mapped with the data from Fig. 13.1 to demonstrate the overall effect of purity and single-crystal structure on resistivity, as shown in Fig. 13.3. This presents a clear overall picture of the effects of grainsize and impurity.

Three main conclusions can be drawn from Figs. 13.2 and 13.3. These relate to grainsize, ionic mobility, and impurities.

Grainsize

Electrical resistivity is much lower in polycrystalline alumina, compared with single-crystal alumina, by several orders of magnitude. In practice, this means that resistivity decreases with decreasing grainsize [551,552] which suggests that the grain boundaries are a region of enhanced conductivity. It is believed that electron conductivity is the parameter that increases with decreasing grainsize, due to hole transport at the grain boundaries [553,554]. This remains a working hypothesis, and there may be deeper factors involved that are not yet fully understood. Moreover, this is complicated by the fact that accurate measurement of the resistivity of alumina is problematic, since the electrical conductivity of high-purity alumina is so low, even at elevated temperatures, that it is in the same order of magnitude as the surface electrical conduction and electrical conduction through the convective surface gas phase [544,545].

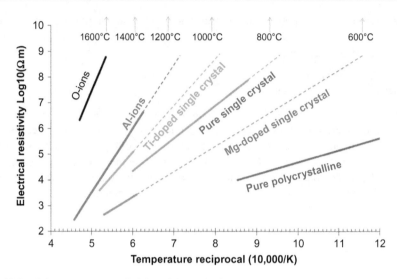

Fig. 13.2 High-temperature resistivity of theoretically pure alumina. The solid lines correspond to published data. The dashed lines are linear extrapolation. *O-ions*: Theoretical resistivity based on oxygen ion diffusion in theoretically pure single-crystal alumina, calculated from self-diffusion data for alumina [546]. *Al-ions*: Theoretical resistivity based on aluminum ion diffusion in theoretically pure single-crystal alumina, calculated from self-diffusion data for alumina [547]. *Pure single crystal*: High-purity single-crystal alumina [548]. *Ti-doped*: Titanium-doped (430 ppm) high-purity single-crystal alumina [549]. *Mg-doped*: Magnesium-doped (10 ppm) high-purity single-crystal alumina [550]. *Pure polycrystalline*: High-purity polycrystalline alumina [551].

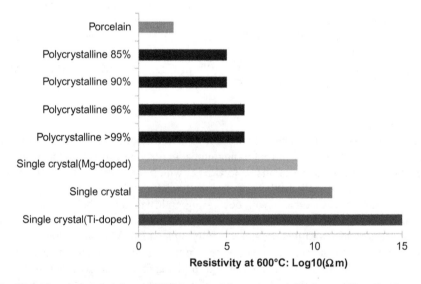

Fig. 13.3 Electrical resistivity at 600°C for porcelain, commercial polycrystalline alumina, and single-crystal alumina.

Alumina as an electrical insulator 421

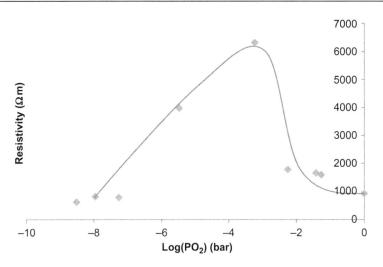

Fig. 13.4 Effect of partial oxygen pressure on resistivity of single crystal ~100% pure alumina at 1650°C [548].

It is further complicated by the fact that truly 100% theoretically pure alumina is very difficult to synthesize, and dopants, even at parts per million (ppm) levels, can have a dramatic effect on conductivity, as discussed in section "Impurities".

The important point, from a commercial perspective, is the fact that grainsize has a dramatic effect on resistivity. This is of major practical significance given alumina's commercial dominance as an electrical insulator, and the commercial impracticality in the extremely high cost of producing single-crystal alumina for mass-market alumina products.

Moreover, Fig. 13.3 clearly demonstrates that grain boundaries are more important than impurities, in diminishing resistivity, since single-crystal high-purity alumina is five orders of magnitude higher in resistivity than polycrystalline high-purity alumina at 600°C, while the presence of magnesium doping reduces this by only two orders of magnitude. Moreover, for impure polycrystalline alumina, the resistivity decreases by only an order of magnitude from >99% pure down to 85%. The sodium content does not dramatically increase between the 99% and 85% alumina grades. It is only with the switch to alkali metal-rich porcelain that this impurity effect becomes substantial, dropping four orders of magnitude. Thus, a very large amount of impurities is required in order to have as dramatic effect on resistivity as does grainsize, unless they are sodium or other monovalent alkali metal impurities. Finally, given that magnesium is the primary sintering aid in high-purity ceramics (Chapter 4), this finding has added significance.

Ionic mobility

The data demonstrate that the aluminum ions are more mobile than the oxygen ions in alumina, and that the theoretical conductivity of alumina, if it was controlled by aluminum and oxygen ions, is many orders of magnitude higher that the resistivity of alumina ceramics, even ultra-high-purity single-crystal alumina. This points to the critical role of impurities in alumina. This can be seen in the context of diffusion and crystal structure issues discussed in Chapter 4, which confirm that aluminum ions are more mobile in the alumina lattice from a crystallographic point of view.

Impurities

The distinction between impurity and dopant is a somewhat blurred one, but in general, a dopant is a deliberately added trace impurity, measured in ppm rarely more than tens or hundreds of ppm, while an impurity is a raw material residual and can be in the order of 1% or more (10,000 ppm). Impurities have a significant effect on alumina resistivity. Fig. 13.2 demonstrates that doping pure single-crystal alumina with 430 ppm titanium enhanced its resistivity by four orders of magnitude, while doping it with 10 ppm magnesium reduced its resistivity by two orders of magnitude.

In order to explain this, a brief digression to semiconductor terminology is called for. A dopant is a trace impurity added to a crystal lattice to alter its properties. In semiconductor terminology, dopants are known as "acceptors" (electron acceptors) or "donors" (electron donors). For example, when silicon (four valence electrons) semiconductors are doped with phosphorous (five valence electrons) the phosphorus is a "donor" dopant and forms a net negative charge or "n-type" semiconductor. When silicon is doped with boron (three valence electrons), the boron is an "acceptor" dopant which form a net positive charge or "p-type" region.

Aluminum has three valence electrons. Therefore, titanium (four valence electrons) is a "donor" dopant and forms a net negative charge in the alumina lattice, while magnesium (two valence electrons) is an "acceptor" dopant and forms a net positive charge in the alumina lattice. Acceptor dopants (e.g., magnesium) in alumina generate positively charged lattice defects. This has the effect of enhancing ionic conductivity. Donor dopants (e.g., titanium) have the opposite effect, and the result is that ionic conductivity is impeded.

This concept can be further expanded to the well-known importance of sodium and the alkali metals (one valence electron) in ionic conductivity in alumina, and the associated dramatic reductions in alumina resistivity. Sodium (one valence electron) is a very strong "acceptor" dopant and forms a strongly net positive charge in the alumina lattice, thereby strongly enhancing ionic conductivity. Thus, sodium and the alkali metals reduce resistivity even more strongly than divalent metals such as magnesium. This explains the dramatically lower resistivity of porcelain compared with the polycrystalline alumina samples of varying purities, as shown in Fig. 13.3. Electrical porcelain has a very high alkali metal content, particularly sodium and potassium, owing to the fact that clay and fluxing additives, such as feldspar, are used in its formulation.

In contrast, the polycrystalline alumina at 85% purity is only one order of magnitude lower in resistivity at 600°C than 95%–99% pure alumina. This is because the primary impurity is silicon in these polycrystalline aluminas in the 85%–99% range. The sodium and alkali earth concentration is not dramatically increased in 85% alumina compared with 95%. In terms of actual sodium/potassium alkali earth content for alumina ceramics, it varies significantly from manufacturer to manufacturer, and also depends on the application. However, as a typical example, a 96% alumina thick-film substrate material, for example, typically contains 0.05%–0.1% Na_2O maximum, that is, 500–1000 ppm alkali earth content. This is a mid-range alumina in purity, between the ultra-high-purity 99.9% extremity and the <90% purity extremity of the alumina ceramic purity range.

In contrast, electrical porcelain has a huge alkali earth content, both Na_2O and K_2O. The precise amount varies from manufacturer to manufacturer, and is mainly dependent on the alkali earth content of the feldspar used in its manufacture. It is, however, possible to estimate the approximate alkali earth content of an electrical porcelain in a generalized sense. It is one of the central tenets of ceramic engineering that a whiteware (porcelain) body formulation typically contains 50% clay, 25% feldspar (or other flux such as nepheline syenite), and 25% flint (silica). Furthermore, a typical clay contains ∼1.5% (Na_2O+K_2O) and a typical feldspar contains about 12% (Na_2O+K_2O). Silica alkali earth content is relatively insignificant compared with feldspar and clay. This means that a typical porcelain contains about 4% (Na_2O+K_2O), or 40,000 ppm. That is around 40–80 times (approximately two orders of magnitude) higher than a 96% alumina thick-film substrate. Fig. 13.3 shows that 96% pure alumina has four orders of magnitude higher resistivity at 600°C than electrical porcelain, while the resistivity difference between 100% and 85% pure alumina is just one order of magnitude at 600°C. The significant effect of alkali earth impurities on resistivity is clearly apparent.

Oxygen partial pressure

The final effect to be discussed is the effect of oxygen partial pressure on the electrical resistivity of alumina. For high $p(O_2)$ values, above 10^{-2} bar and approaching atmospheric pressure, and ultra-low $p(O_2)$ values (below 10^{-6} bar) the resistivity is lowest. For low $p(O_2)$ values, in the order of 10^{-6}–10^{-2} bar, electrical resistivity is maximized. This is shown in Fig. 13.4 [548].

Numerous studies have been made of this phenomenon, and universally the finding has been that the resistivity-$p(O_2)$ correlation has a maximum. This effect is also independent of impurity atoms present, it seems to be inherent to alumina itself. There have been various theories hypothesized to explain this, which generally relate to the inherent nature of alumina which is that it is an ionic conductor at low temperatures and an electron conductor at high temperatures [548,555–557].

While there remains uncertainty as to the scientific mechanisms underlying this phenomenon, the key issue here is that the oxygen partial pressure effect is of substantial practical importance when alumina is used as a high-temperature insulator, whether in a vacuum or inert gas [ultra-low to low $p(O_2)$ conditions associated], or in ambient air [high $p(O_2)$ conditions associated]. Given the importance of alumina in vacuum-feedthrough electrical devices, sometimes at high temperatures, this is of commercial significance.

Alumina resistivity and its commercial significance

It should be stressed that high electrical resistivity is one of the most commercially important properties of alumina, along with its wear resistance, heat resistance, corrosion resistance, and low cost. In terms of commercial applications as an insulator, alumina is dominant in the role of electrical insulator in most applications other than "power pole" urban power transmission systems where the scale of insulator usage dictates porcelain, for cost reasons, with alumina in a niche role. AlN, BN, and BeO are insulators of comparable resistivity to alumina, but not commonly used. AlN is the most commonly used of these, and it is primarily used in microelectronic applications, and only when heatsinking requirements are a stronger imperative than purchasing cost and processing complexity. These issues will be discussed in detail in Section 13.4.

13.1.2 Dielectric loss (AC loss tangent) of alumina

Dielectric loss, commonly known as the AC loss tangent, is a measure of the energy efficiency of an insulator in high-frequency electromagnetic fields. It equates to the frequency-dependent loss of electrical power through electric field attenuation, for a given insulator material. It is commonly reported as either the loss angle δ or the loss tangent $\tan(\delta)$. Its primary importance is in high-frequency applications, for which the most commercially significant is thick-film microelectronics in microwave technology and microwave-transparent windows in microwave generators (particularly gyrotrons).

The study of the dielectric loss tangent of alumina dates back to the 1950s [558]. Therefore, in addition to its pre-eminent resistivity (Section 13.1.1), the low dielectric loss of alumina is the other very commercially important electrical property of alumina in its role as an electrical ceramic, second only to resistivity in its commercial significance.

Loss tangent of insulating ceramics depends on the capacity for ionic migration under the influence of an alternating electric field. Thus, alkali ion content is critical for the loss tangent of alumina, which is why high-purity alumina surpasses lower-purity alumina, and porcelain is lower still. Indeed, high-purity alumina has one of the lowest loss tangents of any known material, which is another primary reason for its commercial dominance as an electrical ceramic.

Table 13.4 demonstrates the preeminence of alumina in terms of its dielectric loss factor $\tan(\delta)$, superior to BeO (five times better) and $BaAl_2Si_2O_8$ (50 times better) and porcelain (125 times better). The data in Tables 13.3 and 13.4 are typical values. There is a significant range of reported loss factors in the literature for ceramics due to the

Table 13.3 **Loss tangent of alumina: purity, frequency, and temperature dependence**

Ceramic	$\tan(\delta)$ 50 Hz 25°C	$\tan(\delta)$ 1 MHz 25°C	$\tan(\delta)$ 1 MHz 500°C
Alumina 99.5%	0.0002	0.001	0.013
Alumina 96%	0.0005	0.001	0.03

Table 13.4 **Loss tangent of alumina: comparison with competitor ceramics**

Ceramic	tan(δ) 50 Hz 25°C	tan δ(competitor)/ tan δ(alumina 99.5%)
Alumina 99.5%	0.0002	1
Alumina 96%	0.0005	2.5
Porcelain	0.025	125
BeO	0.001	5
AlN	0.001[a]	5
$BaAl_2Si_2O_8$	0.01	50

[a]There is significant uncertainty surrounding the loss tangent for AlN which has been reported with values ranging from 0.01 to 0.0005 [559]. Regardless of whether it is 0.01 or 0.0005, AlN is clearly significantly inferior to alumina with regard to loss tangent.

purity, temperature, and frequency dependence of tan(δ), with purity, temperature, and frequency not always reported with the tan(δ) value. Reported loss factors for >99% alumina are as low as 0.0001.

The effects of purity, temperature, and frequency on the loss tangent of alumina are shown in Fig. 13.5. Overall the following trends can be seen:

- Frequency: loss tangent decreases with increasing frequency.
- Purity: loss tangent decreases with increasing purity.
- Temperature: loss tangent decreases with decreasing temperature.

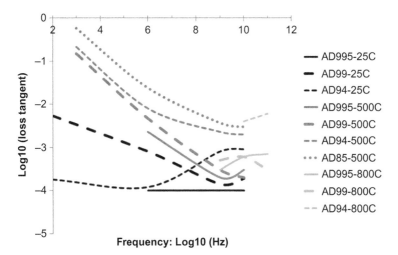

Fig. 13.5 The effects of purity, temperature, and frequency on the loss tangent of alumina of purities 99.5%, 99%, 94%, and 85%, at temperatures of 25°C, 500°C, and 800°C, and within the frequency range 100 Hz (AC power) to 10 GHz (microwaves) [560].

Thus, the ideal conditions for minimal loss tangent for alumina are high purity, high frequency, and low temperature. In microwave windows, an important area of commercial exploitation of the low loss tangent of alumina, the purity is high, the frequency is high, and the temperature depends on the power and configuration of the device.

13.1.3 Relative permittivity (dielectric constant) of alumina

Relative permittivity, also known as dielectric constant, is a measure of the degree of diminution, by the intervening medium, of the coulomb force of an electric field between two point charges in the intervening medium. Most insulating ceramics, including alumina, have relative permittivity values of <10. Therefore, permittivity is not a stand-out property for alumina, unlike resistivity and loss factor. The titanates are the family of materials with extremely high relative permittivity values, which range into the thousands.

Therefore, permittivity has little commercial applicability in a discussion of the commercial applications of alumina as an electrical ceramic. Alumina is sometimes doped with TiO_2 when it is necessary to enhance its permittivity.

13.1.4 Dielectric breakdown voltage

Dielectric breakdown is a threshold voltage above which electrical current flows through an insulator. This can manifest as a transitory electrostatic discharge, or can manifest as a continuous electrical arc. Dielectric breakdown is a principle that has some well-known commercial uses, such as arc welding and arc lamps.

Dielectric breakdown threshold is measured as a voltage per unit thickness of dielectric, with the units $kV\,mm^{-1}$. DC breakdown threshold for a given material tends to be higher than for AC breakdown.

Most insulating ceramics, including alumina, have dielectric breakdown threshold values in the order of $10-25\,kV\,mm^{-1}$. Therefore, dielectric breakdown threshold is not a stand-out property for alumina, unlike resistivity and loss factor. Thus, dielectric breakdown threshold has little commercial applicability in a discussion of the commercial applications of alumina as an electrical ceramic.

13.2 Electrical insulator applications of alumina: Alumina macro-insulators

Alumina is predominantly used as an insulator in electrical applications. Other than insulating hermetic feedthroughs in electrical devices (the subject of Chapter 14), there are four main industrial electrical roles in which alumina is used:

- Macro-insulators: spark plug insulators, power transmission.
- Insulating thin-film substrates in SOI microchips.
- Insulating substrates for thick-film microelectronic devices.
- Specialty electrical applications.

The main commercial use for alumina as a macro-insulator is in spark plug insulators. Its secondary commercial use as a macro-insulator is in insulators for power transmission and management.

13.2.1 Spark plug insulators

The global spark plug industry is a $3 billion industry. Virtually all spark plugs in the world today use a 94% alumina insulator. In the early days, porcelain, steatite, and sillimanite were sometimes used, but today it's primarily 94% alumina. The alumina insulator is the main high-tech high-cost component of the spark plug, and is shown in Fig. 13.6. Therefore, spark plugs represent one of the largest global markets for alumina as an electrical insulator.

Electrical uses of alumina date back to the early 20th century, and arguably the first documented commercial application of alumina was an electrical one—spark plug insulators, which first appeared in the literature about 90 years ago in 1927 as a patent

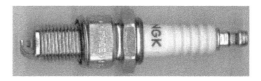

Fig. 13.6 Spark plug. Top: the white alumina insulator clearly seen on the right extends all the way (through the center of the engine-block metal connection sleeve seen on the left) to the electrode tip (small white region on the far left) where it is exposed to temperatures of 500–800°C and corrosive gases in the combustion zone. A significant thermal gradient exists given that the high-tension-lead connection end (right) doesn't reach 100°C in operation. Middle: dissected spark plug showing the whole alumina insulator component. Bottom: close-up of the electrode end of the spark plug, the section that is inside the combustion zone.
Top: Image Author: Sonett72 [114]. Middle: Image Author Industry Shill [561]. Bottom: Image Author: Norris Wong [562]. Reproduced from Wikipedia (Public Domain Image Library).

filed by Hans Reichman on slipcast high-purity alumina ceramic spark plug insulators. Reichman published four patents in the early 1930s on this innovation [110–113]. His original patent [110] was filed in 1927, and granted in 1931.

However, one could equally argue that the first commercial application of alumina was refractory, as per the original alumina patents of 1910, 1012, and 1913 by Graf (Count) Botho Schwerin [17,107,108]. Schwerin, the original alumina ceramic pioneer, saw crucible uses as the purpose for developing slipcast alumina. Regardless of whether one ascribes the first commercial application of alumina as crucibles or as spark plug insulators, it is clear that the work of Reichman follows the work of Schwerin who filed three patents on slipcasting of alumina between 1910 and 1913 [17,107,108].

Schwerin's patents are the first documentation in the literature of alumina ceramics. Much work preceded Schwerin on clay-bonded alumina, but clay-free true alumina ceramics as such, date back to 1910. Very little alumina research and development activity appears to have taken place in the public domain between the outbreak of the World War I in 1914, and the patent filing of Reichman in 1927, 9 years after the cessation of hostilities. Moreover, the patents of Schwerin were processing patents, documenting a slipcasting approach to alumina fabrication. Therefore, the 1927 patent filing of Reichman, while not the first documentation of alumina ceramics as such, was the first documented commercial alumina application in the literature.

Essentially, Reichmann adapted Schwerin's innovation and attempted to commercialize it in the application of spark plug insulators, manufacturing the alumina insulators exactly as per Schwerin's 1910 patent, Reichmann did this by slipcasting (Chapter 4) fine-ground Bayer alumina using acid as a deflocculant, which was successful at the forming stage, but the required sintering temperatures for pure alumina components proved problematic commercially and so Reichmann went on to explore flux additions, up to 5%, compromising the purity of the alumina, and of course its electrical resistivity, but bringing the sintering temperature down to a barely commercially viable 1620–1700°C. Unfortunately for Reichmann, the advent of alumina sintering-aid research, enabling flux to be eliminated and effective sintering with just 0.15 wt% MgO in high-purity alumina and sintering temperatures below 1600°C, was many decades in the future (see Chapter 4). The work of Reichmann, although innovative for its time, was derivative of Schwerin, and largely unsuccessful commercially, although Reichmann certainly advanced the commercialization potential of slipcast alumina much further than Schwerin had.

The next big breakthrough in alumina forming also involved spark plug insulators, and came shortly after the Reichmann patents, in the early 1930s. However, it involved a fundamentally new approach to making alumina ceramics: cold isostatic pressing (Chapter 4), in rubber molds which ultimately became one of the main mass-production processes for alumina spark plug insulators, which were to become a mass market product in the postwar period. This innovation first appeared in the literature in the 1934 patent of Homer Daubenmeyer of the Champion Spark Plug Company [18]. The Champion Spark Plug Company continues to thrive to this day.

The patent of Daubenmeyer, filed in 1933, and granted in 1934, is really the first successful commercial prototype of an alumina ceramic. If one was to nominate the year in which the first commercial alumina ceramic product was prototyped, 1933 would be the best estimate. Alumina spark plug insulators made by isostatic pressing were a relatively pure alumina ceramic that had a ready and ongoing market, and remain with us to this day, including as the Champion brand. However, the first commercial prototype developed is one thing, actual mass-market commercialization of alumina ceramics did not really begin until the 1950s, and it was a very gradual process of market penetration for quite some time.

World War II intervened at this stage, and no significant recorded alumina innovation happened after the Reichmann/Daubenmeyer spark plug innovations of the 1920s and 1930s, until the late 1940s. In 1949, the next major breakthrough happened with the innovation of Karl Schwartzwalder: injection molding of alumina ceramics [19]. It is notable that Schwartwalder was granted a patent on isostatic pressing of spark plug insulators filed in 1938, and granted in 1943 [563]. Injection molding of alumina spark plug insulators and other advanced ceramics was to become a mass-market technology many years later.

Alumina is ideal for spark plug insulators because of its combination of the following essential properties:

- Good mechanical strength.
- High thermal conductivity and therefore good thermal shock resistance.
- Resistance to high-temperature corrosion (a portion of the insulator is exposed to the combustion zone).
- High resistivity at elevated temperatures.

Spark plugs insulate against very high voltages, in the order of 10,000–40,000 V, inside a corrosive combustion zone in the 500–800°C temperature range. It is an application that involves extreme conditions. Alumina continues to dominate the global spark plug insulator industry, and the technology has not changed much since the 1950s: isostatic pressing or injection molding. Generally 94% alumina is used. Only high-performance engines require higher alumina purity, and this is a small proportion of the total market.

As mentioned before, the global spark plug industry is about $3 billion dollars. This is not surprising given that there are about 1 billion cars in the world, about 100 million motorcycles, a large number of piston-powered aircraft, piston-powered boats, and specialty recreational vehicles. The retail cost of spark plugs is in the $10+ range, and spark plugs need replacing several times during the lifetime of an engine. Alumina is the major component of a spark plug in dollar value.

For more details on isostatic pressing and injection molding, see Chapter 4.

13.2.2 Alumina insulators in power transmission and management

As Table 13.1 clearly demonstrates, alumina ceramics are not commonly used in power transmission systems, particularly as regards mass-urban "power-pole" systems. This role is primarily filled by porcelain, both siliceous porcelain and aluminous porcelain. For high-voltage applications, these porcelain insulators are commonly manufactured by plastic forming methods, sintered to zero porosity and glazed. For low-voltage applications, the insulators are either plastic formed or pressed. It is not essential that they be nonporous, although it is preferable that they are nonporous and glazed. It is a question of cost justification for the service voltage involved.

As Section 13.1 demonstrated, sodium and alkali metals in general are problematic for electrical insulation as the alkali metal ions are the principal source of ionic conduction in oxide ceramics. This is the principle underlying ionic conductor beta alumina, which is essentially $Na_2O \cdot 11Al_2O_3$, and is a very effective industrial ionic conductor above about 300°C.

Porcelains have a significant amount of sodium and other alkali ions in their structure of around 4%. Therefore, ideally insulators for power transmission, especially high-voltage power transmission should be high alumina. It is only because of the huge number of these insulators used globally, and the higher cost of alumina compared with porcelain, that porcelain is used in this role.

Nonetheless, in specialty high-voltage applications, and in situations where sodium and alkali content needs to be minimized or eliminated, high alumina ceramics are used in this role. Examples of high-purity alumina electrical insulators (manufactured by Taylor Ceramic Engineering—see Chapter 1, Section 1.3.2.5) are shown in Fig. 13.7. The multiple-flange structures seen in Fig. 13.7 resemble the multiple-flange structures used for regular porcelain electrical insulators. The purpose of the multiple flanges is to maximize the surface conductivity pathway in the event that rain, or other moisture sources, wet the surface of the insulator and thereby create a surface electrolyte conductivity pathway. These flange structures need large-radius rounded curves so as to minimize stress concentrators arising.

13.2.3 Other macro-insulator roles for alumina

Alumina is also used in various other niche roles as a macro-insulator. One example is small low-power fuse bodies. However, cordierite and other aluminosilicates dominate the fuse body field. Alumina is also used in nonferrimagnetic coil formers

Fig. 13.7 Specialist electrical insulators made of high-purity alumina.
Manufactured by Taylor Ceramic Engineering (Images courtesy of Taylor Ceramic Engineering).

for high-frequency coils, due to its outstanding loss tangent. However, most non-ferrimagnetic coil formers are used for low frequency (mains AC current at ~50 Hz) heating coils where thermal expansion is the critical parameter, and they are made of low thermal expansion aluminosilicates such as sillimanite.

13.3 Electrical insulator applications of alumina: Alumina substrates in microelectronic technology

13.3.1 The role of alumina in microelectronics

Other than alumina, there are many other uses for ceramics in the electronics industry. The electronics industry is huge, and comprises a wide range of metals (conductors and semiconductors), polymers (insulators and specialist conductive polymers), and ceramics. In electronics, ceramics have many specialist roles. It is important to see alumina in context, and for this reason, the following nonexhaustive list of ceramic electrical applications is compiled to give some idea of the scope of electrical applications for ceramics. This provides context to show the specific niche that alumina fills in a broader context. In many cases, there are a number of candidates used in each role, however, for brevity, only the dominant ceramic or ceramics are cited each case.

- Insulators (feedthroughs, macro-insulators; microelectronic substrates)—primarily alumina. AlN or BeO when extreme thermal conductivity is required.
- Capacitors—$BaTiO_3$ for its high dielectric constant.
- Resistors—originally carbon. Now numerous ceramics, and even metals, are used for thin-film or thick-film printed resistors.
- Thermistors—$NiMn_2O_4$ (negative temperature coefficient) or $BaTiO_3$ (positive temperature coefficient). Various ceramic thermistors as gas sensors.
- Varistors—primarily ZnO. Formerly SiC.
- Transducers—PLZT (Pb-La-Zr-Ti-O) piezoelectrics.
- Oxygen sensors—semiconducting oxides such as ZrO_2 and TiO_2.
- Magnetic materials—Fe_2O_3-based ferrites.
- Semiconductors—SiC (note silicon the dominant semiconductor, is a metalloid not a ceramic).
- Superconductors—Y-Ba-Cu-O perovskites.

In dollar value, one of the largest uses of alumina as an electrical ceramic is as insulating substrates in *thin-film* and *thick-film* microelectronic devices. A substrate is the surface onto which microelectronic components are deposited, in the construction of microcircuitry. Components for microcircuitry are very small. Thin-film semiconductor components are now approaching 10 nm in size, and are commonly deposited onto a substrate by vapor deposition and lithography. Thick-film microcircuitry components are in the micron size range and are commonly deposited onto a substrate by micro-screen printing technology.

A substrate is usually planar, and can come in the form of a wafer of material, or a thin film of material, commonly deposited onto a wafer. There are essentially three types of substrates:

- *Silicon* wafer semiconducting substrates for microchip manufacture.
- *Alumina "thin-film" wafer* silicon-coated "SOI" substrates for microchip manufacture. Thin-film alumina substrate, typically a few hundred microns thick, with silicon coating. Thin film is a *subtractive technology*, that is, a thin film of silicon is deposited then selectively etched.
- *Alumina "thick-film"* insulating substrate, typically 1 mm thick, used for micro-screen printed, robust, often multilayered, microelectronic devices. Thick film is an *additive technology*, that is, conductor and resistor materials are deposited on the alumina substrate via a screen-printing approach.

While alumina dominates the role of insulating substrates, AlN has a significant niche role in this area, as well as BeO to a lesser extent.

13.3.2 Alumina thin-film semiconductor (microchip) technology: SOI-integrated circuit

A discussion of alumina substrates for microelectronic technology needs to be seen in the context of the SOI semiconductor integrated circuit (IC) industry, the industry that focuses on microchip manufacture. The silicon-substrate-based microchip has dominated the microelectronics industry for several decades, but the microchip is very delicate and sensitive to environment stresses such as heat and corrosion. This creates an important niche for the robust alternative the SOI concept, that is, silicon semiconductor on an alumina thin-film substrate with the alumina functioning in the dual role of insulator and heatsink.

IC electronic devices, as shown in Fig. 13.8, commonly known as "microchips," "microprocessor chips," "microprocessors," or simply "ICs," are miniaturized stand-alone silicon-semiconductor-based microcircuit packages. They are generally made by thin-film technology: vapor deposition of thin films, followed by nanolithography of the thin films to produce nanoscale transistors, diodes, connections, and other componentry of a semiconductor electronic circuit. The size of a thin-film transistor has shrunk over the years and is now in the order of 10 nm.

The microchip was a quantum leap forward in electronic technology that arose in the 1960s, evolved to high levels of sophistication through the late 20th century, and into the 21st century, and essentially underpins the entire electronics industry today. The microchip has its roots in the invention of the transistor in 1947 which gave birth to the microelectronics industry. The transistor was a quantum leap forward from the preceding diode/triode vacuum valve technology, which was a technology quite impractical for miniaturization.

In the 1950s, miniaturized transistorized electronic devices first appeared, such as portable radios and the first wearable pacemaker of Medtronic, invented by Earl Bakken (Chapter 8). The concept of using alumina as the nonconducting base (substrate) for semiconducting coatings was first theoretically explored in 1959 [564]. Some of the early pioneers in the invention of the IC were Werner Jacobi of Siemens who filed the patent for the first simple IC with five transistors in 1949 [565], and Jack Kilby of Texas Instruments who won the Nobel Prize for the invention of the IC, and filed his ground-breaking IC patent in 1959 [566].

Alumina as an electrical insulator 433

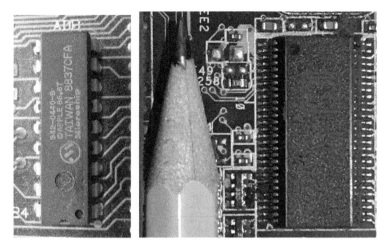

Fig. 13.8 Microprocessor chips [integrated circuits (ICs)]. Left: 1988 vintage microchip as used in a Macintosh SE, one of the early personal computers. Right: contemporary microprocessor, with a pencil for scale (note this microprocessor is substantially more miniaturized). The microchip industry has grown from $12 billion to $500 billion since 1980, with an associated million-fold increase in nanocomponent density.
Left: Image Author Binarysequence [4]. Reproduced from Wikipedia (Public Domain Image Library).

The microchip became widely available in the 1970s and by the late 1970s was underpinning the fledgling computer industry. In this era, popular electronics magazines began to have large advertisements advertising ICs of different types, sold by a growing number of manufacturers, with increasingly competitive pricing. Rudimentary personal computers began to appear as a niche market. Today, four decades later, smart phones, laptops, tablets, smart watches, flash drives, and numerous other microchip devices number in the billions. The microchip industry has seen meteoric growth since 1980 when the electronics industry was around $200 billion and the semiconductor industry was about $12 billion, only 6% of the market value of the electronics industry. Today, the global electronics industry is around $2 trillion. Underpinning this is the microchip (semiconductor) industry, at around $500 billion, or 25% of the market value of the electronics industry. Therefore, the microchip industry has grown by a factor of 40, since 1980 from $12 billion to $500 billion today and is forecast to reach the $trillion mark by about 2030. Silicon wafers, generally supplied as 300 mm wafers, are the dominant substrate in this market.

SOI microchips are commonly produced by depositing a silicon film onto an alumina insulating substrate and then nanolithography is used to engineer the nanometer-scale diodes, transistors, and other microcomponents onto the silicon surface. The only difference between the SOI microchip and the pure silicon microchip is the presence of the alumina substrate, the electrical insulator that greatly enhances heatsinking properties.

On the plus side, silicon technology enables extraordinary miniaturization. Half a century ago, thin-film electronic components were tens of microns in size, now they are around 10 nm, three orders of magnitude decrease linearly, six orders of magnitude in a planar sense. That is a millionfold increase in processor density in half a century. Today, the number of transistors in a microchip surpasses 20 million per square millimeter. On the negative side, microchips are very fragile and delicate and incapable of operating in harsh environments. Increasingly, they do not have the heatsinking capacity to cope with the component density.

Alumina's role in thin-film microchip technology as the "SOI" microchip technology is therefore vital when more heatsinking is required than silicon can deliver. The majority of microchips today (90%) are made from pure silicon substrates (wafers). About 10% are made from SOI wafers.

SOI wafers (alumina thin-film wafers, silicon coated) are primarily used in microelectromechanical systems (MEMS) which represent about 50% of the SOI market, driven primarily by the increased demand for accelerometers and gyroscopes for smart phones, tablets, and cameras. Other SOI markets are power device electronics (mainly in plasma TV drivers), and in the power semiconductor industry. SOI represents a significant proportion of the wafer market. Asia supplies about 75% of all SOI wafers, and consumes about 50% of the global SOI market, making Asia a net SOI exporter.

Since approximately 10% of the global semiconductor market is serviced by SOI substrates (wafers), with the remaining 90% serviced by pure silicon substrates (wafers), the global market in SOI wafers (thin-film alumina, silicon-coated substrates) underpins 10% of the global $500 billion semiconductor industry. That is, SOI technology underpins a $50 billion industry, primarily in MEMS and power device electronics.

13.3.3 Evolution of alumina thick-film microelectronic technology

Alumina thick-film technology arose in the same era as the microchip—the 1960s. In this era, Bell Telephone Laboratories of AT&T first sputtered TaN thin films onto glass, and later onto 99.5% alumina substrates, as resistors and capacitors to interface with transistors [567], and Du Pont developed PdO-Ag thick-film resistor technology [568]. The switch from glass substrates to alumina substrates was an important evolution as alumina has a much higher thermal conductivity and mechanical strength. This proved vital for the miniaturization that lay further down this track. A ground-breaking paper in 1966 reviewed the uses of alumina as a substrate onto which resistor, capacitor, and conductor networks could be deposited in a computer assembly plant, noting the following advantages of alumina in this role [569]:

- High strength
- Chemical inertness
- Smooth surface morphology
- High thermal conductivity
- High electrical resistivity (insulation)

The principles outlined in 1966 remain the key drivers today.

In microelectronic technology, two terms commonly used are thin film (10–500 nm by vapor deposition) and thick film (10–25 μm by screen printing). This is a thickness difference of 50–1000, or approximately two to three orders of magnitude.

Thick-film micro-screen printing methods for making robust, sintered, and sometimes co-fired, electrical devices, developed as a means of printing electrical component precursors from powdered glass (frit) and other functional components, suspended as a slurry in an organic solvent, containing a binder [570]. The screen-printed powder deposit is then consolidated onto the substrate by sintering, generally around 500–900°C. Primarily the functional components thereby produced are conductors, resistors, and dielectrics. They do not contain semiconductor components and therefore they are not microprocessors. They are commonly used to interface with microprocessors. Semiconductors cannot be made by this process, however, they can be added to the top of a co-fired thick-film circuit. Organic field effect transistors (FETs) are an active research area in this regard.

The metal conductor precursor material can comprise powdered metal, usually of a powder <2 μm in size, or can be produced by metallo-organic deposition, which involves metal precursors in the screen-printed deposit [571]. Firing in oxidizing conditions converts the metal precursors into metal oxides and firing in inert gas or reducing conditions converts them into metals. In some cases, the electrical components can be prepared by tapecasting them into thick films, laser machining, laminating, and co-firing with the substrate [572].

The principal advantages of thick-film technology, in contrast to thin-film technology, are:

- Cost and convenience: low capital and operating cost.
- Simple operation.
- Ease of automation.
- Multiple layers can be printed which enhances the capability of the process.

Thick-film technology is also important for applications involving multilayer laminates. Sometimes 50 or more layers are involved, and in most cases alumina-based, but sometimes also incorporating heatsinking AlN or BeO layers.

Thick-film alumina substrates are ideal for harsh environments. Technology applications include transducers, sonar transmitters, amplitude regulators, active filters, microwave devices, high-speed computers, and they see widespread use in the automotive industry, aerospace industry, power electronics, sensors, hybrid microelectronics, and especially microwave devices. There is a clear technology overlap between thick-film circuitry and the high-density multilayer alumina feedthrough technology described in Chapter 10 (bionic eye micro-feedthrough) and Chapter 14 (micro-feedthroughs for conventional electronics). Into the future, the line between multilayered thick-film alumina microelectronic devices and high-channel number alumina micro-feedthroughs is likely to become blurred.

The low loss tangent of alumina makes thick-film alumina microelectronic devices the dominant technology in the huge microwave industry.

13.4 Alumina substrate technology

There are a number of important properties required for the insulating substrate role. While alumina is not the highest performing ceramic on all six criteria, overall it has the most optimal fit to these essential properties, and a huge cost advantage over all the viable alternatives. For example, AlN and mullite are better on thermal expansion. BeO is significantly better on thermal conductivity but toxic. AlN is better on thermal conductivity and nontoxic, but very expensive. Alumina is pre-eminent in dielectric loss tangent, electrical resistivity, and low cost. Across the board, and when cost is taken into consideration, alumina is optimal:

(1) Outstandingly high electrical resistivity (insulating properties)
(2) Outstandingly low dielectric loss tangent
(3) High mechanical strength
(4) Thermal stability
(5) Corrosion resistance
(6) High thermal conductivity
(7) Thermal expansion matching with silicon
(8) Low cost

The importance of the first five properties is self-evident. The sixth property, high thermal conductivity, is important for heat dissipation, that is, heatsinking. Heat generation and dissipation is a key problem in service. The seventh property, thermal expansion matching with silicon, is important so as to minimize thermomechanical interfacial stresses. These requirements reduce the list of possible candidates for insulating substrate to three main oxide ceramics: alumina; mullite ($3Al_2O_3 \cdot 2SiO_2$), and BeO. There are also four nonoxide ceramics used: SiC, Si_3N_4, AlN, and BN.

AlN has emerged in the last two decades as a significant rival for alumina in the microelectronic substrate role. For example, CoorsTek, a leading supplier of alumina substrates supplies a range of AlN substrates [28]. The main reason for the rise of AlN is that while AlN has a similarly high electrical resistivity to alumina (Fig. 13.9) it has five times higher thermal conductivity than alumina, as shown in Fig. 13.10. Thus, AlN is a much better heatsink. This is increasingly important as semiconductor devices trend toward higher power.

AlN has a closer thermal expansion match with silicon, as shown in Fig. 13.11. Note that the thermal expansion coefficients reported in Fig. 13.11 are approximate values for the approximate 20–300°C temperature range. The reality is that dilatometry curves show different expansion coefficients at different temperatures and so thermal expansion matching is a much more complex matter than shown in this simple histogram. Nonetheless, AlN is probably a better match for silicon than alumina or BeO in the operating temperature range of microelectronic devices, which is important as semiconductor devices trend toward larger dimensions.

On the negative side, AlN has a significantly higher dielectric loss factor than alumina (Fig. 13.12). More importantly, AlN is a nonoxide ceramic that is difficult to densify, requiring controlled atmosphere and hot pressing. This AlN densification problem applies not only to the synthesis of AlN substrates for commercial sale (high

Alumina as an electrical insulator 437

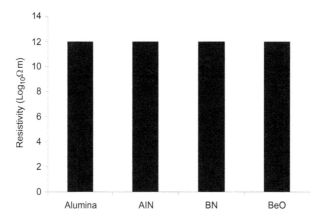

Fig. 13.9 Electrical resistivity at room temperature, as $\text{Log}10(\Omega\text{m})$. for the leading electrical substrate ceramics. $10^{12}\,\Omega\,\text{m}$ is an extremely high electrical resistivity value, and alumina, AlN, BN, and BeO are all around the $10^{12}\,\Omega\text{m}$ order of magnitude on electrical resistivity.

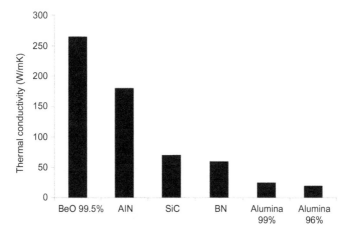

Fig. 13.10 Thermal conductivity of leading electrical ceramics [573]. The higher the thermal conductivity, the better the heatsinking capability of the substrate.

cost of commercially supplied substrates), but also in the sintering or co-firing of multilayer AlN thick-film devices, which translates into higher cost and complexity in AlN multilayer device manufacture. Moreover, the cost of AlN is very much higher than alumina.

In contrast to AlN, Alumina is a low-cost ceramic that can be pressureless sintered to full density in air, or in a controlled atmosphere (if the embedded screen-printed microelectronic components require that). This makes alumina a cost effective substrate to manufacture or purchase, and makes the sintering of alumina multilayer or

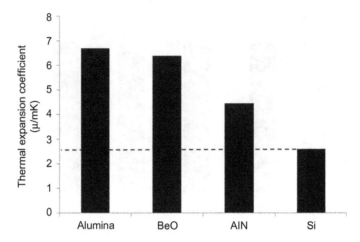

Fig. 13.11 Approximate thermal expansion coefficients, from various sources, in the 0–300°C range for leading electrical insulating substrate materials, with silicon included for context. The ideal scenario is to be just a little higher than silicon. Lower than silicon is going to cause the silicon to fail in tension. If it is too much higher than silicon, the silicon will be placed in higher compressive stress than is wise. Alumina, BeO, and AlN are therefore all suitable on thermal expansion criteria, with AlN the best fit.

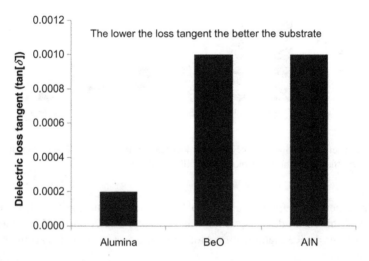

Fig. 13.12 Dielectric loss tangent for alumina, BeO, and AlN, the leading electrical substrate ceramics. The lower the dielectric loss, the better, especially in high-frequency electromagnetic fields such as those in microwave applications.

co-fired devices a simple process. Moreover, the extremely low loss tangent of alumina makes it the material of choice in microwave applications.

BeO has an even higher thermal conductivity than AlN, as shown in Fig. 13.10 but it is toxic and has much lower mechanical strength than alumina. Moreover, it has a higher dielectric loss tangent than alumina. For this reason BeO, always was, and will remain, only a niche product in the thick-film electrical substrates market.

BN meets the resistivity criterion, but is not much higher than alumina in thermal conductivity. Given its high price, this makes BN uncompetitive against alumina as a substrate material.

- *Cost*: alumina 5–10 times lower.
- *Electrical resistivity*: alumina, AlN, BN, and BeO are all similar.
- *Thermal conductivity*: BeO the strongest, closely followed by AlN. Alumina significantly lower.
- *Thermal expansion*: none are a perfect match for silicon, although AlN comes closest.
- *Dielectric loss factor*: alumina by far the strongest.

In Summary:

- *Alumina*: Dominant.
- *AlN*: Used when thermal conductivity (heatsinking) is more important than cost and processing complexity.
- *BeO*: Used in niche applications when the highest thermal conductivity (heatsinking) is essential.
- *BN*: Not competitive against alumina or AlN.

Therefore, it is probable that 96%–99.5% pure alumina will continue to be the low-cost easily-processed dominant substrate material into the future with AlN the high-cost alternative, used when heat dissipation is a stronger imperative than cost. BeO is likely to remain a niche product for applications where thermal conductivity is a very strong imperative. Moreover, given that alumina has a much lower dielectric loss tangent than both AlN and BeO, alumina will always dominate microwave applications. The microwave industry is absolutely huge (see Section 13.5).

13.4.1 The substrates—Thin film versus thick film

There are essentially two types of alumina substrates used in electronics:

- Thick-film alumina substrates
- Thin-film alumina substrates

These are available commercially in alumina, AlN, and BeO, with alumina the dominant substrate of the industry.

Thin-film substrates which are generally silicon, but can be SOI, in which case it is generally alumina coated with silicon. Sometimes SOI can be AlN or BeO substrates coated with silicon. These are used for vapor-deposition/photolithographic deposition of nanometer-scale electronic components, in the manufacture of semiconductor microchips.

Thick-film substrates, which are generally alumina, but can be AlN or BeO, are used for micro-screen printed thick-film microelectronic devices. These involve bonded micron-scale resistors, conductors, and dielectrics, which are sintered, sometimes co-fired in multilayers, to form robust microelectronics devices. They do not contain semiconductor components and therefore they are not microprocessors. They are commonly used to interface with microprocessors.

The terms *thick-film substrates* and *thin-film substrates* don't denote the substrate thickness (albeit thick-film substrates tend to be somewhat thicker) but the purpose for which the substrate is intended: thin-film (vapor-deposited) or thick-film (screen-printed) microelectronic component deposition onto the substrate surface. Thus, purity is higher, grainsize finer, surface finish smoother, and price is higher for thin-film substrates.

A discussion of the thick-film microelectronic technology would be incomplete without overviewing the alumina substrates themselves.

13.4.1.1 Thin-film alumina substrates

Thin-film alumina substrates, primarily used for SOI applications, generally come in two grades: 99.5% pure the economical version, and the higher grade 99.6% pure. They are manufactured as tapecast alumina wafers. Thin-film alumina substrates are generally below 1.5 mm in thickness. CoorsTek, one of the world's leading alumina ceramic manufacturers, offers thin-film alumina substrates as thin as 0.127 mm (0.005 in.) [28].

Thin-film alumina substrates are made to tight thickness and surface morphology tolerances, and commonly laser machined. They can truly be described as wafers whereas thick-film alumina substrates are probably more akin to thin alumina tiles. MgO sintering aid is generally added to the high-purity alumina raw materials so as to help minimize grainsize which is ideally <2 µm, with associated ultra-smooth surface finish, ideally <0.15 µm.

13.4.1.2 Thick-film alumina substrates

Thick-film substrates need to have sufficient glassy phase to enable a bond to form between the substrate and the electronic component precursor materials applied to the substrate surface by screen printing, followed by sintering or co-firing. For this reason, the industry convention for thick-film alumina substrates is 96% alumina, with typically 0.05%–0.1% Na_2O content. This product comprises a small but significant portion of aluminosilicate glassy phase, including alkali metal ions such as sodium. This glassy phase assists in the bonding of conductor and resistor materials to the alumina thick film. Glass-bonded conductors can be bonded to the substrate surface via a glassy interaction between the glass phase in the substrate and the glass phase in the bonded conductor.

It is important to minimize the alkali metal content in the alumina substrate. A high alkali metal content in the substrate can be problematic if there is an alkali metal diffusion interaction between the glassy phase of the substrate and the glassy phase of the bonded electrical conductors/resistors. Alkali metal ions can affect resistivity via their role as ionic conductor species, which is enhanced at elevated temperatures, as for example in beta alumina $Na_2O \cdot 11Al_2O_3$ (see Section 13.1.1). Absolute resistivity can be affected as well as the temperature coefficient of resistance. Generally, Na_2O content is kept below 0.1 wt% in alumina thick-film substrates.

Commercially available thick-film alumina substrates comprise large thin alumina wafers in the order of 1 mm in thickness and many centimeters in size. CoorsTek, a leading global manufacturer of alumina ceramics, are a pioneer in alumina substrate technology and offer thick-film alumina substrates as thick as 3.5 mm, and as thin as 0.25 mm, but typically they are 0.6–1.0 mm in thickness [28].

Thickness determines the fabrication method. Substrates thinner than 1.5 mm or larger than 50×50 mm are generally tapecast. Substrate dimensions of 100 mm and even up to 200 mm are commonly produced. Substrates thicker than 1.5 mm, and smaller than 50×50 mm are generally dry pressed, and sometimes produced by powder roll compaction. Holes and slots can be incorporated during fabrication or by laser machining after sintering.

13.5 Microwave-industry electrical applications of alumina

13.5.1 Microwave technology

Before discussing the large-scale use of alumina in the microwave industry, a general background to the microwave industry is helpful for context.

Microwaves are found between radio waves and infrared waves in the electromagnetic spectrum, that is, the frequency range 300 MHz–300 GHz, which corresponds to the wavelength range 1 m down to 1 mm. There are a number of unique features of microwaves:

- *Focused beams*: unlike radio wave transmissions which permeate the airwaves (multiple transmissions on the same frequency interfere with one another), focused microwave beams of the same frequency can be targeted (using directional antennas) in different directions at the same time in the same location.
- *Line of sight*: unlike radio waves which can permeate over topographic interference, microwave beams travel by line of sight.
- *Omnidirectional antenna*: microwaves can also be transmitted by an omnidirectional antenna, as for example by a cellphone.
- *Heating*: microwave energy couples with water and biomolecules in food and human tissue, and can be used for heating. This microwave effect has widespread use in cooking food, and also use as a weapon.
- *Risk*: microwaves are not mutagenic but there is a suspected carcinogenic effect from long-term tissue exposure.

13.5.2 Industrial uses for microwave technology

The microwave industries are huge in scale:

Ovens: microwave energy couples with water and biomolecules in food and human tissue, generating heat. There are about one billion microwave ovens in the world. The microwave oven market is about 50 million ovens a year.
Radar: essentially radar involves a focused microwave beam targeted at a specific object, with measured return echo. Radar systems are ubiquitous for navigation, object detection, weapons targeting, meteorology, aircraft onboard radar, ground-control radar, weapons radar, and maritime radar. Many thousands of these radar systems are used globally.
Telecommunications: microwave communication is ubiquitous. Short-range low-power links include WiFi, Bluetooth, and Cellphones. Long-range high-power links include cellphone tower to satellite communication, news-van (live-telecast) to satellite communication, and satellite TV receivers. There are over one million cellphone towers in the world today, and millions of cellphone-tower-mounted microwave generators servicing the seven billion cellphones globally (approximately one for every person on the planet).

13.5.3 Types of microwave generators

- *Low energy*: Small solid-state generators such as diode or FET, for example, ~1–5 W cellphone.
- *Medium energy*: Cavity magnetron. The basis of the oven industry, a low-cost 1 kW ~2.5 GHz microwave generator. Klystron and travelling-wave tube (TWT) for telecommunications and radar, where power requirements can range from W to kW range.
- *High energy*: Tens to hundreds of GHz range, KW to MW power range. Gyrotron used for high-intensity industrial heating, nuclear fusion reactors, and "heat ray" weapons systems (human skin heating).

Cavity magnetrons are one of the most widely produced vacuum-tube electrical devices in the world today, with one billion of them out there in ovens. Applications for the Klystron and TWT number in the millions, in the communications and radar realm. Currently, a niche use for gyrotrons is plasma heating in nuclear fusion systems. If the future for humanity is nuclear fusion energy, the gyrotron looks set to become a huge mass-market product also. Currently, it sees significant industrial usage but far less so that oven and communications microwave generators.

The magnetron, the dominant microwave generator globally, was originally invented in 1910. Its modern version, the mass-market cavity magnetron (see Fig. 13.13) that underpins the microwave oven industry, has its origins in the 1946 patent of Percy Spencer [574]. The cavity magnetron, compared with the Klystron, TWT, and gyrotron, is at the bottom level of cost and technology sophistication. A 1 KW microwave magnetron for a kitchen microwave oven needs to be manufactured for less than $50, requiring a low-cost hermetic feedthrough system, and low-cost oven infrastructure and electronics.

Alumina as an electrical insulator 443

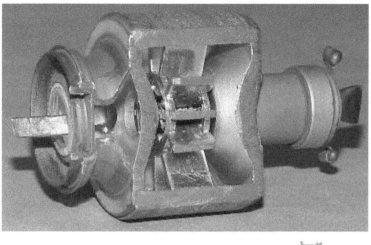

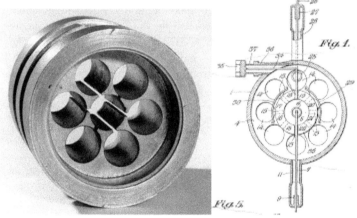

Fig. 13.13 Cavity magnetron. A high-power vacuum device that is used to generate microwaves. In essence, it comprises a vacuum-tube cathode which passes through the center of a cavity anode. Top: cutaway view of an actual cavity magnetron. The cavity anode is the bronze colored center section. Lower left: the cavity anode itself. The cathode passes through the central hole of the cavity anode. Lower right: schematic from the original 1946 patent by inventor Percy Spencer [574]. 1 kW, 2.5 GHz magnetron microwave generators are widely used in the billion ovens globally.
Top: Author: HCRS Home Labor Page [575]. Lower left: Image Author John Cummings, from the Science Museum London [576]. Reproduced from Wikipedia (Public Domain Image Library).

13.5.4 Alumina usage in microwave technology

Alumina is very important in the huge microwave industry due to its extremely low AC loss tangent, in combination with its extreme electrical resistivity, high heat resistance, good thermal conductivity, and when required, its capability for transparency as synthetic sapphire, or translucent polycrystalline alumina (Chapter 5).

13.5.4.1 Microwave-integrated circuits

Microwave-integrated circuits (MICs) are thick-film microelectronic devices that use alumina substrates. MICs take advantage of the extremely low dielectric loss tangent of alumina, and its robustness. Alumina thick-film microelectronic devices are essentially transparent to the extreme high-power high-frequency energy of the microwaves.

Given the enormous size of the global microwave industry, and therefore the huge global demand for microelectronic systems that are able to tolerate intense microwave fields, the alumina MIC is one of the largest applications for alumina in the electrical realm.

13.5.4.2 Microwave windows

Microwave windows depend on the extremely low dielectric loss tangent of alumina. Thus, one particularly important specialty use of alumina is vacuum-seal microwave windows. This is an application that utilizes a microwave transparent window that provides hermetic sealing of a vacuum tube, but allows microwave radiation to pass through the window without significant heating or energy attenuation, where extremely high-energy high-frequency (GHz range) microwave energy is involved. The system is equally applicable to radio frequency (RF) windows. Microwave windows are a special case of feedthrough technology: radiation feedthrough.

In summary:

Technology: Microwave windows:
Important property: Heat resistance. Extremely low dielectric loss tangent. Transparency.

13.6 Alumina as an electrical insulating ceramic. Concluding remarks

The three large markets for alumina as an electrical insulator are all in the $Billion order of magnitude:

Spark plug insulators: In dollar terms, alumina is the majority of the value embodied in a spark plug and is essentially the sole material used in spark plug insulators, the main high-tech component of the spark plug. This was the first commercial application of alumina, dating back to the mid-20th century. Spark plugs are a $3 billion global industry, with alumina a key component of that value. All indications are that alumina's central role in this industry will continue for as long as internal combustion engines remain a dominant global transportation technology.

Thin-film microelectronic devices: Commonly known as "microchips," these are fragile and delicate microelectronic devices that are made by vapor-deposition/nanolithography techniques onto silicon substrates (~90% market share), or SOI silicon-coated alumina thin-film substrates (~10% market share). Thin-film substrates are an essential platform technology underpinning 10% of the $500 billion semiconductor industry. This represents about 10% of global wafer consumption, the other 90% being silicon wafers.

Thick-film microelectronic devices: These are robust microelectronic devices made by screen-printing micron-scale electronic components onto thick-film alumina substrates. They are commonly multilayer devices, sometimes involving >50 interconnected layers.

In contrast to the fragile IC chip technology, thick-film microelectronic devices are not microprocessors, but rather robust electronic devices capable of interfacing with a microprocessor, and capable of withstanding extreme operating conditions, such as severe thermal cycling and corrosive conditions. As such they are an important technology for applications that call for robust microelectronic devices. They see widespread use in the automotive industry, microwave industry, and in sensors. They are also used for specialized applications such as high-speed computers, transducers, amplitude regulators, active filters, and sonar transmitters. The global market for thick-film alumina substrates is not precisely quantified, but it would be similar to the thin-film SOI alumina substrate market.

Other: Alumina has many other boutique applications as an electrical ceramic, the most significant being, fuse bodies, high-frequency (nonmagnetic) coil formers, and ultra-high resistivity insulators for power transmission systems.

In summary: Of the $50 billion SOI market, the extreme heatsinking applications are filled by AlN and BEO, but the majority are the domain of alumina thin-film substrates. The thick-film alumina substrate market size for thick-film microelectronic devices is similar in scale to thin film. Finally, while $50 billion is bigger than the $3 billion spark plug market, alumina is the majority of the value embodied in a spark plug, but a significantly smaller proportion of the value embodied in an SOI microchip, thus the alumina market size in dollar value is comparable in each sector:
- Spark plugs
- SOI microchip substrates
- Thick-film microelectronic substrates

Alumina-metal bonding for electrical feedthroughs   14

The alumina hermetic electrical feedthrough seal is the basis of a very large industry in dollar value. Bionic implants, a $25 billion global industry, exclusively use alumina feedthroughs. Bionic implants are a special case for electrical feedthroughs, for which the following essential criteria apply:

- High biocompatibility.
- High electrical resistivity long term in the warm moist corrosive in vivo environment.
- Chemically stable long-term in the warm moist corrosive in vivo environment.
- Impervious to liquid or gas diffusion through the ceramic, or through the ceramic-platinum interface or ceramic-braze interface in the warm moist corrosive in vivo environment.
- Stable ceramic-platinum bonding long term in the warm moist corrosive in vivo environment.
- Stable ceramic-braze bond long term in the warm moist corrosive in vivo environment.
- High purity and therefore resistance to static fatigue in the warm moist corrosive in vivo environment.

These are a set of extremely stringent requirements and only alumina meets them, in combination with a titanium casing, and platinum channel conductor wires, with a proven track record over the last half-century in the role. The industry for non-biomedical electrical feedthroughs, that is, "electrical-industry feedthroughs" has a much less stringent set of requirements. Moreover, it predates the bionic implant industry by more than a century, with its origins in the 1830s, whereas the alumina/titanium bionic feedthrough was invented in 1970 by David Cowdery (Chapter 8). This chapter outlines the evolution and technology development of the "electrical industry feedthrough," beginning with the lightglobe in the 19th century, and evolving into the alumina feedthrough in the mid 20th century, which diverged into the various manifestations which are found in the world today, all of which are discussed. Alumina feedthroughs for bionic implants are discussed in detail in Chapters 8–10.

In a way, one could say that the bionic feedthrough is one specific case of the very much older parent concept, the electrical feedthrough, but this one special case has grown rapidly in recent decades to be an industry much larger than its parent industry, electrical-industry feedthroughs.

In this book, the two industries are distinguished by the following definitions:

Biomedical: Bionic feedthroughs:
- Ceramic feedthrough insulator: exclusively alumina.
- Metal casing: exclusively titanium.
- Ceramic-casing bond: titanium-based braze or gold-based braze.
- Conductor wires: exclusively platinum or noble-metal alloys.

Electrical-industry feedthroughs (nonbiomedical):
- Ceramic feedthrough insulator: predominantly alumina or glass (amorphous glass commonly, sometimes glass ceramic): other ceramics have been used in the early days.
- Metal casing: various metals.
- Ceramic-casing bond: various metal brazes, or solder glass.
- Conductor wires: various metals, commonly Kovar or copper.

As should be apparent from the above, the electrical-industry feedthrough is a very old technology with many manifestations and quite flexible from a materials point of view. A wide range of materials have been used for electrical-industry feedthroughs, and they continue to be relatively flexible in their requirements. This is because biocompatibility and the warm corrosive human body environment are not requirements for electrical-industry feedthroughs. In contrast, the bionic feedthrough is a special case for the electrical feedthrough, relatively recently developed, and with an extremely stringent and inflexible set of materials requirements.

Secondly, not all feedthroughs are electrical. Feedthroughs come in two main forms:

Electrical feedthrough: Hermetic seal through which electrical conductor wires pass.
- Bionic feedthrough (alumina)
- Electrical-industry feedthrough (glass or alumina)

Nonelectrical feedthrough:
- Microwave window (alumina, in rare cases BeO)
- Vacuum-furnace-viewing window (fused silica or synthetic sapphire)
- Sealed-vessel viewing window (Glass)
- Conduit feedthrough, such as a liquid or gas conduit

Hermetic electrical-industry feedthroughs are *always ceramic-based*, whether that ceramic is glass, a glass ceramic, alumina ceramic, or some kind of glass-ceramic composite. Polymer seals are not hermetic. Metals are not electrical insulators. It has to be ceramic (glass, glass ceramic, alumina ceramic). If it is not glass, it is almost exclusively, alumina. The primary focus in this chapter is on alumina seals, but for context, glass seals will also be briefly discussed. For historical completeness, other ceramics that have been trialed in the past will also be mentioned.

Alumina, as a hermetic insulator, dominates the high-end electrical hermetic feedthrough system applications in electrical-industry feedthroughs and is exclusively used in the biomedical industry (bionic feedthroughs). The main alternative to alumina is glass-metal seals, which are primarily used in low-tech electrical applications where conditions are mild: low-temperature, low-voltage, low-frequency, low-corrosiveness, low-stress, and low-cost requirement.

Electrical-industry feedthroughs are used in many applications including X-ray tubes, microwave generators, cathode ray tubes, triode valves, capacitors, discharge tubes, heatsinks, insulators, and indeed any device, which requires electrical current, fluid conduits, or electromagnetic energy (window) to pass through the wall of a sealed chamber or membrane.

Alumina is ideal for feedthroughs because of its extremely low electrical resistivity, high chemical inertness, capacity for transparency (as synthetic sapphire), and low cost, albeit not as low as glass. Alumina has one other unique quality of importance: the extremely low AC loss tangent of alumina makes it the best choice in microwave systems, radio transmitters, and other systems involving high-frequency electromagnetic fields. Not just potentially in the feedthrough but more importantly in the electrical substrates of the microwave device, for example, thick-film microelectronic systems (Chapter 13). The microwave industry is extremely large (see Chapter 13).

In brief terms, the various alumina electrical feedthrough seal types are as follows:

Alumina feedthroughs	
Bionic implants	High alumina
High-power vacuum metallizing devices	High alumina
Low-power vacuum electrical devices	High alumina
Microwave windows	High alumina, BeO

14.1 Evolution of the hermetic feedthrough concept

Alumina/metal hermetic seals, and indeed ceramic/metal and glass/metal hermetic seals in general, are a long established technology, with origins dating back to the 1830s. They are commonly used for vacuum electronic devices for which the hermeticity is generally measured by the helium leak test in accordance with MIL-STD-883 [378].

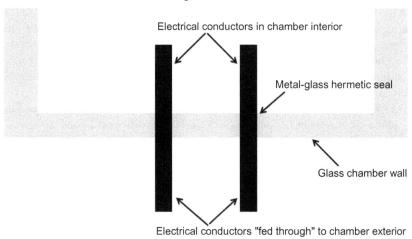

Fig. 14.1 Glass-metal hermetic feedthrough seal of the type used in the light globe and other spin-off technologies. A metal (usually Kovar) with a perfect thermal expansion match to the glass (usually borosilicate) is fused into the wall of the glass chamber, which may be done while the chamber is under vacuum or filled with inert gas, for example, Kovar electrical feedthrough into a glass lightglobe to the (oxidation susceptible) incandescent tungsten filament. This concept is also used in vacuum electrical devices such as triode valves and X-ray tubes.

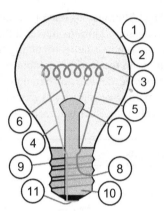

Fig. 14.2 Modern xenon-filled tungsten-halogen light globe schematic: (1) glass globe, (2) xenon or other inert gas inside globe, (3) tungsten filament, (4) contact wire that links to base terminal, (5) contact wire that goes to screw terminal, (6) support wires, (7) glass support post, (8) conductor connection point to screw terminal, (9) screw thread terminal, (10) glass-metal electrical feedthrough through with the two conductor wires pass, and (11) conductor connection point to base terminal.
Image Author Fastfission [577]. Reproduced from Wikipedia (Public Domain Image Library).

The original and simplest example of the hermetic seal is the electric light globe, as shown in Fig. 14.1 as the feedthrough conceptually, and Fig. 14.2. as the actual technical manifestation of the modern lightglobe, which involves a glass/metal hermetic seal, between the inert-gas filled (or evacuated) glass globe and the external environment, in order to protect the incandescent electrical filament from oxidation. The electrical conductors pass through the wall of the glass globe, to which they form a hermetic glass/metal seal, along with the terminal plate. The glass and the metal conductors (e.g., Kovar), sealed to the glass wall, need to have closely matching thermal expansion characteristics across the entire operating temperature range; otherwise, seal failure can occur by cracking at the glass-metal interface.

The concept of the hermetic feedthrough was originally invented by a succession of light globe innovators in the period 1760–1880. This 120-year period of innovation began with the first documented demonstration of heating a wire to incandescence by pioneering British scientist Ebenezer Kinnersley in 1761, and culminated in the optimally engineered, carbon-filament in a vacuum, commercially viable light globe of Thomas Edison in 1880 [578].

While it is Thomas Edison who is generally remembered as the inventor of the commercial light globe device in 1880, shown in Fig. 14.3, he was "standing on the shoulders of giants" in doing so. There is some controversy as to who invented the feedthrough concept for the electric light globe, whether it was James Bowman Lindsay in 1835, Marcellin Jobard in 1838, Warren de la Rue in 1840, or Fredrick de Moleyns who patented the first hermetic light globe in 1841. Indeed the inventor could have been someone else, now undocumented, who inspired one or all of these four documented early pioneers. The fact that all these reports date to the same 6-year

Alumina-metal bonding for electrical feedthroughs 451

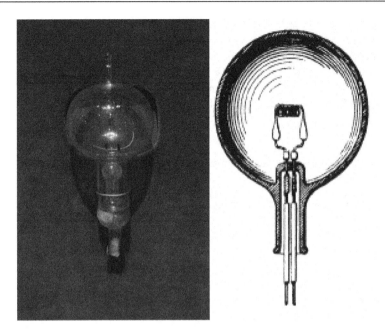

Fig. 14.3 The original carbon-filament/vacuum lightglobe of Thomas Edison. Left: An actual Edison lightglobe. Right: Schematic from the original Edison patent. The electrical feedthrough can clearly be seen with the two conductors passing through the protruding glass membrane at the base of the bulb [578].
Left: Image Author Terren [579]. Reproduced from Wikipedia (Public Domain Image Library).

period suggests that there was one genius inventor, and many competitors who were eager to capitalize on the invention, as is sometimes the case in a technology race where the commercial stakes are high. The identity of that genius, on whose legacy the entire modern electrical-feedthrough industry and the whole concept of electric lighting rests, is unfortunately lost to history.

Bright light requires electrical heating of a filament to an incandescent temperature. A dull red glow is achieved at around 600°C, by 1000°C a bright orange glow, and by about 1500°C the glow approaches a yellow-tinged "white hot" appearance. Around 3000°C, the brilliant white light of a tungsten-halogen filament is achieved, something we are all familiar with. The challenge to be overcome is that at incandescent temperatures, a carbon or metal filament will rapidly burn out in air. The filament therefore needs to be electrically heated either in a vacuum chamber or in a chamber filled with inert gas: the evacuated or inert-gas filled glass globe. This requires a means of getting electrical current through the wall of a sealed chamber and forming a hermetic seal that would be unaffected by the differential thermal expansion of the glass chamber wall, and the electrical conductors passing through that wall, during the heating and cooling cycles as the heat-generating incandescent light is switched on and off. Ideally, this would be achieved by developing a metal/glass combination with matching thermal expansion characteristics over the operating temperature range of the device; otherwise, differential thermal expansion would crack the seal. For a light globe, it also needs to be achieved at a production cost well below $1.

The ultralow-cost incandescent-filament/hermetic-feedthrough lightglobe was the mainstay of domestic lighting from the late 19th century until the early 21st century, when the similarly cheap fluorescent and LED lights began to supplant it for energy efficiency reasons. Fluorescent light tubes also use a low-cost glass-metal electrical feedthrough.

14.1.1 The hermetic feedthrough evolves: Invention of the vacuum valve

The vacuum valve was an evolution of the glass-metal feedthrough technology invented for the lightglobe. It was a vacuum-tube-based electrical component that performed the function of a diode or transistor in the pre-semiconductor era as shown in Figs. 14.4 and 14.5. Largely obsolete today other than in niche uses, the key technical breakthroughs underpinning the invention of the modern electrical feedthrough continue to see widespread use in many applications such as the x-ray tube and cathode ray tube shown in Fig. 14.6. These breakthroughs were:

- The patenting of the tungsten filament lightglobe by Sandor Just and Franjo Hanaman in 1904 [580].

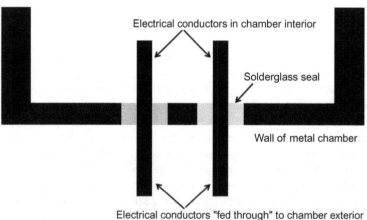

Fig. 14.4 Vacuum-valve solder-glass feedthrough concept. While some vacuum valves use a simple seal of the type shown in Fig. 14.1, the more complex glass-metal seals are of the above concept. Solder-glass seal as used in providing both a hermetic seal and electrical insulation at the hole in the metal chamber wall through which the conductor (usually Kovar) passes. The common scenario is borosilicate solderglass and Kovar (conductor) wires passing through the solder glass seal. The solder glass (gray) fuses to both the Kovar (conductor) wires and chamber wall and provides both a hermetic seal and electrical insulation. Commonly used in electrical component manufacture.

Alumina-metal bonding for electrical feedthroughs

Fig. 14.5 Early glass-metal feedthrough technology. These involve a simple glass-metal feedthrough. Vacuum tubes (also known as the diode valve, triode valve, electron tube, or simply "valve") are used in the electronics industry. While the diode and triode valves and cathode ray tube were gradually supplanted by the transistor and microchip and associated solid-state technology, they continue to be used in many significant niche applications, such as boutique amplifiers.
Image Author Stefan Riepl [584]. Reproduced from Wikipedia (Public Domain Image Library).

- The patenting of the "audion" by Lee De Forest in 1906 who filed two patents with identical titles on the same day [581,582]. This audion technology was the precursor to the triode vacuum valve.
- The filing of the original Kovar patent by Howard Scott in 1929, which was granted in 1936 [583].

The invention of Kovar was a game changer. Kovar is a trademarked name for an iron alloy containing approximately 29% nickel and 17% cobalt. However, Fe-Ni-Co alloys have seen such widespread use in electrical feedthroughs in the last 80 years that the name Kovar has become synonymous with Fe-Ni-Co feedthrough metal alloys.

Kovar was deliberately engineered by inventor Howard Scott [583] to have perfect thermal expansion matching to borosilicate glass [382]. In the 20th century, a wide range of solder glasses was developed for glass-metal sealing of electronic components, again primarily to Kovar [383]. The Kovar concept as the Fe-Ni-Co feedthrough metal bonded to glass originated in the 1920s [384], though it was not called Kovar at that time. From the 1920s to the 1960s, Kovar-glass was the predominant commercial electrical feedthrough system, that is, borosilicate glass, bonded to the electrical metal alloy based on the Kovar concept (iron-nickel-cobalt) [384]. Kovar remains a dominant metal alloy for metal-glass hermetic feedthroughs today, although alumina is now used in all the advanced and high-temperature feedthroughs.

Glass-metal seals remain in common use, in the order of billions of glass-metal seals manufactured a year. The majority of these are low-cost mass-market products in light globes and other low-tech applications, but there are also advanced

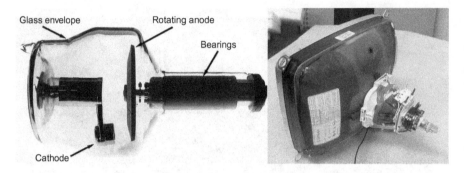

Fig. 14.6 Modern glass-metal feedthrough technology. Left: Rotating anode X-ray tube. Right: cathode-ray tube for a traditional television set. The X-ray tube is a huge mass-market product in medical imaging and security screening. The cathode ray tube is rapidly becoming obsolete, supplanted by LCD and plasma screens.
Left: Image Author Daniel W. Rickey [589]. Right: Image Author Blue tooth7 [590].
Reproduced from Wikipedia (Public Domain Image Library).

applications such as X-ray tubes and cathode-ray tubes where they are used as shown in Fig. 14.6. Their continuing widespread use is due to a number of commercial advantages:

- Low cost compared to ceramic-metal seals.
- Ease of manufacture, on both a small scale (glass workshop) and a large-scale (mass production).
- In conditions involving moderate mechanical and thermal stress, they can provide long-term reliability.
- Glass-metal sealing does not require the advanced equipment, and expertise, such as for welding under a vacuum, required for ceramic-metal sealing.

The state of the art in glass-metal seals is the use of glass ceramics, which have a long history dating back to the late-20th century [585–588].

14.2 Evolution of the alumina feedthrough

Bonding metals to *glass* hermetically is a simple fusion process with thermal expansion matching. The big challenge in the mid-20th century was finding a way to bond metals hermetically to *ceramics*. For this, there are three main criteria:

(1) Thermal expansion matching of ceramic and metal.
(2) Selection of a suitable brazing alloy that is compatible with metal and ceramic and has a melting temperature neither too high for impracticality nor too low for the end-use application.
(3) Metallizing, or in some other way, treating the ceramic surface so as to enable the braze to wet it and bond to it.

Kovar-glass hermetic sealing was sufficient for the majority of hermetic applications in the electrical industry of the early- to mid-20th century and continues to be sufficient in many low-tech applications today. However, there were an increasing number of applications that arose in the late 20th century for which Kovar glass was insufficient.

The focus in the electrical industry is application specific. In most low-temperature applications, Kovar glass is sufficient. When application temperatures exceed a few hundred degrees Celsius, or in other extreme conditions, metal-ceramic bonding technology is required.

Feedthrough applications range from simple low-tech mass market applications, such as insulators, capacitors, heatsinks, incandescent and vapor lamps, triode vacuum valves, and semiconductor housings, such as thick-film multilayer feedthroughs, to feedthroughs of so many kinds from high voltage to high frequency to highly corrosion resistant. Some of these applications are served by glass-metal seals, some require alumina-metal seals.

In general, glass-metal seals service the low-tech mass-market applications, such as lamps in all their many manifestations, while ceramic-metal seals (usually alumina Kovar) service high-tech and severe-environment applications. There are a number of other high-tech applications, with major commercial importance, for which ceramic-metal seals (usually alumina Kovar) might be necessary:

- Thermionic converters.
- Vacuum equipment requiring transfer of electrical signals, energy, or fluids through the chamber wall.
- Sealed equipment (liquid or controlled-gas content) requiring transfer of electrical signals, energy, or fluids through the chamber wall.
- Microwave tube seals.
- High-temperature microwave windows.
- Bionic implants (Chapters 8–10).

Research into ceramic-metal seals began in the 1930s, driven by the need for higher power, higher temperatures, and higher frequencies. A quantum leap forward in feedthrough technology was required, from glass-sealed systems to ceramic-metal bonding systems, where the ceramic evolved to be alumina and the metal evolved to be Kovar. Copper is also relatively common in this role. Ceramics, especially alumina, have many advantages over glass. These include

- Much higher temperature capability (glass is limited to a few hundred degrees Celsius).
- Alumina and ceramics in general are much mechanically stronger than glass.
- Alumina and ceramics in general are more thermal shock resistant than glass.
- Alumina and ceramics in general are less sensitive to flaws and stress concentrators than glass.
- Alumina specifically has much better electrical insulation than glass.
- Alumina specifically has a much better dielectric loss tangent than glass.

Some of the main demands placed on an alumina-metal joint include

- Hermetic sealing as per the helium leak test
- Ductile strain relief at the metal-ceramic interface

- High-temperature strength
- Corrosion resistance

14.2.1 The origin of alumina-metal bonding technology

Three leading German electrical companies and one US company were the drivers of the early ceramic-metal feedthrough development: Siemens, AEG, and Telefunken from Germany and General Electric in the United States. The 1935 Siemens patent of Vatter [591] and the 1939 General Electric patent of Pulfrich [385], both concerning the molybdenum-manganese method, were the result of this drive and represented the first steps toward using oxide ceramics rather than glass for hermetic seals. The Vatter patent [591] involved coating a ceramic with Fe, Cr, Ni, or W (with or without Mn), vacuum sintering, and then bonding to Cu, Ag, or Au or their alloys. The Pulfrich patent [385] was similar but involved coating a ceramic with Mo, Re, or W. Neither of these patents considered alumina ceramics, but rather pyrolusite, baria, calcia, zirconia, and sodium tungstate. At this time, alumina was a virtually unknown ceramic, the subject of a couple of rudimentary US spark plug patents, a couple of pre-World War I concept patents, and little else (Chapter 4).

The first reference to the possibility of using alumina in a ceramic-metal hermetic seal came in the mid-1950s, with the patents of Nolte [38,387], originally filed in 1947, but not granted until 1954. By the 1950s, alumina was starting to receive some attention as an engineering ceramic (Chapter 4). The Nolte patents are primarily about the molybdenum-manganese ceramic-metal bonding method. The ceramics involved are not claimed, although the patent notes "I prefer to use the zirconium silicate or magnesium silicate bodies; however, due to the nature of the reaction, it may be assumed that other materials are satisfactory, providing that they are refractory enough to withstand the prescribed firing temperature. Silicate, titanium dioxide, beryllium oxide, alumina, and others" [387].

In 1954, a key study laid the foundation for metal-ceramic bonding, defining seven methodologies that could be used [592]:

(1) Metal-ceramic bonding by sintering a metal oxide onto the surface of a ceramic and then reducing the oxide to a metal.
(2) Metal-ceramic bonding by sintering laminates comprising metal and ceramic powders in differing ratios (the functionally graded material approach).
(3) Metal-ceramic bonding by hot-pressing in a vacuum or inert atmosphere.
(4) Metal-ceramic bonding by utilizing active metals or metal alloys, or active metal hydrides.
(5) Metal-ceramic bonding by using glazes in an oxidizing atmosphere.
(6) Metal-ceramic bonding by using a combination of glazes and metal powders in a reducing atmosphere.
(7) Metal-ceramic bonding utilizing refractory metal powders (such as molybdenum) with or without manganese (for continuity), nickel or copper (for build-up), and utilizing solder for the final bond.

Of this list, four remain niche applications, and three methods have risen to dominate alumina-metal bonding technology today:

- Method 7 (the Nolte molybdenum-manganese method). The first and the dominant commercialized process.
- Method 4 (active brazing). Commercialized much later.
- Method 5 (solder glass bonding) has continued to be important for regular applications, as it was in the preceramic era. This method is sometimes referred to as brazing with glass.

Method 5 concerns glass bonding, it is low-tech, does not involve alumina, and is therefore beyond the scope of this book.

Methods 4 and 7 are very industrially significant today. Both of these methods involve brazing. Brazing involves joining two surfaces with an intermediate molten metal (braze) that is capable of wetting and bonding to both surfaces. Generally if the joining temperature is significantly below 600°C, and certainly if it is below 400°C, the intermediate metal is called a "solder." If joining with the molten intermediate metal is conducted at 600°C or above, it is called brazing, which is almost always the case for alumina-metal bonding with an intermediate metal.

Brazing is a favored metal-alumina joining method, particularly useful when the alumina and the metal have dissimilar thermal expansion characteristics, as the metal braze interlayer can allow for some ductile yield at the joint, unlike glass-bonding in ceramic-metal joins. Indeed, ductile yield at a metal-ceramic bonding interface is one of the key benefits of brazing with metal alloys, and one of its main advantages over glass bonding, since glass bonds have near-zero capability for yield (near zero failure strain). Alumina is particularly suitable for metal brazing.

Today, the processes of metallization and brazing are widespread. Metallization of the ceramic surface with a metallization film is a standard precursor process to brazing, for the purpose of improving the wetting of the ceramic by the braze, to thereby enhance the braze adherence to the ceramic (method 7). Brazes are generally classified into three main categories:

- Refractory metal braze (e.g., the molybdenum-manganese process).
- Precious metal braze (e.g., gold).
- Active metal braze, generally titanium-based (Method 4).

14.2.1.1 The Nolte molybdenum-manganese method

The molybdenum-manganese (Mo-Mn) method is a very common method used for alumina-metal bonding in the electrical industry. Compared to active brazing, it has the highest firing temperature and therefore provides the greatest high-temperature strength. Thus, it is the method of preference for high-temperature applications. It was invented by Nolte in the late 1940s with his original patent filed in 1947 but not granted until 1954 [386,387], The Nolte patents are primarily about the molybdenum-manganese ceramic-metal bonding method. The ceramics involved are not claimed, although the patent notes alumina as one of various candidate ceramics, and this is the first mention in the literature of alumina in this role, a role it came to dominate by the late-20th century.

The Nolte molybdenum-manganese ceramic-metal bonding method process (Methodology 7 from the above list) was the first approach of the 7 postulated above to be developed into a commercial reality. It was perfected and evolved into an alumina-dominated approach through the 1960s, 1970s, and 1980s and by the 1990s was highly evolved to the industrial process that is widely used today. By the 1990s, the molybdenum-manganese ceramic-metal bonding method, bonding alumina to Kovar, had become commonplace for ceramic-metal hermetic seals in the electrical industry. Alumina emerged as the dominant ceramic used for ceramic-Kovar hermetic seals in the electrical industry, with zirconia and mullite, also used, and in rare cases beryllia. Mullite for its thermal shock resistance, beryllia for its thermal conductivity, and zirconia for its toughness. Zirconia is problematic as a high-temperature insulator given that it needs to be stabilized (Chapter 6), which makes it an ionic conductor (Chapter 15).

The Nolte molybdenum-manganese (Mo-Mn) process remains the standard industry process for alumina-metal bonding in the electrical industry today. The Mo-Mn process in its modern manifestation since the late-20th century is very widespread and extensively covered in the literature [388–390,593–597]. It involves the general steps listed below, but it should be noted that there is a great deal of proprietary industry know-how underlying these simple steps, which involves much fine-tuning and modifications of this general process. Nonetheless, the following general steps summarize the basics of the process:

- Coating the alumina surface with a fine powder of molybdenum and manganese (or their oxides).
- Heat treatment at around 1400–1500°C, commonly under a H_2 reducing atmosphere, to convert the manganese to MnO and preserve the molybdenum as metal.
- The cooled alumina surface is coated with a $\sim 5\,\mu m$ film of nickel to enhance wettability.
- Kovar is usually used as the bonded metal, sometimes copper. In rare cases, stainless steel or nickel.
- Finally, brazing is conducted, usually with a silver-copper alloy.

The MnO reacts with the alumina to produce a glassy phase at the alumina-Mo interface. The glass penetrates the grain boundaries of the alumina and also reacts with MoO to form a chemical bond with the molybdenum. These results in a dense glass-Mo composite interlayer, in the order of a few tens of microns thick, intimately bonded to both alumina and molybdenum. It is a somewhat functionally graded layer, which gives a broad uniform stress distribution across the interface, which is one of the key reasons for the industrial popularity and success of the Mo-Mn process [388–390,593–597]. Finally, the alumina purity, glass content, and composition of its impurities all influence the glass-Mo composite layer, in terms of its formation and its properties and structure.

Nickel coating can be done by coating with NiO paint, which is then heated at $\sim 950°C$ in a H_2-N_2 atmosphere to convert it to the Mo-Mn-alumina surface and reduce it to nickel metal. Coating can also be done using an electrolytic method, or an electroless nickel plating method, and in either case, the subsequent $\sim 950°C$ heat treatment is then carried out. One of the key problems with the Mo-Mn process is the effects of brazing on the nickel layer and alumina-Mo-Mn surface. Pure Cu is commonly used for brazing, which can dissolve and consume the entire Ni coating and then attack the Mo layer. This can result in bond failure at the alumina-Mo interface.

Au-Cu braze is also commonly used that has a similar effect on the Ni coating and alumina-Mo-Mn surface. Ag-based braze, such as the Ag-Cu eutectic, does not easily attack the Ni coating due to the limited solubility of Ni in Ag.

With regard to proprietary fine-tuning of the process, some of the important factors in the alumina metallizing aspect of the process are as follows:

- Alumina grain size has a significant effect on the Mo-Mn process and needs to be precisely managed via exact and reproducible sintering cycles in the sintering of the alumina prior to the Mn-Mo coating process [598]
- Application of the Mo and Mn powders. Various optimized proprietary methods are used involving binders, application method, and the characteristics of the raw powders themselves. Similarly, if electrolytic method or electroless methods are used, various optimized proprietary methods are used.
- The specifics of the sintering cycle for the Mo-Mn coating.
- The specifics and fine tuning of the Ni-coating process.
- The choice of braze and the brazing process.

Essentially the brazing stage of the Mo-Mn process is a metal-metal brazing process: brazing molybdenum to Kovar (or other bonded metal). The nickel coating is primarily there to enhance wetting of the molybdenum by the braze.

In conclusion, the Mo-Mn process is arguably the equal most industrially important alumina-metal bonding process and is well documented in the literature [388–390,593–597]. The other equally most industrially important alumina-metal bonding process is the alumina-titanium process used for bionic implants (see Chapters 8–10).

14.2.1.2 Active metal brazing technology

Active metal brazing, while less commonly used in electrical-industry feedthroughs than the Mo-Mn process, dominates the implantable bionics industry (Chapters 8–10) as an alumina-metal bonding process. It has also seen increasing use in the "electrical-industry feedthrough" in recent decades. Alumina is well suited to active brazing. An active braze is one that is capable of brazing (wetting and bonding) directly and simultaneously onto the metal and native ceramic surfaces. It does not require pre-metallization of the ceramic. An active braze contains metal components that chemically react with the ceramic. This generally means titanium, sometimes zirconium.

Active brazing offers many advantages as an alumina-metal bonding system, not the least of which is the fact that it is a simple one step process, and therefore has the benefits of simplicity and economy. On the negative side, molten titanium requires a protective atmosphere. Therefore, active brazing must be carried out either in a high-vacuum or in a high-purity inert gas atmosphere, such as argon, generally using graphite as a brazing fixture.

The science underlying the active brazing concept concerns redox oxidation potential of the braze metal and ceramic. Cu-Ag braze is one of the most widely used brazes and bonds well to most metals. However, in the case of most ceramics, and specifically alumina, the oxidation potentials of copper and silver are less than the oxidation potential of aluminum, which means that interfacial bonding will not occur. Titanium, on the other hand, has a high oxidation potential, which drives a redox reaction with the alumina surface in which the molten titanium reduces alumina to form TiO_2 and Al_2TiO_5 at the interface, an interlayer that is chemically compatible with both alumina and the brazing alloy.

The first documented accounts of active brazing of metals to ceramics date back to the same era as the Mo-Mn process, that is, the late 1940s, from a patent filed in 1948 by Kelly [599] and a paper by Pearsall in 1949 [600]. Some of the early work in active-metal brazing involved foils of zirconium or titanium as the braze interlayer between the metal and the ceramic to be bonded, as for example the patent of Beggs filed in 1954 [601], which explored the use of titanium shims as an interlayer in solid-state bonding of a ceramic (forsterite) to a metal: various metals were cited including copper and Fe-Co-Ni alloys. Cohen's patent filed in 1957 [602] claimed a similar concept, the use of foils of titanium as an interlayer between ceramic (forsterite) and Kovar-type Fe-Co-Ni alloy called nicosel. Alumina-nicosel bonding was also studied, and this involved a zirconium foil. In the early days, from the 1950s to the 1980s, the Mo-Mn process was the main focus of attention for metal-ceramic bonding in the electrical industry, because it produced more reliable and consistent joining outcomes. Even today the Mo-Mn process remains a primary focus. The key problems with active brazing in the early days were threefold:

- The critical importance of ceramic surface in the braze-ceramic interaction was not well understood.
- Active brazing alloys of adequate ductility were not available in the early developmental era.
- The importance of vacuum quality in vacuum brazing furnaces was not well understood.

Further development in the late-20th century, particularly in the 1980s and 1990s, refined the active brazing process into the highly evolved and effective process that we know it as today [603–606]. The essential process in industry involves active brazing of alumina using a titanium-based braze. In this scenario, both TiO and Ti_2O_3 have been reported as reaction products, as well as a $Cu_2(Ti,Al)_4O$ layer on the alumina surface [607,608]. However, when pure Ti was brazed to sapphire, the main reaction product was Ti_3Al [609].

Essentially, there are four methods used to apply the braze. All involve subsequent heating of the alumina-braze-metal join under a high-vacuum, or in a high-purity inert gas:

- Using a titanium-based braze. Today there is a diversity of active brazes used in active metal brazing in this role, all of which contain titanium. Other common components include copper, nickel, gold, silicon, silver, tin, zirconium, and indium, of which copper, nickel, and silver are the most common.
- Coating the alumina with titanium using powdered titanium as a binder-powder paint or sputter coating and then brazing with a regular brazing alloy.
- Using a powder mixture of titanium and regular brazing alloy, which is applied to the alumina as a binder-powder paint.
- Titanium/braze foil interlayers between alumina and metal surfaces.

The surface condition of the alumina is important to the peel strength of the braze. A ground alumina surface has a much lower peel strength than an as-sintered alumina surface. Most alumina components are ground, after sintering, to precise final dimensions and therefore the component needs to be resintered to $\sim 1600°C$ in air, before brazing, in order to heal grinding surface defects. The other important factor is the alumina purity itself, glass content, and composition of its impurities. These all influence the braze-alumina interaction.

The higher the titanium content of the braze, the greater the reaction interlayer. The strength of the bond is maximized for a moderate interlayer thickness, which means in practice the titanium content of the brazing alloy, for example, a Ti-Cu-Ag braze, needs to be an optimum amount, neither too high nor too low, for maximum bond strength [606]. Indeed, there is much proprietary know-how in the industry regarding optimizing and fine tuning of the active brazing process, particularly with regard to the following:

- Braze composition, especially titanium content.
- Brazing atmosphere, which is ideally a high-vacuum or high-purity inert gas.
- Heating rate that needs to be optimized between the conflicting requirements of being fast enough to minimize undesirable interactions between the braze and metal component (Kovar or copper), but not so fast as to thermally shock the alumina and the joint.
- Brazing temperature, which certainly must be above the solidus and ideally above the liquidus temperature of the braze. This needs to be optimized with respect to braze peel strength, between the conflicting requirements of being high enough for adequate brazing but not so high that the braze becomes too titanium depleted, which reduces peel strength.
- Brazing time, which is typically in the order of 10 min, but like temperature, needs to be optimized with respect to braze peel strength.

Optimization of all of the above is also necessary with respect to not just peel strength but also hermeticity.

A final important point to note about active brazing is the need for careful management of brazing temperature and time. Excess brazing temperature and/or time can lead to severe and damaging penetration of the alumina by the active braze alloy. This can easily become 1 mm deep and can be recognized by a color change in the alumina. Thus, while the active brazing offers the benefits of a good interfacial reaction with the alumina, this comes at the risk, when poorly managed, of excessive and detrimental interfacial reaction with the alumina.

In concluding this section on active brazing, it is important to note that active brazing of alumina-titanium metal-ceramic bonded feedthroughs has been the industry standard in the bionic industry since 1970, using predominantly Ti-Cu-Ni braze, that is, 70Ti-15Cu-15Ni (Chapters 8–10). However, the traditional electrical industry lagged significantly behind the bionics industry in this regard, since it did not have the stringent requirements of the bionics industry; therefore, the imperative was not as strong in the traditional electrical industry. It was not until a couple of decades later that active brazing became significant in the electrical industry. It is also worth noting that precious metal brazing, generally gold-based, is also relatively common in the biomedical industry though less common in the electrical industry (Chapters 10–12).

14.3 Alumina-metal bonded feedthroughs: General discussion

The Alumina-metal feedthrough is one of the most important commercial applications for alumina in the world today. It underpins a $25 billion bionic implant industry that is growing rapidly, and it underpins many areas of the electrical industry, also on a large commercial scale. Alumina-titanium hermetic seals in the biomedical industry are discussed in Chapters 8–10. The alumina-titanium hermetic bionic implant feedthrough technology of the biomedical industry evolved independently of the Kovar-alumina feedthrough technology in the electronics industry. There was no cross-pollination between the biomedical bionic implants industry and the traditional electrical industry. This independent evolution occurred around the same time, in the mid- to late-20th century, while alumina-titanium feedthroughs for bionic implants had their origin in Australia and the British realm, and alumina-Kovar feedthroughs in the electrical industry had their origin in Germany and the United States.

The high-tech feedthrough application in the electronics industry is the alumina-Kovar technology. The principles are the same for nonbiomedical systems but hermetic systems in the biomedical industry (bionic implants) evolved independently of hermetic systems in the electronics industry because bionic feedthroughs are low-power (milliwatt) devices that have different system requirements: specifically biocompatibility and stability in the corrosive moist environment of the body cavity over time periods of many decades. Moreover, with the high bionic implant retail price, feedthrough cost is no object. In contrast, the electronics industry applications that utilize alumina-Kovar feedthroughs (rather than glass-Kovar) tend to involve one or more of the following extreme conditions: high vacuum, high temperature, high voltage, high power up to the kilowatt range, with no requirement for biocompatibility, a dry operating environment, and a strong requirement for low cost.

In short, bionic implant feedthroughs and traditional electrical-industry feedthroughs came from fundamentally different industries, with fundamentally different cost imperatives, fundamentally different operating condition requirements, fundamentally different end-use applications, and no technology cross-pollination. About the only common ground was that electrical and materials engineers were the people behind the innovation in each case.

A key issue with metal-ceramic bonding is thermal expansion matching, particularly if thermal cycling occurs in service. A significant thermomechanical mismatch can produce high residual stress in the join, which can result in delayed sudden catastrophic failure due to cyclic fatigue (thermal cycling). In the absence of thermal cycling, with poor thermal expansion matching, it is possible that the residual thermomechanical stresses are high enough for slow crack growth leading to delayed sudden catastrophic failure by static fatigue via (see Chapter 4). Thermal expansion data are compiled in Fig. 14.7, which demonstrate that Kovar is an excellent match for alumina, as is titanium. Niobium, tantalum, and copper are not such an ideal match, though copper is still very suitable in the role, as discussed below.

Alumina-metal bonding for electrical feedthroughs 463

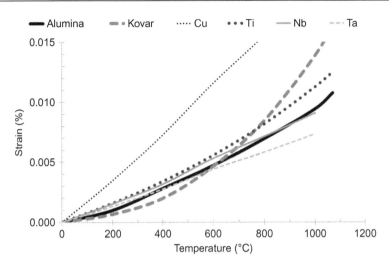

Fig. 14.7 Comparative dilatometry data for the two metals commonly bonded to alumina in electrical feedthroughs (Kovar and copper), as well as the three native metals with the closest thermal expansion match with alumina: Ti, Ta, and Nb [610].

Two other key factors that are important in joint stability are the elastic modulus of the bonded metal and the ductility of the braze. A highly ductile braze is capable of ductile yield for accommodating thermomechanical residual joint stresses. Secondly, a bonded metal that has a low elastic modulus will generate less thermomechanical stress for a given strain mismatch. Thus, a high-thermal expansion mismatch can be compensated for by a low associated thermomechanical stress in the case of a low modulus-bonded metal. Modulus data for the key bonded metals are compiled in Fig. 14.8.

The final factor of importance is that in seal design, wherever possible, the seal should be designed such that the alumina ceramic is placed in mild compression by the metal. This means that the metal ideally has a thermal expansion coefficient slightly higher than alumina, and that the design of the seal is done in such a way that upon cooling, the metal, through a peripheral ring or external lip, places the alumina in residual compression. This is shown schematically in Fig. 14.9. The principles of static fatigue and fracture mechanics (Chapter 4) are also very important. The alumina should be flaw free and designed in such a way as to minimize stress concentrations. Finite element modeling is commonly done to optimize such designs.

Fig. 14.7 demonstrates that titanium, tantalum, and niobium provide the closest thermal expansion match with alumina. However, niobium and tantalum are unsuitable for alumina-metal seals for reasons discussed in Section 14.3.1. Titanium is ideal, but much more expensive to process than Kovar- and copper-based seals. Titanium dominates the $25 billion "money-no-object" bionic-implant seals market. A bionic implant can retail for up to $30,000 for a device a few centimeters in size. So the cost of an alumina/titanium seal is therefore insignificant in that context.

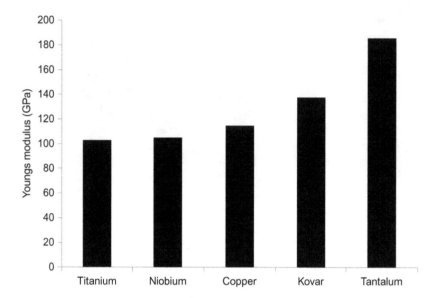

Fig. 14.8 Young's modulus for the key metals used in alumina-metal electrical feedthroughs. Kovar dominates the electrical industry with copper also significant. Titanium is solely used in the biomedical industry. Niobium and tantalum are rarely used commercially due to an unsuitable combination of properties.

Cost of ceramic seals (alumina or glass) is very important in the mass-market, low-priced electrical industry devices. For example, a light globe requires a metal ceramic seal and a production cost well below $1. A magnetron for a kitchen microwave oven needs manufacturing costs pruned to the bone, given that an entire kitchen microwave oven, including 1 kW magnetron, programmable control system, and chamber, can retail for $50 or less. Having said that, high-end microwave generators for telecommunications, radar, and industry can be well into the multiple thousand dollar retail category. In comparison, a small bionic implant can retail for up to $30,000.

Kovar and copper are generally used in alumina-metal feedthrough systems in the electrical industry and generally bonded by the dominant bonding system of the electrical industry—the Mo-Mn process. Clearly Kovar is a closer thermal expansion match to alumina. Indeed Kovar, by microalloying of its constituent components (Fe/Co/Ni), can be engineered to have perfect thermal expansion compatibility to the alumina. Kovar is, after all, a deliberately engineered alloy for thermal expansion matching to ceramics. Copper is also sufficiently well matched for commercial use and being significantly higher in thermal expansion than alumina and is ideal for applications in which the alumina is placed in residual compression, as per Fig. 14.9. Copper has also advantage over Kovar in electrical conductivity. Moreover, as shown in Fig. 14.8, the low elastic modulus of copper also reduces problems associated with thermomechanical mismatch, especially given the very high ductility of copper. Thus, copper is quite a good alternative to Kovar.

Alumina-metal bonding for electrical feedthroughs   465

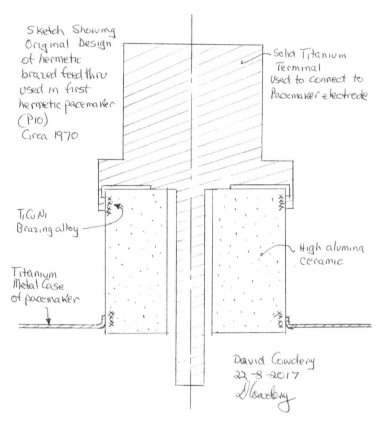

Fig. 14.9 Principles of ceramic-metal seal design so as to place the ceramic in residual compression. In the schematic shown, the two titanium (slightly higher thermal expansion than alumina) components have a lip which places the alumina seal in residual compression on cooling. The lip is also where the brazing is done, as shown, as this helps enhance the seal. Image supplied courtesy of David Cowdery, inventor of the alumina/titanium bionic feedthrough—see Chapter 8.

14.3.1 Titanium, niobium, and tantalum

Titanium and niobium have both been used in alumina-metal seals in bionic implants. Titanium has dominated this field since 1970, while niobium saw limited use in the late-20th century. It is important to note that neither Kovar nor copper can be used in bionic implants because neither metal is a biocompatible material. Therefore, titanium-alumina bonding, using titanium braze, was independently developed for bionic implant feedthroughs (pacemaker) in 1970 (Chapter 8) and has been used in pacemakers and indeed all bionic implants for many years and remains the standard today (Chapter 10) [146,147]. This process is described in detail in Chapter 8. As discussed earlier, the requirements for bionic implant technology are completely different to those of the electrical industry and are essentially twofold:

(1) Biocompatibility of metal and ceramic (titanium and alumina have exemplary biocompatibility).
(2) Long-term durability in a warm (37°C), wet, corrosive environment. The human body is one of the most corrosive operating environments to be found outside of a chemical production plant. Durability and corrosion resistance in the order of many decades are required of the metal-ceramic bond, the alumina, the metal, and the braze.

The alumina-titanium hermetic ceramic-metal bonding technology was developed as the hermetic feedthrough for the world's first implantable bionic pacemaker in 1970 by David Cowdery [146,147]. This was the first documented commercial alumina-titanium bonding system and the first documented alumina-titanium hermetic seal. The bionics industry went on to evolve the titanium-alumina hermetic seal to a very high level of sophistication over the next few decades, as discussed in Chapters 8–10.

The literature contains no documented research on alumina-titanium seals or feedthroughs prior to 1970. Indeed there are only two documented reports of alumina-titanium bonding pre-1970. These did not involve hermetic seals. They were R&D trials from the aerospace industry. A 1960 study was conducted by Boeing (Seattle, USA) into alumina-metal bonding so as to evaluate the potential of alumina-metal bonding (including alumina-titanium) for aerospace missile applications, specifically alumina radomes, IRdomes, nose cones, leading edges of wings and control surfaces, and antenna covers [611,612]. The titanium alloy investigated was the aerospace alloy Ti6Al4V. This was the era in which the US aerospace community was actively developing aerospace applications, particularly in the military sector, based on titanium-alloys, which at a density of $4.5\,g/cm^3$, are much closer to aluminum ($2.7\,g/cm^3$) than steel ($7.9\,g/cm^3$) and Ti6Al4V has a very high specific strength. The all-titanium Mach3+ Lockheed SR71 spy plane is one of the most famous examples of the 1960s. It was within this context that the Boeing alumina-titanium bonding study was conducted.

The Boeing alumina-metal bond-strength research involved a simple study of the tensile strength and shear strength of alumina-metal bonds produced by active metal brazing [611,612]. No seals were made, and therefore hermeticity was not evaluated. The study simply involved joint strengths being mechanically tested in tension and shear from room temperature up to 1000°F (540°C). The metal test pieces were 1.5×0.5 in (38.1×12.7 mm), and the alumina test pieces were $0.5 \times 0.5 \times 0.25$ in ($12.7 \times 12.7 \times 6.4$ mm). Titanium hydride was sprinkled on the alumina surface, and brazing was done in either argon or a vacuum with either with Ag or Ag-Cu braze. This effectively constituted a TiAgCu alloy braze. Over the ensuing decades, experience with the TiCuAg braze showed that while it had initial high strength, it had very poor corrosion resistance. This was to have negative outcomes in both the aerospace industry and bionics industry, where it was trialed.

The Boeing work also contains the first and indeed one of the only documented reports of the use of the Mo-Mn process for bonding alumina to a titanium alloy, specifically to attach a prototype alumina radome to a titanium alloy [612]. No details are provided other than that, "this process requires two or three furnace cycles and the availability of a 2300–2700°F hydrogen or ammo-gas furnace," that is, 1260–1480°C.

Outcomes are not described other than a statement regarding the technical challenges and potential benefits: "this temperature is dangerously near the softening point of the ceramic especially for a large component. The titanium could also be embrittled by the furnace atmosphere. These limitations were offset by the high-temperature capability of the final joints and the assurance of a strong reaction between the ceramic and braze with a resulting increase in reliability and strength of the joint" [398]. It is also worth noting here that in the 1970s, the Mo-Mn process for alumina-metal seals, brazed with AgCu alloy, was trialed in pacemakers. The AgCu braze proved to have a very poor corrosion resistance, which led to failures in service.

As is typical in alumina technology, applications for alumina have evolved independently in their respective industries (refractories, electrical, wear/corrosion resistance, armor, bionics, and the various niche areas such as aerospace) with very little cross-pollination, if any, between these respective industries. Aerospace applications for alumina are discussed further in Chapter 15.

Alumina-tantalum, and alumina-niobium bonding systems appear, based on close thermal expansion matching, to be viable alternatives to titanium, as per Fig. 14.7. A process for bonding tantalum and niobium to alumina was developed by Elssner et al. in the mid-1970s [397] from within the electrical industry. This ultimately saw no significant commercial application in the electrical industry. However, the alumina-niobium bonding technology of Ellsner et al. [397] was adapted by the newly established US pacemaker industry in the late-1970s, from which the world's second bionic pacemaker feedthrough technology was launched in the late-1970s: alumina-niobium. A competitor for the alumina-titanium pacemaker feedthrough technology developed in Australia in 1970.

While niobium is biocompatible, its mechanical property limitations are its Achilles heel. This was a problem in the 1970s when it was first trialed in bionic implants, and it remains a problem today [398]. Moreover, the thermal expansion of niobium, shown in Fig. 14.7, is so close to alumina that niobium is incapable of placing the alumina in significant stable residual compression on cooling, in accordance with the principles of seal design shown in Fig. 14.9. In the 1970s, US pacemaker manufacturers experimented with gold brazing of niobium to alumina. The softness and ductility of the gold, combined with the very close thermal expansion characteristics of alumina and niobium, resulted in minimal interfacial stresses. Moreover, the design needs to be made in such a way that body fluids are not exposed to gold, due to the hazard of gold sensitivity.

Thus, while niobium can be made to work, it never proved to be as successful as titanium in alumina-metal seals. Ultimately, the alumina-niobium technology was supplanted globally by the alumina-titanium technology, as is discussed in Chapters 8–10. Titanium on the other hand has an outstanding combination of excellent mechanical properties, and excellent biocompatibility, with over half a century track record of proven biocompatibility performance in orthopedics, bionics, dentistry, and other implantable medical devices.

Tantalum fails as a suitable metal for alumina-metal seals on two criteria. Firstly, it has too high a modulus to be competitive with titanium in feedthroughs. Titanium has about half the modulus of Tantalum, as shown in Fig. 14.8, and is equal or superior to tantalum in most other respects. Secondly, and more importantly, as Fig. 14.7 shows, tantalum has a thermal expansion coefficient that is slightly lower than alumina. This is fatal in terms of placing the alumina in residual compression on cooling, in accordance with the principles of seal design shown in Fig. 14.9. Tantalum is therefore fatally flawed for the role.

Titanium, which is slightly higher than alumina in thermal expansion coefficient, low in elastic modulus, and has outstanding biocompatibility, is therefore unrivaled in the alumina-metal feedthrough role for bionic implants. The science says this is the case. History has proven it over the last half-century as is discussed in depth in Chapters 8–10.

14.4 Multilayer micro-feedthroughs: Cofired alumina-substrate systems

A major growth area, both in the bionics and conventional electronics industries, is the micro-feedthrough, produced by alumina substrate thick-film technology and cofiring.

Hermetic microchip systems can involve alumina substrate multilayer cofired packages, which can include

- Chip carriers
- Pin-grid arrays
- DIL (dual-in-line) packages

Generally, 90%–94% purity alumina grades are used in multilayer systems. The green alumina undergoes a printed metallization process and is then cofired into the alumina substrate containing internal conductor pathways. The process involves the following steps:

- Typecasting alumina substrates
- Laser machining to produce "via"-holes
- Metallization by screen printing (either tungsten or molybdenum)
- Lamination into a multiple layered structure
- Cofiring at around 1600°C in an inert sintering atmosphere
- After cofiring, connectors are brazed to the appropriate metallization points.

This is shown schematically in Fig. 14.10.

The multilayer alumina substrate with embedded metallized tracks is a hermetic device that can be metallized and brazed to a hermetic metal capsule, as per the metallization technology discussed in Section 14.2.1, so as to interface between a silicon chip and the external environment and has a wide range of commercial applications in microelectronic systems which have hermetic requirements, whether due to environmental conditions involving liquid, gas, corrosion, or other environmental challenges requiring isolation of the microelectronics from the environment.

Alumina, being an outstanding insulator, as well as being impervious, chemically inert, corrosion-resistant, and heat-resistant, is pre-eminent in this role. There are other

Alumina-metal bonding for electrical feedthroughs 469

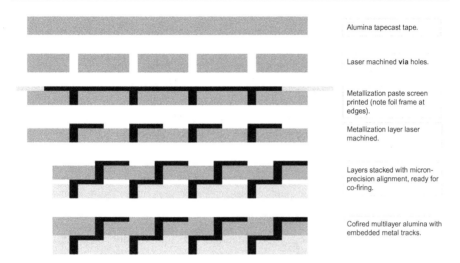

Fig. 14.10 Production process schematic for cofired metalized alumina hermetic multilayer substrates.

less extreme applications for hermetic feedthrough substrates that do not necessarily involve extreme external environmental conditions. Regardless of the specific application requirements for hermeticity in each specific commercial application of this technology, the hermetic feedthrough, as a platform technology for hermetic electronic systems, is one of the most important commercial applications for alumina in the world today.

This technology is essentially a hermetic interface with a huge number of channels (conductor wires)—the logical evolution from the simple light globe with its "two channels" (Kovar electrode connections) passing hermetically through the glass wall of the lightglobe.

This concept has been taken to the next level, in terms of channel number and engineering for biocompatibility, in the alumina/platinum bionic eye hermetic feedthrough. This utilizes biocompatible metallized conductor material (platinum), biocompatible capsule (titanium), and braze (titanium-based) with platinum channel numbers in excess of 1000, as discussed in Chapter 10. The traditional electronics industry metallization brazing metals, and Kovar as the bonded metal, are not biocompatible.

The alumina micro-feedthrough is one of the major growth areas for the future of alumina ceramics.

14.5 Concluding remarks

Alumina-metal feedthrough seals offer a number of advantages over glass-metal feedthrough seals for situations of high temperature or other extreme conditions:

- Much higher temperature capability (glass is limited to a few hundred degrees Celsius).

- Alumina is much mechanically stronger than glass.
- Alumina is more thermal shock resistant than glass.
- Alumina is less sensitive to flaws and stress concentrators than glass.
- Alumina specifically has much better electrical insulation than glass.
- Alumina specifically has a much better dielectric loss tangent than glass.

Therefore, the electronics industry applications that utilize alumina-Kovar feedthroughs (rather than glass-Kovar) tend to involve one or more of the following extreme conditions, of which temperature is the most critical:

- High operating temperature
- Severe thermal cycling
- High vacuum
- High voltage
- High power up to the kilowatt range

There are many applications of alumina/Kovar, alumina/copper, and alumina/specialty-metal hermetic feedthroughs and seals in the electrical and electronics industry. Many of them involve vacuum tubes, or gas-filled tubes, that require internal electrical power. To a significant extent, solid-state technology is increasingly rendering diode and triode valve electronic component technology obsolete in the 21st century. However, these are primarily glass-metal sealed systems, and not commonly utilizing ceramic-metal seals. Therefore, they are a small proportion of the market in electrical alumina-Kovar, alumina-copper hermetic feedthrough seal technology.

Commercial electrical-industry feedthrough applications range from simple low-tech mass market application such as insulators, capacitors, heatsinks, incandescent and vapor lamps, triode vacuum valves, and semiconductor housings such as thick-film multilayer feedthroughs to advanced feedthroughs of many kinds from high voltage to high frequency to highly corrosion resistant. In addition, there are vacuum-chamber and furnace viewports, glass adaptors, thermocouples, electrical breaks, liquid feedthroughs, vacuum-compatible cables and connectors, X-ray tubes, and the various mass-market microwave generators (see below). All of these systems require ceramic seals (alumina or glass), and in most cases an electrical feedthrough. Many of these applications are serviced by glass-metal seals, many require alumina-metal seals.

In general, glass-metal seals service the low-tech mass-market applications, such as lamps in all their many manifestations, triode valves, and cathode ray tubes, while ceramic/metal (usually alumina/Kovar) services high-tech and severe-environment applications.

Alumina is very important in the huge microwave industry due to its extremely low AC loss tangent, in combination with its extreme electrical resistivity, high heat resistance, good thermal conductivity, and when required, its capability for transparency as synthetic sapphire, or translucent polycrystalline alumina (Chapter 5). There is a huge number of microwave generators globally: oven magnetrons (1 billion worldwide) Klystrons and TWTs (millions worldwide), and Gyrotrons (thousands worldwide), see Chapter 13. Microwave generators are vacuum tubes and require an electrical feedthrough. In general, thick-film microwave integrated circuits (MICs) and microwave windows are exclusively alumina, which is essential in this role due to its low loss tangent and its low-processing cost. However, microwave tube feedthroughs, being a relatively low temperature device, need not necessarily be alumina and can be glass or glass ceramic. There is a wide range of manufacturers of microwave generators globally. Among these manufacturers, and between the respective microwave-generator types, and even within each type, there is a diversity of design, materials, and construction approaches used in the electrical feedthrough systems. Various proprietary ceramic feedthrough systems are used, which are generally not disclosed in the public domain.

In summary, the market for hermetically sealed vacuum tube feedthroughs, and other hermetic feedthroughs, in modern industry is large and diverse today. This is mostly used not only in electrical and electronic equipment but also in vacuum, chemical-processing, and high-temperature equipment, where hermetic isolation and electrical or liquid feedthrough is required. Some use glass feedthroughs and some use alumina feedthroughs. In the biomedical sector, alumina feedthroughs are a key enabling technology, and dominant, in the $25 billion bionic feedthrough industry. In the (nonbiomedical) traditional electronics industry, alumina feedthroughs are a key enabling technology for high-end applications in this large and diverse industry and glass feedthroughs for the low-end applications. All indications are that alumina's central role in hermetic electrical feedthroughs will continue long into the future.

Refractory and other specialist industrial applications of alumina

Chapters 12–14 having addressed the major industrial applications of alumina ceramics (wear resistance, corrosion resistance, electrical industry), this chapter will address all other industrial uses for alumina ceramics, of which the most industrially significant is the refractories industry. Refractories are used for heat-containment, primarily in furnace linings, and also as crucibles and kiln furniture. The global refractories industry is one of the largest materials-based industries in the world. It is currently generating revenues of around $50 Billion a year, of which China has 50% global market share. Therefore, refractories are one of the quintessential traditional ceramic products, along with traditional clay-based ceramics from bricks and tiles, to porcelain, and fine China. By their broadest definition, refractories are somewhat outside of the scope of this book. However refractories is such a large industry, and there are so many important points of intersection between alumina ceramics and the refractories industry, that this chapter will dedicate a significant proportion to the refractories topic to highlight these important areas of intersection. While the largest focus in this chapter is on alumina in the refractories industry, important niche applications for alumina will also be discussed including machining/cutting tools, architectural uses, aerospace uses, and oxygen sensors. This chapter contains two important industrial case studies: alumina architectural ceramics with a focus on Taylor Ceramic Engineering (alumina specialty manufacturer for over 40 years); alumina/zirconia oxygen sensors with a focus on Ceramic Oxide Fabricators (AUST) (pioneer of oxygen sensors for over 40 years). Minor niche applications for alumina will also be briefly discussed including gas laser tubes, laboratory instrument tubes, kiln furniture, agricultural-plough tine tips, alumina gauges, jewel bearings in watches, and mortar and pestles. The main commercial applications of alumina can be broadly classified into the following areas:

- Biomedical engineering: orthopedics, bionics, dentistry, scaffolds, body armor, and other specialist biomedical uses (Chapters 5–11).
- Wear/corrosion-resistant linings (Chapter 12).
- Electrical (Chapters 13 and 14).
- Refractory (this chapter).
- Other specialist applications (this chapter).

The focus of this chapter is on refractory and other specialist *industrial* applications. The following list compiles the four *significant* industrial uses of alumina not discussed elsewhere in this book.

- Refractories
- Machining/cutting tools

- Architectural uses
- Aerospace uses
- Oxygen sensors

These will all be discussed in some detail in the following sections:

Refractories: Refractories are the major topic of this chapter. Sections 15.1–15.4 will address the global $50 billion refractories industry and the role alumina has in this large industry.
- 15.1. The refractories industry: A brief overview
- 15.2. The contemporary refractories industry
- 15.3. Contemporary refractory uses of alumina ceramics
- 15.4. High-purity alumina ceramics for heat containment: Labware and industrial uses

Alumina machining tools: This significant niche application will be discussed in Section 15.5.
Architectural applications for alumina: This significant niche application will be discussed in Section 15.6.
Aerospace applications for alumina: This significant industrial application will be discussed in Section 15.7.
Oxygen sensors: Oxygen sensors are an important global technology, in which alumina is an integral component, as discussed in Section 15.8.

Another seven minor niche industrial applications will be discussed in Section 15.8.
- Thermocouple sheaths
- Gas laser tube
- Laboratory instrument tube
- Kiln furniture
- Agricultural-plow tine tip
- Alumina gauge
- Jewel bearings in watches
- Mortar and pestle

Of this list, all are niche applications for alumina in the sense that:

- They do not represent a high-volume market.
- They have the potential to grow into a high-volume market in the future.
- Alumina is an excellent material for this application.

15.1 The refractories industry: A brief overview[1]

15.1.1 Economics of the refractories industry

Refractories are one of the quintessential traditional ceramic products, along with traditional clay-based ceramics from bricks and tiles to porcelain and fine china. Therefore, by their broadest definition, refractories are somewhat outside of the scope of this

---

[1] For some years, the author lectured on refractories. Sections 15.1–15.3 are a distillation of the key principles, as they relate to generic issues (Sections 15.1 and 15.2) and specific alumina-relevant issues (Section 15.3).

Table 15.1 Approximate current refractory consumption of the various industries

Industry	$Billion
Iron & Steel	36
Cement	6.5
Nonferrous Metals	2.3
Energy and Chemical Industries	1.8
Glass	1.3
Ceramics	1.0
Others	0.8

book. However, refractories are such a large industry, and there are so many important points of intersection between alumina ceramics and the refractories industry, that this chapter will dedicate a significant proportion to the refractories topic to highlight these important areas of intersection.

Refractories are used for heat containment, primarily in furnace linings, also crucibles. The global refractories industry is one of the largest materials-based industries in the world. It is currently generating revenues of around $50 billion a year, of which China has 50% global market share. Table 15.1 gives an approximate indication of the current size of the key subsectors of the refractories industry. Clearly the Iron & Steel sector is dominant, representing over 70% of the refractories industry. Cement is also substantial.

15.1.2 Prehistory of the refractories industry

The refractories industry is a huge, very diverse, and very ancient industry. While a detailed history of the refractories industry is beyond the scope of this book, for context, a brief history is helpful. One of the earliest documented uses of the word "refractory" is from 1556 [613].[2] However, the concept of the clay-based firebrick used to line fireplaces, industrial kilns, and metal smelting systems dates back to the dawn of civilization: the brick-lined hearth in homes and buildings, early pottery "kilns," the first commercial utilization of precious metals such as gold, and the beginning of the bronze age.

Recent archeological research suggests that pottery arose around 20,000 years ago, possibly even earlier than advanced stone tools (Neolithic) and that pottery predates agriculture by perhaps as much as 10,000 years [103]. While the precise chronology remains the subject of active research and debate, there is no doubt that pottery is an ancient craft, along with agriculture, and one of the key innovations that drove the rise of early civilization.

---

[2] Georgius Agricola was the author's latinized pseudonym. His actual name was Georg Bauer. This book was one of the earliest printed materials science textbooks in history. Seen for some centuries as an authoritative source, from medieval Germanic Europe (Bohemia and Saxony) on mining and metallurgy.

The refractories industry began with simple pit-kilns dug into the ground for firing pottery. These were dug into soil that was "heat resistant" or refractory, able to combine the dual function of thermal insulation and structural integrity at high temperature. In the Roman era and thereafter, brick constructs were commonly used to fire earthenware products, rather than pits.

Refractories technology began to evolve significantly with the discovery of metal smelting, particularly the onset of the iron age in around 2000 BCE, 4000 years ago. Much higher temperatures and more serious heat containment challenges arrived with the iron age. Iron-smelting refractories made of clay/charcoal blends evolved over many centuries, ultimately evolving into the earliest blast furnaces, which arose in China first, and appeared much later in Medieval Europe around the 1300s, a couple of centuries before the Agricola textbook [613]. By the early 1800s, the first nonclay ceramics (impure as they were), magnesia brick, silica brick, and graphite had evolved as refractories, with chromia appearing late in the 1800s.

15.1.3 A brief history of alumina in the refractories industry

The first industrial applications of alumina "advanced ceramics" were electrical (spark plug insulators) and refractory (crucibles) which date to the early 20th century, although they were only obscure boutique uses in that era. It was in the postwar period, from the 1950s onward, that alumina industrialization began. In the early alumina industrialization era of the mid-20th century, refractory applications for alumina ceramics were a leading alumina application of the time. In the background there was also a fledgling wear resistance and fledgling electrical industry for alumina. Refractory uses remain very important today, though increasingly overshadowed by the 2020 scale of other alumina ceramic applications detailed in other chapters of this book. The alumina landscape of 2020 has changed almost beyond recognition since 1950.

In the history of alumina ceramics, refractory applications evolved gradually, as a subset of the much larger global refractories industry, and largely independently of the other early-evolving alumina industry sectors: electrical and wear resistance. Most alumina-containing refractories are not alumina ceramics as such. Alumina-containing refractories come in six main forms:

1. Multicomponent refractory for which alumina is a component—not an alumina ceramic.
2. Alumina-based refractory, in which alumina is dominant but not the sole component.
3. Calcium aluminate cement, which is the principal binder used in the dominant sector of the contemporary refractories industry: monolithic refractories (calcium aluminate cement-bonded ready-mix).
4. High-purity fibrous alumina—not a ceramic.
5. High-purity porous alumina.
6. High-purity dense alumina—fused-cast glass-contact refractory, crucible, furnace tube, kiln furniture.

Of the above list, only items 2, 5, and 6 could be classified as alumina ceramics, of which the most relevant to this treatise on alumina ceramics is item 6.

Refractories are one of the largest segments of the global ceramics industry and also the oldest commercial application of alumina, generally in a highly impure form. While alumina is one of the major components used in refractories, and a significant component in many refractories, alumina ceramics as such, that is, high-purity alumina, is a relatively small portion of the refractories industry, but it is certainly significant in dollar value all the same. Thus, while alumina is a component in many refractories, alumina as a component or as a monolithic is but one of many refractory materials used in the refractories industry. Indeed, high-purity ceramics of any composition are rare in the refractories industry. *The refractories industry primarily uses natural minerals as raw materials, since natural minerals are much less expensive than highly refined chemical compounds. When higher refractoriness is required, natural minerals of higher refractoriness are generally used rather than highly refined chemical compounds.*

The refractories industry predates by thousands of years the Bayer process of the 1880s [15,16] that made pure alumina ceramics possible, and the first pure alumina ceramic of 1910 [17]. However, calcined bauxite has been a common component for "high alumina" refractories for a long time. As a chemical component of Kaolinite, aluminum oxide has been a significant component of refractories since the emergence of pottery at the dawn of civilization. Much more recently, calcined bauxite began to be used as a source of "high alumina" refractories. Fibrous alumina and porous "high alumina" refractories became commonplace in the 20th century, however, even these are not alumina ceramics.

It should be noted that the term "high alumina" in the advanced ceramics industry, the primary focus of this book, generally means close to 100% pure alumina ceramics, and certainly not below 85% pure. However, the term "high alumina" has a different meaning in the refractories industry. It means 50%–87.5% alumina, and generally refers to refractories made from kaolinitic-bauxite. Thus, "high-alumina refractories" are not alumina ceramics as such. Fused-cast alumina refractories, however, are high-purity alumina ceramics, being within the 90%–100% purity range.

Alumina the ceramic, as the 85%–100% purity alumina ceramic that is the focus of this book, was first used in glass-contact refractories from the middle of the 20th century as fused-cast alumina blocks. Occasionally also as sintered alumina blocks. Previously, glass-contact refractories had been aluminosilicate sinter materials, but since the mid-20th century fused-cast alumina zirconia silicate (AZS) and fused-cast alumina glass-contact refractories became the industry standard.

The scientific research underpinning the development of refractory applications for alumina ceramics dates back to the mid-20th century. Since the 1980s, with much of the relevant fundamental science already published for refractory applications of alumina ceramics, the focus has primarily been on commercial activity. Current industrial practices in the refractories industry are generally underpinned by the fundamental scientific literature in a diverse range of generic areas, from general ceramic engineering to phase equilibrium diagrams, to heat transfer, and thermodynamics, and the study of high-temperature creep. These are well-established traditional fields of learning. Moreover, the industry also rests on an extensive body of proprietary knowledge in the form of industrial knowhow: proprietary formulations, processes, and applications.

15.2 The contemporary refractories industry

While a detailed discussion of the refractories industry is beyond the scope of this book, given the size of the global refractories industry, and the importance of alumina in the industry, both as a component of refractory compositions, and as alumina cement in monolithic refractories, this section will comprise a brief overview of the refractories industry, followed by a discussion of the specific roles for high-purity alumina ceramics in the industry. These roles, though only a portion of the $50 billion refractories industry, are still significant in the context of the global alumina advanced ceramics industry.

15.2.1 Key industry participants: Refractories

Given the huge scale of the global refractories industry, there are many companies involved. The following nonexhaustive list represents some of the key players in the industry:

- Harbison Walker International (formerly ANH Refractories)
- Chosun Refractories
- IFGL Refractories
- Imerys Refractory Minerals
- TRL Krosaki Refractories
- Magnezit Group (Refractories)
- Minteq International
- Morgan Thermal Ceramics
- Puyang Refractories Group
- Refratechnik Ceramics GMBH
- Resco Products
- RHI Magnesita
- Shinagawa Refractories Australasia
- Vesuvius plc

15.2.2 Major uses of refractories

Some of the major uses of refractories are

Furnace linings:
- Ceramic kilns
- Cement kilns
- Metal smelting furnaces
- Metal remelting furnaces (e.g., induction furnaces)
- Other industrial furnaces (e.g., power generation, incinerators)
- High-temperature reactors
- Glass tanks

Heat containment systems:
- Crucibles
- Furnace tubes
- Kiln furniture
- Fuse-casting molds
- Combustion chambers

15.2.3 Key requirements for refractories

The following requirements apply to all, or almost all, refractory applications:

- Heat resistance at operating temperature [pyrometric cone equivalent (PCE)]
- Refractoriness under load (creep resistance)
- Thermal shock/spalling resistance
- Thermal insulation
- Chemical inertness and corrosion resistance

The following requirements apply to some refractory applications, on a case-specific basis:

- Abrasion resistance (dry furnace charge)
- Erosion resistance (molten slag or gas flow)
- Chemical resistance to molten slag
- Carbon monoxide resistance
- Permeability

15.2.4 Classification of refractories

There are essentially two broad classes of refractories: clay and nonclay.

Clay refractories:
- Fireclay: 25%–45% alumina. Low, medium, high, or super-duty depending on alumina content.
- High alumina: Kaolinitic bauxite. 50%–87.5% alumina.

Nonclay refractories:

There is a wide range of formulations used, based on formulations with one or more oxides dominant, of the main six refractory oxides, generally as oxide mixtures, but sometimes as an oxide-nonoxide mixture such as carbon-MgO in the steel industry. The refractory oxides are further defined by their resistance to acid or basic conditions, such as molten slags:

The main six refractory oxides:
- Silica (acidic)
- Zirconia (acidic)
- Alumina (basic/amphoteric)
- Magnesia (basic)
- Calcia (basic)
- Chromia (basic)

In addition, mullite (up to 70% alumina) is a very common oxide refractory comprising mixtures of bauxite, kyanite, sillimanite, andalusite, or other aluminosilicates.

The main refractory nonoxides (reducing conditions):
- Carbon
- Silicon carbide

Refractories are also defined by fusion temperature:
- <1780°C—Normal refractory (e.g., fireclay)
- 1780–2000°C—High refractory (e.g., chromia)
- >2000°C—Superrefractory (e.g., zirconia)

15.2.5 Refractory bricks and monolithic refractories

Over the years, the refractories usage has trended away from bricks toward monolithic refractories. A monolithic refractory is an unshaped ready-mix concept, commonly involving particles of refractory oxides with alumina cement as the binder, used for in situ refractory applications. Just add water and cast, pump, spray, or otherwise apply as an instant refractory liner, or instant liner repair. In short, the trend in refractories has been away from "bricklaying" toward "concreting."

> *Thus calcium aluminate cement is one of the most important materials underpinning the contemporary $50 billion refractories industry. Though calcium aluminate cement is not an alumina ceramic as such, it is an extremely commercially important application of an alumina-based powdered material, comparable to the alumina abrasives industry in scale: $Multibillion order of magnitude.*

Essentially, there are three types of refractories: bricks, monolithics, and fibrous. Bricks (standardized as $9 \times 4.5 \times 2.5$ in – $230 \times 115 \times 64\,mm$):

- Dry-pressed bricks
- Castable bricks
- Fused-cast bricks

Monolithic (ready-mix with binder)—usually requires anchors:

- Ramming
- Gunning
- Plastic forming
- Patching
- Monolithic plastic
- Mortar
- Dry vibration

Binders for monolithic refractories are typically 25%–30%, low cement is 3%–10%, Ultra-low is <3%. Initial drying and firing of castable bricks/linings have a critical bake-out cycle. The main binding systems for monolithics are:

- Alumina cement (dominant)
- Phosphate bonding
- Colloidal silica precursors

Fibrous refractories:
A number of wools, felts, and compressed fiberboard fibrous ceramics are on the market. Aluminosilicate is the most common, for example, aluminosilicate fiber (such as Kaowool), zirconia, and alumina fibrous materials are also used in niche applications.

15.2.6 Ultra-refractories

In the ultra-refractory range, the order of increasing refractoriness is $Al_2O_3 < Cr_2O_3 < La_2O_3 < Y_2O_3 < SrO < BeO < CaO < CeO_2 < ZrO_2 < MgO < HfO_2 < UO_2 < ThO_2$. While alumina is at the bottom of the ultra-refractory scale, it is well ahead of SiO_2

(a widely used refractory oxide), and alumina is one of the most stable oxides in both oxidizing and reducing conditions. Alumina is also one of the lowest cost materials available in the high-purity form. With a melting point of ~2050°C, and a service temperature above 1800°C, alumina is also more than sufficient for the majority of temperatures encountered in the refractories industry.

When reducing conditions are involved, some of the carbides are capable of more extreme temperatures still. Carbon volatilizes at ~3730°C, but the most refractory compound known is tantalum hafnium carbide, which has a melting point of ~3900°C, in the absence of oxygen, but it is rarely used as a refractory. Rocket nozzles/combustion zones commonly utilize carbon-based refractories, with carbon-carbon composites a common choice.

15.3 Contemporary refractory uses of alumina ceramics

High alumina cement is the dominant binder used in monolithic refractories. However, high alumina cement does not fall within the scope of alumina ceramics. High alumina cement is a hydraulic cement, based on calcium aluminate compounds of various stoichiometries. It is therefore a more refractory aluminous equivalent of portland cement (calcium silicate compounds of various stoichiometries).

While alumina is widely used as a component, and even as the dominant component, in refractories, the refractory uses of high-purity alumina ceramics are relatively uncommon. For cost-minimization, most refractories are made from naturally occurring minerals. Bauxite and kaolin are the primary sources of alumina in refractories. Neither mineral is a high-purity alumina precursor.

The primary driver for the use of *high-purity alumina* in the refractories industry is for situations where the maximum precaution must be taken to minimize the risk of contamination. High-purity alumina is ideal in this role as it is inert and relatively insoluble to molten glass or slag at high temperatures. Therefore, the three main niche applications for high-purity alumina ceramics in the refractories industry are as follows:

1. Fused-cast beta-alumina or alpha/beta-alumina blocks (glass tanks).
2. Sintered alumina blocks (glass tanks).
3. High-purity containment systems (furnace tubes, crucibles).

15.3.1 Fused-cast beta-alumina and alpha/beta-alumina

Fuse-casting involves pouring a molten ceramic into a mold. It is a process that is rarely used for making conventional ceramics, which are almost always made by sintering a powder-preform. The primary industrial usage of fused-cast ceramics is for glass-tank refractories. Fused-cast ceramics also see some usage as wear-resistant ceramics. Fused-cast basalt is a commonly used wear-resistant ceramic, fused-cast alumina much less so.

15.3.2 Glass-tank refractories

A glass tank is a furnace used to melt glass on an industrial scale, in the form of a refractory-lined tank, filled with molten glass. The hottest areas of a glass tank typically operate at around 1500°C, with the coldest areas around 1200°C. The key requirement for glass-contact refractories is resistance to corrosion by molten glass. This is a requirement not just for maximizing longevity of the refractory glass-tank lining, but also a requirement for minimizing contamination of the molten glass. Excessive refractory corrosion can produce various glass defects, including

- Stones: crystalline inclusions in the glass.
- Cord or ream: glassy striae in the glass.

Degradation of a refractory by glass involves alkali ions of the molten glass diffusing into the refractory until they reach sufficient concentrations to flux the refractory hotface. For this reason, glass-tank refractories cannot be porous as this enhances glass penetration of the refractory by molten glass. Moreover, the tank needs to be designed in such a way that no refractory is exposed to molten glass on more than one face.

The most severe degradation occurs at the top surface of the glass melt, at the three-phase point where the solid refractory, liquid glass, and gas of the furnace atmosphere all meet. This is known as melt-line corrosion, and minimization of this requires zero porosity refractories. The most severe melt-line corrosion occurs where the surface of the glass melt encounters joins between the refractories.

Secondly, the size of the thermal gradient across the refractory determines the depth of penetration into the glass-contact refractory of alkali ions diffusing into the refractory from the molten glass. For this reason, the rate of degradation of the glass-contact refractory decreases as the thickness of the insulating wall of the glass tank decreases, due to the steeper thermal gradient that results. Therefore, paradoxically, glass-tank refractories have a longer service life when the wall is thin. Convective air or water cooling of the cold face of the refractories is often used to enhance the thermal gradient. This represents a cost in energy loss, which is offset by the cost-saving in prolonged refractory life.

For all of the above reasons, glass-tank refractories need two main attributes:

- Highly impervious
- Highly insoluble in molten glass

These two attributes are best achieved with fuse-cast blocks because these are nonporous refractories with high chemical inertness. Fused-cast AZS (alumina zirconia silicate) is the most common type of glass-tank refractory in conventional soda-lime glass tanks. The ZrO_2 content of the AZS is in the range 35%–40% ZrO_2. Fused-cast zircon is also produced, as well as fused-cast ZrO_2 (up to 95% ZrO_2). If pure zirconia, it needs to be a fully stabilized zirconia, by adding sufficient dopant oxides to stabilize the highest temperature polymorph (cubic) to room temperature (see Chapter 6). This is then stable on heating and cooling all the way to its melting point of ∼2700°C.

Refractory and other specialist industrial applications 483

In fused-cast refractories, the higher the ZrO_2 content, the better the performance. In some extreme cases, fused-cast Chrome-AZS is used. This is a high-performance formulation that can discolor glass. Used for severe conditions such as high-alkali borosilicate insulating fiberglass where chromium contamination is not an issue (chromium contamination can discolor glass).

This begs the question why use alumina at all, whether as fused-cast alumina or dense sintered alumina? Why not purely use AZS? Refractory performance generally correlates with price. Moreover, glass tanks are designed in such a way that the refractory degradation rate is uniform throughout the tank, from the areas of most severe conditions to the most mild areas. This means that a single maintenance event can be scheduled for replacement of all glass-contact refractories, which has the added cost-benefit of minimizing production downtime, as well as optimizing the cost of the refractories. Therefore, AZS is used in the most severe areas of a glass tank, with lower-cost, less-degradation-resistant refractories in the less severe areas, the large zone that alumina services. Commonly alumina is used in the downstream areas of a glass tank. The following alumina-based refractories are commonly used in glass tanks:

Severe glass-tank-severity conditions: e.g., *high-alkali borosilicate insulating fiberglass*
- Fused-cast Chrome-Alumina. Used in combination with Chrome-AZS.

Regular glass-tank-severity conditions: Soda-Lime Glass
- Fused-cast alpha-alumina or alpha/beta-alumina. Used in combination with AZS. Fused-cast alumina has a slightly lower resistance to glass attack than AZS and is less costly than AZS (ZrO_2 is well above alumina in the ultra-refractory range—see Section 15.2).

Mild glass-tank-severity conditions: e.g., *Lead Glass*
- Dense sintered alumina. Dense sintered alumina has lower resistance to glass attack than both AZS and fuse-cast alumina.

The performance of a fuse-casting process is judged by its cast quality (reject rate), and the degree of stone and blister defects caused by the fused-cast refractories in service in the glass melt: stone, cord, and blisters in the resultant glass.

Fused-cast alumina is sometimes also used in wear-resistant applications and as a refractory in other industries such as the iron and steel industry.

15.3.3 Alumina fuse-casting formulations

Commonly fused-cast alumina is produced as beta-alumina or as alpha/beta-alumina. Beta-alumina has the approximate formula $Na_2O \cdot 11Al_2O_3$, which in the case of 100% pure $Na_2O \cdot 11Al_2O_3$ equates to 5.2% Na_2O and 94.8% Al_2O_3 by weight. Of course, being a refractory, it is not 100% pure. Alpha alumina equates to theoretically 100% Al_2O_3. Beta-alumina is best known as an ionic conductor with sodium the conducting species.

Silica content: Generally below 1%.

Sodium content: From 2% to 6.5%. At 6.5% Na_2O, this is defined as beta-alumina, while the lower Na_2O contents are midway between pure alpha-alumina and pure beta-alumina and are called alpha/beta-alumina.
Other impurities: Generally below 1%.
Alumina content: Generally in the range 93%–96.5%.

15.3.4 The fuse-casting process

The alumina melt is heated to temperatures of 2000–2500°C in an electric-arc furnace and poured while molten into a preprepared mold.

Fuse-casting molds are generally made up from ~30 mm thick resin-bonded sand panels, cut and glued into the required dimensions for the block to be fuse-cast. The mold is then placed into a steel box, with a clearance gap of several centimeters between the mold and steel box, which is filled with unbonded sand. If the casting box is also the annealing box, the steel mold box will be larger and the thickness of unbonded sand layer will accordingly be much greater.

When the molten ceramic melt is poured in, the intense heat rapidly breaks down the resin binder in the mold panels. The unbonded sand backing behind the resin-bonded sand panels therefore assists in mold integrity during the critical early cooling stages while chill-crystals are forming in the contact interface between the ceramic melt and the resin-bonded sand panel, while simultaneously the resin binder is rapidly degrading.

In some cases, the fusion casting is left in the casting mold (larger box, thicker unbonded sand layer) to anneal. In other cases, the casting is removed, while hot with a molten core, from the sand-mold when it has formed a skin sufficiently thick for removal (a couple of centimeters), and placed in an annealing box.

Annealing time can be from a few days up to a month, depending on the size of the casting. If the annealing box is filled with diatomite or sand, the long period at high temperature can cause cristobalite and/or tridymite to form. Dust of either causes silicosis. This is why nonsilica products, such as alumina and mullite, are now more commonly used in annealing boxes. Non-optimal annealing cycles, or insulation thickness values, can lead to cracking and other casting defects. A great deal of industrial knowhow, and often finite-element-modeling, ensures optimal annealing cycles.

Given the huge size of the refractories industry, $50 billion, the usage of high-purity alumina ceramics still amounts to a significant volume in dollar value. Glass-industry refractories are a $1.3 billion industry of which fused cast alumina is a sizable proportion, some hundreds of millions of dollars.

15.4 High-purity alumina ceramics for heat containment: Labware and industrial uses

In 1910, when Graf (Count) Botho Schwerin filed the first patent on alumina ceramics [17], the patent involved slipcasting and the application was crucibles. This original alumina crucible application remains a significant application today. High-purity alumina is an ideal crucible material as it has a very high melting point of ~2050°C, is highly chemically inert, and therefore poses a very low contamination risk when used as a crucible up to very high temperatures. Crucibles are a major use of alumina in the world today, both in research and industry: from microcrucibles for differential thermal analysis and other microanalytical methods to larger crucibles for containing heat-treatment experiments, to large industrial crucibles, and a gamut of speciality crucibles, such as those for growing single crystals of sapphire. A range of commercial crucibles is shown in Fig. 15.1.

Other labware alumina containment systems include furnace boats, furnace tubes, and furnace furniture. Some commercial examples are shown in Fig. 15.2.

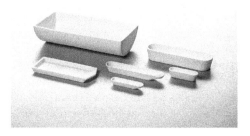

Fig. 15.1 A range of standard and specialty alumina crucibles, lids, and "combustion boats" from milliliter to multiple-liter capacity.
Manufactured by Ceramic Oxide Fabricators (AUST) Pty Ltd. Image courtesy of Ceramic Oxide Fabricators (AUST) Pty Ltd.

Fig. 15.2 Alumina furnace tubes (outer diameter 40–120 mm) and a range of other extruded alumina products.
Manufactured by Ceramic Oxide Fabricators (AUST) Pty Ltd. Image courtesy of Ceramic Oxide Fabricators (AUST) Pty Ltd.

The high-purity heat-containment application services the scientific research market and also various industrial applications.

15.5 Alumina machining tools

Important property: Hardness. Wear resistance.

Tool steel, cemented carbides, and bonded diamond are well known as the ideal machine tool materials. Tool steel and other hard metals for regular machining, cemented carbides for machining hard metals, and diamond for machining ceramics. Cemented carbides such as tungsten carbide have been used in this role for a very long time. Cemented carbides are an obvious fit given their combination of hardness and high toughness, diamond is an obvious fit given its pre-eminent hardness.

Alumina has comparable hardness to cemented carbides, but is less tough, and is well behind diamond in hardness. However, it performs well as a machine tool at high cutting speeds. This is a counter-intuitive situation. Therefore, the machining tool application for alumina warrants a more detailed discussion.

The concept of alumina as a potential machining tool dates back a century to the 1913 patent of the Thomson-Houston company [109]. This patent was also the first example of alumina green machining. A pure alumina body with 10% gum tragacanth binder was pressed into a 6 mm thick 20 mm diameter disk, presintered to 1400°C, green machined into the end use component, then resintered at 1800–2000°C. The proposed use was wear-resistant applications such as tools, bearings, dies, and drills.

The next significant publication on alumina as a machine tool was three decades later with the 1942 patent of one of the early pioneers of alumina, Eugen Ryshkewitch [137]. Entitled "Ceramic Cutting Tool," this patent concerned alumina with chromium oxide added. Chromium oxide addition is now a well-known mechanism for increasing the hardness of alumina, and the Ryshkewitch 1942 patent is the original reference to that effect.

It seems counterintuitive that alumina would make an appropriate choice for a lathing/machining tool, in competition with the comparably hard but tougher cemented carbides. Indeed alumina machine tools were not taken seriously until the 1950s. In the former Soviet Union in the early 1950s, a significant discovery was made to change this perception and identified the important niche for alumina [614–617]: *alumina is superior to cemented carbides in red-hot hardness*, a useful property for a machining tool at high cutting speeds. What alumina loses in toughness, it gains in high-temperature hardness. Given that machine tips routinely reach the red-hot state during machining processes, this was an important discovery. Alumina retains its hardness even at the melting point of steel. Moreover, cemented carbides are about four times the cost of alumina.

This discovery was rapidly taken up by scientists across the globe, and by the late 1950s, the concept of alumina lathing/machining tools was well established, not just for the cost-saving over cemented carbides, but also the hot hardness. The early alumina cutting tools were seen as suitable only for uninterrupted lathing at moderate depth of cut and feed rate and primarily only suitable for machining cast iron. In 1958, a detailed analysis of the performance of alumina cutting tools was published in the United States, which reported that alumina tool performance was metal-dependent in accordance with the following series, from best to worst: nickel, cobalt, gray cast iron, steel, lead, silver, aluminum, beryllium, and titanium [619]. Alumina grain size was found to be critical to tool performance, ideally grain size should be very fine, approaching 1 μm [620].

A great deal of research subsequently took place in the late-20th century, and a plethora of papers were published on the extensive work done in this area. Some of the key outcomes have been:

- Tool design enhancements in terms of edge radius and edge chamfering.
- Zirconia-toughened-alumina (ZTA) tools (increased toughness).
- Hot-pressed alumina-matrix composites containing up to 30% of carbide particles such as TiC (increased hardness).
- Hot-pressed or HIPed SiC whisker-reinforced alumina-matrix composites (increased toughness).
- Hot-isostatic-pressing (HIP) sinter-HIPing, as a postprocessing stage for pressureless sintered tools.

It may seem counterintuitive, but HIPing represents a cost improvement. Sinter-HIPing of presintered ceramics can be cheaper than hot pressing as the presintering by pressureless sintering is low cost, and sinter HIPing requires no encapsulation. Therefore, many presintered tools can be loaded into a sinter HIP, which then follows a simple preprogrammed cycle (Chapter 4). Hot pressing requires molds and is an expensive process that is complex to automate (Chapter 4).

While alumina machine tools remain a niche market in the machining industry, the primary driver for using alumina machine tools has been cost efficiency. Firstly, this is due to the lower cost for the alumina tools compared to cemented carbides. Secondly, the enhanced hot hardness brings economic advantages in long uninterrupted cutting and faster feed rates and cutting speeds. In summary, alumina cutting tools offer the following key benefits:

- Higher hot hardness.
- Lower cost than cemented carbides, in the case of pure alumina. For alumina-matrix composites, or HIPed alumina, the cost is much higher than pure alumina, which can offset the cost advantage.
- Alumina has higher resistance to abrasive wear than cemented carbides.
- Alumina is more chemically inert than cemented carbides.
- Alumina tools can be run at faster cutting speeds.
- Rate of metal removal can be higher for alumina than for cemented carbides.
- Longer tool life than for cemented carbides is possible for alumina, with appropriate usage.

If alumina cutting tools were superior all-rounders compared to cemented carbides, they would be dominant today. Unfortunately, alumina cutting tools have an Achilles heel. At low cutting speeds, below 100 m/min, alumina cutting tool wear rates are very much worse than cemented carbides (Fig. 15.3). Conversely, above 200 m/min, the wear rate of alumina cutting tools is very much better than cemented carbides (Fig. 15.4: the reciprocal of the image on the left). This benefit is not offset by the poor performance of alumina at regular cutting speeds. Therefore, alumina fulfills

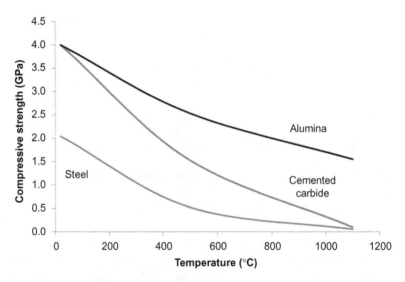

Fig. 15.3 Compressive strength as a function of temperature for steel, cemented carbides, and alumina. This demonstrates the substantial "red-hot hardness" of alumina cutting tools compared to steel and cemented carbide cutting tools [618].

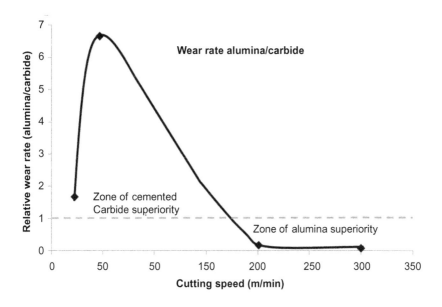

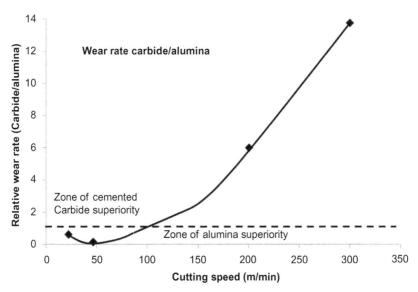

Fig. 15.4 Relative wear rate, as a function of cutting speed, for alumina cutting tools compared with cemented carbides. Above: represented as alumina/carbide; below: (the reciprocal of the image on the left): represented as carbide/alumina. At cutting rates below 100 m/min, cemented carbides are significantly superior. Above 200 m/min alumina is significantly superior. From 100 to 200 m/min is the crossover zone [621].

an excellent niche as a machine tool, but is by no means a strong all-rounder that supplants the alternatives.

15.6 Architectural applications for alumina

It is uncommon for alumina to be used in architectural ceramic applications as it is competing with very low cost alternatives such as glass, terracotta, porcelain, polished cement, and polished stone. However, its extreme chemical durability, high strength, and brilliant white appearance mean that for deluxe architectural applications, alumina has a potential niche role, as for example in the following case study.

The Sydney Opera House, shown in Fig. 15.5, is one of the world's most famous and distinctive buildings. Completed in 1973, and designed by Danish architect Joern Utzon, it is a large 180 × 120 m building with a majestic design involving large white sails rising to a height of 65 m, clad in brilliant white tiles. It is arguably the most remarkable modern building in the Southern Hemisphere.

The original Opera House Tiles were clay tiles and are described by Michael Lewis [623], a Senior Partner of Arup, as follows:

> The tiles were manufactured in Hoganas, Sweden, after extensive research by the architect to achieve a white ceramic tile with glazed finish and an underlying rough texture. There are two types of finish, "matt" and "glazed" which have been arranged in a specific pattern on the lids—although a close scrutiny of the tiles shows a marked difference between the two materials, the difference in reflectiveness creates a subtle

Fig. 15.5 The Sydney Opera House.
Image Author John Hill [622]. Reproduced from Wikipedia (Public Domain Image Library).

pattern on the surface which defines the edges of the tile lids and introduces the underlying anatomy or structural form of the load-bearing elements beneath. Standard tiles are 4¾ × 4¾ in (120 × 120 mm) and 5/8 in (16 mm) thick.

Being on the waterfront, in a windy location, this building is exposed to severe corrosive conditions from sea spray. Furthermore, while most tile-clad buildings involve vertical tile arrangements, the Opera House involves a dome-like tile arrangement that greatly increases the sun exposure. Moreover, Sydney has a hot and sunny climate, with >340 sunny days in the year, and temperatures occasionally reaching 45°C in mid-summer. The Sydney climate also has a relatively wide temperature range from hot days to cool nights. Even in winter, it is not unusual for the waterside temperature to reach 20°C, in sunny conditions, which can cause significant tile heating, but dropping close to 0°C on cold winter nights. The Sydney waterside minimum never drops below 0°C unlike western Sydney, inland of the coast, where morning frosts in winter are standard and minimum temperatures have gone as low as −8°C. Therefore, freeze/thaw cycling is not a problem on the Sydney waterfront.

Thus, the tiles cladding the opera house have some unique and demanding requirements:

- Brilliant white appearance, which needs to be stable in the long term.
- Severe exposure to sea spray: enhanced risk of moisture expansion and chemical corrosion.
- Severe sunlight exposure due to climate conditions and the dome-like tile-clad sails, with cool nights: significant thermal expansion cycling.
- The tiles in the overhanging sections of the sails are more exposed to wind and sea spray, and in the original structure up to two-thirds of the edge tiles cantilever beyond the fixation point, creating enhanced conditions for static fatigue in these tiles: cantilever loads in addition to thermal expansion cycling.

Under these severe conditions, the original clay tiles, particularly those in the overhanging sections of the sails, underwent significant degradation. Therefore, in the mid-1990s, a tender was issued to replace tiles that form the overhanging edge of the sail faces. It was important as part of the specifications of the replacement tiles that the new tiles had significantly improved properties, so as to give them enhanced durability under the service conditions.

This tender had strict criteria for color, mechanical strength, and moisture absorption. A number of companies competed for this tender, including Taylor Ceramic Engineering (TCE), a company that has specialized in alumina ceramics for almost half a century (Chapter 1, Section 1.4.2.5; Chapter 12). Ultimately, TCE was awarded the tender having succeeded in meeting these challenging criteria in a competitive environment.

In 1996, two tile formulations were developed for the tender under the direction of David Taylor (Managing Director/Founder: TCE): (1) an ultra-high-purity alumina tile and (2) a high-alumina tile.

Ultrahigh-purity alumina tile tender submission: The TCE proposal for the ultrahigh-purity-alumina tile, noted in the tender submission that "based on the physical evidence at hand and following an inspection of tiles in situ, a clay based ceramic tile would lack the physical properties and chemical resistance necessary for this harsh application on the edge of Sydney Harbour" [624].

High-alumina-tile tender submission: The second tile in the TCE tender submission was a high-alumina tile of proprietary composition. This had enhanced mechanical properties in comparison to the original Swedish Clay tiles from the 1970s, though obviously lesser than the mechanical properties of the ultrahigh-purity-alumina tile. It was lower in cost than the ultrahigh-purity alumina tile and had a closer color match to the original Swedish clay tiles [625].

Ultimately, the competitive tender was awarded to TCE, for the high-alumina tiles, chosen based on their combination of improved properties compared to the original Opera House Tiles, and close color match. Approximately 10,000 matt high-alumina tiles were supplied by TCE, of approximate dimensions $115 \times 230 \times 24$ mm and weight 1.6 kg, as shown in Fig. 15.6. TCE commenced development of this new ceramic body in the latter half of 1996 and delivered the first test tiles in early April 1997 [625].

A retrospective analysis of the tender submission archive by Alyssa Taylor (Managing Director: TCE) and Julie Taylor (Founder: TCE) states as follows [625]:

In relation to the ultra-high purity alumina proposal: "Ultrahigh purity alumina has a number of advantages over clay based ceramics, those being; a higher physical strength and hence higher Modulus of Rupture (MOR) of up to ten times, a higher density which, coupled with the higher strength, allows the thickness of the tile to be reduced thereby reducing the mass load on the building; a higher resistance to thermal sensitivity (thermal shock); higher chemical corrosion resistance and lower moisture absorption. All of which make Alumina an ideal candidate for replacement tiles."

In relation to the high-alumina tiles that were awarded the tender: "Development and testing continued and improvements to the originally developed body formulation were made which

Fig. 15.6 High-alumina Opera House cladding tiles.
Manufactured by Taylor Ceramic Engineering in 1997. Image courtesy of Taylor Ceramic Engineering.

lowered the water absorption to less than 3% and eventually 0.3% (the criterion was less than 5%). Water absorption (or open porosity) has an effect on physical strength, thermal shock resistance and chemical durability. Not only was TCE involved in the design of the ceramic composition, but they also had input into the mechanical fixation of the tiles to the roof. The original system had up to two thirds of the tile cantilevering beyond the fixation point, whereas the new design not only had the tiles glued to a backing plate, with the said plate protruding or locking into the tile, but these backing plates were then pop-riveted to the roof. Overall TCE's passion for exceptional design solutions and ability to adapt to different engineering situations meant that it was able to fulfil requirements that no other company was able to match."

15.7   Aerospace applications for alumina

Extreme oxidation resistance, up to its melting point, is the primary attribute of alumina for aerospace applications. Offset against this is the modest thermal shock resistance of alumina. To a lesser extent, its moderately low density of $3.96\,g/cm^3$ offers some benefit, though this is nowhere near as low as the density of the dominant aerospace ceramics in space shuttle claddings:

- Fused silica ($2.1\,g/cm^3$), which is oxidation resistant to 1700°C.
- Highly refractory oxidation-susceptible graphite ($2.25\,g/cm^3$) and carbon-carbon composites with their extremely low density (1.6–$2.0\,g/cm^3$) coupled with high toughness.

The 1950s and 1960s were the high point for research in aerospace applications for alumina ceramics, indeed this was the era of rapid development in missiles and rocketry. The primary application explored for alumina was the radome. Rain erosion at speeds >Mach1 is so severe on most materials that only a limited range of materials are able to be used in radomes.[3] Dense alumina proved resistant to rain erosion at such speeds. In the 1950s and 1960s, high alumina and some high performance glasses were seriously considered in these roles [626,627]. Prototype alumina radomes were fabricated by casting in the late 1950s and early 1960s [628].

It was around this time that Boeing (Seattle, USA) took an interest in the field. In 1960, Boeing conducted a study into alumina-metal bonding so as to evaluate the potential of alumina-metal bonding with the following aerospace alloys [608,609]:

- Alumina-titanium alloy (Ti6Al4V)
- Alumina-molybdenum alloy(0.5 Ti)
- Alumina-molybdenum alloy (alloy PH 15-7)
- Alumina-nickel alloy (alloy Rene41)

The purpose was for aerospace and missile applications, specifically alumina radomes, IRdomes, nose cones, leading edges of wings and control surfaces, and antenna covers [608,609]. Of greatest relevance to this book is the titanium alloy, given that titanium later became the dominant metal in alumina-metal seals in bionic

---

[3]The author spent time in the rain erosion facility and the arc-jet facility (atmospheric re-entry simulation) on a 2006 collaborative visit to NASA and US-AFRL.

implants (Chapters 8–10). However, it was not titanium that was investigated by Boeing, but rather the aerospace alloy Ti6Al4V. This was the era in which the US aerospace community was actively developing aerospace applications, particularly in the military sector, based on titanium-alloys, which at a density of 4.5 g/cm^3, are much closer to aluminum (2.7 g/cm^3) than steel (7.9 g/cm^3), and Ti6Al4V has a very high specific strength. The all-titanium Mach3+ Lockheed SR71 spy plane is one of the most famous examples of the 1960s. It was within this context that the Boeing alumina-titanium bonding study was conducted.

The Boeing alumina-metal bond-strength research involved a simple study of the tensile strength and shear strength of alumina-metal bonds produced by active metal brazing [608,609]. No seals were made, and therefore hermeticity was not evaluated. The study simply involved joint strengths being mechanically tested in tension and shear from room temperature up to 1000°F (540°C). The metal test pieces were 1.5 × 0.5 in. (38.1 × 12.7 mm), and the alumina test pieces were 0.5 × 0.5 × 0.25 in. (12.7 × 12.7 × 6.4 mm). Titanium hydride was sprinkled on the alumina surface, and brazing was done in either argon or a vacuum with either with Ag or Ag-Cu braze. This effectively constituted a TiAgCu alloy braze. Over the ensuing decades, experience with the TiCuAg braze showed that while it had initial high strength, it had very poor corrosion resistance. This was to have negative outcomes in both the aerospace industry and bionics industry, where it was trialed.

Active metal brazing is discussed in detail in Chapter 14.

However, it is notable that the Boeing work also contained the first, and indeed one of the only documented reports of the use of the Mo-Mn process (discussed in detail in Chapter 14) for bonding alumina to a titanium alloy, specifically to attach a prototype alumina radome to a titanium alloy [609]. No details are provided other than that "this process requires two or three furnace cycles and the availability of a 2300 − 2700°F hydrogen or ammo-gas furnace," that is, 1260 − 1480°C. Outcomes are not described other than a statement regarding the technical challenges and potential benefits: "this temperature is dangerously near the softening point of the ceramic especially for a large component. The titanium could also be embrittled by the furnace atmosphere. These limitations were offset by the high temperature capability of the final joints and the assurance of a strong reaction between the ceramic and braze with a resulting increase in reliability and strength of the joint" [609].

Since the 1960s, the primary application of alumina in aerospace has been in coatings technology and also advanced alumina-based composite ceramics, particularly SiC-whisker-reinforced alumina. Coatings are beyond the scope of this book. In the composites field, one of the most important developments in aerospace was the functionally graded material approach. From 1997 to 2009, the author was involved in a project developing alumina-based functionally graded materials for aerospace applications, including alumina-aluminum by pore grading and metal infiltration, and by powder-stream blending [629–632]. The role of alumina was twofold: oxidation resistance for re-entry conditions and rain-erosion resistance for lower altitude segments of the reentry flight path [629,632].

Silicon carbide whisker reinforced alumina and alumina matrix composites manufactured by the DIMOX process (directed metal oxidation) also became a research

Refractory and other specialist industrial applications 495

focus some decades later in the aerospace and industry and for other advanced applications. However, as noted in Chapter 1, composites based on alumina are beyond the scope of this book. With all their specialized manufacturing technologies and extraordinary properties and applications, they represent a scope so broad and so different to pure alumina, that the topic really needs to be addressed in a separate dedicated monograph. ZTA, in contrast, is just like a conventional alumina with some particulate zirconia added and is now commercially very important in the orthopedic and dental areas.

15.8 Oxygen sensors

Important property: Electrical insulator. Heat resistant. Gas tight seal.

The zirconia stabilization innovation of Ron Garvie (see Chapter 6, Section 6.4.1.4) had a number of technologically important outcomes:

(1) Partially stabilized zirconia (Chapter 6)
(2) ZTA (Chapter 7)
(3) Solid oxide fuel cells (SOFC)
(4) Stabilized zirconia ionic conductors "solid electrolyte" for oxygen sensors

ZTA is such an important biomaterial in the orthopedics industry that it forms the entire basis of Chapter 7 in this book and is also of importance in Chapter 5 (dental ceramics). Zirconia has many industrial uses, and is important in dentistry, but was problematic as an orthopedic material as discussed in Chapter 6. SOFC are beyond the scope of this book.

The final significant use for alumina is outcome 4 from the list above: oxygen sensors, as shown in Fig. 15.7. The zirconia solid electrolyte was developed by the same CSIRO[4] Materials Science Division in Melbourne, Australia, at which British Physicist Ron Garvie, the inventor of zirconia toughening, perfected his zirconia "ceramic steel" invention in 1972. This CSIRO division became a hothouse for zirconia innovation in the 1970s and 1980s. An oxygen sensor utilizes zirconia solid electrolyte, platinum coated on both sides, and commonly sealed to an alumina support tube. One of the challenges with the early prototype development was containment of the zirconia electrolyte. This was achieved through welding a pellet of stabilized zirconia solid electrolyte in the end of a high-purity alumina tube at around 2000°C, utilizing the 2000°C alumina-zirconia eutectic [633–635]. Harold Kanost, an American Ceramic Engineer, emigrated to Australia in 1970 joining CSIRO to contribute to research on partially stabilized zirconia. In 1971, he started Ceramic Oxide Fabricators Pty Ltd., which CSIRO later licensed to manufacture the patented SIRO2 oxygen sensor.

Ceramic Oxide Fabricators (AUST) Pty Ltd. is an Australian-owned and operated company with >45 years of history in the manufacture and export of alumina and

---

[4]CSIRO—Commonwealth Scientific and Industrial Research Organization. Australia's nationwide Government-funded industrial research organization, with numerous divisions throughout the country.

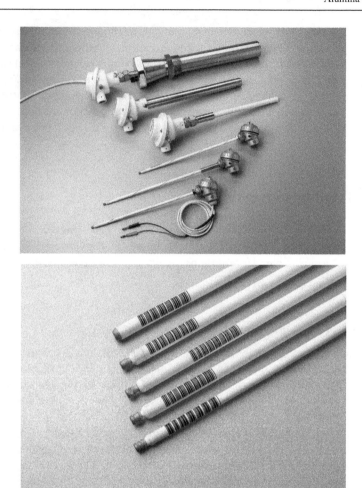

Fig. 15.7 Above: Oxygen sensors manufactured by Ceramic Oxide Fabricators (AUST) Pty Ltd. Below: Oxygen probes manufactured by Australian Oxytrol Systems Pty Ltd.
Image courtesy of Ceramic Oxide Fabricators (AUST) Pty Ltd.

zirconia-based ceramic products. The Company is a world leader in the manufacture of high-temperature oxygen sensors (500–1750°C), used in science and industry (see Fig. 15.7) The oxygen sensor was invented by CSIRO, then commercialized under license by Ceramic Oxide Fabricators for manufacture. Applications include scientific research, measuring fugacity of chemical reactions at elevated temperatures, carburizing furnaces, as well as oxygen measurements in molten metals. Other products of Ceramic Oxide Fabricators (AUST) Pty Ltd. include crucibles, tubes (furnace to multibore), rod, insulators, refractories, and custom fabrication including specialized machining services (see Figs. 15.1, 15.2, and 15.8–15.10). The Company supplies ceramics to many of the world's leading scientific institutions.

Refractory and other specialist industrial applications 497

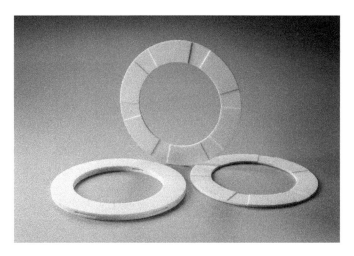

Fig. 15.8 Precision alumina kiln furniture custom manufactured for Ceramic Fuel Cells Limited by Ceramic Oxide Fabricators (AUST) Pty Ltd.
Image courtesy of Ceramic Oxide Fabricators (AUST) Pty Ltd.

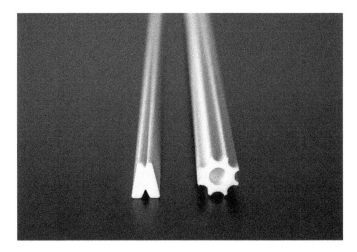

Fig. 15.9 Specialty alumina extrusion. Left: Shearing comb file imparts controlled bluntness to shearing combs allowing shearing of sheep without the sheep bleeding. Right: Precision plasma torch insulator with provision for cooling air flow.
Manufactured by Ceramic Oxide Fabricators (AUST) Pty Ltd. Image courtesy of Ceramic Oxide Fabricators (AUST) Pty Ltd.

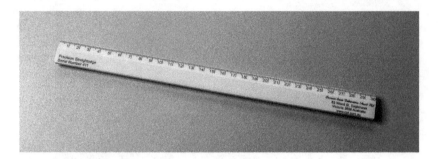

Fig. 15.10 Precision straight edge alumina ruler custom manufactured for Boeing Ltd. by Ceramic Oxide Fabricators (AUST) Pty Ltd.
Image courtesy of Ceramic Oxide Fabricators (AUST) Pty Ltd.

15.9 Specialist industrial applications of alumina not discussed elsewhere

The following list compiles all the significant uses of alumina not discussed elsewhere in this book.

- Gas laser tube
- Laboratory instrument tube
- Kiln furniture
- Agricultural-plow tine tip
- Alumina gauge
- Jewel bearings in watches
- Mortar and pestle

Of this list, all are niche applications for alumina in the sense that:

- They do not represent a high-volume market.
- They have the potential to grow into a high-volume market in the future.
- Alumina is the best choice in this application.

Each of the above niche applications for alumina is an obviously good fit for alumina, with little question as to the suitability of alumina to the role, and for brevity, the descriptions given below are sufficient.

Thermocouple sheaths:

Important property: Electrical insulator. Heat resistant. These are widely used, and essentially function as high-temperature electrical insulators for thermocouple wires, commonly used for operating temperatures up to 1800°C.

Gas laser tubes:

Important property: Chemical inertness. The alumina ceramic CO_2 laser tube technology is a significant improvement over the old metal tube laser technology. CO_2 gas in a metal laser tube reacts significantly with the metal tube, which is problematic for long-term usage, but does not react at all with an alumina tube.

Laboratory instrument tubes:

Important property: Heat resistance. Chemical inertness. With its heat resistance and high chemical inertness, alumina is ideal as instrument tubes and supports for high-temperature analytical instruments, or applications in corrosive environments. This is for exactly the same reasons that alumina makes an excellent analytical-grade crucible.

Kiln furniture:

Important property: Heat resistance. Chemical inertness. With its heat resistance and high chemical inertness, alumina also makes an excellent noncontaminating surface in furnace furniture. This is for exactly the same reasons that alumina makes an excellent analytical-grade crucible.

Agricultural-plow tine tips:

Important property: Wear resistance. Alumina tips have been used as tine tips on agricultural plows. They give much longer tine wear life than conventional plow tips.

Alumina gauges:

Important property: Wear resistance. The high-wear resistance of alumina provides a niche opportunity for alumina precision gauges, such as plug-and-ring gauges, and thread plug gauges operating in high-wear applications.

Jewel bearings in watches:

Important property: Wear resistance. These are commonly made of alumina and called "synthetic sapphire" in the watch trade.

Mortar and pestle:

Important property: Wear resistance. Chemical purity. High-purity alumina is ideal in this role due to its hardness, low wear rate, and minimal contamination risk.

Alumina: The future   16

This chapter summarizes all of the applications for alumina ceramics in the world today, both biomedical and industrial, within the context of the previous 15 chapters which have discussed these applications in detail. This chapter then looks at what the next 30 years is likely to hold in store for the future applications of alumina. In traditional industrial applications, dramatic changes are unlikely. The most dramatic changes are likely to be in the biomedical realm, particularly implantable bionics.

Before forecasting the future for alumina, it is appropriate to summarize the current world market for alumina ceramics. Traditional industrial applications and biomaterials applications for alumina are easily categorized as per the list below. Lightweight body armor uses of alumina, and its spin-offs in lightweight vehicle armor, are neither a traditional industrial application nor a biomaterials application. They do involve protecting the human body, combining materials science and the study of traumatology, and have delivered an enormous humanitarian benefit to society with millions of lives protected, and many saved. Therefore, body armor and its spin-offs have been categorized in this book as belonging to the biomedical engineering category, though not biomaterials.

In summary, the following applications for alumina ceramic have been discussed in this book:

Biomedical engineering applications for alumina (Chapters 5–11)
   $Billion biomaterials markets for alumina
   - Orthopedic hip bearings (Chapters 6 and 7)
   - Bionic feedthroughs (Chapters 8–10)
   Niche biomaterials markets for alumina (Chapter 5)
   - White or translucent orthodontic brackets
   - Dental crowns
   - Dental bridges
   - Dental implants
   - Dental abutments
   - Dental implants
   - Tissue scaffolds
   - Prosthetic eyeball
   - Osteotomy spacers
   $Multimillion biomedical engineering markets for alumina—Armor Realm (Chapter 11)
   - Lightweight body armor
   - Lightweight vehicle/aircraft armor
   - Synthetic sapphire bulletproof windows
Traditional industrial applications for alumina (Chapters 12–15)
   $Multiple hundred million dollar range (collectively not individually) wear- and corrosion-resistance applications market (Chapter 12)
   - Hydrocyclone components and linings (mineral processing)
   - Spigots (wear and corrosion)

- Straight and curved piping (wear and corrosion)
- Reducers (wear and corrosion)
- Orifice and baffle plates (wear, corrosion, papermaking)
- Dart valve plugs and seats (wear and corrosion)
- Tiles for ore chutes (standard, engineered, and weldable)
- Knife-edge blades (conveyors, wine industry)
- Mill linings and milling media (wear and corrosion)
- Thread guides (textiles)
- Wire drawing step cones and dies (wire)
- Brick and heavy clayware (brick cores, sleeves, die boxes, and shaper caps)
- Augers and trough liners (numerous industries)
- Nozzles (wear and corrosion)
- Water-faucet valves
- Rotary seals
- Bearings and sleeves
- Cylinders
- Rods and disks
- Spacers, tubes, and washers
- Tiles (standard, engineered, and weldable)

$Billion (collectively not individually) electrical applications market (Chapters 13 and 14)
- Electrical insulators: high voltage
- Electrical insulators: high temperature
- Electrical insulators: high frequency
- Sparkplug insulators
- Thin-film silicon-on-insulator (SOI) microelectronic substrates
- Thick-film microelectronic substrates
- Microelectronic-insulating heat sinks
- Hermetic electrical feedthroughs
- Fuse bodies
- Heating-coil formers

$Multiple hundred million (collectively not individually) heat containment applications market (Chapter 15)
- General refractories ($50 billion industry, of which alumina is a key component)
- $Multimillion fuse-cast alumina glass-contact refractories
- Crucibles
- Furnace tubes
- Labware

Other niche industrial applications (Chapter 15)
- Machining/cutting tools
- Architectural
- Aerospace applications
- Gas laser tube
- Laboratory instrument tube
- Kiln furniture
- Thermocouple sheath
- Synthetic sapphire microwave window

- Oxygen sensors
- Agricultural-plow tine tips
- Alumina gauge
- Jewel bearings in watches
- Mortar and pestle

Some of the applications in the above list represent $billion markets, and some niche research fields. Moreover, some of these applications are life-saving or life-enhancing technologies, while others are merely traditional engineering technologies. Therefore, before taking a look at the future, it is appropriate to look at the significance of alumina in the year 2020 from an economic viewpoint and also a humanitarian viewpoint.

16.1 Economic impact of alumina—2020

The applications for alumina ceramics in this book were broadly categorized into biomedical and industrial. In pure economic impact, alumina in 2020 is as follows:

Biomedical alumina
- Bionic implants (IPG)
  - $25 billion industry.
  - First used 1971 (pacemaker).
  - Significant broadening from pacemaker to other bionic applications beginning in the 1980s.
  - 14 bionic implant types now on the market, over 25 disease states approved for bionic implant commercial medical use.
  - Alumina feedthrough is the key enabling platform technology for all bionic implants.
- Orthopedics (hip implants)
  - $7 billion hip replacement industry (1.3 million hips PA).
  - FDA-approved alumina orthopedic bearings in 2003.
  - Significant commercialization since 2010.
  - Alumina or zirconia-toughened-alumina (ZTA) bearings with an end-user price of around $1000. About 900,000 alumina bearing components PA currently, which service 55% of global hip implants market (40% alumina-polyethylene; 15% alumina-alumina).
- Dental and other bioinert implants

Little economic impact but promise for the future
  - ZTA is making significant inroads into dentistry.
  - Bioactive-ion-doped alumina tissue scaffolds—new research field.
  - Alumina is the most bioinert biomaterial in current usage.
- Body armor and vehicle armor
  - $Multiple hundred million dollar range.
  - Only elite use from 1965 to 1996.
  - Millions of body armor ceramic plates deployed globally since 1996.
  - Alumina dominates the civilian sector and is one of the two main armor ceramics used in the military sector.

Industrial alumina
- Wear- and corrosion-resistant industrial applications
  - Alumina ceramic consumption $Multiple hundred million dollar range
  - Alumina dominates the wear-resistant linings application in the $500 billion mining industry.
  - Important roles in the following industries: textiles, papermaking, wire drawing, food processing, heavy-clayware manufacture.
  - Used in rotary seals, bearings, pump components, and many niche applications.
- Electrical and electronic applications
  - Alumina is an important heatsink/insulator substrate technology in the global $500 billion microchip industry.
  - Alumina is the standard insulator used in electrical feedthroughs in a wide range of electrical equipment, from magnetrons to vacuum chambers to thick-film electronic devices. This is a huge global industry.
  - Alumina spark plug insulators underpin the $3 billion global spark plug industry.
- Refractory applications
  - Alumina is a significant platform technology in the global $50 billion refractories industry.
  - Fuse-cast pure alumina glass contact refractories a global $300 million industry.
- Specialty industrial applications
  - Cutting tools, laser tubes, oxygen sensors, architectural, many niche uses. Small but growing industry.

16.2   Social impact of alumina—2020

Aside from pure economics, alumina has also proved to be an important material from a social impact point of view.

Life-saving applications of alumina
- Alumina bionic feedthrough technology—the pacemaker
  - Millions of lives saved.
  - First implanted 1971.
  - Over a million pacemakers implanted a year, and many millions since 1971.
- Alumina body armor
  - Millions of people protected, many lives saved.
  - First deployed 1965.
  - Huge global body armor deployment globally since 1996 (millions).

Life-enhancing applications of alumina
- Alumina bionic feedthrough technology—used in all bionic implants
  - Bionic ear
  - Bionic eye
  - Deep-brain stimulator (Parkinsons disease, essential tremor, dystonia, OCD)
  - Spinal cord stimulator (back pain, angina pain, peripheral artery disease pain)
  - Vagal nerve stimulator (epilepsy, depression)
  - Occipital-nerve stimulator (migraine, brain trauma)
  - Sacral-nerve stimulator (incontinence, pelvic pain, ED)
  - Gastric stimulator (obesity, gastroparesis, IBS)

- o Pulmonary stimulator (respiratory support)
- o Functional electrical stimulation/peripheral nerve stimulation (foot drop, mobility for spinal cord-injured patients)
- Orthopedics—alumina bearings in hip implants
  - o 1.3 million hip implants implanted globally PA, 55% use alumina bearings.
  - o Alumina bearings bring greatly enhanced implant longevity.
  - o Alumina bearings show promise for other joints such as the knee.

Community-enhancing applications of alumina
- Mining industry
- Textile industry
- Paper industry
- Food industry
- Wire drawing industry
- Chemical processing industry
- Microchip heatsink/insulator
- Electrical feedthroughs in a wide range of equipment
- Refractories

16.3 Alumina in the future

First, and most importantly, the orthopedics and bionics industries will continue to grow and the usage of alumina in these industries will continue to grow in sales value, in proportion to the growth of these industries.

Similarly, the mining, electronics, refractories, textiles, chemical processing, and other industries in which alumina has a significant role, will continue to grow. The importance of alumina in these industries is unlikely to change in the foreseeable future. The current applications for alumina ceramics in those industries will continue to grow in sales value in proportion to the growth of those industries.

Alumina body armor is likely to continue to be increasingly important in the future. The most significant manifestation of this is likely to be the continued penetration of body armor into the civilian sector. In developing countries, where economics favor low-cost ceramic armor, alumina is the ceramic of choice. Life-saving widespread deployment is really an economic not a technological challenge. This is essentially a humanitarian issue, with enormous potential benefits in lives saved, but only marginal economic benefits to the entrepreneurs involved.

In a technology sense, while the ceramic technology of body armor has evolved very little since 1970, the fabrics used to clad alumina and other ceramic body armor are evolving, for example, spider silk, carbon nanotubes, and enhanced zylon. This is where the future enhancements lie.

16.3.1 New engineering applications from 20th-century alumina science

To a large extent, the era of alumina science was the 20th century and the 21st century is primarily the era of alumina engineering. Therefore, the future for alumina will primarily be a continuation of the present 21st-century engineering applications underpinned by 20th-century alumina science, and an expansion of these current engineering applications.

16.3.1.1 Brain-computer interface

The most world-changing future direction is likely to be brain-computer-interface (BCI) implants, depending as they do on the 1970 alumina feedthrough innovation and 21st-century electrode and microprocessor systems. While experimental short-term prototypes have not necessarily involved alumina feedthrough systems (as indeed experimental prototypes of the bionic ear and bionic eye did not), *long-term BCI implants will depend on alumina feedthroughs*, and probably also inductive coupling systems for transcutaneous power and information transfer, such as those used in the bionic ear and bionic eye. Moreover, rather than the invasive pin-electrode technology (penetrating neural tissue) of the experimental BCI implants, long-term BCI implants are likely to use the noninvasive high-channel density "electrode pad" technology (surface contact with neural tissue) outlined in Chapter 10 or stentrodes.

The BCI will revolutionize many aspects of life in ways too numerous to list, many of which are yet to be imagined. However, some applications are both very likely to eventuate and very likely to have a significant social impact:

- Mobility for spinal cord-injured patients by marriage of BCI and FES technology.
- Communication by synthetic telepathy (military, telephonic, social media).
- Highly evolved exoskeleton robotic frames.
- Novel vehicle/avionic control systems.
- Novel BCI-controlled computer-human interface.

16.3.1.2 Conventional bionics in the future

The bionics realm now involves 14 different types of implantable bionics treating more than 25 disease states. The future for alumina bionics will involve expansion of the existing 21st-century applications:

- More bionic implant types and disease states treated
- New-generation bionic implants with hundreds, even thousands, of channels treating conditions for which large channel numbers are necessary, such as FES-walking implants and FES hand-mobility implants

16.3.1.3 New industrial applications based on 20th-century science

- New wear-resistant industrial applications
- New corrosion-resistant industrial applications
- New innovative applications of alumina in electrical industries, such as the enhanced large-scale Gyrotron if the nuclear fusion industry takes off

16.3.2 New 21st-century alumina science: Engineering applications

There are a number of areas of 21st-century alumina science that promise new alumina applications in the foreseeable future:

- Fiber-reinforced alumina for orthopedics other than convention hip replacements (hip resurfacing, knee bearings, other joints).
- Laser-machined cofired alumina/platinum bionic feedthroughs with thousands of channels (bionic eye enabling near-normal vision).
- Three-dimensional (3D) printing of alumina composites or functional materials.

Thirty years from now in 2050, on the centenary of the commercialization of alumina, it is probable that all of these 21st-century alumina science innovations will have been realized in commercial applications, which could include

- Bionic eye retinal implants, ultraminiaturized, with thousands of channels, enabling near-normal vision.
- Metal/polymer bearings near obsolete in the orthopedic industry. Alumina near 100% market share.

Finally, there will arise new engineering applications for alumina in the future, both new adaptations of 20th-century alumina science, and new applications arising from new 21st-century alumina science.

16.4 Conclusions: Alumina in the year 2050

- In the centenary of alumina ceramic commercialization, bauxite is likely to be as plentiful as today, other than in some countries where resources are small and mining output is large, such as China and parts of Europe.
- The industrial alumina market (wear, corrosion, electrical, refractories) in terms of products and market share is likely to closely resemble the world of industrial alumina in 2020, but proportionally larger in total market size and economic impact.
- There seems little doubt that the important existing industrial and biomedical applications of alumina ceramics discussed in this book will continue to perform the same vital role they do today, well beyond 2050.
- Alumina body armor is unlikely to be significantly different, other than perhaps more common in the developing world than it is today, and in the developed world more highly evolved FRP cladding may be common. Alumina vehicle and aircraft armor may have evolved somewhat.
- It is very likely that new and important applications and industries that depend on alumina ceramics will emerge in the decades to come, especially in implantable bionics (BCI—brain-computer-interface and FES—functional electrical stimulation, in particular), orthopedic implants, and advanced biomedical implants. In this regard, the year 2050 is likely to include
  - A bionic eye capable of near-normal vision, i.e., as highly evolved as the bionic ear was in the year 2000.
  - A BCI implant as highly evolved as the bionic ear was in the early 1980s, and where the bionic eye is now, i.e., functional early commercial technology manifestations, with

synthetic telepathy, thought-controlled robotic exoskeletons, and visual cortex technology three likely manifestations.
- FES/BCI technology enabling unaided walking (no walking frame) for spinal cord-injured patients—at the clinical trial stage.
- Alumina bearings will probably dominate all orthopedic implant systems (both full joint replacements and joint resurfacing), for the hip, knee, ankle, shoulder, elbow, and finger.

16.5 Closing comments

While it may seem a very broad scope to discuss all alumina applications in this book, from those driving $billion markets down to the niche uses of 2018 (notably at the end of Chapter 5 and the end of Chapter 15), on the grounds that all may prove significant in the future, I would like to give the last word to my predecessors.

In the last book published on alumina, which was in 1984 by Erhard Dorre and Heinz Hubner [1], body armor and orthopedic bearings were presented in this context, as niche alumina applications with a possible significant future. It was prophetic of them. In this 2018 book, they are among the leading applications.

Bionic feedthrough applications for alumina, which had obscure fledgling status in 1984, were not mentioned in the 1984 book [1], understandably because almost nobody in the traditional ceramics industry yet knew of the world-changing developments underway in the "sci-fi" bionics field as it was viewed in 1984.

Bionic feedthroughs are the number one commercial application for alumina in 2018, in dollar value. They also formed an important backdrop to my life since the 1990s and the fortunes of the Australian biotechnology industry. Telectronics, the Australian pacemaker company, was Australia's leading medical device company in the 1980s and 1990s, world number 3, and employing thousands, all thanks to David Cowdery's invention of the alumina bionic feedthrough. Telectronics is now long gone, but Cochlear Ltd. the Bionic Ear Company has taken its place. The bionic ear alumina feedthrough technology, adapted by technology transfer from 1980s parent company Telectronics, and with the bionic ear application driven by the innovative vision of Graeme Clark and Jim Patrick, has resulted in Cochlear Ltd., Australia's equally largest Medical Device Company in 2018, employing thousands, and bringing hearing to 250,000 Cochlear implant recipients.

It is difficult to predict the future. 34 years ago in 1984, I did not envisage the world of alumina ceramics in 2018. However, with the benefit of hindsight, and decades of participation in Ceramic Engineering and Biomedical Engineering, it is not too difficult to look 30 years ahead to 2050. In the traditional alumina applications, I believe the world of 2050 will not look greatly different. However, in the bionics and orthopedics realm, I have no doubt it will be a significantly different world from today. History will have the final word on that.

Thus, on that note, it seems appropriate to end with the words of John Fitzgerald Kennedy, whose prescient words from the early 1960s ring true today:

> *Change is the law of life. And those who look only to the past or present are certain to miss the future.*

References

[1] E. Dorre, H. Hubner, Alumina, Processing, Properties, and Applications, Springer-Verlag, Berlin, 1984.
[2] R. Green, T. Guenther, C. Jeschke, A. Jaillon, J. Yu, W. Dueck, W. Lim, W. Henderson, A. Vanhoestenberghe, N.H. Lovell, G.J. Suaning, Integrated electrode and high density feedthrough system for chip-scale implantable bionics, Biomaterials 34 (26) (2013) 6109–6118.
[3] Armslist, Vintage Original Alumina/Fibreglass Lightweight Body Armor Breastplate as Deployed to US Aircrews in the Vietnam War Affectionately Known as "Chicken Plate", As advertised at http://www.armslist.com/posts/3275754/gary-indiana-tactical-gear-for-sale-trade–chicken-plate-body-armor-with-kevlar-helmet, 2018.
[4] Binarysequence, https://en.wikipedia.org/wiki/Microchip_Technology#/media/File:Apple_Desktop_Bus_Microchip.jpg, 2013.
[5] C.S. Hurlbut, C. Klein, Manual of Mineralogy, 20th ed., Wiley, New York, 1985, pp. 300–302.
[6] G.T. Austin, Minerals Yearbook, vol. 1, 1987, pp. 71–84.
[7] P.C. Rickwood, The largest crystals (PDF), Am. Mineral. 66 (1981) 885–907.
[8] LesFacettes, https://commons.wikimedia.org/wiki/File:Geschliffener_blauer_Saphir.jpg, 2011.
[9] Humanfeather, https://commons.wikimedia.org/wiki/File:Ruby_gem.JPG, 2009.
[10] Ra'ike, https://en.wikipedia.org/wiki/Corundum#/media/File:Several_corundum_crystals.jpg, 2006.
[11] JewelsDujour, http://www.jewelsdujour.com/2014/02/ten-most-expensive-sapphires-sold-at-auction/.
[12] Unknown, https://en.wikipedia.org/wiki/Bauxite#/media/File:BauxiteUSGOV.jpg, 2004.
[13] T. Thomson, A System of Chemistry of Inorganic Bodies, Baldwin & Cradock, William Blackwood, London, Edinburgh, 1831.
[14] F. Habashi, Karl Josef Bayer and his time, CIM Bull. 97 (1083) (2004) 62–66.
[15] K.J. Bayer, Process of obtaining alumina, Patent in 3 jurisdictions: Britain 10093; USA 382505; Germany 43977, 1888.
[16] K.J. Bayer, Process of making alumina, Patent in 3 jurisdictions: Britain 5296; USA 515895; Germany 65604, 1894.
[17] B. Schwerin, Method for manufacture of highly resistant objects from naturally nonplastic materials, German Patent 274039, 1910.
[18] H.W. Daubenmeyer, Elastic mold and method of molding material, Champion Spark Plug Company, US Patent 1983602, 1934.
[19] K. Schwartzwalder, Injection moulding of ceramic materials, Am. Ceram. Soc. Bull. 28 (11) (1949) 459–461.
[20] CIA Intelligence Memorandum 344, (Approved for release 1999) 1951.
[21] Saint Gobain Materials, Innovative Materials Sector, https://www.saint-gobain.com/en/group/our-businesses/innovative-materials, 2017.
[22] Saint Gobain Ceramics, Saint Gobain Ceramic Materials, http://www.ceramicmaterials.saint-gobain.com/, 2017.
[23] A. Smallwood, 35 years on. A new look at synthetic opal, Aust. Gemmol. 21 (2003) 438–447.

[24] Kyocera, Characteristics of Kyocera Fine Ceramics, 2017, Product Specification Document. Kyocera Corporation 008/016/1708.
[25] Kyocera, Kyocera Fine Ceramics (Alumina Ceramics), https://global.kyocera.com/prdct/fc/list/material/alumina/alumina.html, 2017.
[26] Sintox, Morgan Advanced Ceramics, Specification Document, http://www.matweb.com/search/datasheettext.aspx?matguid=107b5420f35145f79e56825950b7e09d, 2017.
[27] Morgan, http://www.morganadvancedmaterials.com/, 2017.
[28] CoorsTek, https://www.coorstek.com/, 2018.
[29] CeramTec, https://www.ceramtec.com/, 2018.
[30] CUMI, https://www.cumi-murugappa.com/ceramics/ic/, 2018.
[31] Taylor, Taylor Ceramic Engineering, http://www.taylorceramicengineering.com/, 2018.
[32] AAC, Australian Aluminium Council Ltd, fact sheet, 2010.
[33] Tomago, Tomago Aluminium, http://www.tomago.com.au/about-us/about-aluminium, 2017.
[34] DIIS, Australian Government, Department of Industry Innovation and Science, BAA (Bauxite Alumina and Aluminium), 2017.
[35] K. Davis, Material review: alumina (Al2O3), School of Doctoral Studies European Union J. (2010) 109–114.
[36] AIMR, Bauxite, Mines Atlas, Australian, 2013.
[37] USGS, Mineral Commodity Summaries: Bauxite and Alumina, United States, https://minerals.usgs.gov/minerals/pubs/commodity/bauxite/mcs-2017-bauxi.pdf, 2017.
[38] H.Z. Yang, Bauxite deposits in China, Chin. J. Geochem. 8 (4) (1989) 293–305.
[39] UGS, Mineral Commodity Summaries, Aluminum (PDF), US Geological Survey, 2017.
[40] P.H. Freyssinet, C.R.M. Butt, R.C. Morris, P. Plantone, Ore forming processes related to lateritic weathering, in: Economic Geology One Hundredth Anniversary Volume 1905–2005, Colorado, SEG, 2005.
[41] G. Bardossy, Karst Bauxites, Developments in Economic Geology, vol. 14, Elsevier, Amsterdam, 1982.
[42] G. Bardossy, G.J.J. Aleva, Lateritic Bauxites, Developments in Economic Geology, vol. 27, Elsevier, Amsterdam, 1990.
[43] V.G. Hill, R.J. Robson, in: The classification of bauxites from the Bayer plant standpoint, Proceedings of TMS Light Metals, 1981, pp. 15–27.
[44] Alsglobal, Geochemistry technical note. Bauxite, Alsglobal.com, 2017.
[45] L.R. Liu, L. Aye, Z.W. Lu, P.H. Zhang, Analysis of the overall energy intensity of alumina refinery process using unit process energy intensity and product ratio method, Energy 31 (8–9) (2006) 1167–1176.
[46] J. Doucet, in: Double digestion: technology that leads towards quality and efficiency, Third International Alumina Quality Workshop, Hunter Valley, NSW, 1993, pp. 93–101.
[47] J. Doucet, et al., in: Pressure decantation technology: the Kaiser Gramercy experience, Sixth International Alumina Quality Workshop, Brisbane, QLD, 2002, pp. 94–99.
[48] S. Kumar, B. Hogan, J.-P. Lerredde, J. Laurier, G. Forté, in: Double digestion process: and energy efficient option for treating boehmitic bauxites, Eighth International Alumina Quality Workshop, Darwin, 2008, pp. 44–46.
[49] D.P. Rodda, R.W. Shaw, Extraction of alumina from bauxite by double Digestion— including fast low temp. Digestion of gibbsite fraction, solid/liquid. Sepn. Digestion of boehmite fraction and two-stage post-desilication WO9606043-A, 1996.
[50] S. Bhargava, M. Allen, M. Hollitt, S. Grocott, A. Hartshorn, Thermal activation of bauxite, Chem. Aust. 71 (2004) 6–8.

[51] M. Hollitt, S. Grocott, G. Roe, Treatment of an alumina process feedstock includes controlling the contact time of the solid alumina feedstock during heating, WO200010919-A, 2000.
[52] S. Ostap, Control of silica in the Bayer process used for alumina production, Can. Metall. Q. 25 (2) (1986) 101.
[53] P. Smith, C. Wingate, L. De Silva, in: Mobility of included soda in sodalite, Eighth International Alumina Quality Workshop, Darwin, NT, 2008, pp. 27–30.
[54] B.I. Whittington, in: Quantification and characterisation of hydrogarnet and cancrinite present in desilication product (DSP) by powder X-ray diffraction, Fourth International Alumina Quality Workshop, Darwin, NT, 1996, pp. 413–422.
[55] L.P. Ni, et al., A study of hydrogarnets formed in the Al_2O_3-CaO-Fe_2O_3-Na_2O-SiO_2-H_2O system, Russ. J. Inorg. Chem. 13 (11) (1968) 1585–1587.
[56] L.P. Ni, M.M. Goldman, T.V. Solenko, Processing of Iron-Rich Bauxites, Metallurgica, Moscow, 1979.
[57] S. Gu, in: Chinese alumina industry—development, prospect and challenge in the future, Eighth Alumina Quality Workshop, Darwin, NT, 2008.
[58] H. Zhao, in: Digestion of diasporic bauxite with mass ratio of Al_2O_3/SiO_2 no greater than 7 by Bayer process with an excessive addition of lime, Proceedings of TMS Light Metals, Seattle, WA, 2002, pp. 101–104.
[59] K. Solymar, M. Orban, J. Zoldi, G. Baksa, Methods for reducing NaOH losses in the hungarian alumina plants, Travaux ICSOBA 13 (18) (1983) 377–390.
[60] J. Zoldi, K. Solymar, J. Zambo, K. Jonas, in: Iron hydrogarnets in the Bayer process, Proceedings of TMS Light Metals, 1987, pp. 105–111.
[61] K. Solymar, J. Steiner, J. Zoldi, in: Technical peculiarities and viability of hydrothermal treatment of red mud, Proceedings of TMS Light Metals, 1997, pp. 49–54.
[62] V.V. Medvedev, et al., Method for hydrochemical processing of aluminosilicate feedstock RU2193525-C1, 2002.
[63] V.V. Medvedev, S.N. Akhmedov, V.M. Sizyakov, V.P. Lankin, A.I. Kiselev, Hydrogarnet technology of processing the bauxite raw material as up-to-date alternative to Bayer-sintering technique, Tsvetn. Met. 11 (2003) 58–62.
[64] E.T. Misra, J. White, Crystallisation of Bayer aluminium trihydroxide, J. Cryst. Growth 8 (2) (1971) 172–178.
[65] J. Addai-Mensah, Surface and structural characteristics of gibbsite precipitated from pure, synthetic Bayer liquor, Miner. Eng. 10 (1) (1997) 81–96.
[66] M.Y. Lee, G.M. Parkinson, P.G. Smith, F.J. Lincoln, M.M. Reyhani, Separation and Purification by Crystallisation, ACS Symposium Seriesvol. 667, (1997) 123.
[67] S. Veesler, R. Roure, J. Boistelle, General concepts of hydrargillite Al(OH)$_3$, agglomeration, J. Cryst. Growth 135 (3–4) (1994) 505–512.
[68] S.C. Grocott, S.P. Rosenberg, Proceedings of the 1st International Alumina Quality Workshop, Gladstone, Australia, (1988) 271.
[69] N. Brown, T.J. Cole, The behaviour of sodium oxalate in a Bayer alumina plant, Light Met. (1980) 105–107.
[70] R. Calalo, T. Tran, Effects of sodium oxalate on the precipitation of alumina trihydrate from synthetic sodium aluminate liquors, Light Met. 4 (1993) 125–133.
[71] P.J. The, J.F. Bush, in: Solubility of sodium oxalate in Bayer liquor and a method of control, light metals, Proceedings of Sessions, TMS Annual Meeting, Warrendale, PA, 1987, p. 5.
[72] G.J. Farquharson, S. Gotsis, J.D. Kildea, A.E. Gross, S.C. Grocott, Development of an effective liquor oxalate stabilizer, Light Met. (1995) 95–101.

[73] B. Gnyra, G. Lever, Review of Bayer organics-oxalate control processes, Light Met. (1979) 151–161.
[74] Basu, G.A., Nitowski, P.J., 1986. In: J.E. Dutrizac, A.J. Monhemius (Editions). Proceedings of the International Symposium on Iron Control Hydrometallurgy, 223.
[75] D.J. Cooling, D.J. Glenister, Practical aspects of dry disposal, Light Met. (1992) 25–31.
[76] J.K. Yang, B. Xiao, Development of unsintered construction materials from red mud wastes produced in the sintering alumina process, Constr. Build. Mater. 22 (2008) 2299–2307.
[77] A.I. Zouboulis, K.A. Kydros, Use of red mud for toxic metals removal: the case of nickel, J. Chem. Technol. Biotechnol. 58 (1) (1993) 95–101.
[78] J. Wong, G. Ho, Use of waste gypsum in the revegetation on red mud deposits: a greenhouse study, Waste Manag. Res. 11 (3) (1993) 249–256.
[79] D. Cooling, P.S. Hay, L. Guilfoyle, in: Carbonation of bauxite residue, Sixth International Alumina Quality Workshop, Brisbane, QLD, 2002, pp. 185–190.
[80] R.K. Paramguru, P.C. Rath, V.N. Misra, Trends in red mud utilization—a review, Miner. Process. Extr. Metall. Rev. 26 (2005) 1–29.
[81] D.I. Smirnov, T.V. Molchanova, The investigation of sulphuric acid sorption recovery of scandiumand uraniumfrom the red mud of alumina production, Hydrometallurgy 45 (3) (1997) 249–259.
[82] P. Vachon, R.D. Tyag, J.C. Auclair, K.J. Wilkinson, Chemical and biological leaching of aluminum from red mud, Environ. Sci. Technol. 28 (1994) 26–30.
[83] L. Piga, F. Pochetti, L. Stoppa, Application of thermal analysis techniques to a sample of Red Mud—a by-product of the Bayer process—for magnetic separation, Thermochim. Acta 254 (1995) 337–345.
[84] M. Patel, B.K. Padhi, P. Vidyasagar, A.K. Pattnaik, Extraction of titanium and production of building bricks from red mud, Res. Ind. 37 (1992) 154–158.
[85] T.K. Mukherjee, C.K. Gupta, Extraction of vanadium from and industrial waste, High Temp. Mater. Processes 11 (1–4) (1993) 189–206.
[86] J.K. Yang, D.D. Zhang, J. Hou, et al., Preparation of glass-ceramics from red mud in the aluminium industries, Ceram. Int. 34 (2008) 125–130.
[87] Y. Pontikes, P. Nikolopoulos, G.N. Angelopoulos, Thermal behavior of clay mixtures with bauxite residue for the production of heavy-clay ceramics, J. Eur. Ceram. Soc. 27 (2007) 1645–1649.
[88] T. Kavas, Use of boron waste as a fluxing agent in production of red mud brick, Build. Environ. 41 (12) (2006) 1779–1783.
[89] M. Singh, S.N. Upadhayay, P.M. Prasad, Preparation of special cements from red mud, Waste Manag. 16 (8) (1996) 665–670.
[90] N. Yalçin, V. Sevinç, Utilization of bauxite waste in ceramic glazes, Ceram. Int. 26 (5) (2000) 485–493.
[91] W.K. Biswas, D.J. Cooling, Sustainability assessment of red SandTM as a substitute for virgin sand and crushed limestone, J. Ind. Ecol. 17 (5) (2013) 756–762.
[92] C. Brunori, C. Cremisini, P. Massanisso, V. Pinto, L. Torricelli, Reuse of a treated red mud bauxite waste: studies on environmental compatibility, J. Hazard. Mater. 117 (1) (2005) 55–63.
[93] G. Garau, P. Castaldi, L. Santona, P. Deiana, P. Melis, Influence of red mud, zeolite and lime on heavy metal immobilization, culturable heterotrophic microbial populations and enzyme activities in a contaminated soil, Geoderma 142 (1–2) (2007) 47–57.
[94] F.W. Libbey, Alumina for northwest aluminium plants, Min. Congr. J. 31 (3) (1945) 34–39.
[95] M.Y. Bakr, Extraction of aluminium from Egyptian kaolins and clays. Acid process for recovery of alumina, Sprechsaal 96 (24) (1963) 577.

[96] F.A. Peters, P.W. Johnson, R.C. Kirby, Methods of producing alumina from clay—an evaluation of three sulphuric acid processes, US Bureau of Mines Report Invest. No. 6229, 1963, p. 57.
[97] L.E. Saunders, Process of purifying aluminous materials, US Patent 960712, 1910.
[98] J.B. Glaze, R.R. Ridgway, Aluminous materials and process of preparing the same, US Patent 1719131, 1929.
[99] N.S. Malts, N.S. Shmorgunenko, V.M. Sizyakov, Production of Alumina by Combination Bayer-Sintering Method, Bauxite, Society of Mining Engineers of AIME, New York, NY, Los Angeles, CA, 1984, pp. 775–787.
[100] V. Smirnov, Alumina production in Russia. Part I. Historical background, JOM 48 (8) (1996) 24–26.
[101] J.E. Kogel, N.C. Trivedi, J.M. Barker, S.T. Krukowski, Industrial Minerals & Rocks, seventh ed., Society for Mining, Metallurgy, and Exploration, Littleton, 2006.
[102] D.J. O'Connor, Alumina Extraction From Non-Bauxitic Materials, Aluminium-Verlag, Dusseldorf, 1988.
[103] G. Shelach, On the Invention of Pottery, Science 336 (6089) (2012) 1644–1645.
[104] PHG, https://commons.wikimedia.org/wiki/File:JomonPottery.JPG, 2010.
[105] E.L. Seymour, Improvement in the manufacture of bricks, US Patent 71229, 1867.
[106] M. Buchner, Manufacture of ceramic products, US Patent 700673, 1902.
[107] B. Schwerin, Process of manufacturing porous bodies, US Patent 1027004, 1912.
[108] B. Schwerin, Article for electrolytic purposes, US Patent 1050303, 1913.
[109] Thomson-Houston Company, Ltd., Improvement in and relating to wear resistant bodies, British Patent 4887, 1913.
[110] R. Reichman, Spark Plug, US Patent 1799225, 1931.
[111] R. Reichman, Method of examining the state of crystallisation of calcined aluminium oxide, US Patent 2058178, 1936.
[112] R. Reichman, Shaped bodies of non plastic metallic oxides such as spark plug insulators, US Patent 2031129, 1936.
[113] H. Kohl, R. Reichman, Method of molding nonplastic metallic oxides, US Patent 1934091, 1933.
[114] Sonett72, https://commons.wikimedia.org/wiki/File:Sparkplug.jpg, 2006.
[115] US Machineries, https://commons.wikimedia.org/wiki/File:Spray_Dryer.gif, 2009.
[116] S. Abbas, S. Maleksaeedi, E.C. Kolos, A.J. Ruys, Processing and properties of zirconia toughened alumina prepared by gelcasting, Materials 8 (7) (2015) 4344–4362.
[117] A.J. Ruys, C.C. Sorrell, Production of fibre-reinforced bioceramics by thixotropic casting, in: M.J. Bannister (Ed.), Ceramics: Adding the Value, vol. 1, CSIRO Publications, Melbourne, 1992, pp. 581–585.
[118] S.A. Simpson, A.J. Ruys, C.C. Sorrell, in: S. Bandyopadhyay, A.G. Crosky (Eds.), Thixotropic casting of zirconia ceramic matrix composites, Proceedings of the 3rd Australian Forum on Metal Matrix Composites, IMMA, Sydney, 1992, pp. 113–122.
[119] A.J. Ruys, S.A. Simpson, C. Sorrell, Thixotropic casting of fibre reinforced ceramic matrix composites, J. Mater. Sci. Lett. 13 (16) (1994) 1323–1325.
[120] A.J. Ruys, C.C. Sorrell, Thixotropic casting of ceramic matrix composites, Int. Ceram. Monogr. 1 (1) (1994) 692–700.
[121] N. Ehsani, A.J. Ruys, C.C. Sorrell, Thixotropic casting of fecralloy fibre reinforced hydroxyapatite, in: G.M. Newaz, A.H. Neber, F.H. Wohlbier (Eds.), Metal Matrix Composites. Application and Processing, Trans Tech Publications, Zurich, 1995, pp. 373–380.
[122] J.A. Kerdic, A.J. Ruys, C.C. Sorrell, Thixotropic casting of ceramic metal functionally gradient materials, J. Mater. Sci. 31 (16) (1996) 4347–4355.

[123] N. Ehsani, A.J. Ruys, C.C. Sorrell, in: P.A. Walls, C.C. Sorrell, A.J. Ruys (Eds.), Thixotropic casting of monolithic nonclay ceramics, Symposium 4.1. Second International Meeting of Pacific Rim Ceramic Societies. PacRim 2, Australasian Ceramic Society, Sydney, 1996, pp. 2–6.
[124] A.J. Ruys, C.C. Sorrell, Thixotropic casting. Ceramic monograph 1.4.2.5, in: Handbook of Ceramics, Verlag Schmid GMBH, Freiburg, 1999 (16pp).
[125] E. Ryshkewitch, Cutting tools of sintered alumina, Ber. Dtsch. Keram. Ges. 34 (1) (1957) 3–5.
[126] W. Dawihl, E. Dorre, Strength and deformation properties of sintered aumina bodies as a function of their composition and structure, Ber. Dtsch. Keram. Ges. 41 (2) (1964) 85–96.
[127] C.C. Sorrell, N. Ehsani, A.J. Ruys, O.C. Standard, Absence of microwave effect in ceramics: precise temperature, thermal gradient, and densification determination in a proportional-power microwave furnace, in: A.P. Tomsia, A.M. Glaeser (Eds.), Ceramic Microstructure: Control at the Atomic Level, Plenum Press, New York, 1998, pp. 471–486.
[128] C.C. Sorrell, O.C. Standard, N. Ehsani, C.R.H. Harding, A.J. Ruys, Advanced microwave sintering of ceramics: densification and thermal effects of oxide ceramics in a proportional power microwave furnace, in: R.S.C. Smart, J. Nowotny (Eds.), Ceramic Interfaces. Properties and Applications, IOM Communications Ltd., London, 1998, pp. 333–366.
[129] N. Ehsani, A.J. Ruys, O.C. Standard, C.C. Sorrell, Advanced microwave sintering of ceramics. Temperature measurement, Mater. Eng. 8 (2) (1997) 89–95.
[130] O.C. Standard, A.J. Ruys, C.C. Sorrell, Advanced microwave sintering of ceramics. Temperature calibration, Mater. Eng. 8 (2) (1997) 97–106.
[131] N. Ehsani, A.J. Ruys, C.C. Sorrell, Microwave sintering of Al2O3 fiber-reinforced hydroxyapatite matrix composites, J. Biomimetics Biomater. Tissue Eng. 13 (2012) 91–104.
[132] N. Ehsani, A.J. Ruys, C.C. Sorrell, Microwave sintering of ZrO2 fiber-reinforced hydroxyapatite matrix composites, J. Biomimetics Biomater. Tissue Eng. 14 (2012) 93–101.
[133] A.J. Ruys, E.B. Popov, D. Sun, J.J. Russell, C.C.J. Murray, Functionally graded electrical/thermal ceramic systems, J. Eur. Ceram. Soc. 21 (10) (2001) 2025–2029.
[134] CoorsTek, Ceramic Product Specification Document, 2008.
[135] K.E. Spear, J.P. Dismukes, Synthetic Diamond—Emerging CVD Science and Technology, Wiley, Chichester, 1994.
[136] ASTM E384, Standard Test Method for Microindentation Hardness of Materials, ASTM International, West Conshohocken, 2017.
[137] E. Ryschkewitch, Ceramic cutting tool, US Patent 2270607, 1942.
[138] R.C. Bradt, Cr2O3 solid solution hardening of Al2O3, J. Am. Ceram. Soc. 50 (1967) 54–55.
[139] G.R. Anstis, et al., A critical evaluation of indentation techniques for measuring fracture toughness. Direct crack measurements, J. Am. Ceram. Soc. 64 (9) (1981) 533–538.
[140] M. Rock, Kunstliche ersatzteile fur sas innere and aussere des menschlichen und tierischen korpers (Artificial spare parts for the interior and exterior of the human and animal body), German Patent 583589, Filed 22 April 1932, 1933.
[141] S. Sandhaus, Bone implants and drills and taps for bone surgery, British Patent 1,083,769, 1965.
[142] E.J. Eyring, W. Campbell, Ceramic prostheses, a medical engineering team reports progress in impant research, News Engl. (1969) 4–7.
[143] P. Boutin, L'Alumine et son utilisation en chirurgie de la hanche (Experimental study of alumina and its use in surgery of the hip), Presse Med. 79 (14) (1971) 639–640.

[144] P. Boutin, Arhthroplastie totale de la hanche par prostheses en alumina fritte (Total arthroplasty of the hip by fritted alumina prosthesis), Rev. Chir. Orthop. 58 (3) (1972) 229–246.
[145] O. Aquilina, A brief history of cardiac pacing, Images Paediatr. Cardiol. 8 (2) (2006) 17–81.
[146] V. Bondarew, P. Seligman, The Cochlear Story, CSIRO, Clayton, 2012.
[147] G.G. Wickham, D.J. Cowdery, in: An hermetically sealed implantable cardiac pacemaker, Proceedings of the 9th International Conference on Medical and Biological Engineering, Melbourne, 1971.
[148] P. Boutin, Total hip arthroplasty using a ceramic prosthesis. Pierre Boutin (1924–1989), Clin. Orthop. Relat. Res. (379) (2000) 3–11.
[149] S.F. Hulbert, F.A. Young, R.S. Mathews, J.J. Klawitter, C.D. Talbert, F.H. Stelling, Potential of ceramic materials as permanently implantable skeletal prostheses, J. Biomed. Mater. Res. 4 (1970) 433–456.
[150] S.F. Hulbert, S.J. Morrison, J.J. Klawitter, Tissue reaction to three ceramics of porous and non-porous structures, J. Biomed. Mater. Res. 6 (5) (1972) 347–374.
[151] Hanabishi, https://en.wikipedia.org/wiki/Dental_braces#/media/File:Transparent-Bracket. JPG, 2007.
[152] Suyash, https://en.wikipedia.org/wiki/Dental_braces#/media/File:Teeth_Braces_(1).jpg, 2017.
[153] Y.T. O, J.B. Koo, K.J. Hong, J.S. Park, D.C. Shin, Effect of grain size on transmittance and mechanical strength of sintered alumina, Mater. Sci. Eng. A 374 (2004) 191–195.
[154] R. Apetz, M.P.B. Bruggen, Transparent alumina: a light-scattering model, J. Am. Ceram. Soc. 86 (3) (2003) 480–486.
[155] J.A. Taylor, History of Dentistry: A Practical Treatise for the Use of Dental Students and Practitioners, Lea & Febiger, Philadelphia, PA, 1922, pp. 142–156.
[156] Coronation, https://commons.wikimedia.org/wiki/File:Single_crown_implant.jpg, 2013.
[157] Coronation, https://en.wikipedia.org/wiki/Dental_implant#/media/File:Fractured_implant. jpg, 2013.
[158] W.C. Wagner, T.M. Chu, Biaxial flexural strength and indentation fracture toughness of three new dental core ceramics, J. Prosthet. Dent. 76 (2) (1996) 140–144.
[159] Wagonerj, https://commons.wikimedia.org/wiki/File:Bridge_from_dental_porcelain.jpg, 2005.
[160] Coronation, https://en.wikipedia.org/wiki/Dental_implant#/media/File:Implant_retained_ bridge_model.jpg, 2013.
[161] P.I. Branemark, Osseointegration and its experimental background, J. Prosthet. Dent. 50 (3) (1983) 399–410.
[162] R. Branemark, P.I. Branemark, B. Rydevik, R.R. Myers, Osseointegration in skeletal reconstruction and rehabilitation. A review, J. Rehabil. Res. Dev. 38 (2) (2001) 175–181.
[163] S. Sandhaus, K. Pasche, L'implantologie en un temps chirurgical, Actual Odontostomatol. 199 (1997) 539–558.
[164] J.J. Klawitter, A.M. Weinstein, F.W. Cooke, L.J. Peterson, B.M. Pennel, R.V. Mckinney, An evaluation of porous alumina ceramic dental implants, J. Dent. Res. 56 (7) (1977) 768–776.
[165] G. Heimke, W. Schulte, Dental implant having a biocompatible surface, US Patent 4,185,383, Filed Apr 29 1977, 1980.
[166] J.T. Chess, C.A. Babbush, Restoration of lost dentition using aluminium oxide endosteal implants, Dent. Clin. N. Am. 24 (1980) 521–533.
[167] J.E. Haubenreich, F.G. Robinson, K.P. West, R.Q. Frazer, Did we push dental ceramics too far? A brioef history of ceramic dental implants, J. Long-Term Eff. Med. Implants 15 (2005) 617–628.

[168] H. Kawahara, Materials for hard tissue replacement, in: O.C.C. Lin, E.Y.S. Chao (Eds.), Perspectives on Biomaterials, Elsevier, Amsterdam, 1986, pp. 167–206.
[169] T. Takahashi, T. Sato, R. Hisanaga, O. Miho, Y. Suzuki, M. Tsunoda, K. Nakagawa, Long term observation of porous sapphire dental implants, Bull. Tokyo Dent. Coll. 49 (2008) 23–27.
[170] S. Ban, H. Sato, Y. Suehiro, H. Nakanishi, M. Nawa, Biaxial flexure strength and low temperature degradation of Ce-TZP/Al_2O_3 nanocomposite and Y-TZP as dental restoratives, J. Biomed. Mater. Res. B Appl. Biomater. 87B (2) (2008) 492–498.
[171] R.J. Kohal, M. Wolkewitz, C. Mueller, Alumina-reinforced zirconia implants: survival rate and fracture strength in a masticatory simulation trial, Clin. Oral Implants Res. 21 (12) (2010) 1309–1413.
[172] P. Palmero, M. Fornabaio, L. Montanaro, H. Reveron, C. Esnouf, J. Chevalier, Towards long lasting zirconia-based composites for dental implants. Part I. Innovative synthesis, microstructural characterization and in vitro stability, Biomaterials 50 (2015) 38–46.
[173] B. Andersson, A. Taylor, B.R. Lang, H. Scheller, P. Schärer, J.A. Sorensen, D. Tarnow, Alumina ceramic implant abutments used for single-tooth replacement: a prospective 1- to 3-year multicenter study, Int. J. Prosthodont. 14 (5) (2001) 432–438.
[174] K.H. Soh, A.J. Ruys, P. Parakala, in: Foamed alumina bioceramics for use as tissue scaffolds, 2A3-08 (4 pages), Proceedings ICBMEC 2005. The 12th International Conference on Biomedical Engineering, vol. 12, IFMBE, Singapore, 2005.
[175] E. Soh, A.J. Ruys, Characterisation of foamed porous alumina tissue scaffolds, J. Biomimetics Biomater. Tissue Eng. 4 (2009) 21–26.
[176] K.H. Soh, E. Kolos, A.J. Ruys, Foamed high porosity alumina for use as a bone tissue scaffold, Ceram. Int. 41 (1) (2015) 1031–1047.
[177] K.H. Soh, E.C. Kolos, A.J. Ruys, Cellular response to doping of high porosity foamed alumina with Ca, P, Mg, and Si, Materials 8 (3) (2015) 1074–1088.
[178] M.B. Pabbruwe, O.C. Standard, C.C. Sorrell, C.R. Howlett, Bone formation within alumina tubes: effect of calcium, manganese, and chromium dopants, Biomaterials 25 (2004) 4901–4910.
[179] A. Noiri, F. Hoshi, H. Murakami, K. Sato, S. Kawai, K. Kawai, Biocompatibility of a mobile alumina ceramic orbital implant, Folia Ophthal. Jap. 53 (2002) 476–480.
[180] J.C. Bové, Utilization of a porous alumina ceramic spacer in tibial valgus open-wedge osteotomy: fifty cases at 16 months mean follow-up, Rev. Chir. Orthop. Reparatrice Appar. Mot. 88 (5) (2002) 480–485.
[181] Anon, History of Kevlar, Chm.bris.ac.uk, 2017.
[182] S.M. Kurz, Total hip arthroplasty demand rising on a global level, Orthop. Today (2010).
[183] AJRR American Joint Replacement Registry, Third AJRR annual report on hip and knee arthroplasty data, Ajrr.net, 2016.
[184] R. Dotingan, Number of hip replacement has skyrocketed. US report show, in: HealthDay News for Healthier Living, HealthDay, 2015.
[185] T.P. Schmalzried, J.J. Callaghan, Wear in total hip and knee replacements, J. Bone Joint Surg. Am. 81 (1999) 115–136.
[186] J. Charnley, The bonding of prostheses to bone by cement, J. Bone Joint Surg. 46 (1964) 518–529.
[187] H. Gray, Anatomy of the Human Body, Lee and Febiger, New York, 1918.
[188] C. Fredrik, https://en.wikipedia.org/wiki/Osteoarthritis#/media/File:0910_Oateoarthritis_Hip_A.png, 2015.
[189] C. Fredrik, https://en.wikipedia.org/wiki/Osteoarthritis#/media/File:0910_Oateoarthritis_Hip_B.png, 2015.

[190] NIH, https://en.wikipedia.org/wiki/Hip_replacement#/media/File:Hip_replacement_Image_3684-PH.jpg, 2006.
[191] KimvdLinde, https://commons.wikimedia.org/wiki/File:Hip-replacement.jpg, 2010.
[192] R.A. Brand, M.A. Mont, M. Manning, Biographical sketch: themistocles Gluck (1853-1942), Clin. Orthop. Relat. Res. 469 (6) (2011) 1525–1527.
[193] N.J. Eynon-Lewis, D. Ferry, M.F. Pearse, Themistocles Gluck: an unrecognised genius, BMJ 305 (1992) 1534–1536.
[194] D. Muster, Themistocles Gluck, Berlin 1890: a pioneer of multidisciplinary applied research into biomaterials for endoprostheses, Bull. Hist. Dent. 38 (1990) 306.
[195] E.J. Haboush, A new operation for arthroplasty of the hip based on biomechanics, photoelasticity, fast-setting dental acrylic, and other considerations, Bull. Hosp. Joint Dis. 14 (1953) 242–277.
[196] L.L. Wiltse, R.H. Hall, J.C. Stenehjem, Experimental studies regarding the possible use of self curing acrylic in orthopaedic surgery, J. Bone Joint Surg. 39 (1957) 961–972.
[197] J. Charnley, The classic: the bonding of prostheses to bone by cement, Clin. Orthop. Relat. Res. 468 (12) (2010) 3149–3159.
[198] G. Kuntscher, Die marknagelung von knochenbruchen, Arch. klin. Chir. 200 (1940) 443–455.
[199] V. Surin, The prehistory of total joint. Themistocles Gluck and Jules Emile Pean, HB Valdemar, (2010).
[200] I.D. Learmonth, C. Young, C. Rorabeck, The operation of the century: total hip replacement, Lancet 370 (2007) 1508–1519.
[201] S.R. Knight, R. Auila, S.P. Biswas, Total hip arthroplasty—over 100 years of operative history, Orthop. Rev. 3 (2) (2011) 16.
[202] J.D. Bronzino, Biomedical Engineering Fundamentals, CRC Press, Boca Raton, 2006.
[203] J. Park, R.S. Lakes, Biomaterials: An Introduction, Springer, New York, 1992.
[204] P. Hernigou, S. Quiennec, I. Guissou, Hip hemiarthroplasty: from Venable and Bohlman to Moore and Thompson, Int. Orthop. 38 (3) (2013) 655–681.
[205] A.T. Moore, Metal hip joint. A case report, South. Med. J. 45 (11) (1952) 1015–1019.
[206] J. Howse, Semi-captive cups for hip replacement, in: R. Coombs, A. Gristina, D. Hungerford (Eds.), Joint Replacement, Orthotext, 1990, p. 135.
[207] G. McKee, Total hip replacement—past, present, and future, Biomaterials 3 (1982) 130–135.
[208] S.R. Brown, W.A. Davies, D.H. DeHeer, A.B. Swanson, Long-term survival of Mckee-Farrar total hip prosthesis, Clin. Orthop. Relat. Res. 402 (2002) 157–163.
[209] H. McKellop, S.H. Park, R. Chiesa, In vivo wear of three types of metal on metal hip prostheses during two decades of use, Clin. Orthop. Relat. Res. 329 (1996) 128–140.
[210] W.L. Walter, Orthopaedic Surgeon, Transcript of Interview with Prof Andrew Ruys, 2017.
[211] Heilman, https://en.wikipedia.org/wiki/Hip_replacement#/media/File:MetalonmetalhippreplaceMark.png, 2016.
[212] Walker, https://en.wikipedia.org/wiki/Hip_resurfacing#/media/File:Hipoptions.jpg, 2007.
[213] M.A.R. Freeman, H.U. Cameron, G.C. Brown, Cemented double-cup arthroplasty of the hip. A 5 year experience with ICLS prosthesis, Clin. Orthop. 134 (1978) 45–48.
[214] K. Furuya, H. Tsuchiya, S. Kawachi, Socket cup arthroplasty, Clin. Orthop. (134) (1978) 41–44.
[215] D. McMinn, R. Treacy, K. Lin, P. Pynsent, Metal on metal surface replacement of the hip, Clin. Orthop. Relat. Res. 329 (1996) 89–98.
[216] DePuy Synthes, Depuysynthes.com, 2017.
[217] AJR. Australian Orthopaedic Association, National Joint Replacement Registry, 2017, Accessed 2017, https://aoanjrr.sahmri.com/.

[218] J. Charnley, Arthroplasty of the hip. A new operation, Lancet 277 (7187) (1961) 1129–1132.
[219] J. Older, Low-friction arthroplasty of the hip: a 1–12 year follow-up study, Clin. Orthop. 211 (1986) 36–42.
[220] S.R. Carter, P.B. Pynsent, D.J.W. McMinn, Greater than ten year survivorship of the Chrnley low friction arthroplasty, J. Bone Joint Surg. 73 (1991) 71.
[221] B.M. Wroblewski, G.W. Taylow, P. SIney, Charnley low-friction arthroplasty: 19 to 25 year results, Orthopaedics 15 (1992) 421–424.
[222] K.R. Schulte, J.J. Callaghan, S.S. Kelly, R.C. Johnston, The outcome of Charnley total hip arthroplasty with cement after a minimum twenty year follow-up: the results of one surgeon, J. Bone Joint Surg. 75 (1993) 961–975.
[223] J. Charnley, Anchorage of the femoral head prosthesis o the shaft of the femur, J. Bone Joint Surg. 42 (1960) 961–972.
[224] P. Boutin, Les prostheses totals de la hanche en alumina. L'ancrage direct sans ciment dans 50 cas, Rev. Chir. Orthop. 60 (1974) 233–235.
[225] P. Boutin, Total hip prosthetic apparatus made of non-porous alumina, US Patent 3871031, 1975.
[226] P. Boutin, D. Blanquaert, Le frottement Al/Al en chirurgie de la hanche-1205 arthroplasties totales, Rev. Chir. Orthop. 76 (1981) 279–287.
[227] L. Sedel, L. Kerboull, P. Christel, A. Meunier, J. Witvoet, Alumina—on alumina hip replacement, J. Bone Joint Surg. 72 (1990) 659–663.
[228] T.H. Mittelmeier, A. Walter, The influence of prosthesis design on wear and loosening phenomena, CRC Crit. Rev. Biocompatibility 3 (1987) 19.
[229] M.H. Huo, R.P. Martin, L.E. Zatorski, K.J. Keggi, Total hip replacements using the ceramic Mittelmeier prosthesis, Clin. Orthop. 332 (1996) 143–150.
[230] P. Griss, V.A.H. Werburg, B. Krempien, G. Heimke, Experimental analysis of ceramic-tissue interactions. A morphologic, fluoroescenseoptic, and radiographic study on dense alumina oxide ceramics in various animals, J. Biomed. Mater. Res. 8 (3) (1974) 39–48.
[231] M. Salzer, K. Knahr, H. Locke, N. Stark, Cement-free bioceramics double-cup endoprosthesis of the hip jpint, Clin. Orthop. 134 (1978) 80–86.
[232] M. Salzer, K. Knahr, H. Locke, A bioceramics endoprosthesis for the replacement of the proximal humerus, Arch. Orthop. Trauma Surg. 93 (1979) 169–184.
[233] G. Randelli, F. Lonati, M. D'Imporzano, C. Cuccinello, Wagners bioceramic superficial prosthesis in the hip surgery, in: P. Vincenzini (Ed.), Ceramics in Surgery, Elsevier, Amsterdam, 1983, pp. 341–350.
[234] A.S. Dickinson, M. Browne, K.C. Wilson, J.R.T. Jeffers, A.C. Taylor, Pre-clinical evaluation of ceramic femoral head resurfacing prostheses using computational models and mechanical testing, Proc. Inst. Mech. Eng. H 225 (9) (2011) 866–876.
[235] D. Stock, M. Geissler, The Frialit total hip replacements system. Design considerations and clinical results, in: G. Heimke (Ed.), Osseo-integrated implants, CRC Press, Boca Raton, 1990, pp. 127–152.
[236] M. Salzer, H. Locke, H. Plenk, G. Punzet, N. Stark, K. Zweimuller, Experience with bioceramic endoprosthesis of the hip joint, in: M. Schaidach, D. Hohmann (Eds.), Artificial hip and knee joint technology, Springer, Berlin, 1975, pp. 459–474.
[237] H. Kawahara, M. Hirabayiashi, T. Shikita, Single crystal dental implants for dental and bone screws, J. Biomed. Mater. Res. A 14 (1980) 597–605.
[238] H. Oonishi, Knee and ankle joint, in: G. Heimke (Ed.), Osseo-integrated implants, CRC Press, Boca Raton, 1990, pp. 171–186.

[239] T. Shikati, H. Oonishi, Y. Hashimoto, in: Wear resistance of irradiated UHMWPE polyethylenes to Al_2O_3 ceramics in total hip prostheses, Transactions of the 3rd Annual Meeting of the Society for Biomaterials, 1977, pp. 118–124.
[240] E. Dorre, W. Dawihl, G. Altmeyer, Dauerfestigkeit keramischer huftendo prosthesen, Biomed. Tech. 22 (1977) 3–7.
[241] E. Dorre, W. Dawihl, U. Krohn, G. Altmeyer, M. Semlitsch, Do ceramic components of hip joints maintain their strength in human bodies? in: P. Vincenzini (Ed.), Ceramics in Surgery, Elsevier, Amsterdam, 1983, pp. 61–72.
[242] H.G. Willert, H. Bertram, G.H. Buchorn, Osteolysis in alloarthroplasty of the hip. The role of ultra high molecular weight polyethylene wear particles, Clin. Orthop. 258 (1990) 95–107.
[243] L. Sedel, P. Bizot, R. Nizard, A. Meunier, A perspective on 25 years experience with ceramic-on-ceramic articulation in total hip replacement, Semin. Arthroplast. 9 (1998) 123–134.
[244] J.P. Garino, Modern ceramic-on-ceramic total hip systems in the United States, Clin. Orthop. Relat. Res. 379 (2000) 41–47.
[245] P. Bizot, L. Banallec, L. Sedel, R. Nizard, Alumina-on alumina total hip prostheses in patients 40 years of age or younger, Clin. Orthop. Relat. Res. 379 (2000) 68–76.
[246] R.D. Watson, Zirconium bearings and process of producing same, US Patent 2987352, (Filed 1958), 1961.
[247] J.A. Davidson, Zirconium oxide coated prosthesis for wear and corrosion resistance, US Patent 5037438, (Filed 1989), 1991.
[248] J.C. Haygarth, Processes for producing improved wear resistant coatings on zirconium shapes, US Patent 4671824, (Filed 1984), 1987.
[249] O. Popoola, J.P. Anderson, M.E. Hawkins, T.S. Johnson, H.R. Shetty, Method for producing a zirconia-layered orthopedic implant component, Zimmer Technology, US Patent 20060233944, (Filed 2005), 2006.
[250] A.M. Kop, C. Whitewood, D. Johnston, Damage of oxinium femoral heads subsequent to hip arthroplasty dislocation. Three retrieval case studies, J. Arthroplast. 22 (2007) 775–779.
[251] V. Good, M. Ries, R.L. Barrack, K. Widding, G. Hunter, D. Heuer, Reduced wear with oxidized zirconium femoral heads, J. Bone Joint Surg. Am. 85 (4) (2003) 105–110.
[252] K.L. Garvin, C.W. Hartman, J. Mangla, N. Murdoch, J.M. Martell, Wear analysis in THA utilizing oxidized zirconium and crosslinked polyethylene, Clin. Orthop. Relat. Res. 467 (2009) 141–145.
[253] A. Maichand, Oxinium, for Longer Lasting Knees, Available from: http://archivehealthcare.financialexpress.com/201111/knowledge02.shtml, 2011.
[254] P.S. Walker, G.W. Blunn, P.A. Lilley, Wear testing of materials and surfaces for total knee replacement, J. Biomed. Mater. Res. 33 (1996) 159–175.
[255] B.M. Spector, M.D. Ries, R.B. Bourne, W.S. Sauer, M. Long, G. HunterWear, Performance of ultra-high molecular weight polyethylene on oxidized zirconium total knee femoral components, J. Bone Joint Surg. Am. 83 (2001) 80–86.
[256] G. Hunter, W.M. Jones, M. Spector, Oxidized zirconium, in: J. Bellemans, M.D. Ries, J. Victor (Eds.), Total Knee Arthroplasty, Springer Verlag, Heidelberg, 2005, pp. 370–377.
[257] S. Tsai, J. Sprague, G. Hunter, R. Thomas, A. Salehi, Mechanical testing and finite element analysis of oxidized zirconium femoral components, Trans. Soc. Biomater. 24 (2001) 163.

[258] K. Fukui, A. Kaneuji, T. Sugimori, Wear comparison between a highly cross-linked polyethylene and conventional polyethylene against zirconia femoral head: minimum 5-year follow-up, J. Arthroplasty 26 (2011) 45–49.
[259] I. Nakahara, N. Nakamura, T. Nishii, H. Miki, T. Sakai, N. Sugano, Minimum five-year follow-up wear measurement of longevity highly cross-linked polyethylene cup against cobalt-chromium or zirconia heads, J. Arthroplasty 25 (2010) 1182–1187.
[260] Y.H. Kim, J.S. Kim, S.H. Cho, A comparison of polyethylene wear in hips with cobalt-chrome or zirconia heads. A prospective, randomised study, J. Bone Joint Surg. (Br.) 83 (2001) 742–750.
[261] C. Hui, L. Salmon, S. Maeno, J. Roe, W. Walsh, L. Pinczewski, Five-year comparison of oxidized zirconium and cobalt-chromium femoral components in total knee arthroplasty: a randomized controlled trial, J. Bone Joint Surg. 93 (2011) 624–630.
[262] M. Stilling, K.A. Nielsen, K. Søballe, O. Rahbek, Clinical comparison of polyethylene wear with zirconia or cobalt-chromium femoral heads, Clin. Orthop. Relat. Res. 467 (2009) 2644–2650.
[263] G.T. Evangelista, E. Fulkerson, F. Kummer, P.E. Di Cesare, Surface damage to an oxinium femoral head prosthesis after dislocation, J. Bone Joint Surg. (Br.) 89 (2006) 535–536.
[264] W.L. Jaffe, E.J. Strauss, M. Cardinale, L. Herrera, F.J. Kummer, Surface oxidized zirconium total hip arthroplasty head damage due to closed reduction effects on polyethylene wear, J. Arthroplasty 24 (2009) 898–902.
[265] FDA Reports, FDA Medical Device Report for the Oxinium Knee Implant, Available from:http://www.fda-reports.com/device/oxinium-knee-implant.html, 2013.
[266] C.F. Grain, R.C. Garvie, Mechanism of the Monoclinic to Tetragonal Transformation of Zirconium Dioxide, US Bureau of Mines, Washington, 1965.
[267] R.C. Garvie, The occurrence of metastable tetragonal zirconia as a crystalline size effect, J. Phys. Chem. 69 (4) (1965) 1238–1243.
[268] O. Ruff, F. Ebert, E. Stephan, Beitrage zur keramik hochfeurfester stoffe II. Das system ZrO_2-CaO, Z. Anorg. Allg. Chem. 180 (1) (1929) 215–224.
[269] B. Cotterel, The diversity of materials and their fracture behaviour, in: Fracture and life, Imperial College Press, London, 2010, pp. 296–297 (Chapter 10).
[270] R.C. Garvie, R.H. Hannink, R.T. Pascoe, Ceramic steel, Nature 258 (1975) 703–704.
[271] M. Kuntz, B. Masson, T. Pandorf, Current state of the art of ceramic composite materials Biolox Delta, in: G. Mendes, B. Lago (Eds.), Strength of Materials, Nova Science, Boston, 2009, pp. 133–156.
[272] W. Burger, H.G. Richter, High strength and toughness alumina matrix composites by transformation toughening and "in situ" platelet reinforcement (ZPTA), Key Eng. Mater. 192–195 (2001) 545–548.
[273] G. Willmann, Ceramic femoral head retrieval data, Clin. Orthop. Relat. Res. 379 (2000) 22–28.
[274] K. Kobayashi, H. Kuwajima, T. Masaki, Phase change and mechanical properties of Y_2O_3-ZrO_2 solid electrolyte after aging, Solid State Ionics 3–4 (1981) 489–493.
[275] R.L.K. Matsumoto, Strength recovery in degraded yttria-doped tetragonal zirconia polycrusyals, J. Am. Ceram. Soc. 68 (8) (1985) c213.
[276] M. Yoshimura, Phase stability of zirconia, Am. Ceram. Soc. Bull. 67 (12) (1988) 1950–1955.
[277] J. Chevalier, What future for zirconia as a biomaterial? Biomaterials 27 (2006) 535–543.
[278] M. Pabbruwe, A.J. Ruys, C.C. Sorrell, in: A. Craig (Ed.), Zirconia spherical nanoparticles by the urea method: characterisation by FESEM and TEM, Papers Presented at the First

References

Meeting of the International Union of Microbeam Analysis Societies, ASEM Inc., Sydney, 1996, pp. 140–141.

[279] A.J. Ruys, J.J. Russell, T.M. Lee, in: Werkstoffwoche-Partnerschaft GbR (Ed.), Homogeneous zirconia-toughened-alumina nanocomposites, Materials Week 2001 Proceedings, Werkstoff Informationgesellschaft mBH, Frankfurt, 2001. 6 pp. (CDROM format).

[280] A.J. Ruys, T.M. Lee, J.J. Russell, Processing of homogeneous nanocoposites, Mater. Integr. Jpn. 15 (1) (2002) 51–57.

[281] T.M. Lee, A.J. Ruys, Synthesis and testing of homogeneous ceramic-matrix nanocomposites, in: P. Vincenzini (Ed.), Advances in Science and Technology, 10th International Ceramics Congress CIMTEC 2002, vol. 32, Techna, Faenza, 2003, pp. 603–608.

[282] H.P. Cahoon, C.J. Cristensen, Sintering and grain growth of alpha-alumina, J. Am. Ceram. Soc. 39 (1956) 337–344.

[283] N. Claussen, Fracture toughness of Al_2O_3 with an unstabilised ZrO_2 dispersed phase, J. Am. Ceram. Soc. 59 (1–2) (1976) 49–51.

[284] N. Claussen, J. Jahn, Alumina toughened with PSZ dispersions, Ber. Dtsch. Keram. Ges. 55 (1978) 487.

[285] J. Wang, R. Stevens, Review. Zirconia-toughened alumina (ZTA) ceramics, J. Mater. Sci. 24 (1989) 3421–3440.

[286] A. Mandrino, R. Eloy, B. Moyen, J.L. Lerat, D. Treheux, Base alumina ceramics with dispersoids: mechanical behaviour and tissue response after in vivo implantation, J. Mater. Sci. Mater. Med. 3 (1992) 457–463.

[287] J. Rieu, P. Goeuriot, Ceramic composites for biomedical applications, Clin. Mater. 12 (1993) 211–217.

[288] A. Salomoni, A. Tucci, L. Esposito, L. Stamenkovic, Forming and sintering of multiphase bioceramics, J. Mater. Sci. Mater. Med. 5 (1994) 651–653.

[289] S. Affatato, M. Testoni, G.L. Cacciari, A. Toni, A mixed-oxide prosthetic ceramic ball heads, Biomaterials 20 (1999) 1925–1929.

[290] S. Begand, T. Oberbach, W. Glien, ATZ—a new material with high potential in joint replacement, Key Eng. Mater. 284–286 (2005) 983–986.

[291] A.H. De Aza, J. Chevalier, G. Fantozzi, M. Schehl, R. Torrecillas, Crack growth resistance of alumina, zirconia, and zirconia toughened alumina ceramics for joint prostheses, Biomaterials 23 (2002) 937–945.

[292] J. Chevalier, S. Grandjean, M. Kuntz, G. Pezzotti, On the kinetics and impact of tetragonal to monoclinic transformation in an alumina/zirconia composite for arthroplasty applications, Biomaterials 30 (2009) 5279–5282.

[293] C. Pecharroman, J.F. Bartolome, J. Requena, J.S. Moya, S. Deville, J. Chevalier, G.R. Fantozzi, Percolative mechanism of aging in zirconia-containing ceramics for medical applications, Adv. Mater. 15 (2003) 507–511.

[294] S.M. Kurtz, S. Kocagoz, C. Arnholt, R. Huet, M. Ueno, W.L. Walter, Advances in zirconia toughened alumina biomaterials for total joint replacement, J. Mech. Behav. Biomed. Mater. 31 (3) (2014) 107–116.

[295] G. Pezzotti, T. Saito, G. Padeletti, P. Cassari, K. Yamamoto, Nanoscale topography of bearing surface in advanced alumina/zirconia hip joint before and after severe exposure in water vapor environment, J. Orthop. Res. 28 (2010) 762–766.

[296] S. Deville, J. Chevalier, G. Fantozzi, J.F. Bartoalome, J. Requena, J.S. Moya, Low temperature ageing of zirconia toughened alumina composites, J. Eur. Ceram. Soc. 23 (2003) 2975–2982.

[297] S. Deville, J. Chevalier, C. Dauvergne, J.F. Bartolome, J. Requena, R. Torrecillas, Microstructural investigation of the aging behaviour of Y-TZP toughened alumina composites, J. Am. Ceram. Soc. 88 (2005) 1272–1280.
[298] W.G. Hamilton, J.P. McAuley, D.A. Dennis, J.A. Murphy, T.J. Blumenfeld, J. Politi, THA with Delta ceramic on ceramic: results of a multicenter investigational device exemption trial, Clin. Orthop. Relat. Res. 468 (2) (2010) 358–366.
[299] G. Pezzotti, M.C. Munisso, A.A. Porporati, K. Lessnau, On the role of oxygen vacancies and lattice strain in the tetragonal to monoclinic transformation in alumina/zirconia composites and improved environmental stability, Biomaterials 31 (27) (2010) 6901–6908.
[300] T.D. Stewart, J.L. Tipper, G. Insley, R.M. Streicher, E. Ingham, J. Fisher, Long-term wear of ceramic matrix composite materials for hip prostheses under severe swing phase microseparation, J. Biomed. Mater. Res. B Appl. Biomater. 66 (2) (2003) 567–573.
[301] S.M. Kurtz, K. Ong, Contemporary total hip arthroplasty: hard-on-hard bearings and highly crosslinked UHMWPE, in: UHMWPE Biomaterials Handbook, second ed., Ultra-High Molecular Weight Polyethylene in Total Joint Replacement and Medical Devices, Academic Press, Cambridge, 2009, pp. 55–79.
[302] M. Kuntz, in: Live-time prediction of biolox delta, Proceedings of the 12th BIOLOX®-Symposium 2007, Seoul, 2007, pp. 281–288.
[303] J.M. Dorlot, P. Christel, A. Meunier, Wear analysis of retrieved alumina heads and sockets of hip prostheses, J. Biomed. Mater. Res. 23 (1989) 299–310.
[304] J.M. Dorlot, Long term effects of alumina components in total hip prostheses, Clin. Orthop. 282 (1992) 47–52.
[305] A. Walter, On the material and the tribology of Al/Al coupling for joint prostheses, Clin. Orthop. 282 (1992) 31–46.
[306] M. Kuntz, B. Masson, T. Pandorf, Current state of the art of ceramic composite materials Biolox Delta, in: G. Mendes, B. Lago (Eds.), Strength of Materials, Nova Science Hauppauge, Boston, 2009, pp. 133–156.
[307] M. Semlitsch, M. Lehmann, H. Weber, E. Dorre, H.G. Willert, Neue perspektiven zu verlangerter funktionsdauer kuntslicher huftgelenke durch werkstoffkombinationen polyathyler-aluminiumoxidkeramik-metall, Med. Orthop. Tech. 96 (1976) 152–157.
[308] E. Dorre, H. Beutler, D. Geduldig, Anforderungen an oxidkeramische werkstoffe als biomaterial fur kunstliche gelenke, Arch. Orthop. Unfallchir. 83 (1975) 269–278.
[309] H. McKellop, I. Clarke, K. Markolf, H. Amstutz, Friction and wear properties of polymer, metal, and ceramic prosthrtic joint materials evaluated on a multi-channel screening device, J. Biomed. Mater. Res. 15 (1981) 19–65.
[310] W. Dawihl, H. Mittelmeier, E. Dorre, G. Altmeyer, U. Hanser, Zur tribology von huftgelenk-endoprosthesen aus aluminiumoxidkeramik, Med. Orthop. Tech. 99 (1979) 114–118.
[311] I. Catelas, O. Huk, A. Petit, Flow cytometric analysis of mouse macrophage response to ceramic and polyethylene particles: effects of size concentration and composition, J. Biomed. Mater. Res. 41 (1998) 600–607.
[312] M. Tuke, A. Taylor, A. Roques, C. Maul, 3D liner and volumetric wear measurement on artificial hip joints—validation of a new methodology, Precis. Eng. 34 (4) (2010) 777–783.
[313] C. Esposito, A. Roques, M. Tuke, W. Walter, W. Walsh, Biolox forte versus biolox delta stripe wear: 2 years results, Orthop. Proc. 94 (2012) 143.
[314] I.C. Clarke, D.D. Green, P.A. Williams, K. Kuboc, G. Pezzotti, A. Lombardi, Hip-simulator wear studies of an alumina-matrix composite (AMC) ceramic compared to retrieval studies of AMC balls with 1–7 years follow-up, Wear 267 (2009) 702–709.

[315] A. Wang, C. Stark, J.H. Dumbleton, Role of cclic plastic deformation in the wear of UHMWPE acetabular cups, J. Biomed. Mater. Res. 29 (1995) 619–626.
[316] J.R. Cooper, D. Dowson, J. Fisher, Macroscopic and microscopic wear mechanisms in ultra-high molecular weight polyathylene, Wear 162–164 (1993) 378–384.
[317] M. Jasty, D.D. Goetz, C.R. Bragdon, Wear of polyethylene acetabular components in ttal hip arthroplasty, J. Bone Joint Surg. Am. 79 (1997) 349–358.
[318] A.A. Edidin, L. Pruitt, C.W. Jewett, D. Crane, D. Roberts, S.M. Kurz, Plasticity-induced damage laer is a precursor to wear in radiation-crosslinked UHMWPE acetabular components for total hip replacement, J. Arthroplasty 14 (1999) 616–627.
[319] S.M. Kurtz, C.M. Rimnac, L. Pruitt, C.W. Jewett, V. Goldberg, A.A. Edidin, The relationship between the clinical performance and large deformation mechanical behaviour of retrieved UHMWPE tibial inserts, Biomaterials 21 (2000) 283–291.
[320] R.S. Pascaud, W.T. Evans, P.J. McCullagh, D.P. Fitzpatrick, Influence of gamma-irradiation sterilization and temperature on the fracture toughness of ultra-high-molecular-weight polyethlene, Biomaterials 18 (1997) 727–735.
[321] M. Wrona, M.B. Mayor, J.P. Collier, R.E. Jensen, The correlation between fusion defects and damage in tibial polyethylene bearings, Clin. Orthop. 299 (1994) 92–103.
[322] G.W. Blunn, P.A. Lilley, P.S. Walker, Variability of the wear of ultra high molecular weight polyethlene in simulated TKR, Trans. Orthop. Res. Soc. 19 (1994) 177–181.
[323] Haggstrom, https://en.wikipedia.org/wiki/Hip_replacement#/media/File:Hip_prosthesis_liner_creep_and_wear.png, 2017.
[324] M. Semlitsch, M. Lehmann, H. Weber, New prospects for a prolonged functional life span of artificial joints by using the material combination polyethylene/aluminium oxide cramic/metal, J. Biomed. Mater. Res. 11 (1977) 537–552.
[325] V.O. Saikko, P.O. Paavolainen, P. Slatis, Wear of the polyethylene acetabular cup, Acta Orthop. Scand. 64 (1993) 391–402.
[326] I.C. Clarke, A. Gustafson, Clinical and hip simulator comparisons of ceramic-on-polyethylene and metal-on-polyethylene wear, Clin. Orthop. Relat. Res. 379 (2000) 34–40.
[327] M.C. Hogg, (A.J. Ruys Supervisor), Fretting wear and fatigue of taper junctions in modular orthopaedic implants (Ph.D. thesis), University of Sydney, 2015.
[328] Y. Takami, S. Yamane, K. Makinouchi, G. Otsuka, J. Glueck, R. Benkowski, Y. Nosé, Protein adsorption onto ceramic surfaces, J. Biomed. Mater. Res. A 40 (1) (1998) 24–30.
[329] M. Bohler, K. Knahr, M. Salzer, Long term results of uncemented alumina acetabular impants, J. Bone Joint Surg. 76 (1994) 53–59.
[330] S. Lerouge, O. Huk, L.H. Yahai, L. Sedel, Characterisation of in vivo wear debris from ceramic-ceramic total hip arthroplasties, J. Biomed. Mater. Res. 32 (1996) 627–633.
[331] S. Lerouge, O. Huk, L.H. Yahai, Ceramic-ceramic versus metal-polethlene: a comparison of periprosthetic tissues from loosened total hip arthroplasties, J. Bone Joint Surg. 79 (1997) 135–139.
[332] P. Boutin, P. Christel, J.M. Dorlot, The use of dense alumina-alumina ceramic combination in THR, J. Biomed. Mater. Res. 22 (1988) 1203–1232.
[333] P.Z. Wirganowicz, B.J. Thomas, Massive osteolysis after ceramic on ceramic total hip arthroplasty. A case report, Clin. Orthop. 338 (1997) 100–104.
[334] E. Garcia-Cimbrelo, J.M. Saanes, A. Minuesa, Ceramic-ceramic prothesis after 10 years, J. Arthroplasty 11 (1996) 773–778.
[335] M. Hamadouche, R.S. Nizard, A. Meunier, Cementless bulk alumina socket: preliminary result at 6 years, J. Arthroplast. 14 (1999) 701–707.
[336] J.P. Ivory, C.J. Kershaw, R. Choudry, Autophor cementless total hip arthroplasty for osteoarthritis secondar to congenital hip dysplasia, J. Arthroplasty 9 (1994) 427–433.

[337] E.B. Riska, Ceramic endoprosthesis in total hip arthroplasty, Clin. Orthop. 297 (1993) 87–94.
[338] L. Sedel, L. Kerboull, P. Christel, Alumina-on-alumina hip replacement. Results and survivorship in ounger patients, J. Bone Joint Surg. 72 (1994) 658–663.
[339] I. Catelas, A. Petit, D.J. Zukor, R. Marchand, L. Yahia, O.L. Huk, Induction of macrophage apoptosis by ceramic and polyethylene particles in vitro, Biomaterials 20 (7) (1999) 625–630.
[340] I. Kranz, J.B. Gonzalez, I. Dorfel, M. Gemeinert, M. Griepentrog, D. Klaffke, C. Knabe, W. Osterle, U. Gross, Biological response to micron- and nanometer-sized particles known as potential wear products from artificial hip joints. Part II. Reaction of murine macrophages to corundum particles of different size distributions, J. Biomed. Mater. Res. A 89A (2009) 390–401.
[341] R.K.N. Ryu, E.G. Bovill, H.B. Skinner, W.R. Murray, Soft tissue sarcoma associated with aluminium oxide ceramic total hip arthroplasty. A case report, Clin. Orthop. Relat. Res. 216 (1987) 207–212.
[342] G. Maccauro, G. Bianchino, S. Sangiorgi, G. Magnani, D. Marotta, P.F. Manicone, L. Raffaelli, P. Rossi Iommetti, A. Stewart, A. Cittadini, A. Sgambato, Development of a new zirconia-toughened alumina: promising mechanical properties and absence of in vitro carcinogenicity, Int. J. Immunopathol. Pharmacol. 22 (3) (2009) 773–779.
[343] C. Picconi, R.M. Streicher, Forty years of ceramic-on-ceramic THR bearings, Semin. Arthroplast. 24 (2013) 188–192.
[344] M.B. Cross, C. Esposita, A. Sokolova, R. Jenabzadeh, D. Molloy, S. Munir, B. Zicat, W. K. Walter, W.L. Walter, Fretting and corrosion changes in modular total hip arthroplasty, Orthop. Proc. 95 (2013) 127.
[345] S. Munir, R.A. Oliver, B. Zicat, W.L. Walter, W.K. Walter, W.R. Walsh, The histological and elemental characterisation of corrosion particles from taper junctions, Bone Joint Res. 5 (9) (2016) 370–378.
[346] S. Garbuz, N.V. Greidanus, B.A. Masri, C.P. Duncan, The John Charnley Award: metal-on-metal hip resurfacing versus large diameter head metal-on-metal total hip arthroplasty: a randomized clinica trial, Clin. Orthop. Relat. Res. 468 (2) (2010) 318–325.
[347] B. De Force, H. Pickering, A clearer view of how crevice corrosion works, J. Met. 47 (9) (1995) 22–27.
[348] A.R. Dujovne, J.J. Krygier, D.R. Wilson, C.E. Brooks, in: Fretting at the head/neck interface of modular hip prostheses, Proceedings of the Fourth World Biomaterials Congress, Berlin, 1992, p. 268.
[349] J.R. Goldberg, J.L. Gilbert, Electrochemical response of CoCrMo to high-speed fracture of it metal oxide using an electrochemical scratch test method, J. Biomed. Mater. Res. 37 (3) (1997) 421–431.
[350] A. Toni, S. Stea, A. Bordini, A. Sudanese, Survival analysis of ceramic-on-ceramic coupling vs metal-on-polyethylene. The experience of implant register at Instituto Rizzoli, in: J.P. Garino, G. Wilmann (Eds.), Bioceramics in Joint Arthroplasty, Thieme, Stuttgart, 2002, pp. 75–80.
[351] J. D'Antonio, W. Capello, M. Manley, M. Naughton, K. Sutton, Alumina ceramic bearings for total hip arthroplasty, Clin. Orthop. Relat. Res. 436 (2005) 164–171.
[352] S. Zhenxin, https://commons.wikimedia.org/wiki/File:Hip_joint_aseptic_loosening_ar1938-1.png, 2006.
[353] Rhodes, https://commons.wikimedia.org/wiki/File:Dislocated_hip_replacement.jpg, 2008.

[354] D.H. Owen, N.C. Russell, P.N. Smith, W.L. Walter, An estimation of the incidence of squeaking and revision surgery for squeaking in ceramic-on-ceramic total hip replacement: a meta-analysis and report from the Australian Orthopaedic Association National Joint Registry, Bone Joint J. 96 (2) (2014) 181–187.
[355] Y.D. Levy, S. Munir, S. Donohoo, W.L. Walter, Review on squeaking hips, World J. Orthop. 6 (10) (2015) 812–820.
[356] T. Bernasek, D. Fisher, D. Dalury, M. Levering, K. Dimitris, Is metal-on-metal squeaking related to acetabular angle of inclination, Clin. Orthop. Relat. Res. 469 (9) (2011) 2577–2582.
[357] A. Hothan, G. Huber, C. Weiss, N. Hoffmann, M. Morlock, The influence of component design, bearing clearance and axial load on the squeaking characteristics of ceramic hip articulations, J. Biomech. 44 (5) (2011) 837–841.
[358] N.E. Bishop, A. Hothan, M.M. Morlock, High friction moments in large hard-on-hard hip replacement bearings in conditions of poor lubrication, J. Orthop. Res. 31 (5) (2013) 807–813.
[359] S. Taylor, M.T. Manley, K. Sutton, The role of stripe wear in causing acoustic emissions from alumina ceramic-on-ceramic bearings, J. Arthroplasty 22 (7 Suppl 3) (2007) 47–51.
[360] W.L. Walter, G.M. Insley, W.K. Walter, M.A. Tuke, Edge loading in third generation alumina ceramic-on-ceramic bearings: stripe wear, J. Arthroplasty 19 (4) (2004) 402–413.
[361] W.L. Walter, T.S. Waters, M. Gillies, S. Donohoo, S.M. Kurtz, A.S. Ranawat, W. J. Hozack, M.A. Tuke, Squeaking hips, J. Bone Joint Surg. Am. 90 (4) (2008) 102–111.
[362] A. Toni, F. Traina, S. Stea, A. Sudanese, M. Visentin, B. Bordini, S. Squarzoni, Early diagnosis of ceramic liner fracture. Guidelines based on a twelve-year clinical experience, J. Bone Joint Surg. Am. 88 (4) (2006) 55–63.
[363] M.P. Abdel, T.J. Heyse, M.E. Elpers, D.J. Mayman, E.P. Su, P.M. Pellicci, T.M. Wright, D.E. Padgett, Ceramic liner fractures presenting as squeaking after primary total hip arthroplasty, J. Bone Joint Surg. Am. 96 (2014) 27–31.
[364] C. Restrepo, Z.D. Post, B. Kai, W.J. Hozack, The effect of stem design on the prevalence of squeaking following ceramic-on-ceramic bearing total hip arthroplasty, J. Bone Joint Surg. Am. 92 (3) (2010) 550–557.
[365] S.M. McDonnell, G. Boyce, J. Baré, D. Young, A.J. Shimmin, The incidence of noise generation arising from the large-diameter Delta Motion ceramic total hip bearing, Bone Joint J. 95 (2) (2013) 160–165.
[366] CeramTec, State of the Art Ceramics Manufacturing, Information Briefing Publication, Plochingen, 2005.
[367] G. Kleer, U. Soltesz, D. Siegele, Applicability of the proof-test concept to ceramic hip joint heads, in: P. Christel, A. Meunier, A.J.C. Lee (Eds.), Biological and Biomechanical Performances of Biomaterials. Advances in Biomaterials, Elsevier, Amsterdam, 1986, pp. 489–494.
[368] R. Schaefer, U. Soltesz, D. Siegele, Proof testing of ceramic hip joints, in: H. Heimke (Ed.), Bioceramics, Deutsche Keramische Gesellschaft, Koln, 1990, pp. 172–179.
[369] Statistica, Number of hip replacement surgeries in OECD=Countries as of 2015 (per 100,000 population), (2017).
[370] G. Langer, Ceramic tibial plateau of the 70s, in: J.P. Garino, G. Willman (Eds.), Bioceramics in Joint Arthroplasty, Proceedings of the 7th International Biolox Symposium, Stuttgart, 2002, pp. 128–130.
[371] A.P. Krueger, G. Singh, F.T. Beil, B. Feuerstein, W. Ruether, C.H. Lohmann, Ceramic femoral component fracture in total knee arthroplasty: an analysis using fractography, fourier-transform infrared microscopy, contact radiography and histology, J. Arthroplast. 29 (2014) 1001–1004.

[372] D.J. Cowdery, The pulsed TIG welding process Part 1. Equipment design. The Welding Institute, Res. Bull. 10 (8) (1969) 201–203.
[373] D.J. Cowdery, Welding products division. Pulsed arc MIG welding with variable pulse width, Report No. 672, British Oxygen Company research Laboratory, Cricklewood, 1969.
[374] Zion Research, Cardiac pacemaker market by product (implantable cardiac pacemaker and external cardiac pacemaker) by technology (biventricular, single chambered, and dual chambered): global industry perspective, Comprehensive Analysis and Forecast, 2015–2021, (2016).
[375] T. Guenther, C. Kong, H. Lu, M.J. Svehla, N.H. Lovell, A. Ruys, G.J. Suaning, Pt-Al2O3 interfaces in cofired ceramics for use in miniaturized neuroprosthetic implants, J. Biomed. Mater. Res. B Appl. Biomater. 102 (3) (2014) 500–507.
[376] O. Flandra, The first pacemaker implant in America, Pacing Clin. Electrophysiol. 11 (1988) 1234–1238.
[377] J.G. Davies, H. Siddons, Experience with implanted pacemakers: technical considerations, Thorax 20 (1965) 128–134.
[378] US-DOD, US Department of Defense. Test method standard—microcircuits, 2006. MIL-STD-883G.
[379] US-DOD, US Department of Defense. Test method standard—test methods for semiconductor devices, 2006. MIL-STD-750E.
[380] D.A. Howl, C.A. Mann, The back-pressuring technique of leak-testing, Vacuum 15 (7) (1965) 347–352.
[381] N. Patchett, https://en.wikipedia.org/wiki/Artificial_cardiac_pacemaker#/media/File: PPM.pngmedia/File:PPM.png, 2016.
[382] Elan, Elan Technology, Glass recommendations for Kovar, https://www.elantechnology.com/support/glass-recommendations/kovar-alloy/, 2017.
[383] R.G. Frieser, A review of solder glasses, Electrocompon. Sci. Technol. 2 (1975) 163–199.
[384] W.G. Housekeeper, The art of sealing base metals through glass, J. Am. Inst. Electr. Eng. 42 (1923) 870–876.
[385] H. Pulfrich, Ceramic to metal seals, US Patent 2163407, (Priority 1936), 1939.
[386] H.J. Nolte, Metallized ceramic, US Patent 2.667,432, 1954.
[387] H.J. Nolte, Method of metallizing a ceramic member, US Patent 2,667,427, 1954.
[388] H.G. Pincus, Mechanism of ceramic-to-metal adherence, Ceram. Age 70 (1954) 16–32.
[389] A. Meyer, Zum haftmechanismus von molybdan/maangan-metallisierungsschichten auf korund-keramic, Ber. Dtsch. Keram. Ges. 42 (1965) 405–415.
[390] J.T. Klomp, T.P.J. Botden, Sealing pure alumina ceramics to metals, Am. Ceram. Soc. Bull. 49 (1970) 2014–2211.
[391] ETB, Engineering Toolbox 2017, http://www.engineeringtoolbox.com/linear-expansion-coefficients-d_95.html, 2017.
[392] D.J.C. Cowdery, Transcript of interviews with Prof Andrew Ruys, (2017).
[393] Mathematica, Electrical conductivity of the elements, Mathematica's Element Data, Wolfram Research, 2017.
[394] US-DOD, MIL-STD-202F. Department of Defense Test Method Standard: Electronic and Electrical Component Parts (06 FEB 1998), 1998.
[395] D.J. Cowdery, Titanium covered cardiac pacemaker with elastomer coating and method of applying same, US Patent 3971388, 1976.
[396] Medtronic, Medtronic Ltd. Micra Website 2017, http://www.medtronic.com/us-en/patients/treatments-therapies/pacemakers/our/micra.html, 2017.

[397] G. Elssner, R. Pabst, J. Puhr-Westerheide, Schichtverbundkombinationen aus hochschmelzenden metallen und oxiden, Mater. Werkst. 5 (1974) 61–69.
[398] B. O'Brien, Niobium biomaterials, in: M. Niinomi, T. Narushima, M. Nakai (Eds.), Advances in Metallic Biomaterials, Springer Series in Biomaterials Science and Engineering, vol. 3, Springer, Berlin, 2015, pp. 245–272.
[399] A. Djourno, C. Eyries, Auditory prosthesis by means of a distant electrical stimulation of the sensory nerve with the use of an indwelt coiling, Presse Med. 65 (63) (1957) 1417.
[400] R.J. Fretz, R.P. Fravel, Design and function: a physical and electrical description of the 3M/House cochlear implant system, Ear Hear. 6 (1985) 14–19.
[401] F.B. Simmons, C.J. Mongeon, W.R. Lewis, D.A. Huntington, Electrical stimulation of acoustic nerve and inferior colliculus, Arch. Otolaryngol. 79 (1964) 559–568.
[402] F.B. Simmons, Electrical stimulation of the auditory nerve in man, Arch. Otolaryngol. 84 (1) (1966) 2–54.
[403] F.B. Simmons, J.M. Epley, R.C. Lummis, et al., Auditory nerve: electrical stimulation in man, Science 148 (3666) (1965) 104–106.
[404] R.P. Michelson, Electrical stimulation of the human cochlea: a preliminary report, Arch. Otolaryngol. 93 (3) (1971) 317–323.
[405] C.H. Chouard, P. Mac Leod, La réhabilitation des surdités totales: essai de l'implantation cochleaire par électrodes multiples, Nouv. Press. Med. 2 (44) (1973) 2958.
[406] C.H. Chouard, P. Mac Leod, Implantation of multiple intracochlear electrodes for rehabilitation of total deafness: preliminary report, Laryngoscope 86 (11) (1976) 1743–1751.
[407] C.H. Chouard, P. Mac Leod, B. Meyer, P. Pialoux, Appareillage electronique implanté chirurgicalement pour la réhabilitation des surdités totales et des surdi-mutité, Ann. Otolaryngol. Chir. Cervicofac. 94 (7–8) (1977) 353–363.
[408] G. Henkel, History of the Cochlear Implant, ENT Today. The Triological Society, 2013.
[409] M. Worthing, Graeme Clark. The man who invented the bionic ear, Allen and Unwin, Strawberry Hills, NSW, 2015.
[410] H.J. De Bruin, C.E. Warble, Chemical bonding of metals to ceramic materials, US Patent 4050956, (Filed 1975), 1977.
[411] J. Epstein, The Story of the Bionic Ear, Hyland Publishing House, Melbourne, 1989.
[412] T. Guenther, Miniaturisation of Neuroprosthetic Implants (Ph.D. thesis) in Biomedical Engineering, Supervisors: G. Suaning, (UNSW), Auxiliary Supervisors N. Lovell, (UNSW), C. Kong, (IUNSW), H. Lu, (Cochlear), M. Svehla, (Cochlear), A.J. Ruys, (USYD), University of NSW, 2012.
[413] M. Nayak, Hot-pressing of alumina-platinum retinal implants for the bionic eye (Honours thesis) in Biomedical Engineering, Supervisors: Gregg Suaning (UNSW), Andrew Ruys (USYD), University of Sydney, 2011.
[414] A. Mintri, Optimisation of a Hot-Pressing Cycle for alumina-platinum retinal implants for the bionic eye (Honours thesis) in Biomedical Engineering. Supervisors: G. Suaning, (UNSW), A.J. Ruys, (USYD), University of Sydney, 2011.
[415] D.I. Darley, A.A. McCusker, D. Milojevic, J. Parker, Feedthrough for electrical connectors, US Patent 7396265, (Filed 2002), 2008.
[416] A.J. McAlister, D.J. Kahan, The Al-Pt (aluminium-platinum) system, Bull. Alloy Phase Diagr. 7 (1985) 47–51.
[417] H.A. Wriedt, The Al-O (aluminium-oxygen) system, Bull. Alloy Phase Diagr. 6 (1985) 548–553.
[418] R.V. Allen, W.E. Borbidge, Solid state metal-ceramic bonding of platinum to alumina, J. Mater. Sci. 18 (1983) 2835–2843.

[419] M. De Graef, B.J. Dalgleish, M.R. Turner, A.G. Evans, Interfaces between alumina and platinum: structure, bonding and fracture resistance, Acta Metall. Mater. 40 (1992) S333–S344.
[420] J.C. Yang, Y.C. Lu, S.L. Sass, The influence of crystallography and oxide and metallic types on the structure and properties of metalceramic interfaces, Mater. Sci. Eng. A 162 (1993) 97–106.
[421] H. Lu, M.J. Svehla, M. Skalsky, C. Kong, C.C. Sorrell, Pt-Al2O3 interfacial bonding in implantable hermetic feedthroughs: morphology and orientation, J. Biomed. Mater. Res. 100 B (2012) 817–824.
[422] H.U. Klein, G. Inama, Implantable defibrillators: 30 years of history, G. Ital. Cardiol. 11 (10) (2010) 48s–52s.
[423] D.E. Mann, P.A. Kelly, A.D. Robertson, L. Otto, M.J. Reiter, Significant differences in charge times among currently available implantable cardioverter defibrillators, Pacing Clin. Electrophysiol. 22 (6) (1999) 903–907.
[424] C.N. Shealy, J.T. Mortimer, J.B. Reswick, Electrical inhibition of pain by stimulation of the dorsal columns: preliminary clinical report, Anesth. Analg. 46 (4) (1967) 489–491.
[425] R. Melzack, P.D. Wall, Pain mechanisms: a new theory, Surv. Anesthesiol. 11 (2) (1967) 89–90.
[426] B.E. Ben-Menachem, Vagus-nerve stimulation for the treatment of epilepsy, Lancet Neurol. 1 (8) (2002) 477–482.
[427] P.K. Crumrine, Vagal nerve stimulation in children, Semin Pediatr Neurol 7 (3) (2000) 216–223.
[428] K.W. Hatton, J.T. McLarney, T. Pittman, B.G. Fahy, Vagal nerve stimulation: overview and implications for anesthesiologists, Anesth. Analg. 103 (5) (2006) 1241–1249.
[429] K.M. Peters, et al., Randomized trial of percutaneous tibial nerve stimulation versus sham efficacy in the treatment of overactive bladder syndrome: results from the SUmiT trial, J. Urol. 183 (2010) 1438–1443.
[430] A. Al-zahrani, et al., Long-term outcome and surgical interventions after sacral neuromodulation implant for lower urinary Tract symptoms: 14-year experience at 1 center, J. Urol. 185 (2011) 981–986.
[431] S. Wexner, et al., Sacral nerve stimulation for fecal incontinence: results of a 120-patient prospective multicenter study, Ann. Surg. 251 (3) (2010) 441–449. Lippincott Williams & Wilkins.
[432] C.V. Giaimo, Electrical control of partially denervated muscles, US Patent 2737183, 1956.
[433] W.T. Liberson, H.J. Holmquest, D. Scot, M. Dow, Functional electrotherapy: stimulation of the peroneal nerve synchronized with the swing phase of the gait of hemiplegic patients, Arch. Phys. Med. Rehabil. 42 (1961) 101–105.
[434] C.H. Halpern, et al., Deep brain stimulation for epilepsy, Neurotherapeutics 5 (1) (2008) 59–67.
[435] B.C. Jobst, Electrical stimulation in epilepsy: vagus nerve and brain stimulation, Curr. Treat. Options Neurol. 12 (5) (2010) 443–453.
[436] J.M. Delgado, H. Hamlin, W.P. Chapman, Technique of intracranial electrode implacement for recording and stimulation and its possible therapeutic value in psychotic patient, Confin. Neurol. 12 (5–6) (1952) 315–319.
[437] A.L. Benabid, P. Pollak, A. Louveau, S. Henry, J. De Rougemont, Combined (thalamotomy and stimulation) stereotactic surgery of the VIM thalamic nucleus for bilateral Parkinson disease, Appl. Neurophysiol. 50 (1–6) (1987) 344–346.

[438] R. Coffey, Deep brain stimulation devices: a brief technical history and review, Artif. Organs 33 (3) (2008) 208–220.
[439] J. Laird, Leading the Feedthrough Industry, Medical Design, 2014. medicaldesign.com/author/joyce-laird.
[440] A.K. Ahuja, et al., Blind subjects implanted with the Argus II retinal prosthesis are able to improve performance in a spatial-motor task, Br. J. Ophthalmol. 95 (4) (2011) 539–543.
[441] B. Shaberman, Pixium Vision Reports Progress in Development of Two Advanced Bionic Retina Systems, Foundation Fighting Blindness, 2016.
[442] R. Hornig, et al., A method and technical equipment for an acute human trial to evaluate retinal implant technology, J. Neural Eng. 2 (1) (2005) S129–S134.
[443] R. Hornig, et al., Early clinical experience with a chronic retinal implant system for artificial vision, Invest. Ophthalmol. Vis. Sci. 47 (13) (2006) 3216.
[444] R. Hornig, T. Zehnder, M. Velikay-Parel, T. Laube, M. Feucht, G. Richard, The IMI retinal implant system, in: M.S. Humayun, J.D. Weiland, G. Chader, E. Greenbaum (Eds.), Artificial Sight, Biological and Medical Physics, Biomedical Engineering, Springer, New York, 2007, pp. 111–128.
[445] W. Mokwa, M. Goertz, C. Koch, I. Krisch, H.K. Trieu, P. Walter, in: Intraocular epiretinal prosthesis to restore vision in blind humans, Conference Proceedings IEEE Engineering in Medicine and Biology Society, 2008, pp. 5790–5793.
[446] T. Tanaka, K. Sato, K. Komiya, T. Kobayashi, T. Watanabe, T. Fukushima, H. Tomita, H. Kurino, M. Tamai, M. Koyanagi, in: Fully implantable retinal prosthesis chip with photodetector and stimulus current generator, IEEE International Electron Devices Meeting, Washington, DC, 2007, pp. 1015–1018.
[447] G. Auner, et al., Development of a wireless high-frequency microarray implant for retinal stimulation, in: M.S. Humayun, J.D. Weiland, G. Chader, E. Greenbaum (Eds.), Artificial Sight, Biological and Medical Physics, Biomedical Engineering, Springer, New York, 2007, pp. 169–186.
[448] S.K. Kelly, D.B. Shire, J. Chen, P. Doyle, M.D. Gingerich, W.A. Drohan, L. S. Theogarajan, S.F. Cogan, J.L. Wyatt, J.F. Rizzo, in: Realization of a 15-channel, hermetically-encased wireless subretinal prosthesis for the blind, IEEE Engineering in Medicine and Biology Society, Annual Conference, 2009, pp. 200–203.
[449] E. Zrenner, R. Wilke, H. Sachs, K. Bartz-Schmidt, F. Gekeler, D. Besch, U. Greppmaier, A. Harscher, T. Peters, G. Wrobel, B. Wilhelm, A. Bruckmann, A. Stett, Visual sensations mediated by subretinal microelectrode arrays implanted into blind retinitis pigmentosa patients, Biomed. Technol. 53 (2008) 218–220.
[450] E. Zrenner, K.U. Bartz-Schmidt, K. Benav, D. Besch, A. Bruckmann, V. Gabel, F. Gekeler, U. Greppmaier, A. Harscher, S. Kibbel, J. Koch, A. Kusnyerik, T. Peters, K. Stingl, H. Sachs, A. Stett, P. Szurman, B. Wilhelm, B. Wilke, Subretinal electronic chips allow blind patients to read letters and combine them to words, Proc. R. Soc. B Biol. Sci. 278 (1711) (2011) 1489–1497.
[451] A.Y. Chow, V.Y. Chow, Subretinal artificial silicon retina microchip implantation in retinitis pigmentosa, in: J. Tombran-Tink, C.J. Barnstable, J.F. Rizzo (Eds.), Visual Prosthesis and Ophthalmic Devices, Springer, New York, 2007, pp. 37–54.
[452] K. Stingl, R. Schippert, K.U. Bartz-Schmidt, D. Besch, C.L. Cottriall, T.L. Edwards, F. Gekeler, U. Greppmaier, K. Kiel, A. Koitschev, I. Kühlewein, R.E. MacLaren, J. D. Ramsden, J. Roider, A. Rothermel, H. Sachs, G.S. Schröder, J. Tode, N. Troelenberg, E. Zrenner, Interim results of a multicenter trial with the new electronic subretinal implant alpha AMS in 15 patients blind from inherited retinal degenerations. Front. Neurosci. (2017), https://doi.org/10.3389/fnins.2017.00445.

[453] T. Yagi, Biohybrid visual prosthesis for restoring blindness, Int. J. Biomed. Eng. 2 (1) (2009) 1–5.
[454] Y. Ito, A study on conductive polymer electrodes for stimulating nervous system, Int. J. Appl. Electromagn. Mech. 14 (2001) 347–352.
[455] D. Palanker, A. Vankov, P. Huie, S. Baccus, Design of a high-resolution optoelectronic retinal prosthesis, J. Neural Eng. 2 (1) (2005) S105–S120.
[456] A. Butterwick, P. Huie, B.W. Jones, R.E. Marc, M. Marmor, D. Palanker, Effect of shape and coating of a subretinal prosthesis on its integration with the retina, Exp. Eye Res. 88 (1) (2009) 22–29.
[457] J. Ohta, T. Tokuda, K. Kagawa, S. Sugitani, M. Taniyama, A. Uehara, Y. Terasawa, K. Nakauchi, T. Fujikado, Y. Tano, Laboratory investigation of microelectronics-based stimulators for large-scale Suprachoroidal Transretinal Stimulation (STS), J. Neural Eng. 4 (1) (2007) S85.
[458] E.T. Kim, J.M. Seo, S.J. Woo, J.A. Zhou, H. Chung, S.J. Kim, Fabrication of pillar shaped electrode arrays for artificial retinal implants, Sensors 8 (9) (2008) 5845–5856.
[459] H. Chamtie, The Phoenix 686: A Novel Bionic Eye Prototype (Honours thesis) in Biomedical Engineering, Supervisors: G. Suaning, (UNSW), A.J. Ruys, (USYD), University of Sydney, 2015.
[460] M. Schuettler, J.S. Ordonez, T.S. Santisteban, A. Schatz, J. Wilde, T. Stieglitz, Fabrication and test of a hermetic miniature implant package with 360 electrical feedthroughs, Conf. Proc. IEEE Eng. Med. Biol. Soc. 2010 (2010) 1585.
[461] G.J. Suaning, P. Lavoie, J. Forrester, T. Armitage, N.H. Lovell, Microelectronic retinal prosthesis: a new method for fabrication of high-density hermetic feedthroughs, Conf. Proc. IEEE Eng. Med. Biol. Soc. 1 (2006) 1638–1641.
[462] X.F. Wei, W.M. Grill, Current density distributions, field distributions and impedance analysis of segmented deep brain stimulation electrodes, J. Neural Eng. 2 (4) (2005) 139–147.
[463] B. Howell, W.M. Grill, Evaluation of high-perimeter electrode designs for deep brain stimulation, J. Neural Eng. 11 (4) (2014)046026.
[464] L. Haas, H. Berger, R. Caton, Electroencephalography, J. Neurol. Neurosurg. Psychiatry 74 (1) (2003) 9.
[465] J. Donoghue, N. Hatsopoulos, S. Martel, T. Fofonoff, R. Dyer, I. Hunter, Microstructured arrays for cortex interaction and related methods of manufacture and use, US Patent 7212851, 2007.
[466] L.R. Hochberg, D. Bacher, B. Jarosiewicz, N.Y. Masse, J.D. Simeral, J. Vogel, S. Haddadin, J. Liu, S.S. Cash, P. Smagt, J.P. Donoghue, Reach and grasp by people with tetraplegia using a neurally controlled robotic arm, Nature 485 (7398) (2012) 372–375.
[467] Youtube, Paralyzed Woman Uses Mind To Control Robotic Arm, https://www.youtube.com/watch?v=5s8NsgllTvg, 2012.
[468] E.C. Leuthardt, G. Schalk, D.W. Moran, J.R. Wolpaw, J.G. Ojemann, Brain computer interface, US Patent Number US7120486B2, (Filed 2003), 2006.
[469] P. Wicks, https://en.wikipedia.org/wiki/Brain%E2%80%93computer_interface#/media/File:BrainGate.jpg, 2006.
[470] T.J. Oxley, et al., Minimally invasive endovascular stent-electrode array for high-fidelity, chronic recordings of cortical neural activity, Nat. Biotechnol. 34 (3) (2016) 320–327.
[471] Mattes, https://commons.wikimedia.org/wiki/File:MET_Armures.jpg, 2005.
[472] Unknown, https://en.wikipedia.org/wiki/Armour#/media/File:Elements_of_a_Light-Cavalry_Armor_MET_DT780.jpg, 2005.
[473] Unknown, https://en.wikipedia.org/wiki/Ned_Kelly#/media/File:Nedkellysarmour1882.jpg, 1880.

References 531

[474] Chensiyuan, https://en.wikipedia.org/wiki/Ned_Kelly#/media/File:Ned_kelly_armour_library.JPG, 2008.
[475] B. Dean, Helmets and Body Armor in Modern Warfare, Yale University Press, New Haven, 1920.
[476] R. Laible, Ballistic Materials and Penetration Mechanics, Elsevier, Amsterdam, 1980.
[477] L. Traynor, Tests prove that a bulletproof silk vest could have stopped the first world war, 2014. The Guardian.
[478] M. Bolduc, A. Lazaris, Spider silk-based advanced performance fiber for improved personnel ballistic protection systems, 2002. DRDC Vlcartier Technical Memorandum TM, 2002-2222.
[479] P. Cunniff, S. Fossey, M. Auerbach, J. Song, D. Kaplan, W. Adams, R. Eby, D. Mahoney, D. Vezie, Mechanical and thermal properties of dragline silk from the spider Nephalia Clavipes, Polym. Adv. Technol. 5 (1994) 401–410.
[480] L. King, Lightweight Body Armor, The Quartermaster Review (March–April 1953, Reprinted from the January-February Ordnance), 1953.
[481] History of Kevlar, Chm.bris.ac.uk, 2017.
[482] T. Kitagawa, M. Ishitobi, K. Yabuki, An analysis of deformation process on poly-p-phenylenebenzobisoxazole fiber and a structural study of the new high-modulus type PBO HM+ fiber, J. Polym. Sci. B Polym. Phys. 38 (12) (2000) 1605–1611.
[483] Dyneema, The Dyneema Project, http://www.thedyneemaproject.com/story-of-dyneema/dyneema.html, 2017.
[484] M. Ogilvie, DyneemaR. Brief history of DSM high performance fibres, www.tote.com.au, 2017.
[485] K. Mylvaganam, L. Zhang, Ballistic resistance capacity of carbon nanotubes, Nanotechnology 18 (2007)475701.
[486] I.G. Crouch, The science of armor materials, Elsevier, London, 2017.
[487] AZOM, S-Glass Fibre, www.azom.com, 2017.
[488] A.P. Webster, Diphasic Armor 1. Glass Doron: Status Report, Research Division. Bureau of Medicine and Surgery, Navy Department, 1945.
[489] E.R. Barron, Body armor for aircrewmen, Technical Report 69-43-CE, US Army NATICK Laboratories, 1969.
[490] J.B. Melecker, W.J. Gailus, The light armor testing laboratory and research relating thereto, 1945. Final Report on GMC 30-F Volume 2, Contract No. W44-109-qm-305.
[491] A.L. Alesi, Composite personnel armor, Technical Report CP-5, US Army Natick Laboratories, 1957.
[492] R.J. Eichelberger, Panels for protection of armor against shaped charges, US Patent 3324768, (Filed 1950), 1967.
[493] R.L. Cook, Hard faced ceramic and plastic armor, US Patent 3509833, (Filed1963), 1970.
[494] H.A. King, Lightweight protective armor plate, US Patent 3431818, (Filed 1965), 1969.
[495] R.L. Cook, W.J. Hampshire, R.V. Kolarik, Ballistic armor system, US Patent 4179979, (Filed 1967), 1979.
[496] W.J. Ferguson, Armor plate, US Patent 4131053, (Filed 1965), 1978.
[497] R.A. Alliegro, A.W. Learned, Recomposite ceramic armor with metallic support strip, US Patent 3683828, (Filed 1967), 1972.
[498] E.R. Barron, Disclosure of Invention for Monolithic Armor, US Army, NLABS, Picatinny Arsenal, New Jersey, 1967.
[499] A.A. Bezreh, Interim report on injuries resulting from hostile action against army aircrewmembers in flight, US Army, Vietnam, 1967.
[500] R.A. Green, J.A. Parish, Personnel Armor Handbook, U.S. Naval Weapon Laboratory, Montgomery County, 1971.

[501] W.A. Gooch, Overview of the Development of Ceramic Armor Technology—Past, Present and the Future, Research Gate Publication, 2016. 292400398.
[502] PACRIM, Ceramic armor materials by design, in: J.W. McCauley, A. Crowson, W.A. Gooch, A.M. Rajendran, S.J. Bless, K.V. Logan, M. Normandia, S. Wax (Eds.), Proceedings of the Ceramic Armor Materials by Design Symposium. Held at PACRIM IV, November 4-8 2001, Ceramic Transactions, vol. 134, American Ceramic Society, Wailea, Maui Hawaii, 2001, p. 652.
[503] R.A. Armellino, Lightweight armor and method of fabrication, US Patent 3971072, (Filed 1972), 1976.
[504] M.L. Wilkins, Mechanics of penetration and perforation, Int. J. Eng. Sci. 16 (1978) 793–808.
[505] W.A. Gooch, in: An overview of ceramic armor applications, IDEE 2004. 6th Technical Conference, Trencin Slovakia, 2004.
[506] Z. Rosenberg, Y. Yeshurun, The relationship between the ballistic efficiency of ceramic tiles and their compressive strengths, Int. J. Impact Eng. 7 (1988) 357–362.
[507] MC2, Modern Ceramics Company, 2009. RSSC product specification.
[508] Honeywell, Processing Guidelines for Honeywell Ballistic Materials. Hard Armor Products, Honeywell Advanced Fibers and Composites, Colonial Height, VA, 2013.
[509] J. Craig, http://enacademic.com/pictures/enwiki/67/Craig_Ballistic_Plate.jpg, 2006.
[510] W.N. Horton, E.C. Skaar, Elan 46 LAS Glass Ceramic for Modern Glass-To-Metal Sealing. Advanced Hybrid Glass Ceramic Technology, Elan Technology, 2015. Product Report.
[511] T.J. Volger, W.D. Reinhardt, L.C. Chabildas, Dynamic behaviour of boron carbide, J. Appl. Phys. 95 (8) (2004) 4173–4183.
[512] M. Chen, J.W. McCauley, K.J. Hemker, Shock induced localised amorphisation in boron carbide, Science 299 (5612) (2003) 1563–1566.
[513] D. Taylor, T. Wright, J. McCauley, First principles calculation of stress induced amorphisation in armor ceramics, 2011. Army Research Laboratory Report. ARL-MR-0779.
[514] M. Bengisu, Engineering Ceramics, Springer Verlag, Berlin, 2001.
[515] R. Johnson, P. Biswas, P. Ramavath, R.S. Kumar, G. Padmanabham, Transparent polycrystalline ceramics: an overview, Trans. Indian Ceram. Soc. 71 (2) (2012) 73–85.
[516] R. Klement, S. Rolc, R. Mikulikova, J. Krestan, Transparent armor materials, J. Eur. Ceram. Soc. 28 (2008) 1091–1095.
[517] R. Winzer, Corrosion in crucibles made of pure oxide, Angew. Chem. 45 (1932) 429–431.
[518] W. Dawihl, E. Klingler, Der korrosionwiderstand von aluminiumoxidein-kristallen und von gesinteren werkstoffen auf aluminiumoxidgrundlage gegen anorganische sauren, Ber. Dtsch. Keram. Ges. 44 (1967) 1–4.
[519] M.A. Moore, F.S. King, in: Abrasive wear of brittle solids, Proceedings of the International Conference of the Wear of Materials, Americal Society of Mechanical Engineers, ASME, Dearborn Michigan, MI, 1979, pp. 275–285.
[520] M.G. Gee, E.A. Almond, The affect of surface finish on the sliding wear of alumina, J. Mater. Sci. 25 (1990) 296–310.
[521] S.M. Wiederhorn, B.J. Hockey, Effect of brittle parameters on the erosion resistance of brittle materials, J. Mater. Sci. 18 (1983) 766–780.
[522] A.G. Evans, M.E. Gulden, M. Rosenblatt, Impact damage in brittle materials in the elastic-plastic response regime, Proc. R. Soc. Lond. A 361 (1978) 343–365.
[523] R.W. Rice, B.K. Speronello, Effect of microstructure on rate of machining of ceramics, J. Am. Ceram. Soc. 59 (1976) 330–333.
[524] A.G. Evans, Abrasive wear in ceramics: an assessment, in: B.J. Hockey, R.W. Rice (Eds.), The Science of ceramic machining ans surface finishing, Washington NBS Special Publicationvol. 562, 1979, pp. 1–14.

[525] J.B. Wachtman Jr., T.G. Scuderi, G.W. Cleek, Linear thermal expansion of aluminium oxide and thorium oxide from 100 to 1100K, J. Am. Ceram. Soc. 45 (1962) 319–323.
[526] A.R.C. Westwood, N.H. MacMillan, R.S. Kalyoncu, Environment-sensitive hardness and machinability of Al2O3, J. Am. Ceram. Soc. 56 (1973) 258–262.
[527] R.M. Gruver, H.P. Kirchner, Effect of environment on penetration of surface damage and remaining strength, J. Am. Ceram. Soc. 57 (1974) 220–223.
[528] M.V. Swain, R.M. Latamison, A.R.C. Westwood, Further studies on environment-sensitive hardness and machinability of Al2O3, J. Am. Ceram. Soc. 58 (1975) 372–376.
[529] L.E. Fereira, D.D. Briggs, R.G. Barnhart, Engineering with high-alumina ceramics, Met. Prog. 98 (1970) 78–82.
[530] E. Mayer, Axiale Gleitringdichtungen, VDI, Dusseldorf, 1965.
[531] W. Dawihl, E. Dorre, Adsorption behaviour of high-density alumina ceramics exposed to fluids, in: Evaluation of Biomaterials, Wiley, Chichester, 1980, pp. 239–245.
[532] R.P. Steijn, On the wear of sapphire, J. Appl. Phys. 32 (1961) 1951–1958.
[533] B.J. Hockey, Plastic deformation of aluminium oxide by indentation and abrasion, J. Am. Ceram. Soc. 54 (1971) 223–231.
[534] I.A. Cutter, R. McPherson, Plastic deformation of Al2O3 during abrasion, J. Am. Ceram. Soc. 56 (1973) 266–269.
[535] B.J. Hockey, Pyramidal slip on (1123)(1100) and basal twinning in Al2O3, in: R. C. Bradt, R.E. Tressler (Eds.), Deformation of Ceramic Materials, Plenum Press, New York, 1975, pp. 167–179.
[536] B.J. Hockey, B.R. Lawn, Electron microscopy of microcracking about indentations in aluminium oxide and silicon carbide, J. Mater. Sci. 10 (1975) 1275–1284.
[537] P.F. Becher, Abrasive surface deformation in Sapphire, J. Am. Ceram. Soc. 59 (1976) 143–145.
[538] M.V. Swain, Microcracking associated with the scratching of brittle solids, in: R. C. Bradt, D.P.H. Hasselman, F.F. Lange (Eds.), Fracture Mechanics od Ceramics, vol. 3, Plenum Press, New York, 1978, pp. 257–272.
[539] B.B. Ghate, W.C. Smith, C.H. Kim, D.P.H. Hasselman, G.E. Kane, Effect of chromia alloying on machining performance of alumina ceramic cutting tools, Am. Ceram. Soc. Bull. 54 (1975) 210–215.
[540] R.W. Rice, Machining of ceramics, in: J.J. Burke, A.E. Corum, R.N. Katz (Eds.), Ceramics for High Performance Applications, Brook Hill Publishing, 1974, pp. 287–343.
[541] F.F. Lange, M.R. James, D.J. Green, Determination of residual surface stresses caused by grinding in polycrystalline Al2O3, J. Am. Ceram. Soc. 66 (1983) 16–17.
[542] R.W. Rice, J.J. Mecholsky Jr., P.F. Becher, The effect of grinding direction on flaw character and strength of single crystal and polycrystalline ceramics, J. Mater. Sci. 16 (1981) 853–862.
[543] VanBuren, https://commons.wikimedia.org/wiki/File:Hydrocyclone.png, 2008.
[544] D.W. Peters, L. Feinstein, C. Peltzer, On the high-temperature conductivity of alumina, J. Chem. Phys. 42 (1965) 2345–2346.
[545] J. Yee, F.A. Kroger, Measurements of electromotive force in Al2O3—pitfalls and results, J. Am. Ceram. Soc. 56 (1973) 189–191.
[546] D.J. Reed, B.J. Wuensch, Ion probe measurement of oxygen self-diffusion in single-crystal Al2O3, J. Am. Ceram. Soc. 63 (1980) 88–92.
[547] A.E. Paladino, W.D. KIngery, Aluminium ion diffusion in aluminium oxide, J. Chem. Phys. 37 (1962) 957–962.
[548] K. Kitazawa, R.L. Coble, Electrical conduction in single-crystal and polycrystalline Al2O3 at high temperatures, J. Am. Ceram. Soc. 57 (1974) 245–250.

[549] S.K. Mohapatra, F.A. Kroger, Defect structure of α- Al2O3 doped with titanium, J. Am. Ceram. Soc. 60 (1977) 381–387.
[550] S.K. Mohapatra, F.A. Kroger, Defect structure of α- Al2O3 doped with magnesium, J. Am. Ceram. Soc. 60 (1977) 141–148.
[551] O.T. Ozkan, A.J. Moulson, The electrical conductivity of single-crystal and polycrystalline aluminium oxide, J. Phys. D. Appl. Phys. 3 (1970) 983–987.
[552] H. Dilger, Untersuchungen der mechanischen und elektrischen eigenschaften von Al2O3 mit ZnO und NiO zusatzen bei hohen temperature, II. Elektrische eigenschaften, Ber. Dtsch. Keram. Ges. 51 (1974) 123–126.
[553] L.D. Hou, S.K. Tiku, H.A. Wang, F.A. Kroger, Conductivity and creed in acceptor-dominated polycrystalline Al2O3, J. Mater. Sci. 14 (1979) 1877–1889.
[554] M.M. El-Aiat, L.D. Hou, S.K. Tiku, H.A. Wang, F.A. Kroger, High-temperature conductivity and creep of polycrystalline Al2O3 doped with Fe and/or Ti, J. Am. Ceram. Soc. 64 (1981) 174–182.
[555] J. Pappis, W.D. Kingery, Electrical properties of single-crystal and polycrystalline alumina at high temperatures, J. Am. Ceram. Soc. 44 (1961) 459–464.
[556] W.D. Kingery, G.E. Meiling, Transference number measurements for aluminium oxide, J. Appl. Phys. 32 (1961) 556.
[557] H.W. Hennicke, H.H. Stuhrhahn, Untersuchungen von transportvorgangen im system Al_2O_3-Cr_2O_3, Ber. Dtsch. Keram. Ges. 48 (1971) 394–400.
[558] D.M. Bowie, Microwave dielectric properties of solids for applications at temperatures to 3000oF, IRE Natl. Convent. Rec. 5 (1) (1957) 270–281.
[559] E. Sayrun, V. Nguyen, in: High thermal conductivity aluminium nitride for high power microwave windows—an update, IEEE Vacuum Electronics Conference, Monterey, CA, USA, 2006.
[560] Coors, Properties of the Alumina Product Range, Coors Porcelain Company, Golden, 1970.
[561] Industry Shill, https://en.wikipedia.org/wiki/Spark_plug#/media/File:Spark_plug_insulator.jpeg, 2011.
[562] N. Wong, https://en.wikipedia.org/wiki/Spark_plug#/media/File:Bougie3.jpg, 2011.
[563] K. Schwartzwalder, Spark plug insulator US Patent 2332014, (Filed 1938), 1943.
[564] N.B. Hannay, Semiconductors, Reinhold Publishing, New York, 1959, 767.
[565] W. Jacobi, Halblieterverstarker, German Patent DE 833366, (Filed 1949), 1952.
[566] J.S. Kilby, Miniaturized electronic circuits, US Patent 3138743, (Filed 1959), 1964.
[567] R. Berry, Thin-Film Technology, Van Nostrand, New York, 1968.
[568] L. Hoffman, Precision glaze resistors, Am. Ceram. Soc. Bull. 42 (9) (1963) 490–493.
[569] J.J. Svec, Alumina scores in computer assemblies, Ceram. Ind. 87 (3) (1966) 54–55.
[570] A. Barton, Handbook of Solubility Parameters, CRC Press, Boca Raton, 1983.
[571] C.Y. Kuo, Electrical applications of thin-films produced by metallo-organic deposition, Solid State Technol. 17 (2) (1974) 49–55.
[572] C. Kuo, Hybrid Circuit Technol. 4 (1987) 29–33.
[573] ABI American Beryllia Inc., Beryllium oxide, 2017.
[574] P.L. Spencer, High efficiency magnetron, US Patent 2408235, (Filed 1941), 1946.
[575] HCRS, https://commons.wikimedia.org/wiki/File:Magnetron2.jpg, 2005.
[576] J. Cummings, https://en.wikipedia.org/wiki/Cavity_magnetron#/media/File:Original_cavity_magnetron,_1940_(9663811280).jpg, 2013.
[577] Fastfission, https://commons.wikimedia.org/wiki/File:Incandescent_light_bulb.svg, 2006.
[578] T.A. Edison, Electric lamp, US Patent 223898, (Filed 1879) 1880.

[579] Terren, https://commons.wikimedia.org/wiki/File:Edison_Carbon_Bulb.jpg, 2008.
[580] A. Just, F. Hanaman, Hungarian Patent 34541, 1904.
[581] L. De Forest, Oscillation-responsive device, US Patent 824637, (Filed 1906), 1906.
[582] L. De Forest, Oscillation-responsive device, US Patent 836070, (Filed 1906), 1906.
[583] H. Scott, Glass Metal Seal, US Patent 2062335, (Filed 1929), 1936.
[584] S. Riepl, https://commons.wikimedia.org/wiki/File:Elektronenroehren-auswahl.jpg, 2008.
[585] M.P. Borom, A.M. Turkalo, R.H. Doremus, Strength and microstructure in lithium disilicate glass-ceramics, J. Am. Ceram. Soc. 58 (1975) 385–391.
[586] H.L. McCollister, S.T. Reed, Glass-ceramic seals to Inconel, US Patent 4414282, (Filed 1982), 1983.
[587] R.D. Watkins, R.E. Loehman, Interfacial reactions between a complex lithium silicate glass-ceramic and Inconel718, Adv. Ceram. Mater. 1 (1) (1986) 77–80.
[588] S.C. Kunz, R.E. Loehman, Thermal expansion mismatch produced by interfacial reactions in glass-ceramic to metal seals, Adv. Ceram. Mater. 2 (1) (1987) 69–73.
[589] D.W. Rickey, https://en.wikipedia.org/wiki/X-ray_tube#/media/File:Rotating_anode_x-ray_tube_(labeled).jpg, 2006.
[590] Blue tooth7, https://commons.wikimedia.org/wiki/File:Crt14.jpg, 2011.
[591] H. Vatter, A method for producing vacuum-tight electric vessels after soldering, German Patent 645871, (Filed 1935), 1937.
[592] D.G. Burnside, Ceramic seals of the tungsten-iron type, RCA Rev. 15 (1) (1954) 46–61.
[593] E.P. Denton, H. Rawson, The metallization of high-Al2O3 ceramics, Trans. J. Br. Ceram. Soc. 59 (1960) 25–37.
[594] M. Hirota, Mechanism of Mo-Mn-Ti metallizing and Cu brazing in metal-to-ceramic seals, Trans. Jpn. Inst. Metals 10 (2) (1969) 98–106.
[595] L.W. Bean, The sintering of molybdenum metallizing, Trans. J. Br. Ceram. Soc. 70 (3) (1971) 121–122.
[596] M.E. Twentyman, High temperature metallizing Part 1. The mechanism of glass migration in the production of metal-ceramic seals, J. Mater. Sci. 10 (1976) 765–776.
[597] D.M. Mattox, H.D. Smith, Role of manganese in the metallization of high alumina ceramics, Am. Ceram. Soc. Bull. 64 (10) (1985) 1363–1367.
[598] J.R. Floyd, Effect of composition and crystal size of alumina ceramic on metal-to-ceramic bond strength, Am. Ceram. Soc. Bull. 42 (2) (1963) 65–70.
[599] F.C. Kelly, Metallizing and bonding non-metallic bodies, US Patent 2570248, (Filed 1948) 1951.
[600] C.S. Pearsall, New brazing methods for joining non-metallic material to metals, Mater. Meth. 30 (1949) 61–62.
[601] J.E. Beggs, Metallic bond, US patent 2857663, 1958.
[602] E. Cohen, H. Herbst, D.E.P. Jenkins, K.H. Wilkinson, Sealing metals to ceramic bodies, British Patent 832251, 1960.
[603] H. Mizuhara, K. Mally, Ceramic-to-metal joining with active brazing metal, Weld. J. 64 (10) (1985) 43–51.
[604] A.J. Moorhead, H. Keating, Direct brazing of ceramics for advanced heavy-duty diesels, Weld. J. 65 (10) (1986) 17–31.
[605] A.J. Moorhead, Direct brazing of alumina ceramics, Adv. Ceram. Mater. 2 (3) (1987) 159–166.
[606] M.G. Nicholas, T.M. Valentine, M.J. Waite, The weting of alumina by copper alloyed with titanium and other elements, J. Mater. Sci. 15 (9) (1980) 2197–2206.
[607] C. Peyour, F. Barbier, R. Revelovschi, Characterisation of ceramic/TA6V titanium alloy brazed joints, J. Mater. Res. 5 (1) (1990) 127–135.

[608] F. Barbier, C. Peytour, R. Revelovschi, Microstructure study of the brazed joint between alumina and Ti-6Al-4V alloy, J. Am. Ceram. Soc. 73 (6) (1990) 1582–1586.
[609] J.H. Selverian, F.S. Ohuch, M.R. Notis, Microstructure and kinetics of the interface reaction between titanium thin films and (112) sapphire substrates, Mater. Res. Soc. Symp. Proc. 167 (1990) 335–340.
[610] N.N. Sirota, T.E. Zhabko, X-ray study of the anisotropy of thermal properties in titanium, Phys. Status Solidi 63 (2) (1981) 211–215.
[611] J.P. Sterry, Testing ceramic-metal brazes, Met. Prog. 79 (6) (1961) 109–111.
[612] J.P. Sterry, Ceramic metal brazing for missile structures, Ceram. Ind. Mag. 81 (5) (1963) 69–71.
[613] G. Agricola, De Re Metallica (On the Nature of Metals), in: Medieval Printed Textbook From Germanic Europe (Bohemia and Saxony), 1556.
[614] A.I. Isaev, Use of ceramic materials in cutting metals, Stanki Instrum. 23 (4) (1952) 12–14.
[615] I.I. Kitaigordskii, Corundum microlit and its structure, Dokl. Akad. Nauk SSSR 90 (2) (1953) 225–226.
[616] I.I. Kitaigordskii, N.M. Pavlushkin, Microlit—an artificial superstrong stone, Steklo Keramika 10 (11) (1953) 4–7.
[617] I.I. Kitaigordskii, N.M. Pavlushkin, Characteristics of corundum microlit, Steklo Keramika 12 (11) (1955) 16–21.
[618] E. Dorre, State of development: ceramic cutting tools, Technical Mitt. 64 (1971) 7–9.
[619] M.C. Shaw, P.A. Smith, Workpiece compatibility of ceramic cutting tools, A. S. L. E. Trans. 1 (1958) 336–344.
[620] C.H. Kim, W. Roper, D.P.H. Hassleman, G.E. Kane, Machining performance of sintered versus hot-pressed ceramic cutting tools, Am. Ceram. Soc. Bull. 54 (1975) 589–590.
[621] W. Dawihl, E. Dorre, U. Dworak, Application of ceramic cutting tools in machining steel and cast iron, Powder Metall. Int. 3 (1971) 189–192.
[622] J. Hill, https://en.wikipedia.org/wiki/Sydney_Opera_House#/media/File:Opera_House_and_ferry._Sydney.jpg, 2011.
[623] M. Lewis, Roof cladding of the Sydney Opera House, J. Proc. R. Soc. NSW 106 (1–2) (1973) 18–32.
[624] D. Taylor, Tender Submission for Sydney Opera House Tile Replacement Project, 1996. Taylor Ceramic Engineering.
[625] A. Taylor, J. Taylor, Analysis of the 1996 Opera House Tile Tender Submission Archive, 2017. Alyssa Taylor (Managing Director), Julie Taylor (Founder) Taylor Ceramic Engineering.
[626] N.E. Wahl, Rain erosion of ceramics, Ceram. Age 69 (1) (1957) 28.
[627] H.T. Smyth, Physical principles in radome design, Ceram. Age 69 (4) (1957) 33–35.
[628] E.J. Smoke, Fabrication of dense ceramic radomes by casting, Ceram. Age 69 (4) (1957) 24–26.
[629] D.T. Chavara, A.J. Ruys, in: Fabrication of metal-ceramic functionally graded materials for use as heat shield tiles, Proceedings of the 44th AIAA Aerospace Sciences Meeting and Exhibit, Reno, NV, 2006. 5 pp.
[630] D.T. Chavara, A.J. Ruys, Development of the impeller-dry-blending process for the fabrication of metal-ceramic functionally graded materials, Ceram. Eng. Sci. Proc. 27 (3) (2007) 311–319.
[631] D.T. Chavara, C.X. Wang, A.J. Ruys, Biomimetic functionally graded materials: synthesis by impeller dry blending, J. Biomimetics Biomater. Tissue Eng. 3 (2009) 37–49.

[632] D.T. Chavara, (Ruys, A.J., Supervisor), Metal-Ceramic Functionally Graded Materials (Ph.D. thesis), University of Sydney, 2009.
[633] M.J. Bannister, W.G. Garrett, K.A. Johnston, N.A. McKinnon, R.F. Stringer, H. S. Kanost, The SIRO2 sensor: development, properties and applications. Energy and ceramics, in: P. Vincenzini (Ed.), Proc. 4th Int. Meeting on Modern Ceramics Technologies, Saint-Vincent, Italy, 28–31 May 1979, Materials Science Monographs, vol. 6, Elsevier, Amsterdam, 1980.
[634] J. Mills, T. Yapp, An Economic Evaluation of Three CSIRO Manufacturing Research Projects, CSIRO Institute of Industrial Technologies, 1996. May 1996.
[635] P. Crowhurst, Peter Crowhurst, General Manager, Ceramic Oxide Fabricators (Personal Communication), 2018.
[636] A.J. Ruys, C.C. Sorrell, A slip casting process for small thin walled alumina articles, Mater. Sci. Forum 34 (2) (1988) 875–880.
[637] A.J. Ruys, C.C. Sorrell, A slip casting process for small thin walled alumina articles, Mater. Sci. Forum 34 (2) (1988) 875–880.
[638] A.J. Ruys, C.C. Sorrell, Slip casting of high purity Al2O3 using sodium carboxymethylcellulose as deflocculant/binder, Am. Ceram. Soc. Bull. 69 (5) (1990) 828–832.
[639] A.J. Ruys, C.C. Sorrell, Slip casting of alumina using sodium carboxymethylcellulose, in: Handbook of Ceramics, Ceramic Monograph No. 1.4.2.3Verlag Schmid GMBH, Freiburg, 1996 (15 pp).
[640] A.J. Ruys, C.C. Sorrell, Slip casting of alumina, Am. Ceram. Soc. Bull. 75 (11) (1996) 68–71.
[641] A. El-Hajje, E.C. Kolos, J.K. Wang, S. Maleksaeedi, Z. He, F.E. Wiria, C. Choong, A.J. Ruys, Physical and mechanical characterisation of 3D-printed porous titanium for biomedical applications, J. Mater. Sci. Mater. Med. 25 (11) (2014) 2471–2480.
[642] A. O'Regan, Reinforced Alumina for Ceramic Knee Prostheses (Honours thesis) in Biomedical Engineering, Supervisor Andrew Ruys (USYD), University of Sydney, 2011, 2015.
[643] A.J. Ruys, B.K. Milthorpe, C.C. Sorrell, Preparation of fibre-reinforced hydroxylapatite, Interceram 43 (1) (1994) 7–9.
[644] A.J. Ruys, K.A. Zeigler, A. Brandwood, B.K. Milthorpe, S. Morrey, C.C. Sorrell, Reinforcement of hydroxyapatite with ceramic and metal fibres, in: W. Bonfield, G.W. Hastings, K.E. Tanner (Eds.), Bioceramics, vol. 4, Butterworth-Heinemann, London, 1991.
[645] A.J. Ruys, N. Ehsani, S. Moricca, B.K. Milthorpe, C.C. Sorrell, in: P.A. Walls, C. C. Sorrell, A.J. Ruys (Eds.), Preparation of hydroxyapatite fibre composites by hot isostatic pressing, Symposium 4.3, Second International Meeting of Pacific Rim Ceramic Societies. PacRim 2, Australasian Ceramic Society, Sydney, 1996, pp. 56–61.
[646] A.J. Ruys, N. Ehsani, B.K. Milthorpe, K. Husslein, M. Koschig, C.C. Sorrell, in: P.A. Walls, C.C. Sorrell, A.J. Ruys (Eds.), Fracture toughness of fibre-reinforced hydoxyapatite using a miniaturised compact tension test, Symposium 4.3, Second International Meeting of Pacific Rim Ceramic Societies. PacRim 2, Australasian Ceramic Society, Sydney, 1996, pp. 45–55.
[647] M. Knepper, S. Moricca, A.J. Ruys, K. Schindhelm, in: P.A. Walls, C.C. Sorrell, A. J. Ruys (Eds.), Aspects of the production of fibre-reinforced hydroxyapatite ceramics, Symposium 4.4, Second International Meeting of Pacific Rim Ceramic Societies. PacRim 2, Australasian Ceramic Society, Sydney, 1996, pp. 10–20.
[648] N. Ehsani, A.J. Ruys, C.C. Sorrell, Thixotropic casting of fecralloy fibre reinforced hydroxyapatite, in: G.M. Newaz, H. Neber Aeschbacher, F.H. Wohlbier (Eds.), Metal Matrix Composites. Part 1: Application and Processing, Trans Tech Publications, Zurich, 1995, pp. 373–380.

[649] A.J. Ruys, S.A. Simpson, C.C. Sorrell, Thixotropic casting of fibre reinforced ceramic matrix composites, J. Mater. Sci. Lett. 13 (16) (1994) 1323–1325.
[650] X. Miao, A.J. Ruys, B.K. Milthorpe, Hydroxyapatite-316L fibre composites prepared by vibration assisted slip casting, J. Mater. Sci. 36 (13) (2001) 3323–3332.
[651] N.A. Hamzaid, A. Fornusek, A. Ruys, G.M. Davis, in: Conceptual design of an isokinetic functional electrical stimulation (FES) leg stepping trainer for individuals with neurological disability, Kuala Lumpur International Conference on Biomedical Engineering, 2006.
[652] N.A. Hamzaid, C. Fornusek, A.J. Ruys, G.M. Davis, Mechanical design and driving mechanism of an isokinetic functional electrical stimulation-based leg stepping trainer, Australas. Phys. Eng. Sci. Med. 30 (4) (2007) 323–326.
[653] M.J. Taylor, A.J. Ruys, M.P. Jones, C. Fornusek, Use of a load cell and isokinetic dynamometer to investigate a 6060-T5 aluminium orthosis for FES-evoked isometric exercise, Mater. Forum 38 (2014). Combined Australian Materials Societies 7pp, published on CDROM.
[654] M.J. Taylor, C. Fornusek, A.J. Ruys, M. Bijak, A.E. Bauman, The Vienna FES interview protocol—a mixed-methods protocol to elucidate the opinions of various individuals responsible for the provision of FES exercise, Eur. J. Transl. Myol. 27 (3) (2017) 160–165.
[655] M.J. Taylor, C. Fornusek, P. de Chazal, A.J. Ruys, in: "All talk no torque"—a novel set of metrics to quantify muscle fatigue through isometric dynamometry in functional electrical stimulation (FES) muscle studies, IOP Conf. Series: Materials Science and Engineering, vol. 257, 2017 012018, 11 pp.
[656] N. Amanat, A.F. Nicoll, A.J. Ruys, D.R. McKenzie, N.L. James, Gas permeability reduction in PEEK film: comparison of tetrahedral amorphous carbon and titanium nanofilm coatings, J. Membr. Sci. 378 (1–2) (2011) 265–271.
[657] K. Chen, Q. Li, W. Li, H. Lau, A. Ruys, P. Carter, in: Three-dimensional finite element modeling of cochlear implant induced electrical current flows, IEEE International Conference on Computational Intelligence for Measurement Systems and Applications CIMSA 2009, (IEEE) Institute of Electrical and Electronics Engineers, Piscataway, NJ, 2009.
[658] M. Gomez, Silicone and the Cochlear Implant (Honours thesis) in Biomedical Engineering, Supervisors: A.J. Ruys, (USYD), M. Raje, (Cochlear), University of Sydney, 2007.
[659] S.M. Graham, Plasma Based Treatments for Silicone for the Cochlear Implant (Honours thesis) in Biomedical Engineering. Supervisors: A.J. Ruys, (USYD), M. Svehla, (Cochlear), University of Sydney, 2009.
[660] A. Singh, The Extracochlear Hardball Electrode for the Cochlear Implant (Honours thesis) in Biomedical Engineering. Supervisors: A.J. Ruys, (USYD), N. Eder, (Cochlear), University of Sydney, 2006.
[661] S. Sorour, Moisture and Cochlear Implants (Honours thesis) in Biomedical Engineering, Supervisors: A.J. Ruys, (USYD), J. Dalton, (Cochlear), University of Sydney, 2006.
[662] M. Vozzo, Plasma Polymer Primers for Adhesion and the Cochlear Implant (Honours thesis) in Biomedical Engineering, Supervisors: A.J. Ruys, (USYD), M. Bilek, (USYD), D. Smyth, (Cochlear), University of Sydney, 2010.
[663] M. Wong, Cochlear Implant Fixation to Soft Tissue (Honours thesis) in Biomedical Engineering, Supervisors: A.J. Ruys, (USYD), S. Henshaw, (Cochlear), University of Sydney, 2011.

[664] S. Zhang, F. Awaja, N. James, D.R. McKenzie, A.J. Ruys, A comparison of the strength of autohesion of plasma treated amorphous and semi-crystalline PEEK films, Polym. Adv. Technol. (2010). PAT-10-190.R1.

[665] S. Zhang, F. Awaja, N. James, D.R. McKenzie, A.J. Ruys, A comparison of the strength of autohesion of plasma treated amorphous and semi-crystalline PEEK films, Polym. Adv. Technol. 22 (12) (2011) 2496–2502.

[666] S. Zhang, F. Awaja, N. James, D.R. McKenzie, A.J. Ruys, Autohesion of plasma treated semi-crystalline PEEK: comparative study of argon, nitrogen and oxygen treatments, Colloids Surf. A Physicochem. Eng. Asp. 374 (1–3) (2011) 88–95.

[667] W.X.Y. Zheng, Cochlear Implant Fixation (Honours thesis) in Biomedical Engineering, Supervisors: A.J. Ruys, (USYD), S. Henshaw, (Cochlear), University of Sydney, 2010.

[668] L. Howard, A.J. Ruys, P. Carter, X. Wang, Q. Li, Subject specific modelling of electrical conduction in the body: a case study, J. Biomater. Tissue Eng. 10 (2011) 43–54.

[669] H.K. Lau, A.J. Ruys, P. Carter, X. Wang, Q. Li, in: Spatial correction of echo planar imaging deformation for subject specific diffusion tensor MRI analysis, IEEE Proceedings of ITME 2008 (IEEE International Symposium on IT in Medicine & Education), 2008. 6 pp.

[670] M.G. Lee, Fixation of the Cochlear Implant (Honours thesis) in Biomedical Engineering. Supervisors: A.J. Ruys, (USYD), S. Henshaw, (Cochlear), University of Sydney, 2009.

[671] A.J. Ruys, E.R. McCartney, Progress in developing high strength resorbable bone implants, Mater. Sci. Forum 34 (1) (1988) 399–401.

[672] A.J. Ruys, A feasibility study of silicon doping of hydroxylapatite, Interceram 42 (6) (1993) 372–374.

[673] A.J. Ruys, Silicon-doped hydroxyapatite, J. Aust. Ceram. Soc. 29 (1/2) (1993) 71–80.

[674] L.W. Pullan, Powder Injection Moulding of Transparent Alumina Components With Micro-Features for the Biomedical Industry (Honours thesis) in Biomedical Engineering. Supervisor: A.J. Ruys, (University of Sydney), Auxiliary Supervisors S.X. Zhang, (SIMTech, Singapore) S.W. Leu, (SIMTech, Singapore). University of Sydney, 2012.

Index

Note: Page numbers followed by *f* indicate figures and *t* indicate tables.

A
Abbott Laboratories, 288
Abrasive wear, 195
Activation energy of diffusion, 94
Active metal brazing technology, 459–461
ADA, 367
Adhesive wear, 189, 195, 372
Aerospace applications, 474, 493–495
AGC group, 32–33
Alkaline sinter processing, 69–70
Alumina
  advanced composite materials, 4
  cement, 14
  coatings, 4
  crystalline structure, 103–104, 104*f*
  glass-contact refractories, 24
  machining tools, 474
Alumina-metal bonding, electrical feedthroughs
  active metal brazing technology, 459–461
  alumina-titanium hermetic bionic implant, 462
  bionic feedthroughs, 447
  brazing, 457
  ceramic-metal seal design, 463, 465*f*
  corrosion resistant, 455
  criteria, 454–455
  dilatometry data, 462, 463*f*
  electrical-industry feedthroughs, 448
  glass-metal seals, 453–454, 454*f*
    advantages, 455
  hermetic feedthrough concept
    carbon-filament/vacuum lightglobe, 450–451, 451*f*
    electrical conductors, 450
    glass-metal seal, 449*f*, 450
    thermal expansion, 451–452
    vacuum valve, 452–454, 452–453*f*
    xenon-filled tungsten-halogen light, 450, 450*f*
  Kovar-glass hermetic sealing, 455
  metal-ceramic bonding, 456
  metallization, 457
  molybdenum-manganese (Mo-Mn) method, 457–459
  multilayer micro-feedthroughs, 468–469, 469*f*
  niche applications, 456–457
  thermal expansion, 464
  titanium, niobium and tantalum, 465–468
  Vatter patent, 456
  Young's modulus, 463, 464*f*
Alumina powder preparation
  binder, 78
  characteristics, 80
  coarse powders, 80–81
  comminution process, 81
  deflocculant, 78
  factors, 77
  hydrothermal method, 81
  milling, 79, 81
  nanopowders, 80
  particle shape and size, 80
  plasticizer, 78
  production, 77–78
  purity, 80
  sintering grade alumina powders, 77
  slurry, 78
  sol-gel forming, 77–78
  sol-gel powders, 81
  solids loading, 78
  specific surface area, 80
  spray drying, 79, 79*f*
  spray pyrolysis, 81
  wet-forming methods, 79
Aluminum ions, 95
Aluminum oxide, 14
Aluminum smelting, 74
  bauxite, 46
  industry, 20–21
Alum mineral, 17–18

Architectural applications, 474, 490
  Sydney Opera House, 490, 490f
  high-alumina tiles, 492, 492f
  tile formulations, 491–492
  tiles cladding, 491
  ultra-high purity alumina, 492
Aseptic loosening, 202–205, 206f
Atomic mass units (AMU), 160

B

BAE Advanced Materials, 367
Ball/attrition milling, 81–82
Ballistic armor, 13
Ballistic nylon (Nylon-66) body armor, 327–328, 329t
Bauxite, 17, 17f, 40f. *See also* Bayer process
  alumina wear tiles, 41
  aluminum metal, 40
  aluminum smelting, 46
  applications, 40
  boehmite, 42
  ceramic grade alumina powder, 40
  chemical grade alumina production, 39
  diaspore, 42
  economic demonstrated resources (EDR), 42
  gibbsite, 42
  Hall-Héroult process, 46
  hydrated alumina, 39
  karst, 46
  lateritic, 46
  mineral forms, 42, 46
  nonalumina-ceramic applications, 39–40
  open-cut mining, 42
  producers, 42–44, 43t, 45f
  production quality, 42
  quality of, 44
  raw alumina powder, 41
  red mud
    calcium silicate recovery, 66
    sources, 47
  silica contaminant, sources, 46
  sintering grade alumina powders, 40
  usage of, 43–44
  USA Geological Survey, 42
  value adding, 40
  world economic reserves, 43–44, 44t, 45f
Bayer process, 18, 71
  aluminum hydroxide, 52
  autoclave conditions, 54
  bauxite dissolution, 19
  Bayer liquor, 53
    slurry, 51
    sodium hydroxide solution, 53
  calcination optimization, 57
  chemical principles, 50
  chemical reaction, hold time, 52
  digestion temperatures, 53–54
  digestion time, 54
  energy efficiency, 19–20, 49
  evaporators, 52
  exothermic process, 52
  fluidized bed, 52
  gibbsite/boehmite/diaspore ratio, 57
  gibbsite crystals, 51
  gibbsite precipitation, 57, 61–62
  Hall-Héroult process, 20, 49
  high-purity aluminum hydroxide, 20
  high-temperature refineries, 54
  hydrated alumina minerals, 53
  iron contamination, 54
  lateritic *vs.* karstic bauxites, 55
  low-temperature refineries, 54
  NaOH Bayer liquor, 51
  oxalate removal, 50, 51f, 52
  parameters, 50
  precalcination, 55
  purity, 52
  quartz-contaminated karst bauxite, double digestion, 55
  quartz dissolution, 54
  red mud, 20, 51–52 (*see also* Red mud disposal, Bayer process)
  refining process
    alkaline sinter processing, 69–70
    carbothermic furnace processing, 69
    clay, 68
    wet acid processing, 69, 69t
    wet alkaline processing, 68
  reversible chemical equation, 50
  silica problems
    bauxite impurity, 55
    cancrinite, 60
    desilication products (DSP), 59
    dissolution in, 58–59
    hydrogarnet, 60
    iron hydrogarnet, 61

reactive silica, 56–58
scale forms, 59
sodalite, 60
silica removal, 50, 51f, 52
sodium oxalate impurity cleansing, 52
sodium oxalate problem, 62–63
steam saturation pressure, 54
unreactive silica, 51
Bioceramics conference series, 140
BIOLOX®, 212, 218
Biolox symposium series, 140
Biomaterials applications, 9
Biomaterials markets, 10, 501
Biomedical engineering applications, 2, 10, 501
  billion biomaterials markets, 10
  biomedical material, 124–125
  bionic pacemaker, 123
  bone tissue scaffolds
    bioactive bone scaffold, 136, 136f
    bone cells, in vitro response, 135–136
    calcium phosphate ceramics, 134–135
    limitations, 134–135
    manufacturing process, 135–136
    requirements, 134
  Cowdery alumina-feedthrough/titanium-casing system, 125
  defibrillator, 123
  dental devices, 126 (see also Dental technology)
  epoxy encapsulation, 125
  multimillion biomaterials markets, 10
  niche biomaterials markets, 10
  orbital implants, 137
  orthopedic bearing, 124
  osteotomy spacers, 137
Bionic ear, 283
  all-alumina casing
    concept, 263
    moisture-enhanced static fatigue, 263–264
    problem solution, 264
  "ball electrode," 277
  Cochlear Ltd., 289
    Cochlear C122M, 261, 262f
    Cochlear C1532M, 261, 263f
    Cochlear C124RE, 261, 262f
    Cochlear implant system, 261, 264f
    helical electrode support, 260–261

intra-cochlea 22-electrode-array, 260–261, 260f
    multiple electrode, 260
    soft silicone-coated microcoiled electrode, 260
    thermal expansion coefficient, 261–263
    timeline, 259
    titanium/alumina hermetic encapsulation system, 260
    22-electrodes vs. 1-electrode, 264–265
  cofired platinum-alumina interfaces
    Al-Pt-O system, 278
    focused ion beam milling (FIB), 279
    glass-platinum interactions, 278
    hermetic bonding, 277
    intermetallic compound formation, 279
    laser-patterning/screen-printing approach, 281
    metal-ceramic bonding, 278
    pressure-assisted alumina-platinum bonding experiments, 277–278
    residual stress, 279–280
    selected area electron diffraction (SAED), 280
    thermal expansion coefficients, 280
    titanium, 281
    transmission electron microscopy (TEM) analysis, 278–279
    ultrafine channel wires, 280
  differential diametral contraction, 272
  electrical conductivity of, 270–271
  fluid diffusion, 273–274
  Kuzma concept, 271, 276
  MED-EL, 289
  multiple-channel trials, 257–258
  Nurotron, 289
  Oticon Medical, 289
  platinum channel wires, 273, 276f
  "platinum comb" device, 273, 275f
  powder-injection-molded "ceramic comb" concept, 273, 274f
  quantum leap, hermetic feedthrough technology, 257
  single-channel experimental bionic ear trial, 257
  single-channel implant, 255
  Sonova, 289
  speech processing, 273

Bionic ear *(Continued)*
  Telectronics environment
    electrode side, 266–269, 268f
    electronic components, 269
    facilities of, 265–266
    innovations, 270
    microarray bionic implants, 270
    platinum-channel wires, 266–269, 266f
    platinum tubes, 269
    pressure-assisted presintered-alumina/platinum couples, 269–270
    proto-Cochlear, 265–266
    silicone-mounted platinum strips, 266–269, 268f
    with titanium flange, 266–269, 267f
    welding, Cochlear feedthrough, 266–269, 267f
  Telectronics legacy, 255–256
  thermal expansion coefficients, 271
  titanium casing, 273, 275f
Bionic eye, 13
  features, 285–286
  global market size, 286
  hermetic bionic implant concept, 285–286, 285f
  innovation, 283–284
  object recognition, 296
  physiological systems, 284
  retinal implant
    alumina-focused research initiative, 304
    alumina-platinum feedthrough prototypes, 301, 302f
    1145-channel feedthrough, 301, 302f
    co-fired alumina-platinum feedthrough, 303–304
    concept, 297–298
    electrical stimulation, 301
    electrode number, 300–301, 301f
    FDA-approved system, 300–301
    fragile neural tissue, 300
    global research/commercial teams, 298, 299t
    implantation sites, 298, 300f
    quantum leap (*see* Quantum leap, retinal implant)
    risk-benefit analysis, 298, 298t
    titanium casing, 303

Bionic feedthroughs, 10, 23
Bionic implants, 1, 5–9, 6f
  cardiac pacemaker, 13
  feedthroughs, 25
Bionic pacemaker, 123, 287
  Abbott Laboratories, 288
  alumina/titanium hermetic feedthrough system, 225
  advantage, 242
  alumina ceramic composition, 243
  alumina-metal hermetic sealing, 236–238
  assembling, 246
  biocompatible brazing technology, 245–246
  biomaterials, 238–239
  ceramic feedthroughs, 235
  cobalt chrome, 240
  components, 240, 241f
  electrical conductors, 235
  electrical porcelain, 242–243
  epoxy encapsulation, 233–234
  glass-metal hermetic feedthrough, 236, 237f
  hermetic-bionic-implant concept
    (*see* Hermetic-bionic-implant system)
  hermetic seal, 235
  Kovar/borosilicate glass, 236
  physical properties, 242
  sealing, 247
  system-critical technology, 235
  Telectronics P10, 242, 242f
  thermal expansion coefficients, 239, 243
  titanium casing, 243–244
  titanium electrical terminal feedthrough system, 244–245
  Triode vacuum valves, 236
  bionic, definition, 228
  Biotronik, 288
  Boston Scientific, 288
  Cochlear implant bionic ear device, 226–227
  desktop pacemaker, 229–230
  epoxy encapsulation systems, 225
  events, 229
  feedthrough terminology, 252–253
  global market, 226
  hermetic implantable pacemaker, 231–232
  hermeticity and durability, 225

Index 545

implantable cardioverter defibrillator
  (ICD), 287
LivaNova, 288
Medico, 288
Medtronic, 288
portable pacemaker, 230–231
Shree Pacetronix, 288
technological breakthroughs, 228
Telectronics P10, 247
  (*see also* Telectronics Pacemaker
  Model P10)
transistor and semiconductor
  technology, 229
Biotronik, 288
Birmingham hip, 141, 154
Boehmite, 42, 53
Boston Scientific, 288
Brain computer interface (BCI), 283, 506
  bionic implant, 13
  BrainGate™ cortical implant, 316,
  316–317*f*
  electrocorticography, 316, 317*f*
  PubMed analysis, 315, 316*f*
  Stentrode, 318–319
BrainGate™ cortical implant, 316, 316–317*f*
Breastplate technology, 6, 8*f*
Brewster body armor, 325–326, 326*f*

C

Calcium-doped alumina tissue scaffolds,
  125–126
Calcium phosphate coatings, 160
Calcium silicate recovery, 66
Cancrinite, 59–60
Carbothermic furnace processing, 69
Cardiac pacemakers, 1
"Caustic" content, 53
Cavity magnetrons, 442, 443*f*
Ceramic grade alumina powder, 40
Ceramic injection molding (CIM).
  See Powder injection molding (PIM)
Ceramic-metal seal design, 463, 465*f*
CeramTec, 2, 33–35, 141, 184–186, 187*t*, 367
  alumina biomedical bearings, 24, 24*f*
  BIOLOX®, 212, 218
  flexural strength, 185, 186*f*
  high-purity alumina-on-alumina, 216*f*, 218
  hot isostatic pressing, 214

laser marking, 215
osteolysis, 215–218
polyethylene wear particles, 210–212
proof testing, 214–215
quality assurance, 214
quality manufacturing process, 212–214
timeline of, 212, 213*t*
total hip replacement, revision surgery,
  212, 212*t*
zirconia-toughened-alumina (ZTA)
  alumina-on-alumina, 217*f*, 218
Zyglo crack inspection, 215
CeramTec BIOLOXdelta, 180, 181*f*,
  187–188
CeramTec ZTA BIOLOX™ delta, 210,
  211*f*
CeramTec ZTA biomedical bearings, 24,
  24*f*
Ceraver Osteal alumina-on-alumina-bearing
  system, 162, 163*f*
Champion Spark Plug Company, 75
686 Channel co-fired alumina-platinum
  feedthrough, 25–26, 25*f*
Charnley concept, 160, 201
Chemical alumina, 14
Chemical wear, 189, 372
Chipping, 208, 209*f*
Clay
  ceramic forming techniques, 23, 74
  nanoparticulate powder morphology, 73
  plasticity of, 73
  solid-state sintering, 73
CobaltChrome hip replacement, 1
Cochlear Ltd., 289
Cochlear C122M, 261, 262*f*
Cochlear C1532M, 261, 263*f*
Cochlear C124RE, 261, 262*f*
Cochlear implant system, 261, 264*f*
  electrode side, 266–269, 268*f*
  helical electrode support, 260–261
  intra-cochlea 22-electrode-array, 260–261,
  260*f*
  multiple electrode, 260
  silicone-mounted platinum strips, 266–269,
  268*f*
  soft silicone-coated microcoiled electrode,
  260
  thermal expansion coefficient, 261–263

Cochlear Ltd. *(Continued)*
  timeline, 259
  titanium/alumina hermetic encapsulation system, 260
  22-electrodes vs. 1-electrode, 264–265
  with titanium flange, 266–269, 267f
  welding, 266–269, 267f
CoCr-PE-Ti knee, 222–223, 222f
Cofired alumina-substrate systems, 468–469, 469f
Co-fired platinum channel wires, 293, 294f
Cold isostatic pressing (CIPing), 75
  advantages, 84–85
  low-cost process, 85
  pressurization and depressurization cycles, 85
  spark plug insulators, 84
Comminution process, 81
Community enhancing technologies, 2–3
Compression molding process, 353
CoorsTek, 33–34, 366
  fracture energy, 105, 107t
  mechanical properties, 105, 106t
  thermal properties, 105, 108t
Corning Inc., 36
Corrosive wear, 373
Corundum, 14–15, 16f, 17, 39
Cowdery alumina-feedthrough/titanium-casing system, 125
CUMI Murugappa, 33, 35, 366

D

Deep brain stimulator (DBS), 291–292
Defeat mechanisms, ceramic composite armor
  brittleness, 347
  convex surfaces, 347, 347f
  cost, 348, 349f
  density, 348
  dilatancy, 348, 348f
  edge effect, 348
  fracture conoid, 347
  hardness, 347
  mosaics, 347
  stepped surfaces, 347
Defibrillator, 1, 123
Densification techniques, 24

Dental implants, 24
Dental technology, 30, 503
  abutments, 126, 134
  bridges, 126, 129–131, 130f
  crowns, 126–129, 129f
  implants, 126
    components, 131
    hydrothermal aging, 133–134
    load-bearing prosthetic dental root, 131
    Synthodont, 132
    titanium, 131, 133–134
    Tubigen implant, 132, 133f
    vitallium, 131–132
    ZTA, 132
  orthodontic brackets, 126–127, 128f
De Puy ASR, 154, 155f
Desilication products (DSP), 59
Desktop pacemaker, 229–230
Diaspore, 39, 42, 53
Dielectric loss (AC loss tangent), 424–426, 424–425t, 425f
Die pressing method, 21
Direct manufacture methods, 90–91
Direct-sintered silicon carbide (DSSC), 356–357
Dry forming methods
  advantage, 82
  cold isostatic pressing (CIPing), 84–85
  disadvantage, 82
  uniaxial die pressing, 82–84
Dyneema (DSM) body armor, 330–331, 351–352

E

Economic demonstrated resources (EDR), 42
Economic impact, 503–504
Electrical applications, 2–3, 27, 502
  in-house applications, 12
  insulators, 27, 28f
  in microchip industry, 7, 9f
  process-focussed technology, 12
  sparkplug development trials, 7
  spark plug insulators, 27
Electrical insulator, 502
  global market applications, 414, 415t
  industrial applications, 413
  insulating electrical feedthrough, 414

macro-insulators, 414
  hermetic feedthroughs, 426–427
  power transmission and management, 429–430, 430f
  spark plug insulators, 427–429
  thermal expansion, 430–431
microelectronic technology
  electronics industry, 431
  silicon wafer semiconducting substrates, 432
  "thick-film" insulating substrate, 432
  thick-film microelectronic technology, 434–435
  thin-film and thick-film microelectronic devices, 431
  thin-film semiconductor (microchip) technology, 432–434, 433f
  "thin-film" wafer, 432
micro-insulators, 414
microwave technology
  features, 441–442
  industrial uses, 442
  microwave generators, types, 442, 443f
  microwave-integrated circuits (MICs), 444
  windows, 444
properties
  alkali metal ion conductivity, 415
  dielectric breakdown voltage, 426
  dielectric loss (AC loss tangent), 424–426, 424–425t, 425f
  electrical ceramic, 416
  electrical resistivity, 416–419, 418f, 420f
  electronic conductivity, 419
  grainsize, 419–421
  high-temperature resistivity, 419, 420f
  impurities, 422–423
  International Electrotechnical Commission (IEC) standards, 416
  ionic conduction, 415, 419
  ionic mobility, 421–422
  operating temperature range, 416, 417t
  oxygen partial pressure, 419, 421f, 423
  "power pole" urban power transmission systems, 423–424
  relative permittivity (dielectric constant), 426

substrate technology
  dielectric loss tangent, 436–437, 438f
  electrical resistivity, 436, 437f
  heat dissipation, 439
  properties, 436
  resistivity criterion, 439
  thermal conductivity, 436, 437f
  thermal expansion coefficients, 436, 438f
  thick-film substrates, 440–441
  thin-film substrates, 439–440
wear/corrosion-resistant applications, 413
Electron beam melting (EBM), 91
Electronics implants, 5–9
Electronics industry, 37
Emery, 15
Engineering applications, 4, 4f
Epoxy encapsulation systems, 225
Evaporation condensation mechanisms, 98
Extrusion methods, 90

F

Fatigue wear, 196
"Femoral cap endoprosthesis," 166
Fiber-reinforced polymer (FRP), 351
Ficks first law, 94
Ficks second law, 94
Focused ion beam milling (FIB), 279
Fracture mechanics theory, 113
"Free Caustic" content, 53
Frenkel defect, 94
Fretting wear, 197–198
Functional electrical stimulation (FES), 291
Furnace furniture, 29
Fuse casting process, 93

G

Galvanic corrosion, 142. See also Hip replacements, galvanic and fretting corrosion
Gas diffusion, 94
Gelcasting process, 92
Gibbsite, 42, 53
Glass-contact refractories, 8
Glass-metal seals, 453–454, 454f
  advantages, 455
Global markets, 2
Grain boundary diffusion, 95, 98
Green machining method, 82, 90–91

H

Hall-Héroult process, 21, 46, 49
Heat containment applications, 502
Heavy armored vehicle (tank), 365–366
Hermetic-bionic-implant system, 234*f*
  battery-powered implantable cardiac bionic pacemaker, 233
  electrical signal feedthrough, 232
  features, 233
  short-service-life epoxy-clad bionic cardiac pacemaker, 233
High-density polyethylene, 160
Hip replacements, 5–9, 23
  anatomy
    alumina hip implant, 146, 148*f*
    hemiarthroplasty, 144–145
    modern hip replacement, 146, 147*f*
    prosthesis, components, 145–146
    total hip replacement, X-ray, 144–145, 145*f*
  aseptic loosening, 202–205, 206*f*
  bearings, 5, 7*f*, 25, 140
    ceramic Morse taper, 166
    Ceraver Osteal alumina-on-alumina-bearing system, 162, 163*f*
    clinical trials, 167
    commercialization issues, 167
    "femoral cap endoprosthesis," 166
    "First Generation" alumina, 161
    German innovations, 162
    Mittelmeier hip, 162–164, 164*f*
    parameters, 167
    Regulatory Approval, 167
    "Second Generation" CeramTec alumina, 161
    Shikati alumina-polyethylene hip, 166
    "Third Generation" alumina, 161
    Wagner alumina resurfacing implant, 165, 165*f*
  bioceramics conference series, 140
  Biolox symposium series, 140
  biomaterials innovations, 149
    CoCr, 150
    glass, 150
    stainless steel, 150
    tantalum, 150
    titanium, 150–151
  Birmingham Hip, 141
  cementless implants, 160–161
  CeramTec, 141
  chipping, 208, 209*f*
  chromium metallosis, 142
  CobaltChrome (CoCr) hip bearing, 140
  CoCr-on-polyethylene, 141–142
  complications, 200
  "deluxe" alumina version, 1
  dislocation, 206, 207*f*
  features, 158
  galvanic and fretting corrosion, 142
    acetabular section junction, 201
    alumina femoral head, 202, 204*f*
    alumina-on-alumina wear rate, 200–201
    Charnley concept, 201
    CoCr head metal, 202, 203*f*
    CoCr male taper, 202, 204*f*
    electrochemical process, 202
    femoral section taper junction, 201
    taper junctions, 201
    titanium alloy (Ti6Al4V) taper sleeve, 202, 202–203*f*
    titanium male neck taper, 202, 205*f*
  global market in, 139
  hemiarthroplasty, 151
  human hip joint, 139, 143*f*
  materials, 140–141
  metal-on-metal, 151–152
  metal-on-polyethylene-bearing system, 141
  metal-on-polyethylene revolution
    Charnley concept, 157
    lubrication effect, 156
    market trends, 158
    particle-induced osteolysis, 157
    poly-ether-ether-ketone (PEEK), 158, 159*f*
  osteolysis, 140, 202–205, 206*f*
  polyethylene wear particles, 142
  prosthetic hip
    allografts, 149
    animal testing, 147
    arthroplasty strategy, 149
    biocompatibility, 149
    design and implants, 146
    intramedullary fixation, 148
    joint infection, 146–149
    modular construction, 148
    stabile fixation, 147
    stress shielding, 149

resurfacing, 153f
  benefits, 152
  "Birmingham hip," 154
  complications, 154
  De Puy ASR, 154, 155f
  femoral-preservation procedure, 152
  load-bearing femoral bone stock, 152–153
  long-stem THA, 153
  metal-on-metal implant concept, 155–156
  metal-on-polyethylene implant, 155–156
  polymethyl methacrylate (PMMA) bone cement, 154
  Teflon-on-Teflon resurfacing trial, 153
  traditional THA, 153
squeak
  alumina-on-alumina bearings, 208–209
  benefits, 208
  CeramTec ZTA BIOLOX™ delta, 210, 211f
  factors, 210
  fluid film lubrication, 210
  frictional driving force, 210
  hard-on-hard bearings, 208–209
  metal-on-metal bearings, 209–210
systemic cobalt poisoning, 142
taper-fit articulating acetabular cup (socket), 26, 26f
taper-fit femoral head (ball), 26, 26f
UHMWPE polyethylene acetabular liner, 158–160
zirconia ceramic hip, 140
  catastrophic failure, 174–177, 174t, 178f
  zirconia-toughened-alumina (ZTA), 140
Homemade metal armor, 325, 325f
Hot isostatic pressing (HIPing), 102, 214
Hot-pressed silicon carbide (HPSC), 356
Hot pressing, 102
Hydrocyclones, 393f
  ceramic-coated metals, 394
  components, 394, 395f
  high-throughput coarse/fine particle separation, 393
  liquid slurry, 392–393
  refurbishment, 395f, 396
Hydrogarnet, 59–60
Hydrostatic shock forming, 102–103

Hydrothermal aging
  critical zirconia percolation threshold, 183–184
  encasement/isolation, tetragonal zirconia crystals, 183
  higher matrix modulus, 182–183
  tetragonal zirconia, 182
Hydrothermal method, 81

I

Impact wear, 372
Impingement wear, 373f
  mining and mineral-processing mode, 374
  pipe T-junction, 374, 374f
  relative wear rate, 374, 375f
Implantable bionic medical devices, 225–228, 255
Implantable bionic pacemaker, 1
Implantable cardioverter defibrillator (ICD), 287
Implantable pulse generator (IPG) concept, 283, 293–294
Impurity/dopant ions, 95
Industrial alumina, 76–77
  electrical properties, 105, 120–121
  manufacturing advantages, 104
  mechanical properties, 104
    ceramic strength test configurations, 113, 114f
    compressive strength, 113, 119
    4-point bend test, 112
    fracture mechanics theory, 113
    grain size, 117, 117f, 119
    hardness, 107–111, 109–110f, 119
    internal hydrostatic loading test, 113
    modulus of rupture (MOR), 111, 112f, 119
    porosity, 116–117, 119
    purity and chemical inertness, 118–119, 118f
    strength, definition, 111
    tensile strength, 112–113, 119
    toughness, 114–116, 115f, 119
    Weibull modulus, 114
    Young's modulus, 105–107, 109f, 119
  thermal properties, 105
    coefficient of thermal expansion, 120
    conductivity, 119–120

Industrial alumina *(Continued)*
 refractoriness, 120
 specific heat, 120
 thermal shock resistance, 120
Industrial applications, 2, 9–10, 498, 501
 active metal brazing technology, 499
 alumina gauges, 499
 billion electrical applications market, 11
 billion heat containment applications market, 11
 billion wear- and corrosion-resistance market, 10–11
 gas laser tubes, 498
 jewel bearings, 499
 Kiln furniture, 499
 laboratory instrument tubes, 499
 mortar and pestle, 499
 niche applications, 11–12
 thermocouple sheaths, 498
Industrial ceramics, 2
Ingenieurbüro Deisenroth (IBD), 367
Inkjet 3D printing, 91
International Electrotechnical Commission (IEC) standards, 416
Iron hydrogarnet, 59–61
Iron silico calcium aluminates, 59–60
Isostatic pressing method, 21

J

Japanese Jomon pottery, 72, 72*f*

K

Kaolinite clay (silica), 46, 56, 58
Karstic bauxites, 46–47
Kevlar body armor, 328–330
Kovar/borosilicate glass, 236
Kovar-glass hermetic sealing, 455
Kyocera Corporation, 32, 185, 367

L

Laboratory refractories, 29
Labware, 29
Large ceramic companies, 36–37
Laser marking, 215
Laser-patterning/screen-printing approach, 281
Lasers, 29
LAS glass ceramic, 356

Lateritic bauxites, 46
Lattice diffusion, 95
Le Chatelier process, 18
 alumina refining evolution, 18–19, 19*f*
 high-purity aluminum hydroxide, 19
Lightweight body armor, 1–2, 5–9, 8*f*, 501
 ADA, 367
 aircraft armor, 321
 BAE Advanced Materials, 367
 ballistic applications, 26–27, 27*f*
 ceramic body armor
  acquired/onsold companies, 367–368
  alumina, 355
  autoclave process, 352
  boron carbide (B_4C), 358–359, 359*f*
  breastplates, 353–354, 354*f*
  in civilian market, 343–345
  compression molding process, 353
  design and optimization, 348–350, 350*f*
  Doron innovation, 333
  Dyneema/Spectra, 351–352
  fiber-reinforced polymer (FRP), 351
  Kevlar cladding, 359–360
  LAS glass ceramic, 356
  mass-market technology, 341–343
  military sector, 340–341
  NIJ3, 360–361, 360*t*
  NIJ4, 361–362, 361*t*
  operational use of, 338–340, 339*f*
  properties, 354, 355*t*
  S-Glass, 351–352
  SiC, 356–358, 357*f*
  SpectraShield, 352–353
  toughness and flexural strength, 354
  traumatology benefits, 350–351
 ceramic composite armor, 26–27
  ballistic fiberglass, 334
  brittleness, 347
  catching, 346
  concept, 335, 335*f*
  convex surfaces, 347, 347*f*
  Cook patent, 336–337, 336*f*
  cost, 348, 349*f*
  density, 348
  dilatancy, 348, 348*f*
  edge effect, 348
  erosion, 346
  fracture conoid, 347
  functions, 337

hard-faced enamel coatings, 333
hardness, 347
King cites fiberglass, 337, 337f
mosaics, 347
Picatinny fiberglass, 338
projectile defeat mechanisms, 346
Saint-Gobain, 338
shattering, 345
stepped surfaces, 347
Webster invention, 335
CeramTec, 367
CoorsTek, 366
CUMI Murugappa, 366
Ingenieurbüro Deisenroth (IBD), 367
Kyocera, 367
Morgan Advanced Materials, 367
Saint Gobain, 367
sociocultural context
  Brewster body armor, 325–326, 326f
  homemade metal armor, 325, 325f
  metal body armor, 323–324, 324f
  multihit mosaic panels, 322
  soft body armor, 326–333, 329t
  US Civil war, 324–325
  vehicle armor systems, 326
  workplace health and safety (WHS), 322
3M (ESK), 367
vehicle armor
  civilian lightweight vehicle armor, 363
  classification, 363
  heavy armored vehicle (tank), 365–366
  military lightweight vehicle armor, 363–365
  military secrecy, 362
  mosaic vehicle armor systems, 363, 364f
  multiple-hit armor systems, 363
  steel-armor cast, 362
Liquid diffusion, 94
Liquid-phase sintering, 97
LivaNova, 288
Live-enhancing technologies, 1
Live-saving technologies, 1
Low-density polyethylene, 160

M

Macro-insulators, 414
  hermetic feedthroughs, 426–427
  power transmission and management, 429–430, 430f
  spark plug insulators, 427–429
  thermal expansion, 430–431
M5 body armor, 332
MED-EL, 289
Medico, 288
Medtronic, 288
Medtronic Micra transcatheter pacing system, 251
Melt-line corrosion, 482
Metal body armor, 323–324, 324f
Metallurgical alumina, 14
Metal-on-polyethylene-bearing system, 141
Microelectronic technology
  electronics industry, 431
  silicon wafer semiconducting substrates, 432
  "thick-film" insulating substrate, 432
  thick-film microelectronic technology, 434–435
  thin-film semiconductor (microchip) technology, 432–434, 433f
  "thin-film" wafer, 432
Micro-insulators, 414
Microsphere agglomerates, 82
Microwave-integrated circuits (MICs), 444
Microwave sintering, 102
Microwave technology
  features, 441–442
  industrial uses, 442
  microwave generators, types, 442, 443f
  microwave-integrated circuits (MICs), 444
  windows, 444
Milling process, 79, 81
Mineral processing equipment, 5–9
Mining industry, 2–3
Mining processing equipment, 5–9
Modern spark plug, 75, 76f
Mohs hardness scale, 107
Molybdenum-manganese (Mo-Mn) method, 457–459
Monohydrate alumina (MHA), 46–47, 474
Morgan Advanced Materials, 33–34, 367
Murata Manufacturing Co., Ltd., 36

N

Nanopowders, 80
NaOH concentration, 53
Natural alumina

Natural alumina *(Continued)*
  aluminum-bearing minerals, 14
  aluminum-oxygen phase diagram, 14
  bauxite, 17, 17*f*
  Bayer process, 14
  clay, 17
  conventions of, 14
  corundum, 15, 16*f*, 17
  Earth's crust, 13
  elemental analysis, 16
  emery, 15
  factors, 15
  oxidized state, 14
  ruby, 15, 16*f*
  sapphire, 15, 16*f*
    grade and quality, 17
    heat treatment, 16
    surface oxide coating, 14–15
NIJ3, 360–361, 360*t*
NIJ4, 361–362, 361*t*
The NSG Group, 36
Nurotron, 289

O

Open-cut mining process, 42
Orthodontic brackets, 126
Orthopedics, 1–2, 5, 7*f*, 13, 503. *See also* Hip replacements
  bearings, 10
    adhesive wear, 189
    alumina-on-alumina, 190, 193, 200
    alumina-on-polyethylene wear rates, 197, 200
    biocompatibility, 198–199
    CeramTec, 210–218, 212–213*t*, 216–217*f*
    chemical wear, 189
    CoCr-PE-Ti knee, 222–223, 222*f*
    dual-condyle femoral component, 222–223
    fretting wear, 197–198
    hip replacements, 189
    impact wear, 189
    joint simulators, 189–190
    market impact, 218–222, 219–221*f*
    metal-on-metal, 190–191
    polyethylene, 190 (*see also* Polyethylene wear mechanisms)
    sliding wear, 189
    third-body wear, 189
    titanium tibial plate, 223
    wear-resistance, 188, 194–195
    wear testing, 189, 191–193, 192*f*, 194*f*
    CoCr particles, 179
  hip joint
    anatomy, 143, 144*f*
    pathology, 144
  nonalumina bearings, 179
  oxidized zirconium (OXINIUM$^\diamond$)
    advantages, 169
    biomaterials, 168
    ceramic-coatings technology, 169
    galvanic corrosion and ceramic/ceramic bearings, 170
    "intermix zone," 168–169
    nuclear reactors, 168
    simulator tests, 169
    surface damage, 170
  pure-alumina-bearing technology, 182
  zirconia toughening, 171–174, 172*f*, 179
  zirconium based bearing, 168
  ZTA (*see* Zirconia-toughened alumina (ZTA))
Osmotic-pressure diffusion, 94
Osteolysis, 202–205, 206*f*, 215–218
Oticon Medical, 289
Oxide ceramics, 72
Oxygen ions, 95
Oxygen sensors, 30, 474, 495–497, 496–498*f*

P

Partially stabilized zirconia (PSZ), 172
Particle packing models, 83–84
Plastic flow, 98
Plasticizer, 78
Polycrystalline alumina, 23, 104
Polyethylene wear mechanisms
  abrasive wear, 195
  acetabular liner, 196, 196*f*
  adhesive wear, 195
  fatigue wear, 196
  gamma sterilization, 196–197
  modes, 195
Polymethyl methacrylate (PMMA) bone cement, 154
Portable pacemaker, 230–231

Postprocessing Bayer alumina, 81
Powdered alumina commercial
  technologies, 3
Powdered alumina industry, 30
Powder forming techniques, 23
Powder injection molding (PIM) process, 21,
  75–76
  "ceramic comb" concept, 273, 274f
  Cochlear bionic feedthroughs, 87
  draft, 87
  feedstock pellets, 86
  wall thickness, 87
PPG Industries, 37
Production cost, 5
PubMed analysis, 315, 316f

Q

Quality alumina research, 4
Quality manufacturing process, 212–214
Quantum leap, 295–296
  hermetic feedthrough technology, 257
  retinal implant
    alumina bionic feedthrough technology,
      304–305
    alumina composition, 309
    alumina tapes, 311, 313f
    cell culture testing, 311–314
    ceramic-Pt electrodes, 309, 310f
    686-channel feedthrough, 311–314, 314f
    Cochlear "platinum comb" feedthrough
      concept, 305
    co-fired alumina-platinum feedthrough
      concept, 307
    duplicate layers creation, 310–311
    electrode clustering, 305
    electrode layout, 311–314, 313f
    feedthrough manufacturing process, 311,
      311f
    Kuzma concept, 306
    laser machining, vias, 309
    lumina-platinum interface, 306
    paste printing, powdered platinum, 309
    platinum-alumina Suaning feedthrough,
      307f, 309
    platinum channel pathway, 306, 307f
    platinum paste-printing stencil, 309
    platinum screen-printing paste, 309
    platinum tracks, laser machining, 309
    Suaning concept, 308, 311, 312f
    tape casting, 308–309
    titanium/alumina feedthrough decades,
      305–306
    vias design, 309
Quartz (silica), 46
Quartz-rich karst bauxite, 58

R

Rare-earth elements recovery, 67
Reaction-sintered silicon carbide (RSSC),
  357–358, 357f
Reactive silica, Bayer process
  bauxites, 56
  clays and quartz, 56
  double digestion, 57–58
  high-performance alumina ceramics, 57
  industrial alumina powder, 57
  kaolinite-silica, 56
  moisture-induced static fatigue, 57
  problems with, 57
  sodalite precipitation, 57
  sodium recovery, 56–57
Red mud disposal, Bayer process, 58
  advantages, 67
  calcium silicate recovery, 66
  disposal and environmental cost, 64
  flocculants, 64
  impurity types and concentrations, 63–64
  land reclamation problems, 64–65
  land remediation, 68
  management needs, 65
  rare-earth elements recovery, 67
  sodium recovery, 65–66
  solutions, 67–68
  turbid Bayer liquor, 64
Refractories, 29, 29f
  alpha/beta-alumina, 481
  alumina-based, 476
  applications, 2–3, 8, 473
    in-house applications, 12
    process-focussed technology, 12
  calcined bauxite, 477
  calcium aluminate cement, 476
  clay-based firebrick, 475
  clay refractories, 479
  compressive strength, 488–490, 488f
  cutting tools, 488

Refractories *(Continued)*
  economics of, 474–475, 475t
  furnace linings, 478
  fuse-casting formulations, 483–484
  fuse-casting process, 484
  fused-cast alumina zirconia silicate (AZS), 477
  fused-cast beta-alumina, 481
  fusion temperature, 479
  glass production, 29
  glass-tank refractories, 482–483
  hardness, 486
  heat containment systems, 478
  high-purity alumina, 481
  high-purity dense alumina, 476
  high-purity fibrous alumina, 476
  high-purity porous alumina, 476
  HIPing, 487
  industry participants, 478
  iron-smelting refractories, 476
  labware alumina containment systems, 484–486, 485–486f
  lathing/machining tools, 487
  monolithic, 478
    binding systems, 480
    bricks, 480
    fibrous refractories, 480
  multicomponent, 476
  nonclay refractories, 479
  nonoxides, 479
  oxides, 479
  relative wear rate, 488–490, 489f
  requirements, 479
  thermal insulation and structural integrity, 476
  ultra-refractories, 480–481
  wear resistance, 486
Refractory high-purity oxide ceramics, 73
RHI AG, 37
Rotary seals, 30
Ruby, 15, 16f

S

Sacral-nerve stimulator (SNS), 290
Saint Gobain, 31–32, 338, 367
Sapphire, 14–15, 16f
  grade and quality, 17
  heat treatment, 16
  orthodontic brackets, 127

Schott AG, 37
Selected area electron diffraction (SAED), 280
Selective laser sintering (SLS), 91
Self-propagating high-temperature synthesis (SHS), 102
Semiconductor industry, 2–3
S-Glass, 351–352
Shikati alumina-polyethylene hip, 166
Shree Pacetronix, 288
Silico calcium (SiC) aluminates, 59, 356–358, 357f
Silicone coating implants, 247–248, 248–249f
Silicon on insulator (SOI), 2–3
Silk body armor, 326–327
Sintering process
  activation energy of diffusion, 94
  agglomerates, 97
  alkali earth/silicate fluxes, 100
  "apparent porosity," 98
  atmosphere, 97
  cooling rate, 97
  diffusion
    coefficient, 94
    electrical and ionic conductivity, 94
    extrinsic and intrinsic defects, 95
    factors, 94–95
    grain boundary, 95
    ions, types, 95
    lattice, 95
    magnesium, 95
    properties, 94
    surface, 95
    titanium, 95
    types, 94
  factors, 96–97
  Ficks first law, 94
  Ficks second law, 94
  Frenkel defect, 94
  grain size, 97
  green density, 97
  heating rate, 97
  hot isostatic pressing (HIPing), 102
  hot pressing, 102
  hydrostatic shock forming, 102–103
  liquid-phase sintering, 97
  mass-transport mechanisms, 98, 99f
  microwave sintering, 102
  particle size, 96

porosity, 97
pressure, 97
pressureless sintering, 96
secondary recrystallization, 101
self-propagating high-temperature synthesis (SHS), 102
shrinkage, 97
sol-gel methods, 97–98
solid-state sintering, 96
spark plasma sintering (SPS), 102–103
stages, 98–100, 99f
temperature, 96
time, 96
Sintox alumina armor ceramics, 24
Sliding wear
  abrasion wear rate, 377–378, 377f
  abrasion wear resistance, 376f, 377
  applications, 378
  cost-effective performance, 375
  fuse-cast basalt, 377
  relative wear rate, 375, 376f
  test parameters, 375–377
  tungsten carbide (WC), 378
Slipcasting process, 74
  advantages, 90
  flowchart of, 89–90, 89f
  polyelectrolytes, 89
  slurry requirements, 88–89
Slurry processing, plumbing, 396
  curved pipe sections, 397, 398f
  cylinders/tubing, 397, 398f
  knife-edge blades, 399–400, 400f
  mill linings, 400, 401f
  ore chutes, 397–399, 399f
  spigots, 396, 397f
  valves and seats, 397, 398f
Slurry pump components, 30
Small- to medium-sized companies (SME), 36
Social impact, 504–505
"Soda" content, 53
Sodalite, 59–60
Sodium aluminosilicates, 59
Sodium calcium aluminosilicates, 59
Sodium recovery, 65–66
Soft body armor
  ballistic nylon (Nylon-66), 327–328, 329t
  ballistic weight efficiency, 332
  cost-effective armor ceramic, 332–333

Dyneema (DSM), 330–331
Kevlar, 328–330
M5, 332
silk, 326–327
Zylon, 331
SOI-integrated circuit, 432–434, 433f
Sol-gel forming, 77–78
Sol-gel methods, 97–98
Sol-gel powders, 81
Solid-state diffusion, 94
Solid-state sintering, 72–73, 77, 96
Sonova, 289
Spark plasma sintering (SPS), 102–103
Spark plug insulators, 75, 427f, 444
  application, 427–428
  Champion Spark Plug Company, 428
  clay-bonded alumina, 428
  high-purity alumina and sintering temperatures, 428
  injection molding, 429
  properties, 429
SpectraShield, 352–353
Speech recognition, 295
Spinal cord stimulator (SCS), 288–289
Spray drying, 79, 79f
Spray pyrolysis, 81
Stentrode, 318–319
Surface diffusion, 95, 98
Synthetic sapphire, 24, 30
Synthodont implant, 132

T

Tapecasting process, 21, 87–88, 308–309
Taylor Ceramic Engineering (TCE), 33, 35–36, 248–250, 250f
Telectronics Pacemaker Model P10
  evolutions, 251
  failure mechanisms, 248–250
  high-purity alumina components, 248–250
  medical-grade silicone elastomer, 247–248
  Medtronic Micra transcatheter pacing system, 251
  niobium alloy, 251
  silicone coating implants, 247–248, 248–249f
  Taylor Ceramic Engineering (TCE), 248–250, 250f
  titanium feedthrough channels, 247

Tetragonal zirconia precipitate (TZP), 182–183
Textile industry, 2
Thick-film microelectronics
  devices, 444
  technology, 434–435
Thin and thick film alumina wafers, 3
Thin-film microelectronic devices, 444
Thin-film semiconductor (microchip) technology, 432–434, 433f
Third-body wear, 373
Thixotropic casting process, 92–93
Thomson-Houston company, 74
3D printing method, 82, 91
3M (ESK), 367
Timeline, alumina ceramics, 21–23, 22t
Titanium/alumina gold-brazed feedthrough, 292, 293f
Titanium/alumina-platinum hermetic system, 286, 287f
Titanium casing, 292
Titanium diboride armor ceramics, 340
Titanium hermetic feedthrough system, 225
  advantage, 242
  alumina ceramic composition, 243
  alumina-metal hermetic sealing, 236–238
  assembling, 246
  biocompatible brazing technology, 245–246
  biomaterials, 238–239
  ceramic feedthroughs, 235
  cobalt chrome, 240
  components, 240, 241f
  electrical conductors, 235
  electrical porcelain, 242–243
  epoxy encapsulation, 233–234
  glass-metal hermetic feedthrough, 236, 237f
  hermetic-bionic-implant concept (see Hermetic-bionic-implant system)
  hermetic seal, 235
  Kovar/borosilicate glass, 236
  physical properties, 242
  sealing, 247
  system-critical technology, 235
  Telectronics P10, 242, 242f
  thermal expansion coefficient, 239
  thermal expansion coefficients, 243
  titanium casing, 243–244
  titanium electrical terminal feedthrough system, 244–245
  Triode vacuum valves, 236
Titanium osseointegration, 161
Total aluminum content, 53
"Total Caustic" content, 53
Total hip arthroplasty (THA), 53, 144–145, 145f
  dislocation, 206, 207f
Total hip replacement, revision surgery, 212, 212t
Total knee replacement (TKR), 223
Transformation-induced plasticity (TRIP) steels, 172
Transformation toughening, 171–174, 172f
Trihydrate alumina (THA) component, 46–47
Tubigen implant, 132, 133f

U

Ultrahigh-molecular-weight-polyethylene (UHMWPE)
  body armor, 160, 330–331
  polyethylene acetabular liner, 158–160
Ultrapure alumina nanopowders, 81
Uniaxial die pressing
  applications, 82–83
  benefits, 83
  disadvantages, 83
  "green" ceramic component, 83
  particle packing models, 83–84
  production cost, 83
  QA process, cracks detection, 83
  spray-dried nanopowders, 83
  vibratory press, 83–84
Unreactive silica, 51
USA Geological Survey, 42

V

Vagus nerve stimulator (VNS), 290
Valence electrons, titanium, 422
Vapor deposition methods, 24
Vesuvius PLC, 36
Vickers test geometry, 110, 110f, 378–380, 379f
Volume diffusion mechanisms, 98

Index 557

W

Wagner alumina resurfacing implant, 165, 165*f*
Wear-resistant industrial ceramics, 2–3, 27, 504
  alumina machining, 388–389
  applications, 74
  bearings, sleeves, and shafts, 409, 409*f*
  brick and heavy clayware industry, 405, 406*f*
  ceramic-coated metals, 390
  commercial use of, 389
  components, 27–29, 28*f*, 391–392
  and corrosion resistance, 12, 370
  dry pin-on-disk study, 371
  engineering operating principles, 369–370
  flame-spray coating, 390
  food industry, 404–405
  material properties
    chemical inertness and corrosion resistance, 383
    chemisorption, 384
    grain size, 382–383, 383*f*
    hardness, 378–380, 379*f*
    porosity, 382
    strength, 382
    thermal conductivity, 383–384
    toughness, 380–381, 381*f*
  mechanisms
    abrasion "scratch," 386*f*, 387
    ballistic impact mechanisms, 385, 388
    crack depth and load parameter, 387, 387*f*
    hardness testing, 385–386
    material properties, 385
    micro-spallation, 387
    sliding wear, 386
    stage process, 386
  minerals productions, 389–390
  mining and mineral processing industries, 27–29
    advantages, 392
    applications, 392
    hydrocyclone, 392–396, 393*f*, 395*f*
    plumbing (*see* Slurry processing, plumbing)
    silicon carbide, 392
  modes
    adhesive wear, 372
    chemical wear, 372
    corrosive wear, 373
    impact wear, 372
    impingement wear, 373–375*f*, 374
    sliding abrasion, 373
    sliding wear, 372, 375–378, 376–377*f*
    third-body wear, 373
  nozzles, 406, 407*f*
  paper industry, 403–404, 405*f*
  seals and pump components, 6, 8*f*, 406–409, 407*f*
  spacers, tubes, washers, 410, 410*f*
  TCE, 391
  textile industry, 400–402
  tribological parameters, 371, 372*t*
  tungsten carbide/ceramic-coated metals, 370
  wear testing, 371
  wet-chemical corrosion, 369
  wire drawing industry, 402–403, 402*f*
Weibull modulus, 114
Wet acid processing, 69, 69*t*
Wet alkaline processing, 68
Wet forming methods, 79
  benefits, 85–86
  drying shrinkage, 86
  extrusion, 90
  powder injection molding (PIM), 86–87
  slipcasting, 88–90, 89*f*
  tapecasting, 87–88
Wood-fired fireplace/brick-furnace construct (kiln), 72

Y

Young's modulus, 105–107, 109*f*, 119, 463, 464*f*

Z

Zirconia ceramic hip, 140
  catastrophic failure, 174–177, 174*t*, 178*f*
Zirconia-toughened alumina (ZTA), 3, 140, 179, 378
  alumina-on-alumina, 217*f*, 218
  alumina-toughened zirconia (ATZ), 184
  biomedical-grade
    benefits, 186–187
    burst strength, 187, 188*f*
    ceramic femoral heads, 187
    CeramTec, 184–186, 187*t*

Zirconia-toughened alumina (ZTA) *(Continued)*
   flexural strength, 185, 186f
   hardness, 186
   Kyocera, 185
   microstructure, 185
   parameters, 186
   wear resistance, 187
   "ceramic steel," 180
   CeramTec BIOLOXdelta, 180, 181f, 187–188
   concept of, 182
   hydrothermal aging
      critical zirconia percolation threshold, 183–184
      encasement/isolation, tetragonal zirconia crystals, 183
      higher matrix modulus, 182–183
      tetragonal zirconia, 182
   silver lining, 180
Zyglo crack inspection, 215
Zylon body armor, 331

CPSIA information can be obtained
at www.ICGtesting.com
Printed in the USA
LVHW060604301018
595207LV00008B/58/P